Active Calculus

Active Calculus

Matthew Boelkins
Grand Valley State University

Contributing Authors

David Austin
Grand Valley State University

Steven Schlicker
Grand Valley State University

August 15, 2017

Acknowledgements

This text began as my sabbatical project in the winter semester of 2012, during which I wrote the preponderance of the materials for the first four chapters. For the sabbatical leave, I am indebted to Grand Valley State University for its support of the project and the time to write, as well as to my colleagues in the Department of Mathematics and the College of Liberal Arts and Sciences for their endorsement of the project as a valuable undertaking.

The beautiful full-color .eps graphics in the text are only possible because of David Austin of GVSU and Bill Casselman of the University of British Columbia. Building on their collective longstanding efforts to develop tools for high quality mathematical graphics, David wrote a library of Python routines that build on Bill's PiScript program, and David's routines are so easy to use that even I could generate graphics like the professionals that he and Bill are. I am deeply grateful to them both.

For the print-on-demand version of the 2016 text, I am thankful for the support of Lon Mitchell and Orthogonal Publishing L3C. For the first print edition of the text, Lon provided considerable guidance on LaTeX and related typesetting issues, and volunteered his time throughout the production process. I met Lon at a special session devoted to open textbooks at the 2014 Joint Mathematics Meetings; I am grateful as well to the organizers of that session and subsequent ones who are part of a growing community of mathematicians committed to free and open texts. You can start to learn more about this work at the American Institute of Mathematics' open textbook site.

The current .html version of the text is possible only because of the amazing work of Rob Beezer and his development of the original Mathbook XML, soon to be known as PreTeXt. My ability to take advantage of Rob's work is largely due to the support of the American Institute of Mathematics, which funded me for a weeklong workshop in Mathbook XML in San Jose, CA, in April 2016. David Farmer's conversion script saved me hundreds of hours of work by taking my original LaTeX source and converting it to MBX. Alex Jordan of Portland Community College has also been a tremendous help, and it is through Alex's fantastic work that live WeBWorK exercises are not only possible, but also included in the 2017 version. Alex has also contributed substantially to the .pdf version of the 2017 text.

Over my more than 15 years at GVSU, many of my colleagues have shared with me ideas and resources for teaching calculus. I am particularly indebted to David Austin, Will Dickinson, Paul Fishback, Jon Hodge, and Steve Schlicker for their contributions that improved my teaching of and thinking about calculus, including materials that I have modified and used over many different semesters with students. Parts of these ideas can be found throughout this text. In addition, Will Dickinson and Steve Schlicker provided me access to a large number of their electronic notes and activities from teaching of differential and integral calculus, and those ideas and materials have similarly impacted my work and writing in positive ways, with some of their problems and approaches finding parallel presentation here.

Shelly Smith of GVSU and Matt Delong of Taylor University both provided extensive comments on the first few chapters of early drafts, feedback that was immensely helpful in improving the text. As more and more people use the text, I am grateful to everyone who reads, edits, and uses this book, and hence contributes to its improvement through ongoing discussion.

Finally, I am grateful for all that my students have taught me over the years. Their responses and feedback have helped to shape me as a teacher, and I appreciate their willingness to wholeheartedly engage in the activities and approaches I've tried in class, to let me know how those affect their learning, and to help me learn and grow as an instructor. Early on, they provided useful editorial feedback on this text.

Any and all remaining errors or inconsistencies are mine. I will gladly take reader and user feedback to correct them, along with other suggestions to improve the text.

Contributors

A large and growing number of people have generously contributed to the development or improvement of the text. Contributing authors David Austin and Steven Schlicker have each written drafts of at least one full chapter of the text.

The following contributing editors have offered significant feedback that includes information about typographical errors or suggestions to improve the exposition.

DAVID AUSTIN
GVSU

ALLAN BICKLE
GVSU

DAVID CLARK
GVSU

WILL DICKINSON
GVSU

CHARLES FORTIN
Champlain Regional College
St-Lambert, Quebec, Canada

MARCIA FROBIS
GVSU

PATTI HUNTER
Westmont College

MITCH KELLER
Washington and Lee University

DAVE KUNG
St. Mary's College of Maryland

PAUL LATIOLAIS
Portland State University

HUGH MCGUIRE
GVSU

RAY ROSENTRATER
Westmont College

LUIS SANJUAN
Conservatorio Profesional de Musica de Avila
Spain

STEVEN SCHLICKER
GVSU

BRIAN STANLEY
Foothill Community College

AMY STONE
GVSU

ROBERT TALBERT
GVSU

GREG THULL
GVSU

SUE VAN HATTUM
Contra Costa College

Active Calculus: Our Goals

Several fundamental ideas in calculus are more than 2000 years old. As a formal subdiscipline of mathematics, calculus was first introduced and developed in the late 1600s, with key independent contributions from Sir Isaac Newton and Gottfried Wilhelm Leibniz. Mathematicians agree that the subject has been understood rigorously since the work of Augustin Louis Cauchy and Karl Weierstrass in the mid 1800s when the field of modern analysis was developed, in part to make sense of the infinitely small quantities on which calculus rests. As a body of knowledge, calculus has been completely understood for at least 150 years. The discipline is one of our great human intellectual achievements: among many spectacular ideas, calculus models how objects fall under the forces of gravity and wind resistance, explains how to compute areas and volumes of interesting shapes, enables us to work rigorously with infinitely small and infinitely large quantities, and connects the varying rates at which quantities change to the total change in the quantities themselves.

While each author of a calculus textbook certainly offers her own creative perspective on the subject, it is hardly the case that many of the ideas she presents are new. Indeed, the mathematics community broadly agrees on what the main ideas of calculus are, as well as their justification and their importance; the core parts of nearly all calculus textbooks are very similar. As such, it is our opinion that in the 21st century—an age where the internet permits seamless and immediate transmission of information—no one should be required to purchase a calculus text to read, to use for a class, or to find a coherent collection of problems to solve. Calculus belongs to humankind, not any individual author or publishing company. Thus, a main purpose of this work is to present a new calculus text that is *free*. In addition, instructors who are looking for a calculus text should have the opportunity to download the source files and make modifications that they see fit; thus this text is *open-source*. Since August 2013, *Active Calculus* has been endorsed by the American Institute of Mathematics and its Open Textbook Initiative.

In *Active Calculus*, we endeavor to actively engage students in learning the subject through an activity-driven approach in which the vast majority of the examples are completed by students. Where many texts present a general theory of calculus followed by substantial collections of worked examples, we instead pose problems or situations, consider possibilities, and then ask students to investigate and explore. Following key activities or examples, the presentation normally includes some overall perspective and a brief synopsis of general trends or properties, followed by formal statements of rules or theorems. While we often offer plausibility arguments for such results, rarely do we include formal proofs. It is not the intent of this text for the instructor or author to *demonstrate* to students that the ideas of calculus are coherent and true, but rather for students to *encounter* these ideas in a supportive, leading manner that enables them to begin to understand for themselves why calculus is both coherent and true. This approach is consistent with the growing body of scholarship that calls for students to be interactively engaged in class.

Moreover, this approach is consistent with the following goals:

- To have students engage in an active, inquiry-driven approach, where learners strive to construct solutions and approaches to ideas on their own, with appropriate support through questions posed, hints, and guidance from the instructor and text.

- To build in students intuition for why the main ideas in calculus are natural and true. Often, we do this through consideration of the instantaneous position and velocity of a moving object, a scenario that is common and familiar.

- To challenge students to acquire deep, personal understanding of calculus through reading the text and completing preview activities on their own, through working on activities in small groups in class, and through doing substantial exercises outside of class time.

- To strengthen students' written and oral communicating skills by having them write about and explain aloud the key ideas of calculus.

Features of the Text

Instructors and students alike will find several consistent features in the presentation, including:

Motivating Questions At the start of each section, we list 2–3 *motivating questions* that provide motivation for why the following material is of interest to us. One goal of each section is to answer each of the motivating questions.

Preview Activities Each section of the text begins with a short introduction, followed by a *preview activity*. This brief reading and the preview activity are designed to foreshadow the upcoming ideas in the remainder of the section; both the reading and preview activity are intended to be accessible to students *in advance* of class, and indeed to be completed by students before a day on which a particular section is to be considered.

Activities A typical section in the text has three *activities*. These are designed to engage students in an inquiry-based style that encourages them to construct solutions to key examples on their own, working individually or in small groups.

Exercises There are dozens of calculus texts with (collectively) tens of thousands of exercises. Rather than repeat standard and routine exercises in this text, we recommend the use of WeBWorK with its access to the Open Problem Library and around 20,000 calculus problems. In this text, each section includes a small number of anonymous WeBWorK exercises, as well as 3–4 challenging problems per section. The WeBWorK exercises are best completed in the .html version of the text. Almost every non-WeBWorK problem has multiple parts, requires the student to connect several key ideas, and expects that the student will do at least a modest amount of writing to answer the questions and explain their findings. For instructors interested in a more conventional source of exercises, consider the freely available *APEX Calculus* text by Greg Hartmann et al., available from www.apexcalculus.com.

Graphics As much as possible, we strive to demonstrate key fundamental ideas visually, and to encourage students to do the same. Throughout the text, we use full-color[1] graphics to exemplify and magnify key ideas, and to use this graphical perspective alongside both numerical and algebraic representations of calculus.

Links to interactive graphics Many of the ideas of calculus are best understood dynamically; java applets offer an often ideal format for investigations and demonstrations. Relying primarily on the work of David Austin of Grand Valley State University and Marc Renault of Shippensburg University, each of whom has developed a large library of applets for calculus, we frequently point the reader (through active links in the electronic versions of the text) to applets that are relevant for key ideas under con-

---

[1]To keep cost low, the graphics in the print-on-demand version are in black and white. When the text itself refers to color in images, one needs to view the .html or .pdf electronically.

sideration.

Summary of Key Ideas Each section concludes with a summary of the key ideas encountered in the preceding section; this summary normally reflects responses to the motivating questions that began the section.

How to Use this Text

Because the text is free, any professor or student may use the electronic version of the text for no charge. For reading on laptops or mobile devices, the best electronic version to use is the .html one at gvsu.edu/s/0uo, but you may also download a full .pdf copy of the text from gvsu.edu/s/0vM, where there is also a link to a print-on-demand option for purchasing a bound, softcover version for under $25. Other ancillary materials, such as WeBWorK .def files, an activities-only workbook, and sample computer laboratory activities are available upon direct request to the author via email at boelkinm@gvsu.edu. Furthermore, because the text is open-source, any instructor may acquire the full set of source files, which are available on GitHub.

This text may be used as a stand-alone textbook for a standard first semester college calculus course or as a supplement to a more traditional text. Chapters 1–4 address the typical topics for differential calculus, while Chapters 5–8 provide the standard topics of integral calculus, including a chapter on differential equations (Chapter 7) and on infinite series (Chapter 8).

Electronic Edition Because students and instructors alike have access to the book in electronic format, there are several advantages to the text over a traditional print text. One is that the text may be projected on a screen in the classroom (or even better, on a whiteboard) and the instructor may reference ideas in the text directly, add comments or notation or features to graphs, and indeed write right on the text itself. Students can do likewise, choosing to print only whatever portions of the text are needed for them. In addition, the electronic versions of the text includes live .html links to java applets, so student and instructor alike may follow those links to additional resources that lie outside the text itself. Finally, students can have access to a copy of the text anywhere they have a computer. The .html version is far superior to the .pdf version; this is especially true for viewing on a smartphone.

Note. In the .pdf version, there is not an obvious visual indicator of the live .html links, so some availalable information is suppressed. If you are using the text electronically in a setting with internet access, please know that it is assumed you are using the .html version.

Activities Workbook Each section of the text has a preview activity and at least three in-class activities embedded in the discussion. As it is the expectation that students will complete all of these activities, it is ideal for them to have room to work on them adjacent to the problem statements themselves. As a separate document, we have compiled a workbook of activities that includes only the individual activity prompts, along with space provided for students to write their responses. This workbook is the one printing expense that students will almost certainly have to undertake, and is available upon request.

Community of Users Because this text is free and open-source, we hope that as people use the text, they will contribute corrections, suggestions, and new material. The best way to communicate such feedback is by email to Matt Boelkins. I also have a blog at open-calculus.wordpress.com, at which we post new developments, other free resources, feedback, and other points of discussion.

Contents

Contents

CHAPTER 1

Understanding the Derivative

1.1 How do we measure velocity?

Motivating Questions

- How is the average velocity of a moving object connected to the values of its position function?

- How do we interpret the average velocity of an object geometrically with regard to the graph of its position function?

- How is the notion of instantaneous velocity connected to average velocity?

Calculus can be viewed broadly as the study of change. A natural and important question to ask about any changing quantity is "how fast is the quantity changing?" It turns out that in order to make the answer to this question precise, substantial mathematics is required.

We begin with a familiar problem: a ball being tossed straight up in the air from an initial height. From this elementary scenario, we will ask questions about how the ball is moving. These questions will lead us to begin investigating ideas that will be central throughout our study of differential calculus and that have wide-ranging consequences. In a great deal of our thinking about calculus, we will be well-served by remembering this first example and asking ourselves how the various (sometimes abstract) ideas we are considering are related to the simple act of tossing a ball straight up in the air.

Preview Activity 1.1.1. Suppose that the height s of a ball (in feet) at time t (in seconds) is given by the formula $s(t) = 64 - 16(t - 1)^2$.

a. Construct an accurate graph of $y = s(t)$ on the time interval $0 \leq t \leq 3$. Label at least six distinct points on the graph, including the three points that correspond to when the ball was released, when the ball reaches its highest point, and when the ball lands.

b. In everyday language, describe the behavior of the ball on the time interval $0 <$

$t < 1$ and on time interval $1 < t < 3$. What occurs at the instant $t = 1$?

c. Consider the expression

$$AV_{[0.5,1]} = \frac{s(1) - s(0.5)}{1 - 0.5}.$$

Compute the value of $AV_{[0.5,1]}$. What does this value measure geometrically? What does this value measure physically? In particular, what are the units on $AV_{[0.5,1]}$?

1.1.1 Position and average velocity

Any moving object has a *position* that can be considered a function of *time*. When this motion is along a straight line, the position is given by a single variable, and we usually let this position be denoted by $s(t)$, which reflects the fact that position is a function of time. For example, we might view $s(t)$ as telling the mile marker of a car traveling on a straight highway at time t in hours; similarly, the function s described in Preview Activity 1.1.1 is a position function, where position is measured vertically relative to the ground.

Not only does such a moving object have a position associated with its motion, but on any time interval, the object has an *average velocity*. Think, for example, about driving from one location to another: the vehicle travels some number of miles over a certain time interval (measured in hours), from which we can compute the vehicle's average velocity. In this situation, average velocity is the number of miles traveled divided by the time elapsed, which of course is given in *miles per hour*. Similarly, the calculation of $AV_{[0.5,1]}$ in Preview Activity 1.1.1 found the average velocity of the ball on the time interval $[0.5, 1]$, measured in feet per second.

In general, we make the following definition: for an object moving in a straight line whose position at time t is given by the function $s(t)$, the *average velocity of the object on the interval from $t = a$ to $t = b$*, denoted $AV_{[a,b]}$, is given by the formula

$$AV_{[a,b]} = \frac{s(b) - s(a)}{b - a}.$$

Note well: the units on $AV_{[a,b]}$ are "units of s per unit of t," such as "miles per hour" or "feet per second."

Activity 1.1.2. The following questions concern the position function given by $s(t) = 64 - 16(t - 1)^2$, which is the same function considered in Preview Activity 1.1.1.

a. Compute the average velocity of the ball on each of the following time intervals: $[0.4, 0.8], [0.7, 0.8], [0.79, 0.8], [0.799, 0.8], [0.8, 1.2], [0.8, 0.9], [0.8, 0.81], [0.8, 0.801]$. Include units for each value.

b. On the provided graph in Figure 1.1.1, sketch the line that passes through the points $A = (0.4, s(0.4))$ and $B = (0.8, s(0.8))$. What is the meaning of the slope of this line? In light of this meaning, what is a geometric way to interpret each of the values computed in the preceding question?

c. Use a graphing utility to plot the graph of $s(t) = 64 - 16(t - 1)^2$ on an interval containing the value $t = 0.8$. Then, zoom in repeatedly on the point $(0.8, s(0.8))$. What do you observe about how the graph appears as you view it more and more closely?

d. What do you conjecture is the velocity of the ball at the instant $t = 0.8$? Why?

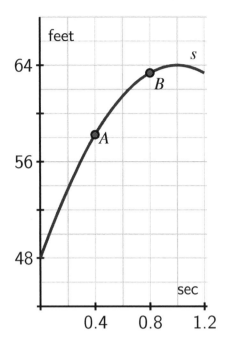

Figure 1.1.1: A partial plot of $s(t) = 64 - 16(t - 1)^2$.

1.1.2 Instantaneous Velocity

Whether driving a car, riding a bike, or throwing a ball, we have an intuitive sense that any moving object has a velocity at any given moment – a number that measures how fast the object is moving *right now*. For instance, a car's speedometer tells the driver what appears to be the car's velocity at any given instant. In fact, the posted velocity on a speedometer is really an average velocity that is computed over a very small time interval (by computing how many revolutions the tires have undergone to compute distance traveled), since velocity fundamentally comes from considering a change in position divided by a change in time. But if we let the time interval over which average velocity is computed become shorter and shorter, then we can progress from average velocity to *instantaneous* velocity.

Informally, we define the *instantaneous velocity* of a moving object at time $t = a$ to be the value that the average velocity approaches as we take smaller and smaller intervals of time containing $t = a$ to compute the average velocity. We will develop a more formal definition of this momentarily, one that will end up being the foundation of much of our work in first semester calculus. For now, it is fine to think of instantaneous velocity this way: take average velocities on smaller and smaller time intervals, and if those average velocities approach a single number, then that number will be the instantaneous velocity at that point.

> **Activity 1.1.3.** Each of the following questions concern $s(t) = 64 - 16(t-1)^2$, the position function from Preview Activity 1.1.1.
>
>   a. Compute the average velocity of the ball on the time interval $[1.5, 2]$. What is different between this value and the average velocity on the interval $[0, 0.5]$?
>
>   b. Use appropriate computing technology to estimate the instantaneous velocity of the ball at $t = 1.5$. Likewise, estimate the instantaneous velocity of the ball at $t = 2$. Which value is greater?
>
>   c. How is the sign of the instantaneous velocity of the ball related to its behavior at a given point in time? That is, what does positive instantaneous velocity tell you the ball is doing? Negative instantaneous velocity?
>
>   d. Without doing any computations, what do you expect to be the instantaneous velocity of the ball at $t = 1$? Why?

At this point we have started to see a close connection between average velocity and instantaneous velocity, as well as how each is connected not only to the physical behavior of the moving object but also to the geometric behavior of the graph of the position function. In order to make the link between average and instantaneous velocity more formal, we will introduce the notion of *limit* in Section 1.2. As a preview of that concept, we look at a way to consider the limiting value of average velocity through the introduction of a parameter. Note that if we desire to know the instantaneous velocity at $t = a$ of a moving object with position function s, we are interested in computing average velocities on the interval $[a, b]$ for smaller and smaller intervals. One way to visualize this is to think of the value b as being $b = a + h$, where h is a small number that is allowed to vary. Thus, we observe that the average velocity of the object on the interval $[a, a + h]$ is

$$AV_{[a,a+h]} = \frac{s(a+h) - s(a)}{h},$$

with the denominator being simply h because $(a + h) - a = h$. Initially, it is fine to think of h being a small positive real number; but it is important to note that we allow h to be a small negative number, too, as this enables us to investigate the average velocity of the moving object on intervals prior to $t = a$, as well as following $t = a$. When $h < 0$, $AV_{[a,a+h]}$ measures the average velocity on the interval $[a + h, a]$.

To attempt to find the instantaneous velocity at $t = a$, we investigate what happens as the value of h approaches zero. We consider this further in the following example.

Example 1.1.2 (Computing instantaneous velocity for a falling ball). For a falling ball whose position function is given by $s(t) = 16 - 16t^2$ (where s is measured in feet and t in seconds), find an expression for the average velocity of the ball on a time interval of the form $[0.5, 0.5 + h]$ where $-0.5 < h < 0.5$ and $h \neq 0$. Use this expression to compute the average velocity on

[0.5, 0.75] and [0.4, 0.5], as well as to make a conjecture about the instantaneous velocity at $t = 0.5$.

Solution. We make the assumptions that $-0.5 < h < 0.5$ and $h \neq 0$ because h cannot be zero (otherwise there is no interval on which to compute average velocity) and because the function only makes sense on the time interval $0 \leq t \leq 1$, as this is the duration of time during which the ball is falling. Observe that we want to compute and simplify

$$AV_{[0.5, 0.5+h]} = \frac{s(0.5 + h) - s(0.5)}{(0.5 + h) - 0.5}.$$

The most unusual part of this computation is finding $s(0.5 + h)$. To do so, we follow the rule that defines the function s. In particular, since $s(t) = 16 - 16t^2$, we see that

$$s(0.5 + h) = 16 - 16(0.5 + h)^2$$
$$= 16 - 16(0.25 + h + h^2)$$
$$= 16 - 4 - 16h - 16h^2$$
$$= 12 - 16h - 16h^2.$$

Now, returning to our computation of the average velocity, we find that

$$AV_{[0.5, 0.5+h]} = \frac{s(0.5 + h) - s(0.5)}{(0.5 + h) - 0.5}$$
$$= \frac{(12 - 16h - 16h^2) - (16 - 16(0.5)^2)}{0.5 + h - 0.5}$$
$$= \frac{12 - 16h - 16h^2 - 12}{h}$$
$$= \frac{-16h - 16h^2}{h}.$$

At this point, we note two things: first, the expression for average velocity clearly depends on h, which it must, since as h changes the average velocity will change. Further, we note that since h can never equal zero, we may further simplify the most recent expression. Removing the common factor of h from the numerator and denominator, it follows that

$$AV_{[0.5, 0.5+h]} = -16 - 16h.$$

Now, for any small positive or negative value of h, we can compute the average velocity. For instance, to obtain the average velocity on [0.5, 0.75], we let $h = 0.25$, and the average velocity is $-16 - 16(0.25) = -20$ ft/sec. To get the average velocity on [0.4, 0.5], we let $h = -0.1$, which tells us the average velocity is $-16 - 16(-0.1) = -14.4$ ft/sec. Moreover, we can even explore what happens to $AV_{[0.5, 0.5+h]}$ as h gets closer and closer to zero. As h approaches zero, $-16h$ will also approach zero, and thus it appears that the instantaneous velocity of the ball at $t = 0.5$ should be -16 ft/sec.

Activity 1.1.4. For the function given by $s(t) = 64 - 16(t - 1)^2$ from Preview Activity 1.1.1, find the most simplified expression you can for the average velocity of the ball on the interval $[2, 2 + h]$. Use your result to compute the average velocity on $[1.5, 2]$ and to estimate the instantaneous velocity at $t = 2$. Finally, compare your earlier work in Activity 1.1.2.

Summary

- The average velocity on $[a, b]$ can be viewed geometrically as the slope of the line between the points $(a, s(a))$ and $(b, s(b))$ on the graph of $y = s(t)$, as shown in Figure 1.1.3.

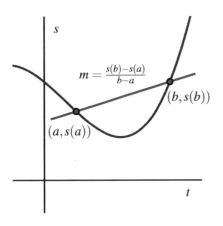

- Given a moving object whose position at time t is given by a function s, the average velocity of the object on the time interval $[a, b]$ is given by $AV_{[a,b]} = \frac{s(b)-s(a)}{b-a}$. Viewing the interval $[a, b]$ as having the form $[a, a + h]$, we equivalently compute average velocity by the formula $AV_{[a,a+h]} = \frac{s(a+h)-s(a)}{h}$.

- The instantaneous velocity of a moving object at a fixed time is estimated by considering average velocities on shorter and shorter time intervals that contain the instant of interest.

Figure 1.1.3: The graph of position function s together with the line through $(a, s(a))$ and $(b, s(b))$ whose slope is $m = \frac{s(b)-s(a)}{b-a}$. The line's slope is the average rate of change of s on the interval $[a, b]$.

Exercises

1. Consider a car whose position, s, is given by the table

t (s)	0	0.2	0.4	0.6	0.8	1
s (ft)	0	0.45	1.7	3.8	6.5	9.6

Find the average velocity over the interval $0 \le t \le 0.2$.

average velocity = [   ] help (units)

Estimate the velocity at $t = 0.2$.

velocity = [   ] help (units)

2. The table below shows the number of calories used per minute as a function of an individual's body weight for three sports:

Activity	100 lb	120 lb	150 lb	170 lb	200 lb	220 lb
Walking	2.7	3.2	4	4.6	5.4	5.9
Bicycling	5.4	6.5	8.1	9.2	10.8	11.9
Swimming	5.8	6.9	8.7	9.8	11.6	12.7

a) Determine the number of calories that a 200 lb person uses in one half-hour of walking .

| | calories

b) Who uses more calories, a 120 lb person swimming for one hour, or a 220 lb person bicycling for a half-hour?

c) Does the number of calories of a person swimming increase or decrease as weight increases?

3. Let $f(x) = 36 - x^2$.

a) Compute each of the following expressions and interpret each as an average rate of change:

(i) $\frac{f(4)-f(0)}{4-0} =$

(ii) $\frac{f(6)-f(4)}{6-4} =$

(iii) $\frac{f(6)-f(0)}{6-0} =$

b) Based on the graph sketched below, match each of your answers in (i) - (iii) with one of the lines labeled A - F. Type the corresponding letter of the line segment next to the appropriate formula. Clearly not all letters will be used.

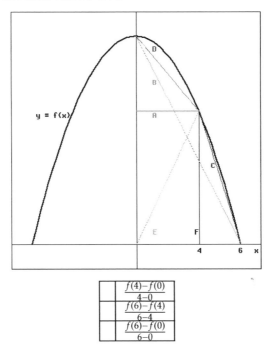

	$\frac{f(4)-f(0)}{4-0}$
	$\frac{f(6)-f(4)}{6-4}$
	$\frac{f(6)-f(0)}{6-0}$

4. Consider the graphs of $f(x)$ and $g(x)$ below:

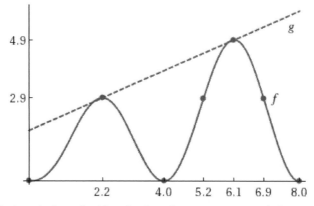

For each interval given below, decide whether the average rate of change of $f(x)$ or $g(x)$ is greater over that particular interval.

Interval	Which function has GREATER average rate of change?
$0 \le x \le 2.2$	[Choose: f | g | both have an equal rate of change]
$5.2 \le x \le 6.1$	[Choose: f | g | both have an equal rate of change]
$5.2 \le x \le 8$	[Choose: f | g | both have an equal rate of change]
$2.2 \le x \le 4$	[Choose: f | g | both have an equal rate of change]
$2.2 \le x \le 6.1$	[Choose: f | g | both have an equal rate of change]

5. A car is driven at a constant speed, starting at noon. Which of the following could be a graph of the distance the car has traveled as a function of time past noon?

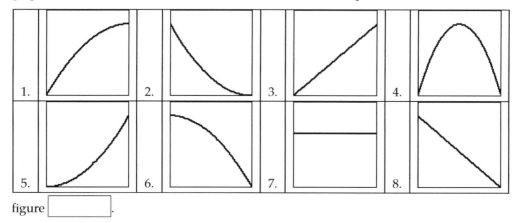

figure [＿＿＿].

6. A bungee jumper dives from a tower at time $t = 0$. Her height h (measured in feet) at time t (in seconds) is given by the graph in Figure 1.1.4. In this problem, you may base your answers on estimates from the graph or use the fact that the jumper's height function is given by $s(t) = 100\cos(0.75t) \cdot e^{-0.2t} + 100$.

   a. What is the change in vertical position of the bungee jumper between $t = 0$ and $t = 15$?

b. Estimate the jumper's average velocity on each of the following time intervals: $[0, 15]$, $[0, 2]$, $[1, 6]$, and $[8, 10]$. Include units on your answers.

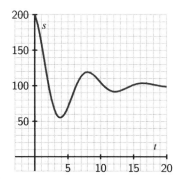

c. On what time interval(s) do you think the bungee jumper achieves her greatest average velocity? Why?

d. Estimate the jumper's instantaneous velocity at $t = 5$. Show your work and explain your reasoning, and include units on your answer.

e. Among the average and instantaneous velocities you computed in earlier questions, which are positive and which are negative? What does negative velocity indicate?

Figure 1.1.4: A bungee jumper's height function.

7. A diver leaps from a 3 meter springboard. His feet leave the board at time $t = 0$, he reaches his maximum height of 4.5 m at $t = 1.1$ seconds, and enters the water at $t = 2.45$. Once in the water, the diver coasts to the bottom of the pool (depth 3.5 m), touches bottom at $t = 7$, rests for one second, and then pushes off the bottom. From there he coasts to the surface, and takes his first breath at $t = 13$.

a. Let $s(t)$ denote the function that gives the height of the diver's feet (in meters) above the water at time t. (Note that the "height" of the bottom of the pool is -3.5 meters.) Sketch a carefully labeled graph of $s(t)$ on the provided axes in Figure 1.1.5. Include scale and units on the vertical axis. Be as detailed as possible.

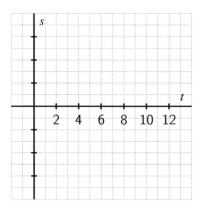

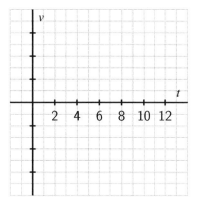

Figure 1.1.5: Axes for plotting $s(t)$ in part (a).

Figure 1.1.6: Axes for plotting $v(t)$ in part (c).

b. Based on your graph in (a), what is the average velocity of the diver between $t = 2.45$ and $t = 7$? Is his average velocity the same on every time interval within $[2.45, 7]$?

c. Let the function $v(t)$ represent the *instantaneous vertical velocity* of the diver at time t (i.e. the speed at which the height function $s(t)$ is changing; note that velocity in the upward direction is positive, while the velocity of a falling object is negative). Based on your understanding of the diver's behavior, as well as your graph of the position function, sketch a carefully labeled graph of $v(t)$ on the axes provided in Figure 1.1.6. Include scale and units on the vertical axis. Write several sentences that explain how you constructed your graph, discussing when you expect $v(t)$ to be zero, positive, negative, relatively large, and relatively small.

d. Is there a connection between the two graphs that you can describe? What can you say about the velocity graph when the height function is increasing? decreasing? Make as many observations as you can.

8. According to the U.S. census, the population of the city of Grand Rapids, MI, was 181,843 in 1980; 189,126 in 1990; and 197,800 in 2000.

a. Between 1980 and 2000, by how many people did the population of Grand Rapids grow?

b. In an average year between 1980 and 2000, by how many people did the population of Grand Rapids grow?

c. Just like we can find the average velocity of a moving body by computing change in position over change in time, we can compute the average rate of change of any function f. In particular, the *average rate of change* of a function f over an interval $[a, b]$ is the quotient

$$\frac{f(b) - f(a)}{b - a}.$$

What does the quantity $\frac{f(b) - f(a)}{b - a}$ measure on the graph of $y = f(x)$ over the interval $[a, b]$?

d. Let $P(t)$ represent the population of Grand Rapids at time t, where t is measured in years from January 1, 1980. What is the average rate of change of P on the interval $t = 0$ to $t = 20$? What are the units on this quantity?

e. If we assume the population of Grand Rapids is growing at a rate of approximately 4% per decade, we can model the population function with the formula

$$P(t) = 181843(1.04)^{t/10}.$$

Use this formula to compute the average rate of change of the population on the intervals $[5, 10]$, $[5, 9]$, $[5, 8]$, $[5, 7]$, and $[5, 6]$.

f. How fast do you think the population of Grand Rapids was changing on January 1, 1985? Said differently, at what rate do you think people were being added to the population of Grand Rapids as of January 1, 1985? How many additional people should the city have expected in the following year? Why?

1.2 The notion of limit

Motivating Questions

- What is the mathematical notion of *limit* and what role do limits play in the study of functions?

- What is the meaning of the notation $\lim_{x \to a} f(x) = L$?

- How do we go about determining the value of the limit of a function at a point?

- How do we manipulate average velocity to compute instantaneous velocity?

Functions are at the heart of mathematics: a function is a process or rule that associates each individual input to exactly one corresponding output. Students learn in courses prior to calculus that there are many different ways to represent functions, including through formulas, graphs, tables, and even words. For example, the squaring function can be thought of in any of these ways. In words, the squaring function takes any real number x and computes its square. The formulaic and graphical representations go hand in hand, as $y = f(x) = x^2$ is one of the simplest curves to graph. Finally, we can also partially represent this function through a table of values, essentially by listing some of the ordered pairs that lie on the curve, such as $(-2, 4)$, $(-1, 1)$, $(0, 0)$, $(1, 1)$, and $(2, 4)$.

Functions are especially important in calculus because they often model important phenomena — the location of a moving object at a given time, the rate at which an automobile is consuming gasoline at a certain velocity, the reaction of a patient to the size of a dose of a drug — and calculus can be used to study how these output quantities change in response to changes in the input variable. Moreover, thinking about concepts like average and instantaneous velocity leads us naturally from an initial function to a related, sometimes more complicated function. As one example of this, think about the falling ball whose position function is given by $s(t) = 64 - 16t^2$ and the average velocity of the ball on the interval $[1, x]$. Observe that

$$AV_{[1,x]} = \frac{s(x) - s(1)}{x - 1} = \frac{(64 - 16x^2) - (64 - 16)}{x - 1} = \frac{16 - 16x^2}{x - 1}.$$

Two things are essential to note: this average velocity depends on x ($AV_{[1,x]}$ is a function of x), and our most focused interest in this function occurs near $x = 1$, which is where the function is not defined. Said differently, the function $g(x) = \frac{16 - 16x^2}{x - 1}$ tells us the average velocity of the ball on the interval from $t = 1$ to $t = x$, and if we are interested in the instantaneous velocity of the ball when $t = 1$, we'd like to know what happens to $g(x)$ as x gets closer and closer to 1. At the same time, $g(1)$ is not defined, because it leads to the quotient $0/0$.

This is where the idea of *limits* comes in. By using a limit, we'll be able to allow x to get arbitrarily close, but not equal, to 1 and fully understand the behavior of $g(x)$ near this value. We'll develop key language, notation, and conceptual understanding in what follows, but for now we consider a preliminary activity that uses the graphical interpretation of a function to explore points on a graph where interesting behavior occurs.

Preview Activity 1.2.1. Suppose that g is the function given by the graph below. Use the graph to answer each of the following questions.

a. Determine the values $g(-2)$, $g(-1)$, $g(0)$, $g(1)$, and $g(2)$, if defined. If the function value is not defined, explain what feature of the graph tells you this.

b. For each of the values $a = -1$, $a = 0$, and $a = 2$, complete the following sentence: "As x gets closer and closer (but not equal) to a, $g(x)$ gets as close as we want to _____."

c. What happens as x gets closer and closer (but not equal) to $a = 1$? Does the function $g(x)$ get as close as we would like to a single value?

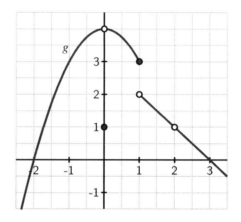

Figure 1.2.1: Graph of $y = g(x)$ for Preview Activity 1.2.1.

1.2.1 The Notion of Limit

Limits can be thought of as a way to study the tendency or trend of a function as the input variable approaches a fixed value, or even as the input variable increases or decreases without bound. We put off the study of the latter idea until further along in the course when we will have some helpful calculus tools for understanding the end behavior of functions. Here, we focus on what it means to say that "a function f has limit L as x approaches a." To begin, we think about a recent example.

In Preview Activity 1.2.1, we saw that for the given function g, as x gets closer and closer (but not equal) to 0, $g(x)$ gets as close as we want to the value 4. At first, this may feel counterintuitive, because the value of $g(0)$ is 1, not 4. By their very definition, limits regard the behavior of a function *arbitrarily close to* a fixed input, but the value of the function *at* the fixed input does not matter. More formally[1], we say the following.

---

[1]What follows here is not what mathematicians consider the formal definition of a limit. To be completely precise, it is necessary to quantify both what it means to say "as close to L as we like" and "sufficiently close to a." That can be accomplished through what is traditionally called the epsilon-delta definition of limits. The definition

Definition 1.2.2. Given a function f, a fixed input $x = a$, and a real number L, we say that f **has limit** L **as** x **approaches** a, and write

$$\lim_{x \to a} f(x) = L$$

provided that we can make $f(x)$ as close to L as we like by taking x sufficiently close (but not equal) to a. If we cannot make $f(x)$ as close to a single value as we would like as x approaches a, then we say that f **does not have a limit as** x **approaches** a.

For the function g pictured in Figure 1.2.1, we can make the following observations:

$$\lim_{x \to -1} g(x) = 3, \ \lim_{x \to 0} g(x) = 4, \ \text{and} \ \lim_{x \to 2} g(x) = 1,$$

but g does not have a limit as $x \to 1$. When working graphically, it suffices to ask if the function approaches a single value from each side of the fixed input, while understanding that the function value right at the fixed input is irrelevant. This reasoning explains the values of the first three stated limits. In a situation such as the jump in the graph of g at $x = 1$, the issue is that if we approach $x = 1$ from the left, the function values tend to get as close to 3 as we'd like, but if we approach $x = 1$ from the right, the function values get as close to 2 as we'd like, and there is no single number that all of these function values approach. This is why the limit of g does not exist at $x = 1$.

For any function f, there are typically three ways to answer the question "does f have a limit at $x = a$, and if so, what is the limit?" The first is to reason graphically as we have just done with the example from Preview Activity 1.2.1. If we have a formula for $f(x)$, there are two additional possibilities: (1) evaluate the function at a sequence of inputs that approach a on either side, typically using some sort of computing technology, and ask if the sequence of outputs seems to approach a single value; (2) use the algebraic form of the function to understand the trend in its output as the input values approach a. The first approach only produces an approximation of the value of the limit, while the latter can often be used to determine the limit exactly. The following example demonstrates both of these approaches, while also using the graphs of the respective functions to help confirm our conclusions.

Example 1.2.3 (Limits of Two Functions). For each of the following functions, we'd like to know whether or not the function has a limit at the stated a-values. Use both numerical and algebraic approaches to investigate and, if possible, estimate or determine the value of the limit. Compare the results with a careful graph of the function on an interval containing the points of interest.

a. $f(x) = \frac{4-x^2}{x+2}; a = -1, a = -2$ b. $g(x) = \sin\left(\frac{\pi}{x}\right); a = 3, a = 0$

Solution. We first construct a graph of f along with tables of values near $a = -1$ and $a = -2$.

---

presented here is sufficient for the purposes of this text.

x	$f(x)$		x	$f(x)$
−0.9	2.9		−1.9	3.9
−0.99	2.99		−1.99	3.99
−0.999	2.999		−1.999	3.999
−0.9999	2.9999		−1.9999	3.9999
−1.1	3.1		−2.1	4.1
−1.01	3.01		−2.01	4.01
−1.001	3.001		−2.001	4.001
−1.0001	3.0001		−2.0001	4.0001

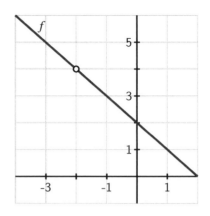

Table 1.2.4: Table of f values near $x = -1$.   **Table 1.2.5:** Table of f values near $x = -2$.   **Figure 1.2.6:** Plot of $f(x)$ on $[-4, 2]$.

From Table 1.2.4, it appears that we can make f as close as we want to 3 by taking x sufficiently close to −1, which suggests that $\lim_{x \to -1} f(x) = 3$. This is also consistent with the graph of f. To see this a bit more rigorously and from an algebraic point of view, consider the formula for f: $f(x) = \frac{4-x^2}{x+2}$. The numerator and denominator are each polynomial functions, which are among the most well-behaved functions that exist. Formally, such functions are *continuous*[2], which means that the limit of the function at any point is equal to its function value. Here, it follows that as $x \to -1$, $(4 - x^2) \to (4 - (-1)^2) = 3$, and $(x + 2) \to (-1 + 2) = 1$, so as $x \to -1$, the numerator of f tends to 3 and the denominator tends to 1, hence $\lim_{x \to -1} f(x) = \frac{3}{1} = 3$.

The situation is more complicated when $x \to -2$, due in part to the fact that $f(-2)$ is not defined. If we attempt to use a similar algebraic argument regarding the numerator and denominator, we observe that as $x \to -2$, $(4 - x^2) \to (4 - (-2)^2) = 0$, and $(x + 2) \to (-2 + 2) = 0$, so as $x \to -2$, the numerator of f tends to 0 and the denominator tends to 0. We call 0/0 an *indeterminate form* and will revisit several important issues surrounding such quantities later in the course. For now, we simply observe that this tells us there is somehow more work to do. From Table 1.2.5 and Figure 1.2.6, it appears that f should have a limit of 4 at $x = -2$. To see algebraically why this is the case, let's work directly with the form of $f(x)$. Observe that

$$\lim_{x \to -2} f(x) = \lim_{x \to -2} \frac{4 - x^2}{x + 2}$$
$$= \lim_{x \to -2} \frac{(2 - x)(2 + x)}{x + 2}.$$

At this point, it is important to observe that since we are taking the limit as $x \to -2$, we are considering x values that are close, but not equal, to −2. Since we never actually allow x to equal −2, the quotient $\frac{2+x}{x+2}$ has value 1 for every possible value of x. Thus, we can simplify

---

[2]See Section 1.7 for more on the notion of continuity.

the most recent expression above, and now find that

$$\lim_{x \to -2} f(x) = \lim_{x \to -2} 2 - x.$$

Because $2 - x$ is simply a linear function, this limit is now easy to determine, and its value clearly is 4. Thus, from several points of view we've seen that $\lim_{x \to -2} f(x) = 4$.

Next we turn to the function g, and construct two tables and a graph.

x	$g(x)$
2.9	0.84864
2.99	0.86428
2.999	0.86585
2.9999	0.86601
3.1	0.88351
3.01	0.86777
3.001	0.86620
3.0001	0.86604

x	$g(x)$
−0.1	0
−0.01	0
−0.001	0
−0.0001	0
0.1	0
0.01	0
0.001	0
0.0001	0

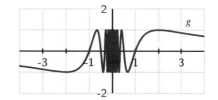

Table 1.2.7: Table of g values near $x = 3$.   **Table 1.2.8:** Table of g values near $x = 0$.   **Figure 1.2.9:** Plot of $g(x)$ on $[-4, 4]$.

First, as $x \to 3$, it appears from the data (and the graph) that the function is approaching approximately 0.866025. To be precise, we have to use the fact that $\frac{\pi}{x} \to \frac{\pi}{3}$, and thus we find that $g(x) = \sin(\frac{\pi}{x}) \to \sin(\frac{\pi}{3})$ as $x \to 3$. The exact value of $\sin(\frac{\pi}{3})$ is $\frac{\sqrt{3}}{2}$, which is approximately 0.8660254038. Thus, we see that

$$\lim_{x \to 3} g(x) = \frac{\sqrt{3}}{2}.$$

As $x \to 0$, we observe that $\frac{\pi}{x}$ does not behave in an elementary way. When x is positive and approaching zero, we are dividing by smaller and smaller positive values, and $\frac{\pi}{x}$ increases without bound. When x is negative and approaching zero, $\frac{\pi}{x}$ decreases without bound. In this sense, as we get close to $x = 0$, the inputs to the sine function are growing rapidly, and this leads to wild oscillations in the graph of g. It is an instructive exercise to plot the function $g(x) = \sin\left(\frac{\pi}{x}\right)$ with a graphing utility and then zoom in on $x = 0$. Doing so shows that the function never settles down to a single value near the origin and suggests that g does not have a limit at $x = 0$.

How do we reconcile this with the righthand table above, which seems to suggest that the limit of g as x approaches 0 may in fact be 0? Here we need to recognize that the data misleads us because of the special nature of the sequence $\{0.1, 0.01, 0.001, \ldots\}$: when we evaluate $g(10^{-k})$, we get $g(10^{-k}) = \sin\left(\frac{\pi}{10^{-k}}\right) = \sin(10^k \pi) = 0$ for each positive integer value of k. But if we take a different sequence of values approaching zero, say $\{0.3, 0.03, 0.003, \ldots\}$, then we find that

$$g(3 \cdot 10^{-k}) = \sin\left(\frac{\pi}{3 \cdot 10^{-k}}\right) = \sin\left(\frac{10^k \pi}{3}\right) = \frac{\sqrt{3}}{2} \approx 0.866025.$$

That sequence of data would suggest that the value of the limit is $\frac{\sqrt{3}}{2}$. Clearly the function cannot have two different values for the limit, and this shows that g has no limit as $x \to 0$.

An important lesson to take from Example 1.2.3 is that tables can be misleading when determining the value of a limit. While a table of values is useful for investigating the possible value of a limit, we should also use other tools to confirm the value, if we think the table suggests the limit exists.

> **Activity 1.2.2.** Estimate the value of each of the following limits by constructing appropriate tables of values. Then determine the exact value of the limit by using algebra to simplify the function. Finally, plot each function on an appropriate interval to check your result visually.
>
> a. $\lim_{x \to 1} \frac{x^2 - 1}{x - 1}$ b. $\lim_{x \to 0} \frac{(2 + x)^3 - 8}{x}$ c. $\lim_{x \to 0} \frac{\sqrt{x + 1} - 1}{x}$

This concludes a rather lengthy introduction to the notion of limits. It is important to remember that our primary motivation for considering limits of functions comes from our interest in studying the rate of change of a function. To that end, we close this section by revisiting our previous work with average and instantaneous velocity and highlighting the role that limits play.

1.2.2 Instantaneous Velocity

Suppose that we have a moving object whose position at time t is given by a function s. We know that the average velocity of the object on the time interval $[a, b]$ is $AV_{[a,b]} = \frac{s(b) - s(a)}{b - a}$. We define the *instantaneous velocity* at a to be the limit of average velocity as b approaches a. Note particularly that as $b \to a$, the length of the time interval gets shorter and shorter (while always including a). In Section 1.3, we will introduce a helpful shorthand notation to represent the instantaneous rate of change. For now, we will write $IV_{t=a}$ for the instantaneous velocity at $t = a$, and thus

$$IV_{t=a} = \lim_{b \to a} AV_{[a,b]} = \lim_{b \to a} \frac{s(b) - s(a)}{b - a}.$$

Equivalently, if we think of the changing value b as being of the form $b = a + h$, where h is some small number, then we may instead write

$$IV_{t=a} = \lim_{h \to 0} AV_{[a,a+h]} = \lim_{h \to 0} \frac{s(a + h) - s(a)}{h}.$$

Again, the most important idea here is that to compute instantaneous velocity, we take a limit of average velocities as the time interval shrinks. Two different activities offer the opportunity to investigate these ideas and the role of limits further.

Activity 1.2.3. Consider a moving object whose position function is given by $s(t) = t^2$, where s is measured in meters and t is measured in minutes.

   a. Determine the most simplified expression for the average velocity of the object on the interval $[3, 3 + h]$, where $h > 0$.

   b. Determine the average velocity of the object on the interval $[3, 3.2]$. Include units on your answer.

   c. Determine the instantaneous velocity of the object when $t = 3$. Include units on your answer.

The closing activity of this section asks you to make some connections among average velocity, instantaneous velocity, and slopes of certain lines.

Activity 1.2.4. For the moving object whose position s at time t is given by the graph below, answer each of the following questions. Assume that s is measured in feet and t is measured in seconds.

   a. Use the graph to estimate the average velocity of the object on each of the following intervals: $[0.5, 1]$, $[1.5, 2.5]$, $[0, 5]$. Draw each line whose slope represents the average velocity you seek.

   b. How could you use average velocities or slopes of lines to estimate the instantaneous velocity of the object at a fixed time?

   c. Use the graph to estimate the instantaneous velocity of the object when $t = 2$. Should this instantaneous velocity at $t = 2$ be greater or less than the average velocity on $[1.5, 2.5]$ that you computed in (a)? Why?

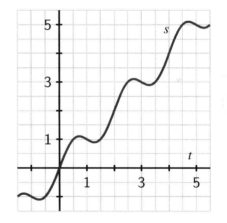

Figure 1.2.10: Plot of the position function $y = s(t)$ in Activity 1.2.4.

Summary

- Limits enable us to examine trends in function behavior near a specific point. In particular, taking a limit at a given point asks if the function values nearby tend to approach a particular fixed value.

- When we write $\lim_{x \to a} f(x) = L$, we read this as saying "the limit of f as x approaches a is L," and this means that we can make the value of $f(x)$ as close to L as we want by taking x sufficiently close (but not equal) to a.

- If we desire to know $\lim_{x \to a} f(x)$ for a given value of a and a known function f, we can estimate this value from the graph of f or by generating a table of function values that result from a sequence of x-values that are closer and closer to a. If we want the exact value of the limit, we need to work with the function algebraically and see if we can use familiar properties of known, basic functions to understand how different parts of the formula for f change as $x \to a$.

- The instantaneous velocity of a moving object at a fixed time is found by taking the limit of average velocities of the object over shorter and shorter time intervals that all contain the time of interest.

Exercises

1. Use the figure below, which gives a graph of the function $f(x)$, to give values for the indicated limits. If a limit does not exist, enter none.

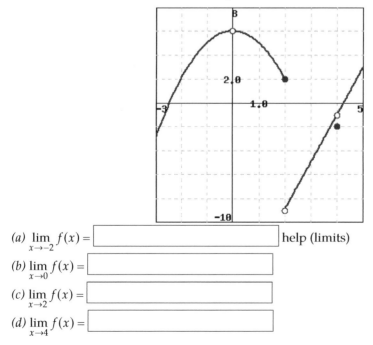

(a) $\displaystyle \lim_{x \to -2} f(x) =$ [   ] help (limits)

(b) $\displaystyle \lim_{x \to 0} f(x) =$ [   ]

(c) $\displaystyle \lim_{x \to 2} f(x) =$ [   ]

(d) $\displaystyle \lim_{x \to 4} f(x) =$ [   ]

2. Use a graph to estimate the limit

$$\lim_{\theta \to 0} \frac{\sin(5\theta)}{\theta}.$$

Note: θ is measured in radians. All angles will be in radians in this class unless otherwise specified.

$\lim_{\theta \to 0} \frac{\sin(5\theta)}{\theta} =$ [   ]

3. For the function

$$f(x) = \begin{cases} x^2 - 4, & 0 \le x < 1 \\ 2, & x = 1 \\ 2x - 5, & 1 < x \end{cases}$$

use algebra to find each of the following limits:

$\lim_{x \to 1^+} f(x) =$ [   ]

$\lim_{x \to 1^-} f(x) =$ [   ]

$\lim_{x \to 1} f(x) =$ [   ]

(For each, enter DNE if the limit does not exist.)
Sketch a graph of $f(x)$ to confirm your answers.

4. Evaluate the limit

$$\lim_{x \to -6} \frac{x^2 - 36}{x + 6}$$

If the limit does not exist enter DNE.
Limit = [   ]

5. Consider the function whose formula is $f(x) = \frac{16 - x^4}{x^2 - 4}$.

  a. What is the domain of f?

  b. Use a sequence of values of x near $a = 2$ to estimate the value of $\lim_{x \to 2} f(x)$, if you think the limit exists. If you think the limit doesn't exist, explain why.

  c. Use algebra to simplify the expression $\frac{16 - x^4}{x^2 - 4}$ and hence work to evaluate $\lim_{x \to 2} f(x)$ exactly, if it exists, or to explain how your work shows the limit fails to exist. Discuss how your findings compare to your results in (b).

  d. True or false: $f(2) = -8$. Why?

  e. True or false: $\frac{16 - x^4}{x^2 - 4} = -4 - x^2$. Why? How is this equality connected to your work above with the function f?

  f. Based on all of your work above, construct an accurate, labeled graph of $y = f(x)$ on the interval $[1, 3]$, and write a sentence that explains what you now know about $\lim_{x \to 2} \frac{16 - x^4}{x^2 - 4}$.

6. Let $g(x) = -\frac{|x+3|}{x+3}$.

    a. What is the domain of g?

    b. Use a sequence of values near $a = -3$ to estimate the value of $\lim_{x \to -3} g(x)$, if you think the limit exists. If you think the limit doesn't exist, explain why.

    c. Use algebra to simplify the expression $\frac{|x+3|}{x+3}$ and hence work to evaluate $\lim_{x \to -3} g(x)$ exactly, if it exists, or to explain how your work shows the limit fails to exist. Discuss how your findings compare to your results in (b). (Hint: $|a| = a$ whenever $a \geq 0$, but $|a| = -a$ whenever $a < 0$.)

    d. True or false: $g(-3) = -1$. Why?

    e. True or false: $-\frac{|x+3|}{x+3} = -1$. Why? How is this equality connected to your work above with the function g?

    f. Based on all of your work above, construct an accurate, labeled graph of $y = g(x)$ on the interval $[-4, -2]$, and write a sentence that explains what you now know about $\lim_{x \to -3} g(x)$.

7. For each of the following prompts, sketch a graph on the provided axes of a function that has the stated properties.

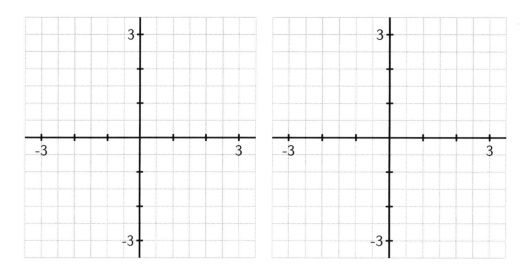

Figure 1.2.11: Axes for plotting $y = f(x)$ in (a) and $y = g(x)$ in (b).

    a. $y = f(x)$ such that
- $f(-2) = 2$ and $\lim_{x \to -2} f(x) = 1$
- $f(-1) = 3$ and $\lim_{x \to -1} f(x) = 3$
- $f(1)$ is not defined and $\lim_{x \to 1} f(x) = 0$

  - $f(2) = 1$ and $\lim_{x \to 2} f(x)$ does not exist.

b. $y = g(x)$ such that

  - $g(-2) = 3$, $g(-1) = -1$, $g(1) = -2$, and $g(2) = 3$
  - At $x = -2, -1, 1$ and 2, g has a limit, and its limit equals the value of the function at that point.
  - $g(0)$ is not defined and $\lim_{x \to 0} g(x)$ does not exist.

8. A bungee jumper dives from a tower at time $t = 0$. Her height s in feet at time t in seconds is given by $s(t) = 100 \cos(0.75t) \cdot e^{-0.2t} + 100$.

a. Write an expression for the average velocity of the bungee jumper on the interval $[1, 1 + h]$.

b. Use computing technology to estimate the value of the limit as $h \to 0$ of the quantity you found in (a).

c. What is the meaning of the value of the limit in (b)? What are its units?

1.3 The derivative of a function at a point

Motivating Questions

- How is the average rate of change of a function on a given interval defined, and what does this quantity measure?

- How is the instantaneous rate of change of a function at a particular point defined? How is the instantaneous rate of change linked to average rate of change?

- What is the derivative of a function at a given point? What does this derivative value measure? How do we interpret the derivative value graphically?

- How are limits used formally in the computation of derivatives?

An idea that sits at the foundations of calculus is the *instantaneous rate of change* of a function. This rate of change is always considered with respect to change in the input variable, often at a particular fixed input value. This is a generalization of the notion of instantaneous velocity and essentially allows us to consider the question "how do we measure how fast a particular function is changing at a given point?" When the original function represents the position of a moving object, this instantaneous rate of change is precisely velocity, and might be measured in units such as feet per second. But in other contexts, instantaneous rate of change could measure the number of cells added to a bacteria culture per day, the number of additional gallons of gasoline consumed by going one mile per additional mile per hour in a car's velocity, or the number of dollars added to a mortgage payment for each percentage increase in interest rate. Regardless of the presence of a physical or practical interpretation of a function, the instantaneous rate of change may also be interpreted geometrically in connection to the function's graph, and this connection is also foundational to many of the main ideas in calculus.

In what follows, we will introduce terminology and notation that makes it easier to talk about the instantaneous rate of change of a function at a point. In addition, just as instantaneous velocity is defined in terms of average velocity, the more general instantaneous rate of change will be connected to the more general average rate of change. Recall that for a moving object with position function s, its average velocity on the time interval $t = a$ to $t = a + h$ is given by the quotient

$$AV_{[a,a+h]} = \frac{s(a + h) - s(a)}{h}.$$

In a similar way, we make the following definition for an arbitrary function $y = f(x)$.

Definition 1.3.1. For a function f, the **average rate of change** of f on the interval $[a, a + h]$ is given by the value

$$AV_{[a,a+h]} = \frac{f(a + h) - f(a)}{h}.$$

Equivalently, if we want to consider the average rate of change of f on $[a, b]$, we compute

$$AV_{[a,b]} = \frac{f(b) - f(a)}{b - a}.$$

It is essential to understand how the average rate of change of f on an interval is connected to its graph.

Preview Activity 1.3.1. Suppose that f is the function given by the graph below and that a and $a + h$ are the input values as labeled on the x-axis. Use the graph in Figure 1.3.2 to answer the following questions.

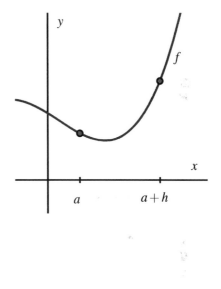

a. Locate and label the points $(a, f(a))$ and $(a + h, f(a + h))$ on the graph.

b. Construct a right triangle whose hypotenuse is the line segment from $(a, f(a))$ to $(a + h, f(a + h))$. What are the lengths of the respective legs of this triangle?

c. What is the slope of the line that connects the points $(a, f(a))$ and $(a + h, f(a + h))$?

d. Write a meaningful sentence that explains how the average rate of change of the function on a given interval and the slope of a related line are connected.

Figure 1.3.2: Plot of $y = f(x)$ for Preview Activity 1.3.1.

1.3.1 The Derivative of a Function at a Point

Just as we defined instantaneous velocity in terms of average velocity, we now define the instantaneous rate of change of a function at a point in terms of the average rate of change of the function f over related intervals. In addition, we give a special name to "the instantaneous rate of change of f at a," calling this quantity "the *derivative* of f at a," with this value being represented by the shorthand notation $f'(a)$. Specifically, we make the following definition.

Definition 1.3.3. Let f be a function and $x = a$ a value in the function's domain. We define the **derivative of** f **with respect to** x **evaluated at** $x = a$, denoted $f'(a)$, by the formula

$$f'(a) = \lim_{h \to 0} \frac{f(a+h) - f(a)}{h},$$

provided this limit exists.

Aloud, we read the symbol $f'(a)$ as either "f-prime at a" or "the derivative of f evaluated at $x = a$." Much of the next several chapters will be devoted to understanding, computing, applying, and interpreting derivatives. For now, we observe the following important things.

Note 1.3.4.

- The derivative of f at the value $x = a$ is defined as the limit of the average rate of change of f on the interval $[a, a+h]$ as $h \to 0$. It is possible for this limit not to exist, so not every function has a derivative at every point.

- We say that a function that has a derivative at $x = a$ is *differentiable* at $x = a$.

- The derivative is a generalization of the instantaneous velocity of a position function: when $y = s(t)$ is a position function of a moving body, $s'(a)$ tells us the instantaneous velocity of the body at time $t = a$.

- Because the units on $\frac{f(a+h)-f(a)}{h}$ are "units of f per unit of x," the derivative has these very same units. For instance, if s measures position in feet and t measures time in seconds, the units on $s'(a)$ are feet per second.

- Because the quantity $\frac{f(a+h)-f(a)}{h}$ represents the slope of the line through $(a, f(a))$ and $(a+h, f(a+h))$, when we compute the derivative we are taking the limit of a collection of slopes of lines, and thus the derivative itself represents the slope of a particularly important line.

While all of the above ideas are important and we will add depth and perspective to them through additional time and study, for now it is most essential to recognize how the derivative of a function at a given value represents the slope of a certain line. Thus, we expand upon the last bullet item above.

As we move from an average rate of change to an instantaneous one, we can think of one point as "sliding towards" another. In particular, provided the function has a derivative at $(a, f(a))$, the point $(a+h, f(a+h))$ will approach $(a, f(a))$ as $h \to 0$. Because this process of taking a limit is a dynamic one, it can be helpful to use computing technology to visualize what the limit is accomplishing. While there are many different options, one of the best is a java applet in which the user is able to control the point that is moving. For a helpful collection of examples, consider the work of David Austin of Grand Valley State University, and this particularly relevant example. For applets that have been built in Geogebra[1], see Marc Renault's library via Shippensburg University, with this example being especially fitting for our work in this section.

---

[1]You can even consider building your own examples; the fantastic program Geogebra is available for free download and is easy to learn and use.

In Figure 1.3.5, we provide a sequence of figures with several different lines through the points $(a, f(a))$ and $(a + h, f(a + h))$ that are generated by different values of h. These lines (shown in the first three figures in magenta), are often called *secant lines* to the curve $y = f(x)$. A secant line to a curve is simply a line that passes through two points that lie on the curve. For each such line, the slope of the secant line is $m = \frac{f(a+h)-f(a)}{h}$, where the value of h depends on the location of the point we choose. We can see in the diagram how, as $h \to 0$, the secant lines start to approach a single line that passes through the point $(a, f(a))$. In the situation where the limit of the slopes of the secant lines exists, we say that the resulting value is the slope of the *tangent line* to the curve. This tangent line (shown in the right-most figure in green) to the graph of $y = f(x)$ at the point $(a, f(a))$ is the line through $(a, f(a))$ whose slope is $m = f'(a)$.

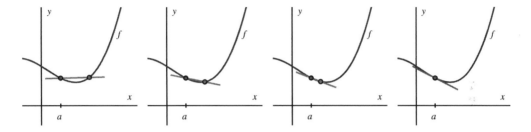

Figure 1.3.5: A sequence of secant lines approaching the tangent line to f at $(a, f(a))$.

As we will see in subsequent study, the existence of the tangent line at $x = a$ is connected to whether or not the function f looks like a straight line when viewed up close at $(a, f(a))$, which can also be seen in Figure 1.3.6, where we combine the four graphs in Figure 1.3.5 into the single one on the left, and then we zoom in on the box centered at $(a, f(a))$, with that view expanded on the right (with two of the secant lines omitted). Note how the tangent line sits relative to the curve $y = f(x)$ at $(a, f(a))$ and how closely it resembles the curve near $x = a$.

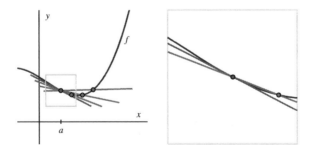

Figure 1.3.6: A sequence of secant lines approaching the tangent line to f at $(a, f(a))$. At right, we zoom in on the point $(a, f(a))$. The slope of the tangent line (in green) to f at $(a, f(a))$ is given by $f'(a)$.

Note 1.3.7. The instantaneous rate of change of f with respect to x at $x = a$, $f'(a)$, also measures the slope of the tangent line to the curve $y = f(x)$ at $(a, f(a))$.

The following example demonstrates several key ideas involving the derivative of a function.

Example 1.3.8 (Using the limit definition of the derivative). For the function given by $f(x) = x - x^2$, use the limit definition of the derivative to compute $f'(2)$. In addition, discuss the meaning of this value and draw a labeled graph that supports your explanation.

Solution.   From the limit definition, we know that

$$f'(2) = \lim_{h \to 0} \frac{f(2 + h) - f(2)}{h}.$$

Now we use the rule for f, and observe that $f(2) = 2 - 2^2 = -2$ and $f(2+h) = (2+h) - (2+h)^2$. Substituting these values into the limit definition, we have that

$$f'(2) = \lim_{h \to 0} \frac{(2 + h) - (2 + h)^2 - (-2)}{h}.$$

With h in the denominator and our desire to let $h \to 0$, we have to wait to take the limit (that is, we wait to actually let h approach 0). Thus, we do additional algebra. Expanding and distributing in the numerator,

$$f'(2) = \lim_{h \to 0} \frac{2 + h - 4 - 4h - h^2 + 2}{h}.$$

Combining like terms, we have

$$f'(2) = \lim_{h \to 0} \frac{-3h - h^2}{h}.$$

Next, we observe that there is a common factor of h in both the numerator and denominator, which allows us to simplify and find that

$$f'(2) = \lim_{h \to 0} (-3 - h).$$

Finally, we are able to take the limit as $h \to 0$, and thus conclude that $f'(2) = -3$.

Now, we know that $f'(2)$ represents the slope of the tangent line to the curve $y = x - x^2$ at the point $(2, -2)$; $f'(2)$ is also the instantaneous rate of change of f at the point $(2, -2)$. Graphing both the function and the line through $(2, -2)$ with slope $m = f'(2) = -3$, we indeed see that by calculating the derivative, we have found the slope of the tangent line at this point, as shown in Figure 1.3.9.

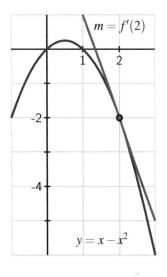

Figure 1.3.9: The tangent line to $y = x - x^2$ at the point $(2, -2)$.

The following activities will help you explore a variety of key ideas related to derivatives.

Activity 1.3.2. Consider the function f whose formula is $f(x) = 3 - 2x$.

    a. What familiar type of function is f? What can you say about the slope of f at every value of x?

    b. Compute the average rate of change of f on the intervals $[1, 4]$, $[3, 7]$, and $[5, 5+h]$; simplify each result as much as possible. What do you notice about these quantities?

    c. Use the limit definition of the derivative to compute the exact instantaneous rate of change of f with respect to x at the value $a = 1$. That is, compute $f'(1)$ using the limit definition. Show your work. Is your result surprising?

    d. Without doing any additional computations, what are the values of $f'(2)$, $f'(\pi)$, and $f'(-\sqrt{2})$? Why?

Activity 1.3.3. A water balloon is tossed vertically in the air from a window. The balloon's height in feet at time t in seconds after being launched is given by $s(t) = -16t^2 + 16t + 32$. Use this function to respond to each of the following questions.

a. Sketch an accurate, labeled graph of s on the axes provided in Figure 1.3.10. You should be able to do this without using computing technology.

b. Compute the average rate of change of s on the time interval $[1, 2]$. Include units on your answer and write one sentence to explain the meaning of the value you found.

c. Use the limit definition to compute the instantaneous rate of change of s with respect to time, t, at the instant $a = 1$. Show your work using proper notation, include units on your answer, and write one sentence to explain the meaning of the value you found.

d. On your graph in (a), sketch two lines: one whose slope represents the average rate of change of s on $[1, 2]$, the other whose slope represents the instantaneous rate of change of s at the instant $a = 1$. Label each line clearly.

e. For what values of a do you expect $s'(a)$ to be positive? Why? Answer the same questions when "positive" is replaced by "negative" and "zero."

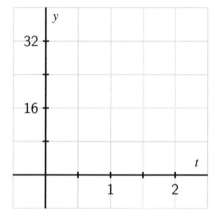

Figure 1.3.10: Axes for plotting $y = s(t)$ in Activity 1.3.3.

Activity 1.3.4. A rapidly growing city in Arizona has its population P at time t, where t is the number of decades after the year 2010, modeled by the formula $P(t) = 25000e^{t/5}$. Use this function to respond to the following questions.

a. Sketch an accurate graph of P for $t = 0$ to $t = 5$ on the axes provided in Figure 1.3.11. Label the scale on the axes carefully.

b. Compute the average rate of change of P between 2030 and 2050. Include units on your answer and write one sentence to explain the meaning (in everyday language) of the value you found.

c. Use the limit definition to write an expression for the instantaneous rate of change of P with respect to time, t, at the instant $a = 2$. Explain why this limit is difficult to evaluate exactly.

d. Estimate the limit in (c) for the instantaneous rate of change of P at the instant $a = 2$ by using several small h values. Once you have determined an accurate estimate of $P'(2)$, include units on your answer, and write one sentence (using everyday language) to explain the meaning of the value you found.

e. On your graph above, sketch two lines: one whose slope represents the average rate of change of P on $[2, 4]$, the other whose slope represents the instantaneous rate of change of P at the instant $a = 2$.

f. In a carefully-worded sentence, describe the behavior of $P'(a)$ as a increases in value. What does this reflect about the behavior of the given function P?

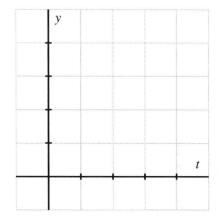

Figure 1.3.11: Axes for plotting $y = P(t)$ in Activity 1.3.4.

Summary

- The average rate of change of a function f on the interval $[a, b]$ is $\frac{f(b)-f(a)}{b-a}$. The units on the average rate of change are units of f per unit of x, and the numerical value of the average rate of change represents the slope of the secant line between the points $(a, f(a))$ and $(b, f(b))$ on the graph of $y = f(x)$. If we view the interval as being $[a, a+h]$ instead of $[a, b]$, the meaning is still the same, but the average rate of change is now computed by $\frac{f(a+h)-f(a)}{h}$.

- The instantaneous rate of change with respect to x of a function f at a value $x = a$ is denoted $f'(a)$ (read "the derivative of f evaluated at a" or "f-prime at a") and is defined by the formula

$$f'(a) = \lim_{h \to 0} \frac{f(a + h) - f(a)}{h},$$

provided the limit exists. Note particularly that the instantaneous rate of change at $x = a$ is the limit of the average rate of change on $[a, a + h]$ as $h \to 0$.

- Provided the derivative $f'(a)$ exists, its value tells us the instantaneous rate of change of f with respect to x at $x = a$, which geometrically is the slope of the tangent line to the curve $y = f(x)$ at the point $(a, f(a))$. We even say that $f'(a)$ is the "slope of the curve" at the point $(a, f(a))$.

- Limits are the link between average rate of change and instantaneous rate of change: they allow us to move from the rate of change over an interval to the rate of change at a single point.

Exercises

1. Consider the function $y = f(x)$ graphed below.

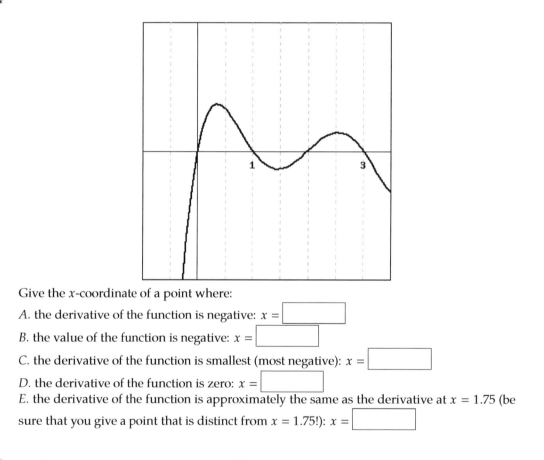

Give the x-coordinate of a point where:

A. the derivative of the function is negative: $x =$ ⬚

B. the value of the function is negative: $x =$ ⬚

C. the derivative of the function is smallest (most negative): $x =$ ⬚

D. the derivative of the function is zero: $x =$ ⬚

E. the derivative of the function is approximately the same as the derivative at $x = 1.75$ (be sure that you give a point that is distinct from $x = 1.75$!): $x =$ ⬚

2. The figure below shows a function $g(x)$ and its tangent line at the point $B = (2.6, 8)$. If the point A on the tangent line is $(2.55, 8.06)$, fill in the blanks below to complete the statements about the function g at the point B.

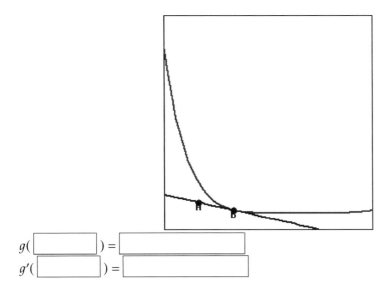

$g\left(\boxed{}\right) = \boxed{}$

$g'\left(\boxed{}\right) = \boxed{}$

3. Consider the graph of the function $f(x)$ shown below.

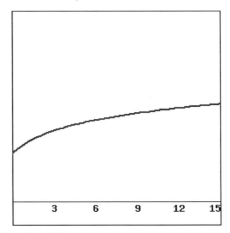

Using this graph, for each of the following pairs of numbers decide which is larger. *Be sure that you can explain your answer.*

A. $f(9)$ [Choose: < | = | >] $f(12)$

B. $f(9) - f(6)$ [Choose: < | = | >] $f(6) - f(3)$

C. $\frac{f(6)-f(3)}{6-3}$ [Choose: < | = | >] $\frac{f(9)-f(3)}{9-3}$

D. $f'(3)$ [Choose: < | = | >] $f'(12)$

4. Find the derivative of $g(t) = t^2 + 6t$ at $t = 3$ algebraically.

$g'(3) = \boxed{}$

5. Estimate $f'(3)$ for $f(x) = 3^x$. Be sure your answer is accurate to within 0.1 of the actual value.

$f'(3) \approx \boxed{}$

Be sure that you can explain your reasoning.

6. Consider the graph of $y = f(x)$ provided in Figure 1.3.12.

a. On the graph of $y = f(x)$, sketch and label the following quantities:
  - the secant line to $y = f(x)$ on the interval $[-3, -1]$ and the secant line to $y = f(x)$ on the interval $[0, 2]$.
  - the tangent line to $y = f(x)$ at $x = -3$ and the tangent line to $y = f(x)$ at $x = 0$.

b. What is the approximate value of the average rate of change of f on $[-3, -1]$? On $[0, 2]$? How are these values related to your work in (a)?

c. What is the approximate value of the instantaneous rate of change of f at $x = -3$? At $x = 0$? How are these values related to your work in (a)?

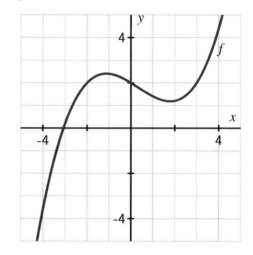

Figure 1.3.12: Plot of $y = f(x)$.

7. For each of the following prompts, sketch a graph on the provided axes in Figure 1.3.13 of a function that has the stated properties.

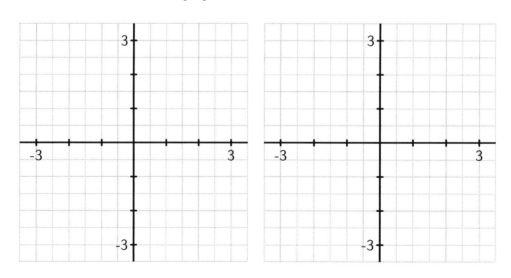

Figure 1.3.13: Axes for plotting $y = f(x)$ in (a) and $y = g(x)$ in (b).

a. $y = f(x)$ such that
- the average rate of change of f on $[-3,0]$ is -2 and the average rate of change of f on $[1,3]$ is 0.5, and
- the instantaneous rate of change of f at $x = -1$ is -1 and the instantaneous rate of change of f at $x = 2$ is 1.

b. $y = g(x)$ such that
- $\frac{g(3)-g(-2)}{5} = 0$ and $\frac{g(1)-g(-1)}{2} = -1$, and
- $g'(2) = 1$ and $g'(-1) = 0$

8. Suppose that the population, P, of China (in billions) can be approximated by the function $P(t) = 1.15(1.014)^t$ where t is the number of years since the start of 1993.

a. According to the model, what was the total change in the population of China between January 1, 1993 and January 1, 2000? What will be the average rate of change of the population over this time period? Is this average rate of change greater or less than the instantaneous rate of change of the population on January 1, 2000? Explain and justify, being sure to include proper units on all your answers.

b. According to the model, what is the average rate of change of the population of China in the ten-year period starting on January 1, 2012?

c. Write an expression involving limits that, if evaluated, would give the exact instantaneous rate of change of the population on today's date. Then estimate the value of this limit (discuss how you chose to do so) and explain the meaning (including units) of the value you have found.

d. Find an equation for the tangent line to the function $y = P(t)$ at the point where the t-value is given by today's date.

9. The goal of this problem is to compute the value of the derivative at a point for several different functions, where for each one we do so in three different ways, and then to compare the results to see that each produces the same value.

For each of the following functions, use the limit definition of the derivative to compute the value of $f'(a)$ using three different approaches: strive to use the algebraic approach first (to compute the limit exactly), then test your result using numerical evidence (with small values of h), and finally plot the graph of $y = f(x)$ near $(a, f(a))$ along with the appropriate tangent line to estimate the value of $f'(a)$ visually. Compare your findings among all three approaches; if you are unable to complete the algebraic approach, still work numerically and graphically.

a. $f(x) = x^2 - 3x$, $a = 2$

b. $f(x) = \frac{1}{x}$, $a = 1$

c. $f(x) = \sqrt{x}$, $a = 1$

d. $f(x) = 2 - |x - 1|$, $a = 1$

e. $f(x) = \sin(x)$, $a = \frac{\pi}{2}$

1.4 The derivative function

Motivating Questions

- How does the limit definition of the derivative of a function f lead to an entirely new (but related) function f'?

- What is the difference between writing $f'(a)$ and $f'(x)$?

- How is the graph of the derivative function $f'(x)$ connected to the graph of $f(x)$?

- What are some examples of functions f for which f' is not defined at one or more points?

Given a function $y = f(x)$, we now know that if we are interested in the instantaneous rate of change of the function at $x = a$, or equivalently the slope of the tangent line to $y = f(x)$ at $x = a$, we can compute the value $f'(a)$. In all of our examples to date, we have arbitrarily identified a particular value of a as our point of interest: $a = 1$, $a = 3$, etc. But it is not hard to imagine that we will often be interested in the derivative value for more than just one a-value, and possibly for many of them. In this section, we explore how we can move from computing simply $f'(1)$ or $f'(3)$ to working more generally with $f'(a)$, and indeed $f'(x)$. Said differently, we will work toward understanding how the so-called process of "taking the derivative" generates a new function that is derived from the original function $y = f(x)$. The following preview activity starts us down this path.

> **Preview Activity 1.4.1.** Consider the function $f(x) = 4x - x^2$.
>
> a. Use the limit definition to compute the derivative values: $f'(0)$, $f'(1)$, $f'(2)$, and $f'(3)$.
>
> b. Observe that the work to find $f'(a)$ is the same, regardless of the value of a. Based on your work in (a), what do you conjecture is the value of $f'(4)$? How about $f'(5)$? (Note: you should *not* use the limit definition of the derivative to find either value.)
>
> c. Conjecture a formula for $f'(a)$ that depends only on the value a. That is, in the same way that we have a formula for $f(x)$ (recall $f(x) = 4x - x^2$), see if you can use your work above to guess a formula for $f'(a)$ in terms of a.

1.4.1 How the derivative is itself a function

In your work in Preview Activity 1.4.1 with $f(x) = 4x - x^2$, you may have found several patterns. One comes from observing that $f'(0) = 4$, $f'(1) = 2$, $f'(2) = 0$, and $f'(3) = -2$. That sequence of values leads us naturally to conjecture that $f'(4) = -4$ and $f'(5) = -6$. Even more than these individual numbers, if we consider the role of 0, 1, 2, and 3 in the process of computing the value of the derivative through the limit definition, we observe that the particular number has very little effect on our work. To see this more clearly, we compute $f'(a)$, where a represents a number to be named later. Following the now standard process of using the limit definition of the derivative,

$$f'(a) = \lim_{h \to 0} \frac{f(a+h) - f(a)}{h} = \lim_{h \to 0} \frac{4(a+h) - (a+h)^2 - (4a - a^2)}{h}$$

$$= \lim_{h \to 0} \frac{4a + 4h - a^2 - 2ha - h^2 - 4a + a^2}{h} = \lim_{h \to 0} \frac{4h - 2ha - h^2}{h}$$

$$= \lim_{h \to 0} \frac{h(4 - 2a - h)}{h} = \lim_{h \to 0}(4 - 2a - h).$$

Here we observe that neither 4 nor $2a$ depend on the value of h, so as $h \to 0$, $(4 - 2a - h) \to (4 - 2a)$. Thus, $f'(a) = 4 - 2a$.

This observation is consistent with the specific values we found above: e.g., $f'(3) = 4-2(3) = -2$. And indeed, our work with a confirms that while the particular value of a at which we evaluate the derivative affects the value of the derivative, that value has almost no bearing on the process of computing the derivative. We note further that the letter being used is immaterial: whether we call it a, x, or anything else, the derivative at a given value is simply given by "4 minus 2 times the value." We choose to use x for consistency with the original function given by $y = f(x)$, as well as for the purpose of graphing the derivative function, and thus we have found that for the function $f(x) = 4x - x^2$, it follows that $f'(x) = 4 - 2x$.

Because the value of the derivative function is linked to the graphical behavior of the original function, it makes sense to look at both of these functions plotted on the same domain.

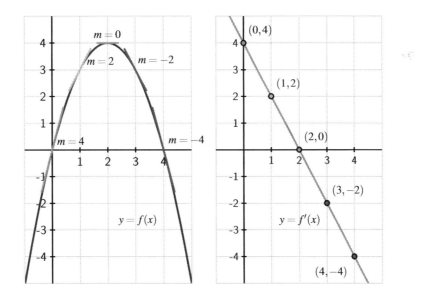

Figure 1.4.1: The graphs of $f(x) = 4x - x^2$ (at left) and $f'(x) = 4 - 2x$ (at right). Slopes on the graph of f correspond to heights on the graph of f'.

In Figure 1.4.1, on the left we show a plot of $f(x) = 4x - x^2$ together with a selection of tangent lines at the points we've considered above. On the right, we show a plot of $f'(x) = 4 - 2x$ with emphasis on the heights of the derivative graph at the same selection of points. Notice the connection between colors in the left and right graph: the green tangent line on the original graph is tied to the green point on the right graph in the following way: *the slope of the tangent line* at a point on the lefthand graph is the same as the *height* at the corresponding point on the righthand graph. That is, at each respective value of x, the slope of the tangent line to the original function at that x-value is the same as the height of the derivative function at that x-value. Do note, however, that the units on the vertical axes are different: in the left graph, the vertical units are simply the output units of f. On the righthand graph of $y = f'(x)$, the units on the vertical axis are units of f per unit of x.

Of course, this relationship between the graph of a function $y = f(x)$ and its derivative is a dynamic one. An excellent way to explore how the graph of $f(x)$ generates the graph of $f'(x)$ is through a java applet. See, for instance, the applets at http://gvsu.edu/s/5C or http://gvsu.edu/s/5D, via the sites of Austin and Renault[1].

In Section 1.3 when we first defined the derivative, we wrote the definition in terms of a value a to find $f'(a)$. As we have seen above, the letter a is merely a placeholder, and it often makes more sense to use x instead. For the record, here we restate the definition of the derivative.

Definition 1.4.2. Let f be a function and x a value in the function's domain. We define the **derivative of f with respect to x at the value** x, denoted $f'(x)$, by the formula $f'(x) = \lim_{h \to 0} \frac{f(x+h) - f(x)}{h}$, provided this limit exists.

We now may take two different perspectives on thinking about the derivative function: given a graph of $y = f(x)$, how does this graph lead to the graph of the derivative function $y = f'(x)$? and given a formula for $y = f(x)$, how does the limit definition of the derivative generate a formula for $y = f'(x)$? Both of these issues are explored in the following activities.

> **Activity 1.4.2.** For each given graph of $y = f(x)$, sketch an approximate graph of its derivative function, $y = f'(x)$, on the axes immediately below. The scale of the grid for the graph of f is 1×1; assume the horizontal scale of the grid for the graph of f' is identical to that for f. If necessary, adjust and label the vertical scale on the axes for f'.

---

[1]David Austin, http://gvsu.edu/s/5r; Marc Renault, http://gvsu.edu/s/5p.

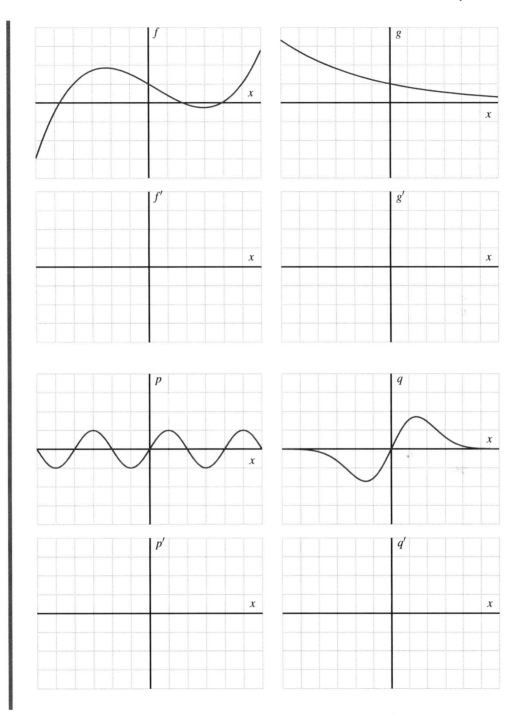

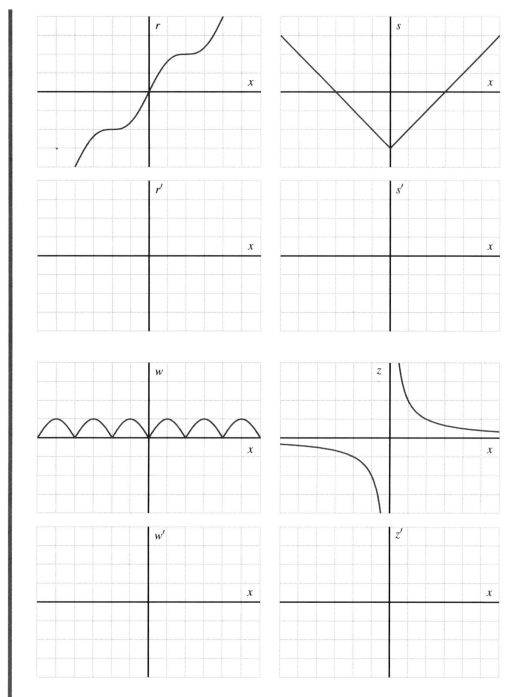

When you are finished with all 8 graphs, write several sentences that describe your overall process for sketching the graph of the derivative function, given the graph the original function. What are the values of the derivative function that you tend to identify first? What do you do thereafter? How do key traits of the graph of the derivative function exemplify properties of the graph of the original function?

For a dynamic investigation that allows you to experiment with graphing f' when given the graph of f, see http://gvsu.edu/s/8y.[2]

Now, recall the opening example of this section: we began with the function $y = f(x) = 4x - x^2$ and used the limit definition of the derivative to show that $f'(a) = 4 - 2a$, or equivalently that $f'(x) = 4-2x$. We subsequently graphed the functions f and f' as shown in Figure 1.4.1. Following Activity 1.4.2, we now understand that we could have constructed a fairly accurate graph of $f'(x)$ *without* knowing a formula for either f or f'. At the same time, it is ideal to know a formula for the derivative function whenever it is possible to find one.

In the next activity, we further explore the more algebraic approach to finding $f'(x)$: given a formula for $y = f(x)$, the limit definition of the derivative will be used to develop a formula for $f'(x)$.

Activity 1.4.3. For each of the listed functions, determine a formula for the derivative function. For the first two, determine the formula for the derivative by thinking about the nature of the given function and its slope at various points; do not use the limit definition. For the latter four, use the limit definition. Pay careful attention to the function names and independent variables. It is important to be comfortable with using letters other than f and x. For example, given a function $p(z)$, we call its derivative $p'(z)$.

a. $f(x) = 1$

b. $g(t) = t$

c. $p(z) = z^2$

d. $q(s) = s^3$

e. $F(t) = \frac{1}{t}$

f. $G(y) = \sqrt{y}$

Summary

- The limit definition of the derivative, $f'(x) = \lim_{h \to 0} \frac{f(x+h)-f(x)}{h}$, produces a value for each x at which the derivative is defined, and this leads to a new function whose formula is $y = f'(x)$. Hence we talk both about a given function f and its derivative f'. It is especially important to note that taking the derivative is a process that starts with a given function (f) and produces a new, related function (f').

- There is essentially no difference between writing $f'(a)$ (as we did regularly in Section 1.3) and writing $f'(x)$. In either case, the variable is just a placeholder that is used to define the rule for the derivative function.

- Given the graph of a function $y = f(x)$, we can sketch an approximate graph of its

[2]Marc Renault, Calculus Applets Using Geogebra.

derivative $y = f'(x)$ by observing that *heights* on the derivative's graph correspond to *slopes* on the original function's graph.

- In Activity 1.4.2, we encountered some functions that had sharp corners on their graphs, such as the shifted absolute value function. At such points, the derivative fails to exist, and we say that f is not differentiable there. For now, it suffices to understand this as a consequence of the jump that must occur in the derivative function at a sharp corner on the graph of the original function.

Exercises

1. Consider the function $f(x)$ shown in the graph below.

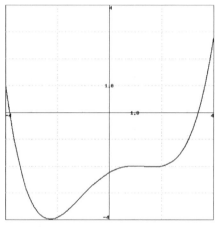

Carefully sketch the derivative function of the given function (you will want to estimate values on the derivative function at different x values as you do this). Use your derivative function graph to estimate the following values on the derivative function.

at $x =$	-3	-1	1	3
the derivative is				

2. Find a formula for the derivative of the function $g(x) = 3x^2 - 4$ using difference quotients:

$g'(x) = \lim\limits_{h \to 0} [($ _____ $)/h]$

$=$ _____ .

(In the first answer blank, fill in the numerator of the difference quotient you use to evaluate the derivative. In the second, fill out the derivative you obtain after completing the limit calculation.)

3. For the function $f(x)$ shown in the graph below, sketch a graph of the derivative. You will then be picking which of the following is the correct derivative graph, but should be sure to first sketch the derivative yourself.

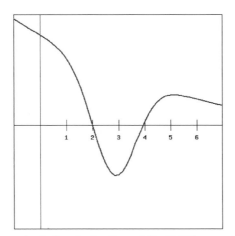

Which of the following graphs is the derivative of $f(x)$? [Choose: 1 | 2 | 3 | 4 | 5 | 6 | 7 | 8]

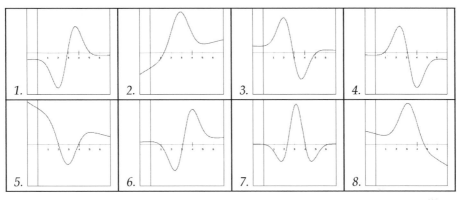

1. 2. 3. 4.

5. 6. 7. 8.

4. The graph of a function f is shown below.

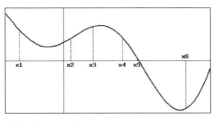

At which of the labeled x-values is
$f(x)$ least? x = [Choose: x1 | x2 | x3 | x4 | x5 | x6]
$f(x)$ greatest? x = [Choose: x1 | x2 | x3 | x4 | x5 | x6]
$f'(x)$ least? x = [Choose: x1 | x2 | x3 | x4 | x5 | x6]
$f'(x)$ greatest? x = [Choose: x1 | x2 | x3 | x4 | x5 | x6]

5. Let

$$f(x) = \frac{1}{x - 3}$$

41

Find

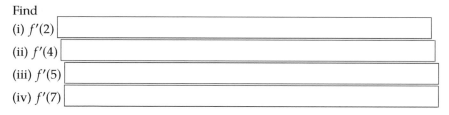

(i) $f'(2)$

(ii) $f'(4)$

(iii) $f'(5)$

(iv) $f'(7)$

6. Let f be a function with the following properties: f is differentiable at every value of x (that is, f has a derivative at every point), $f(-2) = 1$, and $f'(-2) = -2$, $f'(-1) = -1$, $f'(0) = 0$, $f'(1) = 1$, and $f'(2) = 2$.

    a. On the axes provided at left in Figure 1.4.3, sketch a possible graph of $y = f(x)$. Explain why your graph meets the stated criteria.

    b. On the axes at right in Figure 1.4.3, sketch a possible graph of $y = f'(x)$. What type of curve does the provided data suggest for the graph of $y = f'(x)$?

    c. Conjecture a formula for the function $y = f(x)$. Use the limit definition of the derivative to determine the corresponding formula for $y = f'(x)$. Discuss both graphical and algebraic evidence for whether or not your conjecture is correct.

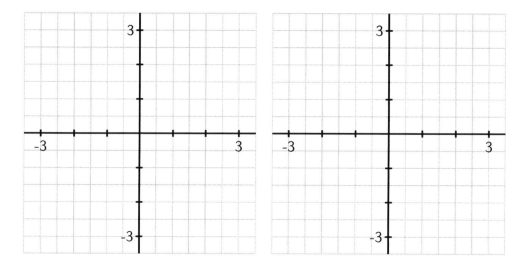

Figure 1.4.3: Axes for plotting $y = f(x)$ in (a) and $y = f'(x)$ in (b).

7. Consider the function $g(x) = x^2 - x + 3$.

    a. Use the limit definition of the derivative to determine a formula for $g'(x)$.

    b. Use a graphing utility to plot both $y = g(x)$ and your result for $y = g'(x)$; does your formula for $g'(x)$ generate the graph you expected?

    c. Use the limit definition of the derivative to find a formula for $p'(x)$ where $p(x) = 5x^2 - 4x + 12$.

d. Compare and contrast the formulas for $g'(x)$ and $p'(x)$ you have found. How do the constants 5, 4, 12, and 3 affect the results?

8. Let g be a continuous function (that is, one with no jumps or holes in the graph) and suppose that a graph of $y = g'(x)$ is given by the graph on the right in Figure 1.4.4.

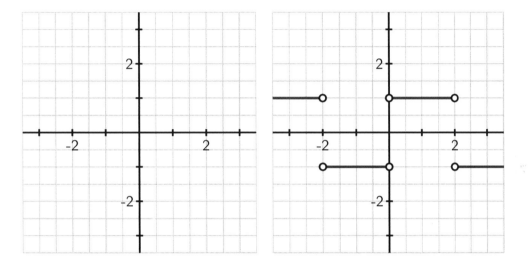

Figure 1.4.4: Axes for plotting $y = g(x)$ and, at right, the graph of $y = g'(x)$.

a. Observe that for every value of x that satisfies $0 < x < 2$, the value of $g'(x)$ is constant. What does this tell you about the behavior of the graph of $y = g(x)$ on this interval?

b. On what intervals other than $0 < x < 2$ do you expect $y = g(x)$ to be a linear function? Why?

c. At which values of x is $g'(x)$ not defined? What behavior does this lead you to expect to see in the graph of $y = g(x)$?

d. Suppose that $g(0) = 1$. On the axes provided at left in Figure 1.4.4, sketch an accurate graph of $y = g(x)$.

9. For each graph that provides an original function $y = f(x)$ in Figure 1.4.5, your task is to sketch an approximate graph of its derivative function, $y = f'(x)$, on the axes immediately below. View the scale of the grid for the graph of f as being 1×1, and assume the horizontal scale of the grid for the graph of f' is identical to that for f. If you need to adjust the vertical scale on the axes for the graph of f', you should label that accordingly.

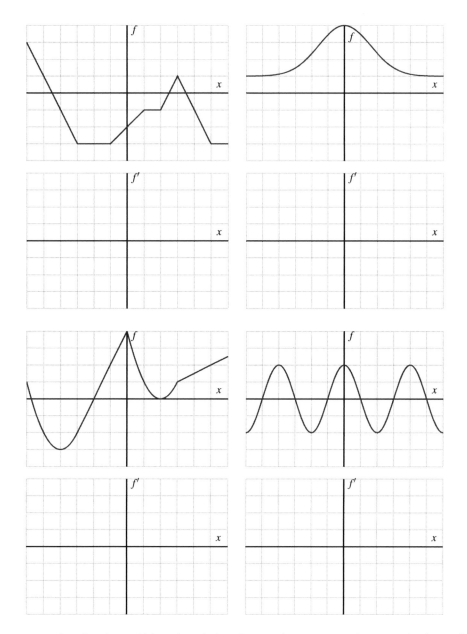

Figure 1.4.5: Graphs of $y = f(x)$ and grids for plotting the corresponding graph of $y = f'(x)$.

1.5 Interpreting, estimating, and using the derivative

Motivating Questions

- In contexts other than the position of a moving object, what does the derivative of a function measure?

- What are the units on the derivative function f', and how are they related to the units of the original function f?

- What is a central difference, and how can one be used to estimate the value of the derivative at a point from given function data?

- Given the value of the derivative of a function at a point, what can we infer about how the value of the function changes nearby?

An interesting and powerful feature of mathematics is that it can often be thought of both in abstract terms and in applied ones. For instance, calculus can be developed almost entirely as an abstract collection of ideas that focus on properties of arbitrary functions. At the same time, calculus can also be very directly connected to our experience of physical reality by considering functions that represent meaningful processes. We have already seen that for a position function $y = s(t)$, say for a ball being tossed straight up in the air, the ball's velocity at time t is given by $v(t) = s'(t)$, the derivative of the position function. Further, recall that if $s(t)$ is measured in feet at time t, the units on $v(t) = s'(t)$ are feet per second.

In what follows in this section, we investigate several different functions, each with specific physical meaning, and think about how the units on the independent variable, dependent variable, and the derivative function add to our understanding. To start, we consider the familiar problem of a position function of a moving object.

> **Preview Activity 1.5.1.** One of the longest stretches of straight (and flat) road in North America can be found on the Great Plains in the state of North Dakota on state highway 46, which lies just south of the interstate highway I-94 and runs through the town of Gackle. A car leaves town (at time $t = 0$) and heads east on highway 46; its position in miles from Gackle at time t in minutes is given by the graph of the function in Figure 1.5.1. Three important points are labeled on the graph; where the curve looks linear, assume that it is indeed a straight line.
>
> a. In everyday language, describe the behavior of the car over the provided time interval. In particular, discuss what is happening on the time intervals $[57, 68]$ and $[68, 104]$.
>
> b. Find the slope of the line between the points $(57, 63.8)$ and $(104, 106.8)$. What are the units on this slope? What does the slope represent?

c. Find the average rate of change of the car's position on the interval $[68, 104]$. Include units on your answer.

d. Estimate the instantaneous rate of change of the car's position at the moment $t = 80$. Write a sentence to explain your reasoning and the meaning of this value.

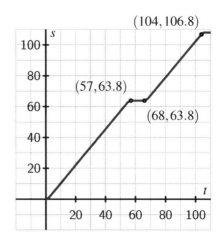

Figure 1.5.1: The graph of $y = s(t)$, the position of the car along highway 46, which tells its distance in miles from Gackle, ND, at time t in minutes.

1.5.1 Units of the derivative function

As we now know, the derivative of the function f at a fixed value x is given by

$$f'(x) = \lim_{h \to 0} \frac{f(x + h) - f(x)}{h},$$

and this value has several different interpretations. If we set $x = a$, one meaning of $f'(a)$ is the slope of the tangent line at the point $(a, f(a))$.

In alternate notation, we also sometimes equivalently write $\frac{df}{dx}$ or $\frac{dy}{dx}$ instead of $f'(x)$, and these notations helps us to further see the units (and thus the meaning) of the derivative as it is viewed as *the instantaneous rate of change of f with respect to x*. Note that the units on the slope of the secant line, $\frac{f(x+h)-f(x)}{h}$, are "units of f per unit of x." Thus, when we take the limit to get $f'(x)$, we get these same units on the derivative $f'(x)$: units of f per unit of x. Regardless of the function f under consideration (and regardless of the variables being used), it is helpful to remember that the units on the derivative function are "units of output per unit of input," in terms of the input and output of the original function.

For example, say that we have a function $y = P(t)$, where P measures the population of a city (in thousands) at the start of year t (where $t = 0$ corresponds to 2010 AD), and we are told that $P'(2) = 21.37$. What is the meaning of this value? Well, since P is measured in thousands and t is measured in years, we can say that the instantaneous rate of change of the city's population with respect to time at the start of 2012 is 21.37 thousand people per year. We therefore expect that in the coming year, about 21,370 people will be added to the city's population.

1.5.2 Toward more accurate derivative estimates

It is also helpful to recall, as we first experienced in Section 1.3, that when we want to estimate the value of $f'(x)$ at a given x, we can use the *difference quotient* $\frac{f(x+h)-f(x)}{h}$ with a relatively small value of h. In doing so, we should use both positive and negative values of h in order to make sure we account for the behavior of the function on both sides of the point of interest. To that end, we consider the following brief example to demonstrate the notion of a *central difference* and its role in estimating derivatives.

Example 1.5.2. Suppose that $y = f(x)$ is a function for which three values are known: $f(1) = 2.5$, $f(2) = 3.25$, and $f(3) = 3.625$. Estimate $f'(2)$.

Solution. We know that $f'(2) = \lim_{h\to 0} \frac{f(2+h)-f(2)}{h}$. But since we don't have a graph for $y = f(x)$ nor a formula for the function, we can neither sketch a tangent line nor evaluate the limit exactly. We can't even use smaller and smaller values of h to estimate the limit. Instead, we have just two choices: using $h = -1$ or $h = 1$, depending on which point we pair with $(2, 3.25)$.

So, one estimate is

$$f'(2) \approx \frac{f(1) - f(2)}{1 - 2} = \frac{2.5 - 3.25}{-1} = 0.75.$$

The other is

$$f'(2) \approx \frac{f(3) - f(2)}{3 - 2} = \frac{3.625 - 3.25}{1} = 0.375.$$

Since the first approximation looks only backward from the point $(2, 3.25)$ and the second approximation looks only forward from $(2, 3.25)$, it makes sense to average these two values in order to account for behavior on both sides of the point of interest. Doing so, we find that

$$f'(2) \approx \frac{0.75 + 0.375}{2} = 0.5625.$$

The intuitive approach to average the two estimates found in Example 1.5.2 is in fact the best possible estimate to $f'(2)$ when we have just two function values for f on opposite sides of the point of interest.

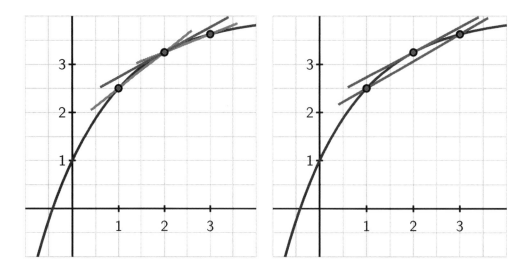

Figure 1.5.3: At left, the graph of $y = f(x)$ along with the secant line through $(1, 2.5)$ and $(2, 3.25)$, the secant line through $(2, 3.25)$ and $(3, 3.625)$, as well as the tangent line. At right, the same graph along with the secant line through $(1, 2.5)$ and $(3, 3.625)$, plus the tangent line.

To see why, we think about the diagram in Figure 1.5.3, which shows a possible function $y = f(x)$ that satisfies the data given in Example 1.5.2. On the left, we see the two secant lines with slopes that come from computing the *backward difference* $\frac{f(1)-f(2)}{1-2} = 0.75$ and from the *forward difference* $\frac{f(3)-f(2)}{3-2} = 0.375$. Note how the first such line's slope over-estimates the slope of the tangent line at $(2, f(2))$, while the second line's slope underestimates $f'(2)$. On the right, however, we see the secant line whose slope is given by the *central difference*

$$\frac{f(3) - f(1)}{3 - 1} = \frac{3.625 - 2.5}{2} = \frac{1.125}{2} = 0.5625.$$

Note that this central difference has the exact same value as the average of the forward difference and backward difference (and it is straightforward to explain why this always holds), and moreover that the central difference yields a very good approximation to the derivative's value, in part because the secant line that uses both a point before and after the point of tangency yields a line that is closer to being parallel to the tangent line.

In general, the central difference approximation to the value of the first derivative is given by

$$f'(a) \approx \frac{f(a + h) - f(a - h)}{2h},$$

and this quantity measures the slope of the secant line to $y = f(x)$ through the points $(a - h, f(a - h))$ and $(a + h, f(a + h))$. Anytime we have symmetric data surrounding a point at which we desire to estimate the derivative, the central difference is an ideal choice for so doing.

The following activities will further explore the meaning of the derivative in several contexts while viewing the derivative from graphical, numerical, and algebraic perspectives.

Activity 1.5.2. A potato is placed in an oven, and the potato's temperature F (in degrees Fahrenheit) at various points in time is taken and recorded in the following table. Time t is measured in minutes.

t	0	15	30	45	60	75	90
$F(t)$	70	180.5	251	296	324.5	342.8	354.5

Table 1.5.4: Temperature data in degrees Fahrenheit.

a. Use a central difference to estimate the instantaneous rate of change of the temperature of the potato at $t = 30$. Include units on your answer.

b. Use a central difference to estimate the instantaneous rate of change of the temperature of the potato at $t = 60$. Include units on your answer.

c. Without doing any calculation, which do you expect to be greater: $F'(75)$ or $F'(90)$? Why?

d. Suppose it is given that $F(64) = 330.28$ and $F'(64) = 1.341$. What are the units on these two quantities? What do you expect the temperature of the potato to be when $t = 65$? when $t = 66$? Why?

e. Write a couple of careful sentences that describe the behavior of the temperature of the potato on the time interval $[0, 90]$, as well as the behavior of the instantaneous rate of change of the temperature of the potato on the same time interval.

Activity 1.5.3. A company manufactures rope, and the total cost of producing r feet of rope is $C(r)$ dollars.

a. What does it mean to say that $C(2000) = 800$?

b. What are the units of $C'(r)$?

c. Suppose that $C(2000) = 800$ and $C'(2000) = 0.35$. Estimate $C(2100)$, and justify your estimate by writing at least one sentence that explains your thinking.

d. Do you think $C'(2000)$ is less than, equal to, or greater than $C'(3000)$? Why?

e. Suppose someone claims that $C'(5000) = -0.1$. What would the practical meaning of this derivative value tell you about the approximate cost of the next foot of rope? Is this possible? Why or why not?

Activity 1.5.4. Researchers at a major car company have found a function that relates gasoline consumption to speed for a particular model of car. In particular, they have determined that the consumption C, in liters per kilometer, at a given speed s, is given by a function $C = f(s)$, where s is the car's speed in *kilometers per hour*.

a. Data provided by the car company tells us that $f(80) = 0.015$, $f(90) = 0.02$, and $f(100) = 0.027$. Use this information to estimate the instantaneous rate of change of fuel consumption with respect to speed at $s = 90$. Be as accurate as possible, use proper notation, and include units on your answer.

b. By writing a complete sentence, interpret the meaning (in the context of fuel consumption) of "$f(80) = 0.015$."

c. Write at least one complete sentence that interprets the meaning of the value of $f'(90)$ that you estimated in (a).

In Section 1.4, we learned how use to the graph of a given function f to plot the graph of its derivative, f'. It is important to remember that when we do so, not only does the scale on the vertical axis often have to change to accurately represent f', but the units on that axis also differ. For example, suppose that $P(t) = 400 - 330e^{-0.03t}$ tells us the temperature in degrees Fahrenheit of a potato in an oven at time t in minutes. In Figure 1.5.5, we sketch the graph of P on the left and the graph of P' on the right.

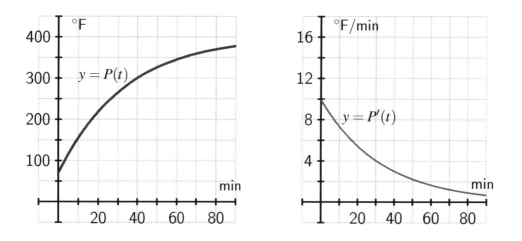

Figure 1.5.5: Plot of $P(t) = 400 - 330e^{-0.03t}$ at left, and its derivative $P'(t)$ at right.

Note how not only are the vertical scales different in size, but different in units, as the units of P are °F, while those of P' are °F/min. In all cases where we work with functions that have an applied context, it is helpful and instructive to think carefully about units involved and how they further inform the meaning of our computations.

Summary

- Regardless of the context of a given function $y = f(x)$, the derivative always measures the instantaneous rate of change of the output variable with respect to the input variable.

- The units on the derivative function $y = f'(x)$ are units of f per unit of x. Again, this measures how fast the output of the function f changes when the input of the function changes.

- The central difference approximation to the value of the first derivative is given by

$$f'(a) \approx \frac{f(a + h) - f(a - h)}{2h},$$

  and this quantity measures the slope of the secant line to $y = f(x)$ through the points $(a - h, f(a - h))$ and $(a + h, f(a + h))$. The central difference generates a good approximation of the derivative's value any time we have symmetric data surrounding a point of interest.

- Knowing the derivative and function values at a single point enables us to estimate other function values nearby. If, for example, we know that $f'(7) = 2$, then we know that at $x = 7$, the function f is increasing at an instantaneous rate of 2 units of output for every one unit of input. Thus, we expect $f(8)$ to be approximately 2 units greater than $f(7)$. The value is approximate because we don't know that the rate of change stays the same as x changes.

Exercises

1. The temperature, H, in degrees Celsius, of a cup of coffee placed on the kitchen counter is given by $H = f(t)$, where t is in minutes since the coffee was put on the counter.
(a) Is $f'(t)$ positive or negative? [Choose: positive | negative]
(*Be sure that you are able to give a reason for your answer.*)
(b) What are the units of $f'(35)$? [_____] help (units)
Suppose that $|f'(35)| = 1.2$ and $f(35) = 52$. Fill in the blanks (including units where needed) and select the appropriate terms to complete the following statement about the temperature of the coffee in this case.
At [____] minutes after the coffee was put on the counter, its [Choose: derivative | temperature | change in temperature] is [_____] and will [Choose: increase | decrease] by about [_____] in the next 90 seconds.

2. The cost, C (in dollars) to produce g gallons of ice cream can be expressed as $C = f(g)$.
(a) In the expression $f(300) = 350$,
what are the units of 300? [Choose: ? | dollars | gallons | dollars*gallons | dollars/gallon | gallons/dollar]
what are the units of 350? [Choose: ? | dollars | gallons | dollars*gallons | dollars/gallon | gallons/dollar]

(b) In the expression $f'(300) = 1.2$,

what are the units of 300? [Choose: ? | dollars | gallons | dollars*gallons | dollars/gallon | gallons/dollar]

what are the units of 1.2? [Choose: ? | dollars | gallons | dollars*gallons | dollars/gallon | gallons/dollar]

(Be sure that you can carefully put into words the meanings of each of these statement in terms of ice cream and money.)

3. A laboratory study investigating the relationship between diet and weight in adult humans found that the weight of a subject, W, in pounds, was a function, $W = f(c)$, of the average number of Calories, c, consumed by the subject in a day.

(a) In the statement $f(1500) = 150$

what are the units of 1500? [Choose: lb | cal | day | lb/cal | cal/lb | cal/day | lb/day | day/lb | day/cal]

what are the units of 150? [Choose: lb | cal | day | lb/cal | cal/lb | cal/day | lb/day | day/lb | day/cal]

(Think about what this statement means in terms of the weight of the subject and the number of calories that the subject consumes.)

(b) In the statement $f'(2000) = 0$,

what are the units of 2000? [Choose: lb | cal | day | lb/cal | cal/lb | cal/day | lb/day | day/lb | day/cal]

what are the units of 0? [Choose: lb | cal | day | lb/cal | cal/lb | cal/day | lb/day | day/lb | day/cal]

(Think about what this statement means in terms of the weight of the subject and the number of calories that the subject consumes.)

(c) In the statement $f^{-1}(157) = 2300$,

what are the units of 157? [Choose: lb | cal | day | lb/cal | cal/lb | cal/day | lb/day | day/lb | day/cal]

what are the units of 2300? [Choose: lb | cal | day | lb/cal | cal/lb | cal/day | lb/day | day/lb | day/cal]

(Think about what this statement means in terms of the weight of the subject and the number of calories that the subject consumes.)

(d) What are the units of $f'(c) = dW/dc$? [Choose: lb | cal | day | lb/cal | cal/lb | cal/day | lb/day | day/lb | day/cal]

(e) Suppose that Sam reads about f' in this study and draws the following conclusion: If Sam increases her average calorie intake from 2800 to 2840 calories per day, then her weight will increase by approximately 0.8 pounds.

Fill in the blanks below so that the equation supports her conclusion.

$f'\left(\rule{8cm}{0.4pt}\right) = \rule{6cm}{0.4pt}$

4. The displacement (in meters) of a particle moving in a straight line is given by

$$s = t^2 - 5t + 12,$$

where t is measured in seconds.

(A)

(i) Find the average velocity over the time interval [3,4].

Average Velocity = [   ]
(ii) Find the average velocity over the time interval [3.5,4].

Average Velocity = [   ]
(iii) Find the average velocity over the time interval [4,5].

Average Velocity = [   ]
(iv) Find the average velocity over the time interval [4,4.5].

Average Velocity = [   ]
(B) Find the instantaneous velocity when $t = 4$.

Instantaneous velocity = [   ]

5. A cup of coffee has its temperature F (in degrees Fahrenheit) at time t given by the function $F(t) = 75 + 110e^{-0.05t}$, where time is measured in minutes.

   a. Use a central difference with $h = 0.01$ to estimate the value of $F'(10)$.

   b. What are the units on the value of $F'(10)$ that you computed in (a)? What is the practical meaning of the value of $F'(10)$?

   c. Which do you expect to be greater: $F'(10)$ or $F'(20)$? Why?

   d. Write a sentence that describes the behavior of the function $y = F'(t)$ on the time interval $0 \leq t \leq 30$. How do you think its graph will look? Why?

6. The temperature change T (in Fahrenheit degrees), in a patient, that is generated by a dose q (in milliliters), of a drug, is given by the function $T = f(q)$.

   a. What does it mean to say $f(50) = 0.75$? Write a complete sentence to explain, using correct units.

   b. A person's sensitivity, s, to the drug is defined by the function $s(q) = f'(q)$. What are the units of sensitivity?

   c. Suppose that $f'(50) = -0.02$. Write a complete sentence to explain the meaning of this value. Include in your response the information given in (a).

7. The velocity of a ball that has been tossed vertically in the air is given by $v(t) = 16 - 32t$, where v is measured in feet per second, and t is measured in seconds. The ball is in the air from $t = 0$ until $t = 2$.

   a. When is the ball's velocity greatest?

   b. Determine the value of $v'(1)$. Justify your thinking.

   c. What are the units on the value of $v'(1)$? What does this value and the corresponding units tell you about the behavior of the ball at time $t = 1$?

   d. What is the physical meaning of the function $v'(t)$?

8. The value, V, of a particular automobile (in dollars) depends on the number of miles, m, the car has been driven, according to the function $V = h(m)$.

   a. Suppose that $h(40000) = 15500$ and $h(55000) = 13200$. What is the average rate of change of h on the interval $[40000, 55000]$, and what are the units on this value?

b. In addition to the information given in (a), say that $h(70000) = 11100$. Determine the best possible estimate of $h'(55000)$ and write one sentence to explain the meaning of your result, including units on your answer.

c. Which value do you expect to be greater: $h'(30000)$ or $h'(80000)$? Why?

d. Write a sentence to describe the long-term behavior of the function $V = h(m)$, plus another sentence to describe the long-term behavior of $h'(m)$. Provide your discussion in practical terms regarding the value of the car and the rate at which that value is changing.

1.6 The second derivative

Motivating Questions

- How does the derivative of a function tell us whether the function is increasing or decreasing at a point or on an interval?

- What can we learn by taking the derivative of the derivative (to achieve the *second* derivative) of a function f?

- What does it mean to say that a function is concave up or concave down? How are these characteristics connected to certain properties of the derivative of the function?

- What are the units of the second derivative? How do they help us understand the rate of change of the rate of change?

Given a differentiable function $y = f(x)$, we know that its derivative, $y = f'(x)$, is a related function whose output at a value $x = a$ tells us the slope of the tangent line to $y = f(x)$ at the point $(a, f(a))$. That is, heights on the derivative graph tell us the values of slopes on the original function's graph. Therefore, the derivative tells us important information about the function f.

At any point where $f'(x)$ is positive, it means that the slope of the tangent line to f is positive. On an interval where $f'(x)$ is positive, we therefore say the function f is increasing (or rising). Similarly, if $f'(x)$ is negative on an interval, we know that the graph of f is decreasing (or falling).

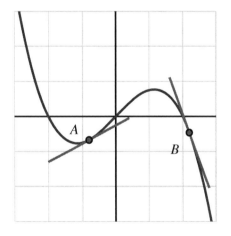

Figure 1.6.1: Two tangent lines on a graph.

In the next part of our study, we work to understand not only *whether* the function f is increasing or decreasing on an interval, but also *how* the function f is increasing or decreasing. Comparing the two tangent lines shown in Figure 1.6.1, we see that at point A, the value of $f'(x)$ is positive and relatively close to zero, which coincides with the graph rising slowly. By contrast, at point B, the derivative is negative and relatively large in absolute value, which is tied to the fact that f is decreasing rapidly at B. It also makes sense to not only ask whether the value of the derivative function is positive or negative and whether the derivative is large or small, but also to ask "how is the derivative changing?"

We also now know that the derivative, $y = f'(x)$, is itself a function. This means that we can consider taking its derivative — the derivative of the derivative — and therefore ask questions like "what does the derivative of the derivative tell us about how the original function behaves?" As we have done regularly in our work to date, we start with an investigation of a familiar problem in the context of a moving object.

Preview Activity 1.6.1.

The position of a car driving along a straight road at time t in minutes is given by the function $y = s(t)$ that is pictured in Figure 1.6.2. The car's position function has units measured in thousands of feet. For instance, the point $(2, 4)$ on the graph indicates that after 2 minutes, the car has traveled 4000 feet.

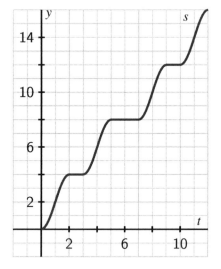

Figure 1.6.2: The graph of $y = s(t)$, the position of the car (measured in thousands of feet from its starting location) at time t in minutes.

a. In everyday language, describe the behavior of the car over the provided time interval. In particular, you should carefully discuss what is happening on each of the time intervals $[0, 1]$, $[1, 2]$, $[2, 3]$, $[3, 4]$, and $[4, 5]$, plus provide commentary overall on what the car is doing on the interval $[0, 12]$.

b. On the lefthand axes provided in Figure 1.6.3, sketch a careful, accurate graph of $y = s'(t)$.

c. What is the meaning of the function $y = s'(t)$ in the context of the given problem? What can we say about the car's behavior when $s'(t)$ is positive? when $s'(t)$ is zero? when $s'(t)$ is negative?

d. Rename the function you graphed in (b) to be called $y = v(t)$. Describe the behavior of v in words, using phrases like "v is increasing on the interval ..." and "v is constant on the interval ...."

e. Sketch a graph of the function $y = v'(t)$ on the righthand axes provide in Figure 1.6.3. Write at least one sentence to explain how the behavior of $v'(t)$ is connected to the graph of $y = v(t)$.

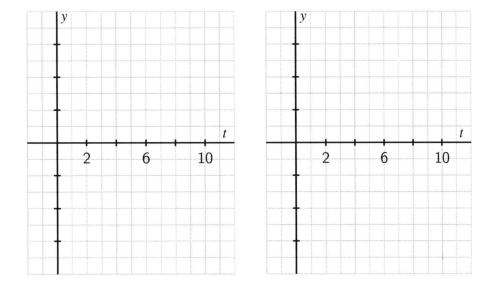

Figure 1.6.3: Axes for plotting $y = v(t) = s'(t)$ and $y = v'(t)$.

1.6.1 Increasing, decreasing, or neither

When we look at the graph of a function, there are features that strike us naturally, and common language can be used to name these features. In many different settings so far, we have intuitively used the words *increasing* and *decreasing* to describe a function's graph. Here we connect these terms more formally to a function's behavior on an interval of input values.

Definition 1.6.4. Given a function $f(x)$ defined on the interval (a, b), we say that f **is increasing on** (a, b) provided that for all x, y in the interval (a, b), if $x < y$, then $f(x) < f(y)$. Similarly, we say that f **is decreasing on** (a, b) provided that for all x, y in the interval (a, b), if $x < y$, then $f(x) > f(y)$.

Simply put, an increasing function is one that is rising as we move from left to right along the graph, and a decreasing function is one that falls as the value of the input increases. For a function that has a derivative, we can use the sign of the derivative to determine whether or not the function is increasing or decreasing.

Let f be a function that is differentiable on an interval (a, b). We say that f is increasing on (a, b) if and only if $f'(x) > 0$ for every x such that $a < x < b$; similarly, f is decreasing on (a, b) if and only if $f'(x) < 0$. If $f'(a) = 0$, then we say f is neither increasing nor decreasing at $x = a$.

For example, the function pictured in Figure 1.6.5 is increasing on the entire interval $-2 < x < 0$. Note that at both $x = \pm 2$ and $x = 0$, we say that f is neither increasing nor decreasing, because $f'(x) = 0$ at these values.

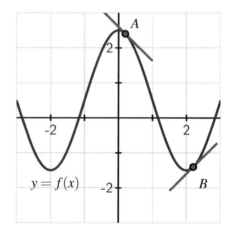

Figure 1.6.5: A function that is decreasing on the intervals $-3 < x < -2$ and $0 < x < 2$ and increasing on $-2 < x < 0$ and $2 < x < 3$.

1.6.2 The Second Derivative

For any function, we are now accustomed to investigating its behavior by thinking about its derivative. Given a function f, its derivative is a new function, one that is given by the rule

$$f'(x) = \lim_{h \to 0} \frac{f(x + h) - f(x)}{h}.$$

Because f' is itself a function, it is perfectly feasible for us to consider the derivative of the derivative, which is the new function $y = [f'(x)]'$. We call this resulting function *the second derivative* of $y = f(x)$, and denote the second derivative by $y = f''(x)$. Due to the presence of multiple possible derivatives, we will sometimes call f' "the first derivative" of f, rather than simply "the derivative" of f. Formally, the second derivative is defined by the limit definition of the derivative of the first derivative:

$$f''(x) = \lim_{h \to 0} \frac{f'(x+h) - f'(x)}{h}.$$

We note that all of the established meaning of the derivative function still holds, so when we compute $y = f''(x)$, this new function measures slopes of tangent lines to the curve $y = f'(x)$, as well as the instantaneous rate of change of $y = f'(x)$. In other words, just as the first derivative measures the rate at which the original function changes, the second derivative measures the rate at which the first derivative changes. This means that the second derivative tracks the instantaneous rate of change of the instantaneous rate of change of f. That is, the second derivative will help us to understand how the rate of change of the original function is itself changing.

1.6.3 Concavity

In addition to asking *whether* a function is increasing or decreasing, it is also natural to inquire *how* a function is increasing or decreasing. To begin, there are three basic behaviors that an increasing function can demonstrate on an interval, as pictured in Figure 1.6.6: the function can increase more and more rapidly, increase at the same rate, or increase in a way that is slowing down. Fundamentally, we are beginning to think about how a particular curve bends, with the natural comparison being made to lines, which don't bend at all. More than this, we want to understand how the bend in a function's graph is tied to behavior characterized by the first derivative of the function.

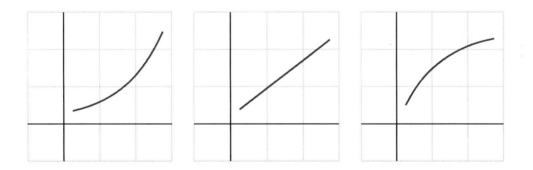

Figure 1.6.6: Three functions that are all increasing, but doing so at an increasing rate, at a constant rate, and at a decreasing rate, respectively.

For the leftmost curve in Figure 1.6.6, picture a sequence of tangent lines to the curve. As we move from left to right, the slopes of those tangent lines will increase. Therefore, the rate of change of the pictured function is increasing, and this explains why we say this function is *increasing at an increasing rate*. For the rightmost graph in Figure 1.6.6, observe that as x increases, the function increases but the slope of the tangent line decreases, hence this function is *increasing at a decreasing rate*.

Of course, similar options hold for how a function can decrease. Here we must be extra careful with our language, since decreasing functions involve negative slopes, and negative numbers present an interesting situation in the tension between common language and mathematical language. For example, it can be tempting to say that "-100 is bigger than -2." But we must remember that when we say one number is greater than another, this describes how the numbers lie on a number line: $x < y$ provided that x lies to the left of y. So of course, -100 is less than -2. Informally, it might be helpful to say that "-100 is more negative than -2." This leads us to note particularly that when a function's values are negative, and those values subsequently get more negative, the function must be decreasing.

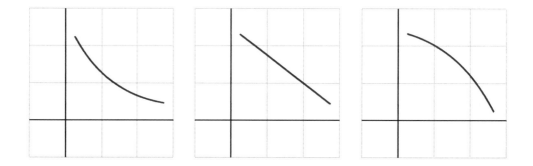

Figure 1.6.7: From left to right, three functions that are all decreasing, but doing so in different ways.

Now consider the three graphs shown in Figure 1.6.7. Clearly the middle graph demonstrates the behavior of a function decreasing at a constant rate. If we think about a sequence of tangent lines to the first curve that progress from left to right, we see that the slopes of these lines get less and less negative as we move from left to right. That means that the values of the first derivative, while all negative, are increasing, and thus we say that the leftmost curve is *decreasing at an increasing rate*.

This leaves only the rightmost curve in Figure 1.6.7 to consider. For that function, the slope of the tangent line is negative throughout the pictured interval, but as we move from left to right, the slopes get more and more negative. Hence the slope of the curve is decreasing, and we say that the function is *decreasing at a decreasing rate*.

This leads us to introduce the notion of *concavity* which provides simpler language to describe some of these behaviors. Informally, when a curve opens up on a given interval, like the upright parabola $y = x^2$ or the exponential growth function $y = e^x$, we say that the curve is *concave up* on that interval. Likewise, when a curve opens down, such as the parabola $y = -x^2$ or the opposite of the exponential function $y = -e^x$, we say that the function is *concave down*. This behavior is linked to both the first and second derivatives of the function.

In Figure 1.6.8, we see two functions along with a sequence of tangent lines to each. On the lefthand plot where the function is concave up, observe that the tangent lines to the curve always lie below the curve itself and that, as we move from left to right, the slope of the tangent line is increasing. Said differently, the function f is concave up on the interval shown because its derivative, f', is increasing on that interval. Similarly, on the righthand plot in Figure 1.6.8, where the function shown is concave down, there we see that the tangent lines alway lie above the curve and that the value of the slope of the tangent line is decreasing as we move from left to right. Hence, what makes f concave down on the interval is the fact that its derivative, f', is decreasing.

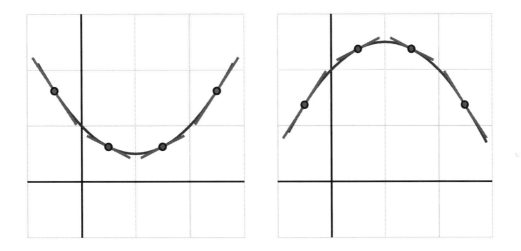

Figure 1.6.8: At left, a function that is concave up; at right, one that is concave down.

We state these most recent observations formally as the definitions of the terms *concave up* and *concave down*.

Definition 1.6.9. Let f be a differentiable function on an interval (a, b). Then f is **concave up** on (a, b) if and only if f' is increasing on (a, b); f is **concave down** on (a, b) if and only if f' is decreasing on (a, b).

The following activities lead us to further explore how the first and second derivatives of a function determine the behavior and shape of its graph. We begin by revisiting Preview Activity 1.6.1.

Activity 1.6.2. The position of a car driving along a straight road at time t in minutes is given by the function $y = s(t)$ that is pictured in Figure 1.6.10. The car's position function has units measured in thousands of feet. Remember that you worked with this function and sketched graphs of $y = v(t) = s'(t)$ and $y = v'(t)$ in Preview Activity 1.6.1.

a. On what intervals is the position function $y = s(t)$ increasing? decreasing? Why?

b. On which intervals is the velocity function $y = v(t) = s'(t)$ increasing? decreasing? neither? Why?

c. *Acceleration* is defined to be the instantaneous rate of change of velocity, as the acceleration of an object measures the rate at which the velocity of the object is changing. Say that the car's acceleration function is named $a(t)$. How is $a(t)$ computed from $v(t)$? How is $a(t)$ computed from $s(t)$? Explain.

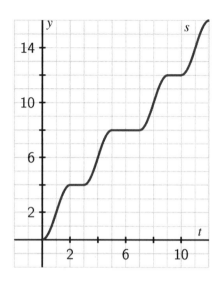

Figure 1.6.10: The graph of $y = s(t)$, the position of the car (measured in thousands of feet from its starting location) at time t in minutes.

d. What can you say about s'' whenever s' is increasing? Why?

e. Using only the words *increasing, decreasing, constant, concave up, concave down,* and *linear,* complete the following sentences. For the position function s with velocity v and acceleration a,

- on an interval where v is positive, s is _____.

- on an interval where v is negative, s is _____.

- on an interval where v is zero, s is _____.

- on an interval where a is positive, v is _____.

- on an interval where a is negative, v is _____.

- on an interval where a is zero, v is _____.

- on an interval where a is positive, s is _____.

- on an interval where a is negative, s is _____.

- on an interval where a is zero, s is _____.

The context of position, velocity, and acceleration is an excellent one in which to understand how a function, its first derivative, and its second derivative are related to one another. In Activity 1.6.2, we can replace s, v, and a with an arbitrary function f and its derivatives f'

and f'', and essentially all the same observations hold. In particular, note that f' is increasing if and only if f is concave up, and similarly f' is increasing if and only if f'' is positive. Likewise, f' is decreasing if and only if f is concave down, and f' is decreasing if and only if f'' is negative.

Activity 1.6.3. A potato is placed in an oven, and the potato's temperature F (in degrees Fahrenheit) at various points in time is taken and recorded in the following table. Time t is measured in minutes. In Activity 1.5.2, we computed approximations to $F'(30)$ and $F'(60)$ using central differences. Those values and more are provided in the second table below, along with several others computed in the same way.

t	$F(t)$
0	70
15	180.5
30	251
45	296
60	324.5
75	342.8
90	354.5

t	$F'(t)$
0	NA
15	6.03
30	3.85
45	2.45
60	1.56
75	1.00
90	NA

Table 1.6.11: Select values of $F(t)$.   **Table 1.6.12:** Select values of $F'(t)$.

a. What are the units on the values of $F'(t)$?

b. Use a central difference to estimate the value of $F''(30)$.

c. What is the meaning of the value of $F''(30)$ that you have computed in (b) in terms of the potato's temperature? Write several careful sentences that discuss, with appropriate units, the values of $F(30)$, $F'(30)$, and $F''(30)$, and explain the overall behavior of the potato's temperature at this point in time.

d. Overall, is the potato's temperature increasing at an increasing rate, increasing at a constant rate, or increasing at a decreasing rate? Why?

Activity 1.6.4. This activity builds on our experience and understanding of how to sketch the graph of f' given the graph of f.

In Figure 1.6.13, given the respective graphs of two different functions f, sketch the corresponding graph of f' on the first axes below, and then sketch f'' on the second set of axes. In addition, for each, write several careful sentences in the spirit of those in Activity 1.6.2 that connect the behaviors of f, f', and f''. For instance, write something such as

f' is _____ on the interval _____, which is connected to the fact that
f is _____ on the same interval _____, and f'' is _____ on the
interval as well

but of course with the blanks filled in. Throughout, view the scale of the grid for the graph of f as being 1×1, and assume the horizontal scale of the grid for the graph of f' is identical to that for f. If you need to adjust the vertical scale on the axes for the graph of f' or f'', you should label that accordingly.

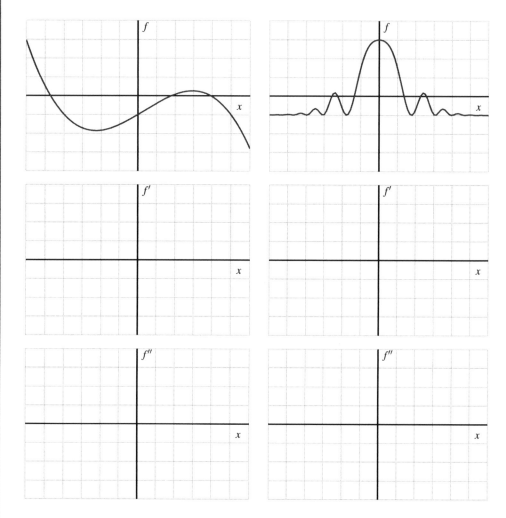

Figure 1.6.13: Two given functions f, with axes provided for plotting f' and f'' below.

Summary

- A differentiable function f is increasing at a point or on an interval whenever its first derivative is positive, and decreasing whenever its first derivative is negative.

- By taking the derivative of the derivative of a function f, we arrive at the second derivative, f''. The second derivative measures the instantaneous rate of change of the first derivative, and thus the sign of the second derivative tells us whether or not the slope of the tangent line to f is increasing or decreasing.

- A differentiable function is concave up whenever its first derivative is increasing (or equivalently whenever its second derivative is positive), and concave down whenever its first derivative is decreasing (or equivalently whenever its second derivative is negative). Examples of functions that are everywhere concave up are $y = x^2$ and $y = e^x$; examples of functions that are everywhere concave down are $y = -x^2$ and $y = -e^x$.

- The units on the second derivative are "units of output per unit of input per unit of input." They tell us how the value of the derivative function is changing in response to changes in the input. In other words, the second derivative tells us the rate of change of the rate of change of the original function.

Exercises

1. Consider the function $f(x)$ graphed below.

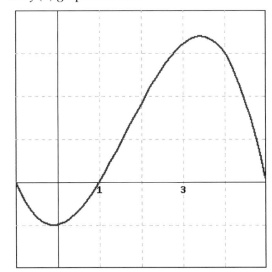

For this function, are the following nonzero quantities positive or negative?

$f(4)$ is [Choose: positive | negative]

$f'(4)$ is [Choose: positive | negative]

$f''(4)$ is [Choose: positive | negative]

(Because this is a multiple choice problem, it will not show which parts of the problem are correct or incorrect when you submit it.)

2. At exactly two of the labeled points in the figure below, which shows a function f, the derivative f' is zero; the second derivative f'' is not zero at any of the labeled points. Select the correct signs for each of f, f' and f'' at each marked point.

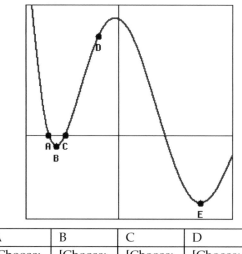

Point	A	B	C	D	E
f	[Choose: positive | zero | negative]	[Choose: positive | zero | negative]	[Choose: positive | zero | negative]	[Choose: positive | zero | negative]	[Choose: positive | zero | negative]
f'	[Choose: positive | zero | negative]	[Choose: positive | zero | negative]	[Choose: positive | zero | negative]	[Choose: positive | zero | negative]	[Choose: positive | zero | negative]
f''	[Choose: positive | zero | negative]	[Choose: positive | zero | negative]	[Choose: positive | zero | negative]	[Choose: positive | zero | negative]	[Choose: positive | zero | negative]

3. Suppose that an accelerating car goes from 0 mph to 61.4 mph in five seconds. Its velocity is given in the following table, converted from miles per hour to feet per second, so that all time measurements are in seconds. (Note: 1 mph is 22/15 ft/sec.) Find the average acceleration of the car over each of the first two seconds.

t (s)	0	1	2	3	4	5
$v(t)$ (ft/s)	0.00	30.68	53.18	69.55	81.82	90.00

average acceleration over the first second =
help (units)

average acceleration over the second second =
help (units)

4. Let $P(t)$ represent the price of a share of stock of a corporation at time t. What does each of the following statements tell us about the signs of the first and second derivatives of $P(t)$?

(*a*) The price of the stock is rising slower and slower.
The first derivative of $P(t)$ is [Choose: positive | zero | negative]
The second derivative of $P(t)$ is [Choose: positive | zero | negative]
(*b*) The price of the stock is just past where it bottomed out.
The first derivative of $P(t)$ is [Choose: positive | zero | negative]
The second derivative of $P(t)$ is [Choose: positive | zero | negative]

5. The graph of f' (*not* f) is given below.

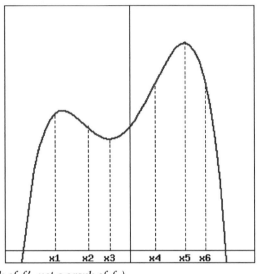

(*Note that this is a graph of f', not a graph of f.*)
At which of the marked values of x is

A. $f(x)$ greatest? $x =$

B. $f(x)$ least? $x =$

C. $f'(x)$ greatest? $x =$

D. $f'(x)$ least? $x =$

E. $f''(x)$ greatest? $x =$

F. $f''(x)$ least? $x =$

6. Suppose that $y = f(x)$ is a differentiable function for which the following information is known: $f(2) = -3$, $f'(2) = 1.5$, $f''(2) = -0.25$.

   a. Is f increasing or decreasing at $x = 2$? Is f concave up or concave down at $x = 2$?

   b. Do you expect $f(2.1)$ to be greater than -3, equal to -3, or less than -3? Why?

   c. Do you expect $f'(2.1)$ to be greater than 1.5, equal to 1.5, or less than 1.5? Why?

   d. Sketch a graph of $y = f(x)$ near $(2, f(2))$ and include a graph of the tangent line.

7. For a certain function $y = g(x)$, its derivative is given by the function pictured in Figure 1.6.14.

a. What is the approximate slope of the tangent line to $y = g(x)$ at the point $(2, g(2))$?

b. How many real number solutions can there be to the equation $g(x) = 0$? Justify your conclusion fully and carefully by explaining what you know about how the graph of g must behave based on the given graph of g'.

c. On the interval $-3 < x < 3$, how many times does the concavity of g change? Why?

d. Use the provided graph to estimate the value of $g''(2)$.

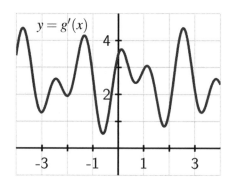

Figure 1.6.14: The graph of $y = g'(x)$.

8. A bungee jumper's height h (in feet) at time t (in seconds) is given in part by the table:

t	0.0	0.5	1.0	1.5	2.0	2.5	3.0	3.5	4.0	4.5	5.0
$h(t)$	200	184.2	159.9	131.9	104.7	81.8	65.5	56.8	55.5	60.4	69.8

t	5.5	6.0	6.5	7.0	7.5	8.0	8.5	9.0	9.5	10.0
$h(t)$	81.6	93.7	104.4	112.6	117.7	119.4	118.2	114.8	110.0	104.7

a. Use the given data to estimate $h'(4.5)$, $h'(5)$, and $h'(5.5)$. At which of these times is the bungee jumper rising most rapidly?

b. Use the given data and your work in (a) to estimate $h''(5)$.

c. What physical property of the bungee jumper does the value of $h''(5)$ measure? What are its units?

d. Based on the data, on what approximate time intervals is the function $y = h(t)$ concave down? What is happening to the velocity of the bungee jumper on these time intervals?

9. For each prompt that follows, sketch a possible graph of a function on the interval $-3 < x < 3$ that satisfies the stated properties.

a. $y = f(x)$ such that f is increasing on $-3 < x < 3$, concave up on $-3 < x < 0$, and concave down on $0 < x < 3$.

b. $y = g(x)$ such that g is increasing on $-3 < x < 3$, concave down on $-3 < x < 0$, and concave up on $0 < x < 3$.

c. $y = h(x)$ such that h is decreasing on $-3 < x < 3$, concave up on $-3 < x < -1$, neither concave up nor concave down on $-1 < x < 1$, and concave down on $1 < x < 3$.

d. $y = p(x)$ such that p is decreasing and concave down on $-3 < x < 0$ and is increasing and concave down on $0 < x < 3$.

1.7 Limits, Continuity, and Differentiability

Motivating Questions

- What does it mean graphically to say that f has limit L as $x \to a$? How is this connected to having a left-hand limit at $x = a$ and having a right-hand limit at $x = a$?

- What does it mean to say that a function f is continuous at $x = a$? What role do limits play in determining whether or not a function is continuous at a point?

- What does it mean graphically to say that a function f is differentiable at $x = a$? How is this connected to the function being locally linear?

- How are the characteristics of a function having a limit, being continuous, and being differentiable at a given point related to one another?

In Section 1.2, we learned about how the concept of limits can be used to study the trend of a function near a fixed input value. As we study such trends, we are fundamentally interested in knowing how well-behaved the function is at the given point, say $x = a$. In this present section, we aim to expand our perspective and develop language and understanding to quantify how the function acts and how its value changes near a particular point. Beyond thinking about whether or not the function has a limit L at $x = a$, we will also consider the value of the function $f(a)$ and how this value is related to $\lim_{x \to a} f(x)$, as well as whether or not the function has a derivative $f'(a)$ at the point of interest. Throughout, we will build on and formalize ideas that we have encountered in several settings.

We begin to consider these issues through the following preview activity that asks you to consider the graph of a function with a variety of interesting behaviors.

> **Preview Activity 1.7.1.** A function f defined on $-4 < x < 4$ is given by the graph in Figure 1.7.1. Use the graph to answer each of the following questions. Note: to the right of $x = 2$, the graph of f is exhibiting infinite oscillatory behavior similar to the function $\sin\left(\frac{\pi}{x}\right)$ that we encountered in the key example early in Section 1.2.
>
> a. For each of the values $a = -3, -2, -1, 0, 1, 2, 3$, determine whether or not $\lim_{x \to a} f(x)$ exists. If the function has a limit L at a given point, state the value of the limit using the notation $\lim_{x \to a} f(x) = L$. If the function does not have a limit at a given point, write a sentence to explain why.
>
> b. For each of the values of a from part (a) where f has a limit, determine the value of $f(a)$ at each such point. In addition, for each such a value, does $f(a)$ have the same value as $\lim_{x \to a} f(x)$?
>
> c. For each of the values $a = -3, -2, -1, 0, 1, 2, 3$, determine whether or not $f'(a)$

exists. In particular, based on the given graph, ask yourself if it is reasonable to say that f has a tangent line at $(a, f(a))$ for each of the given a-values. If so, visually estimate the slope of the tangent line to find the value of $f'(a)$.

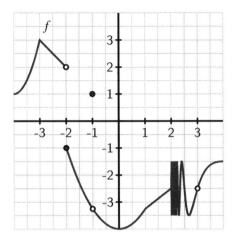

Figure 1.7.1: The graph of $y = f(x)$.

1.7.1 Having a limit at a point

In Section 1.2, we first encountered limits and learned that we say that f has limit L as x approaches a and write $\lim_{x \to a} f(x) = L$ provided that we can make the value of $f(x)$ as close to L as we like by taking x sufficiently close (but not equal to) a. Here, we expand further on this definition and focus in more depth on what it means for a function not to have a limit at a given value.

Essentially there are two behaviors that a function can exhibit at a point where it fails to have a limit. In Figure 1.7.2, at left we see a function f whose graph shows a jump at $a = 1$. In particular, if we let x approach 1 from the left side, the value of f approaches 2, while if we let x go to 1 from the right, the value of f tends to 3. Because the value of f does not approach a single number as x gets arbitrarily close to 1 from both sides, we know that f does not have a limit at $a = 1$.

Since f does approach a single value on each side of $a = 1$, we can introduce the notion of *left* and *right* (or *one-sided*) limits. We say that *f has limit L_1 as x approaches a from the left* and write

$$\lim_{x \to a^-} f(x) = L_1$$

provided that we can make the value of $f(x)$ as close to L_1 as we like by taking x sufficiently close to a while always having $x < a$. In this case, we call L_1 the left-hand limit of f as x approaches a. Similarly, we say L_2 is the right-hand limit of f as x approaches a and write

$$\lim_{x \to a^+} f(x) = L_2$$

provided that we can make the value of $f(x)$ as close to L_2 as we like by taking x sufficiently close to a while always having $x > a$. In the graph of the function f in Figure 1.7.2, we see that

$$\lim_{x \to 1^-} f(x) = 2 \text{ and } \lim_{x \to 1^+} f(x) = 3$$

and precisely because the left and right limits are not equal, the overall limit of f as $x \to 1$ fails to exist.

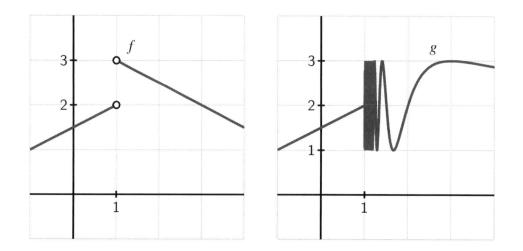

Figure 1.7.2: Functions f and g that each fail to have a limit at $a = 1$.

For the function g pictured at right in Figure 1.7.2, the function fails to have a limit at $a = 1$ for a different reason. While the function does not have a jump in its graph at $a = 1$, it is still not the case that g approaches a single value as x approaches 1. In particular, due to the infinitely oscillating behavior of g to the right of $a = 1$, we say that the right-hand limit of g as $x \to 1^+$ does not exist, and thus $\lim_{x \to 1} g(x)$ does not exist.

To summarize, anytime either a left- or right-hand limit fails to exist or the left- and right-hand limits are not equal to each other, the overall limit will not exist. Said differently,

A function f has limit L as $x \to a$ if and only if

$$\lim_{x \to a^-} f(x) = L = \lim_{x \to a^+} f(x).$$

That is, a function has a limit at $x = a$ if and only if both the left- and right-hand limits at $x = a$ exist and share the same value.

In Preview Activity 1.7.1, the function f given in Figure 1.7.1 only fails to have a limit at two values: at $a = -2$ (where the left- and right-hand limits are 2 and -1, respectively) and at $x = 2$, where $\lim_{x \to 2^+} f(x)$ does not exist). Note well that even at values like $a = -1$ and $a = 0$ where there are holes in the graph, the limit still exists.

Activity 1.7.2. Consider a function that is piecewise-defined according to the formula

$$f(x) = \begin{cases} 3(x+2)+2 & \text{for } -3 < x < -2 \\ \frac{2}{3}(x+2)+1 & \text{for } -2 \leq x < -1 \\ \frac{2}{3}(x+2)+1 & \text{for } -1 < x < 1 \\ 2 & \text{for } x = 1 \\ 4-x & \text{for } x > 1 \end{cases}$$

Use the given formula to answer the following questions.

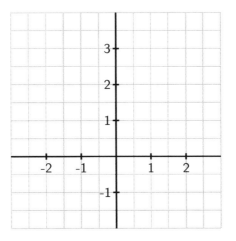

Figure 1.7.3: Axes for plotting the function $y = f(x)$ in Activity 1.7.2.

a. For each of the values $a = -2, -1, 0, 1, 2$, compute $f(a)$.

b. For each of the values $a = -2, -1, 0, 1, 2$, determine $\lim_{x \to a^-} f(x)$ and $\lim_{x \to a^+} f(x)$.

c. For each of the values $a = -2, -1, 0, 1, 2$, determine $\lim_{x \to a} f(x)$. If the limit fails to exist, explain why by discussing the left- and right-hand limits at the relevant a-value.

d. For which values of a is the following statement true?

$$\lim_{x \to a} f(x) \neq f(a)$$

e. On the axes provided in Figure 1.7.3, sketch an accurate, labeled graph of $y = f(x)$. Be sure to carefully use open circles (○) and filled circles (●) to represent key points on the graph, as dictated by the piecewise formula.

1.7.2 Being continuous at a point

Intuitively, a function is continuous if we can draw it without ever lifting our pencil from the page. Alternatively, we might say that the graph of a continuous function has no jumps or holes in it. We first consider three specific situations in Figure 1.7.4 where all three functions have a limit at $a = 1$, and then work to make the idea of continuity more precise.

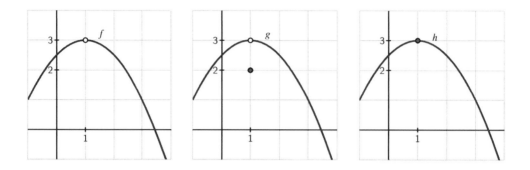

Figure 1.7.4: Functions f, g, and h that demonstrate subtly different behaviors at $a = 1$.

Note that $f(1)$ is not defined, which leads to the resulting hole in the graph of f at $a = 1$. We will naturally say that f is *not continuous at* $a = 1$. For the next function g in in Figure 1.7.4, we observe that while $\lim_{x\to1} g(x) = 3$, the value of $g(1) = 2$, and thus the limit does not equal the function value. Here, too, we will say that g is *not continuous*, even though the function is defined at $a = 1$. Finally, the function h appears to be the most well-behaved of all three, since at $a = 1$ its limit and its function value agree. That is,

$$\lim_{x\to1} h(x) = 3 = h(1).$$

With no hole or jump in the graph of h at $a = 1$, we desire to say that h is *continuous* there. More formally, we make the following definition.

Definition 1.7.5. A function f is **continuous at** $x = a$ provided that

  a. f has a limit as $x \to a$,

  b. f is defined at $x = a$, and

  c. $\lim_{x\to a} f(x) = f(a)$.

Conditions (a) and (b) are technically contained implicitly in (c), but we state them explicitly to emphasize their individual importance. In words, (c) essentially says that a function is continuous at $x = a$ provided that its limit as $x \to a$ exists and equals its function value at $x = a$. If a function is continuous at every point in an interval $[a, b]$, we say the function is "continuous on $[a, b]$." If a function is continuous at every point in its domain, we simply say the function is "continuous." Thus, continuous functions are particularly nice: to evaluate the limit of a continuous function at a point, all we need to do is evaluate the function.

For example, consider $p(x) = x^2 - 2x + 3$. It can be proved that every polynomial is a continuous function at every real number, and thus if we would like to know $\lim_{x\to 2} p(x)$, we simply compute

$$\lim_{x\to 2}(x^2 - 2x + 3) = 2^2 - 2 \cdot 2 + 3 = 3.$$

This route of substituting an input value to evaluate a limit works anytime we know the function being considered is continuous. Besides polynomial functions, all exponential functions and the sine and cosine functions are continuous at every point, as are many other familiar functions and combinations thereof.

Activity 1.7.3. This activity builds on your work in Preview Activity 1.7.1, using the same function f as given by the graph that is repeated in Figure 1.7.6.

    a. At which values of a does $\lim_{x\to a} f(x)$ not exist?

    b. At which values of a is $f(a)$ not defined?

    c. At which values of a does f have a limit, but $\lim_{x\to a} f(x) \neq f(a)$?

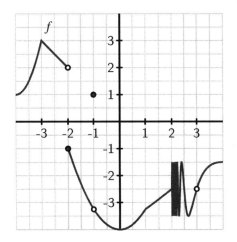

    c. State all values of a for which f is not continuous at $x = a$.

    d. Which condition is stronger, and hence implies the other: f has a limit at $x = a$ or f is continuous at $x = a$? Explain, and hence complete the following sentence: "If f _____ at $x = a$, then f _____ at $x = a$," where you complete the blanks with *has a limit* and *is continuous*, using each phrase once.

Figure 1.7.6: The graph of $y = f(x)$ for Activity 1.7.3.

1.7.3 Being differentiable at a point

We recall that a function f is said to be differentiable at $x = a$ whenever $f'(a)$ exists. Moreover, for $f'(a)$ to exist, we know that the function $y = f(x)$ must have a tangent line at the point $(a, f(a))$, since $f'(a)$ is precisely the slope of this line. In order to even ask if f has a tangent line at $(a, f(a))$, it is necessary that f be continuous at $x = a$: if f fails to have a limit at $x = a$, if $f(a)$ is not defined, or if $f(a)$ does not equal the value of $\lim_{x \to a} f(x)$, then it doesn't even make sense to talk about a tangent line to the curve at this point.

Indeed, it can be proved formally that if a function f is differentiable at $x = a$, then it must be continuous at $x = a$. So, if f is not continuous at $x = a$, then it is automatically the case that f is not differentiable there. For example, in Figure 1.7.4 from our early discussion of continuity, both f and g fail to be differentiable at $x = 1$ because neither function is continuous at $x = 1$. But can a function fail to be differentiable at a point where the function is continuous?

In Figure 1.7.7, we revisit the situation where a function has a sharp corner at a point, something we encountered several times in Section 1.4. For the pictured function f, we observe that f is clearly continuous at $a = 1$, since $\lim_{x \to 1} f(x) = 1 = f(1)$.

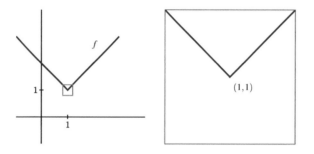

Figure 1.7.7: A function f that is continuous at $a = 1$ but not differentiable at $a = 1$; at right, we zoom in on the point $(1, 1)$ in a magnified version of the box in the left-hand plot.

But the function f in Figure 1.7.7 is not differentiable at $a = 1$ because $f'(1)$ fails to exist. One way to see this is to observe that $f'(x) = -1$ for every value of x that is less than 1, while $f'(x) = +1$ for every value of x that is greater than 1. That makes it seem that either $+1$ or -1 would be equally good candidates for the value of the derivative at $x = 1$. Alternately, we could use the limit definition of the derivative to attempt to compute $f'(1)$, and discover that the derivative does not exist. A similar problem will be investigated in Activity 1.7.4. Finally, we can also see visually that the function f in Figure 1.7.7 does not have a tangent line. When we zoom in on $(1, 1)$ on the graph of f, no matter how closely we examine the function, it will always look like a "V", and never like a single line, which tells us there is no possibility for a tangent line there.

To make a more general observation, if a function does have a tangent line at a given point, when we zoom in on the point of tangency, the function and the tangent line should appear essentially indistinguishable[1]. Conversely, if we have a function such that when we zoom in on a point the function looks like a single straight line, then the function should have a tangent line there, and thus be differentiable. Hence, a function that is differentiable at $x = a$ will, up close, look more and more like its tangent line at $(a, f(a))$, and thus we say that a

---

[1]See, for instance, http://gvsu.edu/s/6J for an applet (due to David Austin, GVSU) where zooming in shows the increasing similarity between the tangent line and the curve.

function is differentiable at $x = a$ is *locally linear*.

To summarize the preceding discussion of differentiability and continuity, we make several important observations.

- If f is differentiable at $x = a$, then f is continuous at $x = a$. Equivalently, if f fails to be continuous at $x = a$, then f will not be differentiable at $x = a$.

- A function can be continuous at a point, but not be differentiable there. In particular, a function f is not differentiable at $x = a$ if the graph has a sharp corner (or *cusp*) at the point $(a, f(a))$.

- If f is differentiable at $x = a$, then f is locally linear at $x = a$. That is, when a function is differentiable, it looks linear when viewed up close because it resembles its tangent line there.

Activity 1.7.4. In this activity, we explore two different functions and classify the points at which each is not differentiable. Let g be the function given by the rule $g(x) = |x|$, and let f be the function that we have previously explored in Preview Activity 1.7.1, whose graph is given again in Figure 1.7.8.

a. Reasoning visually, explain why g is differentiable at every point x such that $x \neq 0$.

b. Use the limit definition of the derivative to show that $g'(0) = \lim_{h \to 0} \frac{|h|}{h}$.

c. Explain why $g'(0)$ fails to exist by using small positive and negative values of h.

d. State all values of a for which f is not differentiable at $x = a$. For each, provide a reason for your conclusion.

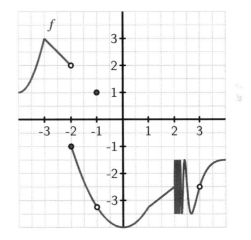

Figure 1.7.8: The graph of $y = f(x)$ for Activity 1.7.4.

e. True or false: if a function p is differentiable at $x = b$, then $\lim_{x \to b} p(x)$ must exist. Why?

Summary

- A function f has limit L as $x \to a$ if and only if f has a left-hand limit at $x = a$, has a right-hand limit at $x = a$, and the left- and right-hand limits are equal. Visually, this means that there can be a hole in the graph at $x = a$, but the function must approach the same single value from either side of $x = a$.

- A function f is continuous at $x = a$ whenever $f(a)$ is defined, f has a limit as $x \to a$, and the value of the limit and the value of the function agree. This guarantees that there is not a hole or jump in the graph of f at $x = a$.

- A function f is differentiable at $x = a$ whenever $f'(a)$ exists, which means that f has a tangent line at $(a, f(a))$ and thus f is locally linear at the value $x = a$. Informally, this means that the function looks like a line when viewed up close at $(a, f(a))$ and that there is not a corner point or cusp at $(a, f(a))$.

- Of the three conditions discussed in this section (having a limit at $x = a$, being continuous at $x = a$, and being differentiable at $x = a$), the strongest condition is being differentiable, and the next strongest is being continuous. In particular, if f is differentiable at $x = a$, then f is also continuous at $x = a$, and if f is continuous at $x = a$, then f has a limit at $x = a$.

Exercises

1. Use the figure below, which gives a graph of the function $f(x)$, to give values for the indicated limits. If a limit does not exist, enter none.

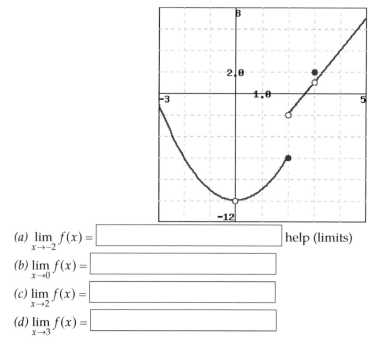

(a) $\lim\limits_{x \to -2} f(x) =$ [_____] help (limits)

(b) $\lim\limits_{x \to 0} f(x) =$ [_____]

(c) $\lim\limits_{x \to 2} f(x) =$ [_____]

(d) $\lim\limits_{x \to 3} f(x) =$ [_____]

2. For the function

$$f(x) = \begin{cases} 2x - 5, & 0 \leq x < 3 \\ 4, & x = 3 \\ x^2 - 6x + 10, & 3 < x \end{cases}$$

use algebra to find each of the following limits:

$\lim\limits_{x \to 3^+} f(x) =$

$\lim\limits_{x \to 3^-} f(x) =$

$\lim\limits_{x \to 3} f(x) =$

(For each, enter DNE *if the limit does not exist.)*
Sketch a graph of $f(x)$ to confirm your answers.

3. Consider the function graphed below.

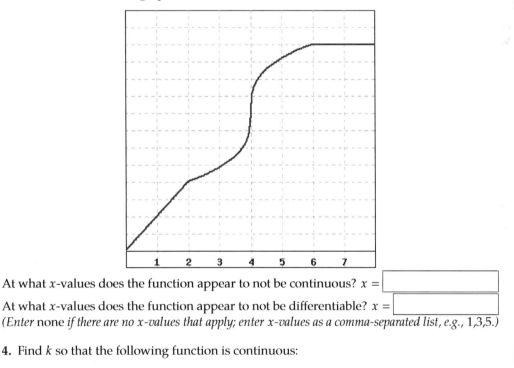

At what x-values does the function appear to not be continuous? $x =$

At what x-values does the function appear to not be differentiable? $x =$

(Enter none *if there are no x-values that apply; enter x-values as a comma-separated list, e.g., 1,3,5.)*

4. Find k so that the following function is continuous:

$$f(x) = \begin{cases} kx & \text{if } 0 \leq x < 2 \\ 9x^2 & \text{if } 2 \leq x. \end{cases}$$

$k =$

5. Consider the graph of the function $y = p(x)$ that is provided in Figure 1.7.9. Assume that each portion of the graph of p is a straight line, as pictured.

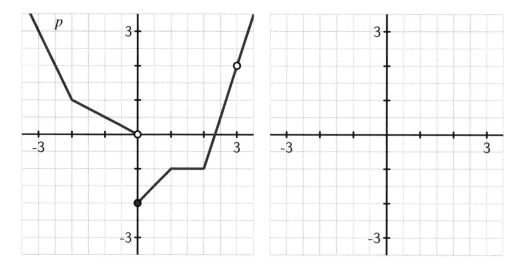

Figure 1.7.9: At left, the piecewise linear function $y = p(x)$. At right, axes for plotting $y = p'(x)$.

  a. State all values of a for which $\lim_{x\to a} p(x)$ does not exist.

  b. State all values of a for which p is not continuous at a.

  c. State all values of a for which p is not differentiable at $x = a$.

  d. On the axes provided in Figure 1.7.9, sketch an accurate graph of $y = p'(x)$.

6. For each of the following prompts, give an example of a function that satisfies the stated criteria. A formula or a graph, with reasoning, is sufficient for each. If no such example is possible, explain why.

  a. A function f that is continuous at $a = 2$ but not differentiable at $a = 2$.

  b. A function g that is differentiable at $a = 3$ but does not have a limit at $a = 3$.

  c. A function h that has a limit at $a = -2$, is defined at $a = -2$, but is not continuous at $a = -2$.

  d. A function p that satisfies all of the following:
   - $p(-1) = 3$ and $\lim_{x\to-1} p(x) = 2$
   - $p(0) = 1$ and $p'(0) = 0$
   - $\lim_{x\to1} p(x) = p(1)$ and $p'(1)$ does not exist

7. Let $h(x)$ be a function whose derivative $y = h'(x)$ is given by the graph on the right in Figure 1.7.10.

  a. Based on the graph of $y = h'(x)$, what can you say about the behavior of the function $y = h(x)$?

b. At which values of x is $y = h'(x)$ not defined? What behavior does this lead you to expect to see in the graph of $y = h(x)$?

c. Is it possible for $y = h(x)$ to have points where h is not continuous? Explain your answer.

d. On the axes provided at left, sketch at least two distinct graphs that are possible functions $y = h(x)$ that each have a derivative $y = h'(x)$ that matches the provided graph at right. Explain why there are multiple possibilities for $y = h(x)$.

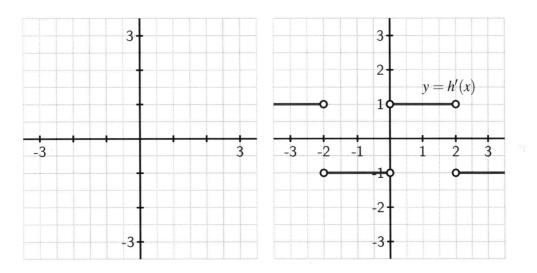

Figure 1.7.10: Axes for plotting $y = h(x)$ and, at right, the graph of $y = h'(x)$.

8. Consider the function $g(x) = \sqrt{|x|}$.

a. Use a graph to explain visually why g is not differentiable at $x = 0$.

b. Use the limit definition of the derivative to show that

$$g'(0) = \lim_{h \to 0} \frac{\sqrt{|h|}}{h}.$$

c. Investigate the value of $g'(0)$ by estimating the limit in (b) using small positive and negative values of h. For instance, you might compute $\frac{\sqrt{|-0.01|}}{0.01}$. Be sure to use several different values of h (both positive and negative), including ones closer to 0 than 0.01. What do your results tell you about $g'(0)$?

d. Use your graph in (a) to sketch an approximate graph of $y = g'(x)$.

1.8 The Tangent Line Approximation

Motivating Questions

- What is the formula for the general tangent line approximation to a differentiable function $y = f(x)$ at the point $(a, f(a))$?

- What is the principle of local linearity and what is the local linearization of a differentiable function f at a point $(a, f(a))$?

- How does knowing just the tangent line approximation tell us information about the behavior of the original function itself near the point of approximation? How does knowing the second derivative's value at this point provide us additional knowledge of the original function's behavior?

Among all functions, linear functions are simplest. One of the powerful consequences of a function $y = f(x)$ being differentiable at a point $(a, f(a))$ is that, up close, the function $y = f(x)$ is locally linear and looks like its tangent line at that point. In certain circumstances, this allows us to approximate the original function f with a simpler function L that is linear: this can be advantageous when we have limited information about f or when f is computationally or algebraically complicated. We will explore all of these situations in what follows.

It is essential to recall that when f is differentiable at $x = a$, the value of $f'(a)$ provides the slope of the tangent line to $y = f(x)$ at the point $(a, f(a))$. By knowing both a point on the line and the slope of the line we are thus able to find the equation of the tangent line. Preview Activity 1.8.1 will refresh these concepts through a key example and set the stage for further study.

> **Preview Activity 1.8.1.** Consider the function $y = g(x) = -x^2 + 3x + 2$.
>
> a. Use the limit definition of the derivative to compute a formula for $y = g'(x)$.
>
> b. Determine the slope of the tangent line to $y = g(x)$ at the value $x = 2$.
>
> c. Compute $g(2)$.
>
> d. Find an equation for the tangent line to $y = g(x)$ at the point $(2, g(2))$. Write your result in point-slope form[a].
>
> e. On the axes provided in Figure 1.8.1, sketch an accurate, labeled graph of $y = g(x)$ along with its tangent line at the point $(2, g(2))$.

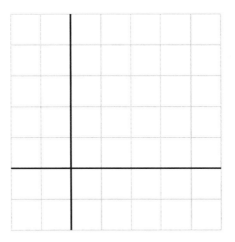

Figure 1.8.1: Axes for plotting $y = g(x)$ and its tangent line to the point $(2, g(2))$.

---

[a]Recall that a line with slope m that passes through (x_0, y_0) has equation $y - y_0 = m(x - x_0)$, and this is the *point-slope form* of the equation.

1.8.1 The tangent line

Given a function f that is differentiable at $x = a$, we know that we can determine the slope of the tangent line to $y = f(x)$ at $(a, f(a))$ by computing $f'(a)$. The resulting tangent line through $(a, f(a))$ with slope $m = f'(a)$ has its equation in point-slope form given by

$$y - f(a) = f'(a)(x - a),$$

which we can also express as $y = f'(a)(x - a) + f(a)$. Note well: there is a major difference between $f(a)$ and $f(x)$ in this context. The former is a constant that results from using the given fixed value of a, while the latter is the general expression for the rule that defines the function. The same is true for $f'(a)$ and $f'(x)$: we must carefully distinguish between these expressions. Each time we find the tangent line, we need to evaluate the function and its derivative at a fixed a-value.

In Figure 1.8.2, we see a labeled plot of the graph of a function f and its tangent line at the point $(a, f(a))$. Notice how when we zoom in we see the local linearity of f more clearly highlighted as the function and its tangent line are nearly indistinguishable up close. This can also be seen dynamically in the java applet at http://gvsu.edu/s/6J.

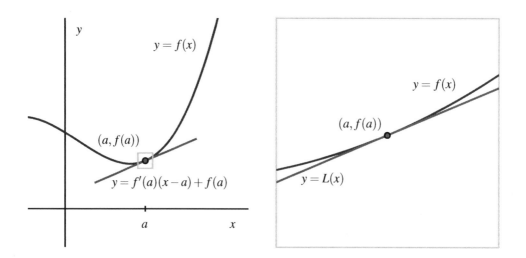

Figure 1.8.2: A function $y = f(x)$ and its tangent line at the point $(a, f(a))$: at left, from a distance, and at right, up close. At right, we label the tangent line function by $y = L(x)$ and observe that for x near a, $f(x) \approx L(x)$.

1.8.2 The local linearization

A slight change in perspective and notation will enable us to be more precise in discussing how the tangent line to $y = f(x)$ at $(a, f(a))$ approximates f near $x = a$. Taking the equation for the tangent line and solving for y, we observe that the tangent line is given by

$$y = f'(a)(x - a) + f(a)$$

and moreover that this line is itself a function of x. Replacing the variable y with the expression $L(x)$, we call

$$L(x) = f'(a)(x - a) + f(a)$$

the *local linearization of f* at the point $(a, f(a))$. In this notation, it is particularly important to observe that $L(x)$ is nothing more than a new name for the tangent line, and that for x close to a, we have that $f(x) \approx L(x)$.

Say, for example, that we know that a function $y = f(x)$ has its tangent line approximation given by $L(x) = 3 - 2(x - 1)$ at the point $(1, 3)$, but we do not know anything else about the function f. If we are interested in estimating a value of $f(x)$ for x near 1, such as $f(1.2)$, we can use the fact that $f(1.2) \approx L(1.2)$ and hence

$$f(1.2) \approx L(1.2) = 3 - 2(1.2 - 1) = 3 - 2(0.2) = 2.6.$$

Again, much of the new perspective here is only in notation since $y = L(x)$ is simply a new name for the tangent line function. In light of this new notation and our observations above, we note that since $L(x) = f(a) + f'(a)(x - a)$ and $L(x) \approx f(x)$ for x near a, it also follows that we can write

$$f(x) \approx f(a) + f'(a)(x - a) \text{ for } x \text{ near } a.$$

The next activity explores some additional important properties of the local linearization $y = L(x)$ to a function f at given a-value.

Activity 1.8.2. Suppose it is known that for a given differentiable function $y = g(x)$, its local linearization at the point where $a = -1$ is given by $L(x) = -2 + 3(x + 1)$.

a. Compute the values of $L(-1)$ and $L'(-1)$.

b. What must be the values of $g(-1)$ and $g'(-1)$? Why?

c. Do you expect the value of $g(-1.03)$ to be greater than or less than the value of $g(-1)$? Why?

d. Use the local linearization to estimate the value of $g(-1.03)$.

e. Suppose that you also know that $g''(-1) = 2$. What does this tell you about the graph of $y = g(x)$ at $a = -1$?

f. For x near -1, sketch the graph of the local linearization $y = L(x)$ as well as a possible graph of $y = g(x)$ on the axes provided in Figure 1.8.3.

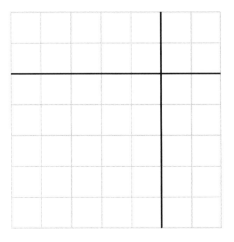

Figure 1.8.3: Axes for plotting $y = L(x)$ and $y = g(x)$.

As we saw in the example provided by Activity 1.8.2, the local linearization $y = L(x)$ is a linear function that shares two important values with the function $y = f(x)$ that it is derived from. In particular, observe that since $L(x) = f(a) + f'(a)(x - a)$, it follows that $L(a) = f(a)$. In addition, since L is a linear function, its derivative is its slope. Hence, $L'(x) = f'(a)$ for every value of x, and specifically $L'(a) = f'(a)$. Therefore, we see that L is a linear function that has both the same value and the same slope as the function f at the point $(a, f(a))$.

In situations where we know the linear approximation $y = L(x)$, we therefore know the original function's value and slope at the point of tangency. What remains unknown, however, is the shape of the function f at the point of tangency. There are essentially four possibilities, as enumerated in Figure 1.8.4.

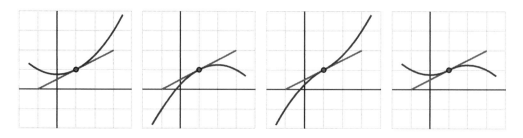

Figure 1.8.4: Four possible graphs for a nonlinear differentiable function and how it can be situated relative to its tangent line at a point.

These stem from the fact that there are three options for the value of the second derivative: either $f''(a) < 0$, $f''(a) = 0$, or $f''(a) > 0$. If $f''(a) > 0$, then we know the graph of f is concave up, and we see the first possibility on the left, where the tangent line lies entirely below the curve. If $f''(a) < 0$, then we find ourselves in the second situation (from left) where f is concave down and the tangent line lies above the curve. In the situation where $f''(a) = 0$ and f'' changes sign at $x = a$, the concavity of the graph will change, and we will see either the third or fourth option[1]. A fifth option (that is not very interesting) can occur, which is where the function f is linear, and so $f(x) = L(x)$ for all values of x.

The plots in Figure 1.8.4 highlight yet another important thing that we can learn from the concavity of the graph near the point of tangency: whether the tangent line lies above or below the curve itself. This is key because it tells us whether or not the tangent line approximation's values will be too large or too small in comparison to the true value of f. For instance, in the first situation in the leftmost plot in Figure 1.8.4 where $f''(a) > 0$, since the tangent line falls below the curve, we know that $L(x) \leq f(x)$ for all values of x near a.

We explore these ideas further in the following activity.

> **Activity 1.8.3.** This activity concerns a function $f(x)$ about which the following information is known:
>
> - f is a differentiable function defined at every real number x
> - $f(2) = -1$
> - $y = f'(x)$ has its graph given in Figure 1.8.5

---

[1]It is possible to have $f''(a) = 0$ and have f'' not change sign at $x = a$, in which case the graph will look like one of the first two options.

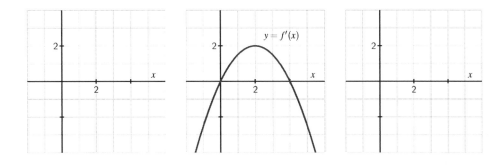

Figure 1.8.5: At center, a graph of $y = f'(x)$; at left, axes for plotting $y = f(x)$; at right, axes for plotting $y = f''(x)$.

Your task is to determine as much information as possible about f (especially near the value $a = 2$) by responding to the questions below.

a. Find a formula for the tangent line approximation, $L(x)$, to f at the point $(2, -1)$.

b. Use the tangent line approximation to estimate the value of $f(2.07)$. Show your work carefully and clearly.

c. Sketch a graph of $y = f''(x)$ on the righthand grid in Figure 1.8.5; label it appropriately.

d. Is the slope of the tangent line to $y = f(x)$ increasing, decreasing, or neither when $x = 2$? Explain.

e. Sketch a possible graph of $y = f(x)$ near $x = 2$ on the lefthand grid in Figure 1.8.5. Include a sketch of $y = L(x)$ (found in part (a)). Explain how you know the graph of $y = f(x)$ looks like you have drawn it.

f. Does your estimate in (b) over- or under-estimate the true value of $f(2.07)$? Why?

The idea that a differentiable function looks linear and can be well-approximated by a linear function is an important one that finds wide application in calculus. For example, by approximating a function with its local linearization, it is possible to develop an effective algorithm to estimate the zeroes of a function. Local linearity also helps us to make further sense of certain challenging limits. For instance, we have seen that a limit such as

$$\lim_{x \to 0} \frac{\sin(x)}{x}$$

is indeterminate because both its numerator and denominator tend to 0. While there is no algebra that we can do to simplify $\frac{\sin(x)}{x}$, it is straightforward to show that the linearization of $f(x) = \sin(x)$ at the point $(0, 0)$ is given by $L(x) = x$. Hence, for values of x near 0, $\sin(x) \approx x$. As such, for values of x near 0,

$$\frac{\sin(x)}{x} \approx \frac{x}{x} = 1,$$

87

which makes plausible the fact that

$$\lim_{x \to 0} \frac{\sin(x)}{x} = 1.$$

These ideas and other applications of local linearity will be explored later on in our work.

Summary

- The tangent line to a differentiable function $y = f(x)$ at the point $(a, f(a))$ is given in point-slope form by the equation

$$y - f(a) = f'(a)(x - a).$$

- The principle of local linearity tells us that if we zoom in on a point where a function $y = f(x)$ is differentiable, the function should become indistinguishable from its tangent line. That is, a differentiable function looks linear when viewed up close. We rename the tangent line to be the function $y = L(x)$ where $L(x) = f(a) + f'(a)(x - a)$ and note that $f(x) \approx L(x)$ for all x near $x = a$.

- If we know the tangent line approximation $L(x) = f(a) + f'(a)(x - a)$, then because $L(a) = f(a)$ and $L'(a) = f'(a)$, we also know both the value and the derivative of the function $y = f(x)$ at the point where $x = a$. In other words, the linear approximation tells us the height and slope of the original function. If, in addition, we know the value of $f''(a)$, we then know whether the tangent line lies above or below the graph of $y = f(x)$ depending on the concavity of f.

Exercises

1. Use linear approximation to approximate $\sqrt{25.3}$ as follows.
Let $f(x) = \sqrt{x}$. The equation of the tangent line to $f(x)$ at $x = 25$ can be written in the form $y = mx + b$. Compute m and b.

$m = $ _____

$b = $ _____

Using this find the approximation for $\sqrt{25.3}$.

Answer: _____

2. The figure below shows $f(x)$ and its local linearization at $x = a$, $y = 2x - 3$. (The local linearization is shown in blue.)

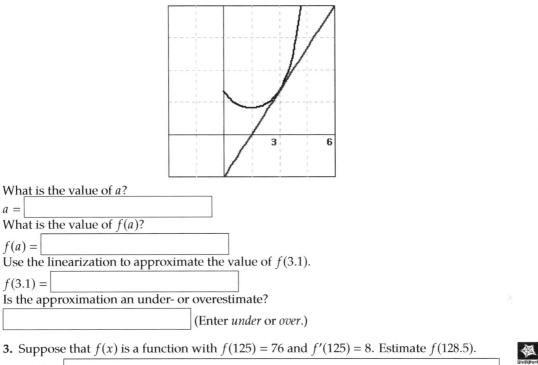

What is the value of a?

$a =$ [   ]

What is the value of $f(a)$?

$f(a) =$ [   ]

Use the linearization to approximate the value of $f(3.1)$.

$f(3.1) =$ [   ]

Is the approximation an under- or overestimate?

[   ] (Enter *under* or *over*.)

3. Suppose that $f(x)$ is a function with $f(125) = 76$ and $f'(125) = 8$. Estimate $f(128.5)$.

$f(128.5) =$ [   ]

4. The temperature, H, in degrees Celsius, of a cup of coffee placed on the kitchen counter is given by $H = f(t)$, where t is in minutes since the coffee was put on the counter.

(*a*) Is $f'(t)$ positive or negative? [Choose: positive | negative]

(*Be sure that you are able to give a reason for your answer.*)

(*b*) What are the units of $f'(25)$? [   ] help (units)

Suppose that $|f'(25)| = 0.6$ and $f(25) = 65$. Fill in the blanks (including units where needed) and select the appropriate terms to complete the following statement about the temperature of the coffee in this case.

At [   ] minutes after the coffee was put on the counter, its [Choose: derivative | temperature | change in temperature] is [   ] and will [Choose: increase | decrease] by about [   ] in the next 60 seconds.

5. A certain function $y = p(x)$ has its local linearization at $a = 3$ given by $L(x) = -2x + 5$.

   a. What are the values of $p(3)$ and $p'(3)$? Why?

   b. Estimate the value of $p(2.79)$.

   c. Suppose that $p''(3) = 0$ and you know that $p''(x) < 0$ for $x < 3$. Is your estimate in (b) too large or too small?

   d. Suppose that $p''(x) > 0$ for $x > 3$. Use this fact and the additional information above to sketch an accurate graph of $y = p(x)$ near $x = 3$. Include a sketch of $y = L(x)$ in your work.

6. A potato is placed in an oven, and the potato's temperature F (in degrees Fahrenheit) at various points in time is taken and recorded in the following table. Time t is measured in minutes.

t	$F(t)$
0	70
15	180.5
30	251
45	296
60	324.5
75	342.8
90	354.5

Table 1.8.6: Temperature data for the potato.

   a. Use a central difference to estimate $F'(60)$. Use this estimate as needed in subsequent questions.

   b. Find the local linearization $y = L(t)$ to the function $y = F(t)$ at the point where $a = 60$.

   c. Determine an estimate for $F(63)$ by employing the local linearization.

   d. Do you think your estimate in (c) is too large or too small? Why?

7. An object moving along a straight line path has a differentiable position function $y = s(t)$; $s(t)$ measures the object's position relative to the origin at time t. It is known that at time $t = 9$ seconds, the object's position is $s(9) = 4$ feet (i.e., 4 feet to the right of the origin). Furthermore, the object's instantaneous velocity at $t = 9$ is -1.2 feet per second, and its acceleration at the same instant is 0.08 feet per second per second.

   a. Use local linearity to estimate the position of the object at $t = 9.34$.

   b. Is your estimate likely too large or too small? Why?

   c. In everyday language, describe the behavior of the moving object at $t = 9$. Is it moving toward the origin or away from it? Is its velocity increasing or decreasing?

8. For a certain function f, its derivative is known to be $f'(x) = (x - 1)e^{-x^2}$. Note that you do not know a formula for $y = f(x)$.

   a. At what x-value(s) is $f'(x) = 0$? Justify your answer algebraically, but include a graph of f' to support your conclusion.

   b. Reasoning graphically, for what intervals of x-values is $f''(x) > 0$? What does this tell you about the behavior of the original function f? Explain.

   c. Assuming that $f(2) = -3$, estimate the value of $f(1.88)$ by finding and using the tangent line approximation to f at $x = 2$. Is your estimate larger or smaller than the true value of $f(1.88)$? Justify your answer.

CHAPTER 2

Computing Derivatives

2.1 Elementary derivative rules

Motivating Questions

- What are alternate notations for the derivative?

- How can we sometimes use the algebraic structure of a function $f(x)$ to easily compute a formula for $f'(x)$?

- What is the derivative of a power function of the form $f(x) = x^n$? What is the derivative of an exponential function of form $f(x) = a^x$?

- If we know the derivative of $y = f(x)$, how is the derivative of $y = kf(x)$ computed, where k is a constant?

- If we know the derivatives of $y = f(x)$ and $y = g(x)$, how is the derivative of $y = f(x) + g(x)$ computed?

In Chapter 1, we developed the concept of the derivative of a function. We now know that the derivative f' of a function f measures the instantaneous rate of change of f with respect to x as well as the slope of the tangent line to $y = f(x)$ at any given value of x. To date, we have focused primarily on interpreting the derivative graphically or, in the context of functions in a physical setting, as a meaningful rate of change. To actually calculate the value of the derivative at a specific point, we have typically relied on the limit definition of the derivative.

In this present chapter, we will investigate how the limit definition of the derivative,

$$f'(x) = \lim_{h \to 0} \frac{f(x+h) - f(x)}{h},$$

leads to interesting patterns and rules that enable us to quickly find a formula for $f'(x)$ based on the formula for $f(x)$ *without* using the limit definition directly. For example, we already know that if $f(x) = x$, then it follows that $f'(x) = 1$. While we could use the limit definition of the derivative to confirm this, we know it to be true because $f(x)$ is a linear function with slope 1 at every value of x. One of our goals is to be able to take standard functions, say ones such as $g(x) = 4x^7 - \sin(x) + 3e^x$, and, based on the algebraic form of the function, be able to apply shortcuts to almost immediately determine the formula for $g'(x)$.

> **Preview Activity 2.1.1.** Functions of the form $f(x) = x^n$, where $n = 1, 2, 3, \ldots$, are often called **power functions**. The first two questions below revisit work we did earlier in Chapter 1, and the following questions extend those ideas to higher powers of x.
>
> a. Use the limit definition of the derivative to find $f'(x)$ for $f(x) = x^2$.
>
> b. Use the limit definition of the derivative to find $f'(x)$ for $f(x) = x^3$.
>
> c. Use the limit definition of the derivative to find $f'(x)$ for $f(x) = x^4$. (Hint: $(a+b)^4 = a^4 + 4a^3b + 6a^2b^2 + 4ab^3 + b^4$. Apply this rule to $(x+h)^4$ within the limit definition.)
>
> d. Based on your work in (a), (b), and (c), what do you conjecture is the derivative of $f(x) = x^5$? Of $f(x) = x^{13}$?
>
> e. Conjecture a formula for the derivative of $f(x) = x^n$ that holds for any positive integer n. That is, given $f(x) = x^n$ where n is a positive integer, what do you think is the formula for $f'(x)$?

2.1.1 Some Key Notation

In addition to our usual f' notation for the derivative, there are other ways to symbolically denote the derivative of a function, as well as the instruction to take the derivative. We know that if we have a function, say $f(x) = x^2$, that we can denote its derivative by $f'(x)$, and we write $f'(x) = 2x$. Equivalently, if we are thinking more about the relationship between y and x, we sometimes denote the derivative of y with respect to x with the symbol

$$\frac{dy}{dx}$$

which we read "dee-y dee-x." This notation comes from the fact that the derivative is related to the slope of a line, and slope is measured by $\frac{\Delta y}{\Delta x}$. Note that while we read $\frac{\Delta y}{\Delta x}$ as "change in y over change in x," for the derivative symbol $\frac{dy}{dx}$, we view this is a single symbol, not a quotient of two quantities[1]. For example, if $y = x^2$, we'll write that the derivative is $\frac{dy}{dx} = 2x$.

Furthermore, we use a variant of $\frac{dy}{dx}$ notation to convey the instruction to take the derivative of a certain quantity with respect to a given variable. In particular, if we write

$$\frac{d}{dx}[\Box]$$

this means "take the derivative of the quantity in $\Box$ with respect to x." To continue our example above with the squaring function, here we may write $\frac{d}{dx}[x^2] = 2x$.

---

[1]That is, we do *not* say "dee-y over dee-x."

It is important to note that the independent variable can be different from x. If we have $f(z) = z^2$, we then write $f'(z) = 2z$. Similarly, if $y = t^2$, we can say $\frac{dy}{dt} = 2t$. And changing the variable and derivative notation once more, it is also true that $\frac{d}{dq}[q^2] = 2q$. This notation may also be applied to second derivatives: $f''(z) = \frac{d}{dz}\left[\frac{df}{dz}\right] = \frac{d^2 f}{dz^2}$.

In what follows, we'll be working to widely expand our repertoire of functions for which we can quickly compute the corresponding derivative formula

2.1.2 Constant, Power, and Exponential Functions

So far, we know the derivative formula for two important classes of functions: constant functions and power functions. For the first kind, observe that if $f(x) = c$ is a constant function, then its graph is a horizontal line with slope zero at every point. Thus, $\frac{d}{dx}[c] = 0$. We summarize this with the following rule.

> **Constant Functions**
>
> For any real number c, if $f(x) = c$, then $f'(x) = 0$.

Thus, if $f(x) = 7$, then $f'(x) = 0$. Similarly, $\frac{d}{dx}[\sqrt{3}] = 0$.

For power functions, from your work in Preview Activity 2.1.1, you have conjectured that for any positive integer n, if $f(x) = x^n$, then $f'(x) = nx^{n-1}$. Not only can this rule be formally proved to hold for any positive integer n, but also for any nonzero real number (positive or negative).

> **Power Functions**
>
> For any nonzero real number, if $f(x) = x^n$, then $f'(x) = nx^{n-1}$.

This rule for power functions allows us to find derivatives such as the following: if $g(z) = z^{-3}$, then $g'(z) = -3z^{-4}$. Similarly, if $h(t) = t^{7/5}$, then $\frac{dh}{dt} = \frac{7}{5}t^{2/5}$; likewise, $\frac{d}{dq}[q^\pi] = \pi q^{\pi-1}$.

As we next turn to thinking about derivatives of combinations of basic functions, it will be instructive to have one more type of basic function whose derivative formula we know. For now, we simply state this rule without explanation or justification; we will explore why this rule is true in one of the exercises at the end of this section, plus we will encounter graphical reasoning for why the rule is plausible in Preview Activity 2.2.1.

> **Exponential Functions**
>
> For any positive real number a, if $f(x) = a^x$, then $f'(x) = a^x \ln(a)$.

For instance, this rule tells us that if $f(x) = 2^x$, then $f'(x) = 2^x \ln(2)$. Similarly, for $p(t) = 10^t$, $p'(t) = 10^t \ln(10)$. It is especially important to note that when $a = e$, where e is the base of the natural logarithm function, we have that

$$\frac{d}{dx}[e^x] = e^x \ln(e) = e^x$$

since $\ln(e) = 1$. This is an extremely important property of the function e^x: its derivative function is itself!

Finally, note carefully the distinction between power functions and exponential functions: in power functions, the variable is in the base, as in x^2, while in exponential functions, the variable is in the power, as in 2^x. As we can see from the rules, this makes a big difference in the form of the derivative.

The following activity will check your understanding of the derivatives of the three basic types of functions noted above.

> **Activity 2.1.2.** Use the three rules above to determine the derivative of each of the following functions. For each, state your answer using full and proper notation, labeling the derivative with its name. For example, if you are given a function $h(z)$, you should write "$h'(z) =$" or "$\frac{dh}{dz} =$" as part of your response.
>
> a. $f(t) = \pi$
>
> b. $g(z) = 7^z$
>
> c. $h(w) = w^{3/4}$
>
> d. $p(x) = 3^{1/2}$
>
> e. $r(t) = (\sqrt{2})^t$
>
> f. $\frac{d}{dq}[q^{-1}]$
>
> g. $m(t) = \frac{1}{t^3}$

2.1.3 Constant Multiples and Sums of Functions

Of course, most of the functions we encounter in mathematics are more complicated than being simply constant, a power of a variable, or a base raised to a variable power. In this section and several following, we will learn how to quickly compute the derivative of a function constructed as an algebraic combination of basic functions. For instance, we'd like to be able to understand how to take the derivative of a polynomial function such as $p(t) = 3t^5 - 7t^4 + t^2 - 9$, which is a function made up of constant multiples and sums of powers of t. To that end, we develop two new rules: the Constant Multiple Rule and the Sum Rule.

Say we have a function $y = f(x)$ whose derivative formula is known. How is the derivative of $y = kf(x)$ related to the derivative of the original function? Recall that when we multiply a function by a constant k, we vertically stretch the graph by a factor of $|k|$ (and reflect the graph across $y = 0$ if $k < 0$). This vertical stretch affects the slope of the graph, making the slope of the function $y = kf(x)$ be k times as steep as the slope of $y = f(x)$. In terms of the derivative, this is essentially saying that when we multiply a function by a factor of k, we change the value of its derivative by a factor of k as well. Thus[2], the Constant Multiple Rule holds:

The Constant Multiple Rule

For any real number k, if $f(x)$ is a differentiable function with derivative $f'(x)$, then $\frac{d}{dx}[kf(x)] = kf'(x)$.

---

[2]The Constant Multiple Rule can be formally proved as a consequence of properties of limits, using the limit definition of the derivative.

In words, this rule says that "the derivative of a constant times a function is the constant times the derivative of the function." For example, if $g(t) = 3 \cdot 5^t$, we have $g'(t) = 3 \cdot 5^t \ln(5)$. Similarly, $\frac{d}{dz}[5z^{-2}] = 5(-2z^{-3})$.

Next we examine what happens when we take a sum of two functions. If we have $y = f(x)$ and $y = g(x)$, we can compute a new function $y = (f + g)(x)$ by adding the outputs of the two functions: $(f + g)(x) = f(x) + g(x)$. Not only does this result in the value of the new function being the sum of the values of the two known functions, but also the slope of the new function is the sum of the slopes of the known functions. Therefore[3], we arrive at the following Sum Rule for derivatives:

The Sum Rule

If $f(x)$ and $g(x)$ are differentiable functions with derivatives $f'(x)$ and $g'(x)$ respectively, then $\frac{d}{dx}[f(x) + g(x)] = f'(x) + g'(x)$.

In words, the Sum Rule tells us that "the derivative of a sum is the sum of the derivatives." It also tells us that any time we take a sum of two differentiable functions, the result must also be differentiable. Furthermore, because we can view the difference function $y = (f - g)(x) = f(x) - g(x)$ as $y = f(x) + (-1 \cdot g(x))$, the Sum Rule and Constant Multiple Rules together tell us that $\frac{d}{dx}[f(x) + (-1 \cdot g(x))] = f'(x) - g'(x)$, or that "the derivative of a difference is the difference of the derivatives." Hence we can now compute derivatives of sums and differences of elementary functions. For instance, $\frac{d}{dw}(2^w + w^2) = 2^w \ln(2) + 2w$, and if $h(q) = 3q^6 - 4q^{-3}$, then $h'(q) = 3(6q^5) - 4(-3q^{-4}) = 18q^5 + 12q^{-4}$.

Activity 2.1.3. Use only the rules for constant, power, and exponential functions, together with the Constant Multiple and Sum Rules, to compute the derivative of each function below with respect to the given independent variable. Note well that we do not yet know any rules for how to differentiate the product or quotient of functions. This means that you may have to do some algebra first on the functions below before you can actually use existing rules to compute the desired derivative formula. In each case, label the derivative you calculate with its name using proper notation such as $f'(x)$, $h'(z)$, dr/dt, etc.

a. $f(x) = x^{5/3} - x^4 + 2^x$

b. $g(x) = 14e^x + 3x^5 - x$

c. $h(z) = \sqrt{z} + \frac{1}{z^4} + 5^z$

d. $r(t) = \sqrt{53}\, t^7 - \pi e^t + e^4$

e. $s(y) = (y^2 + 1)(y^2 - 1)$

f. $q(x) = \frac{x^3 - x + 2}{x}$

g. $p(a) = 3a^4 - 2a^3 + 7a^2 - a + 12$

In the same way that we have shortcut rules to help us find derivatives, we introduce some language that is simpler and shorter. Often, rather than say "take the derivative of f," we'll

---
[3]Like the Constant Multiple Rule, the Sum Rule can be formally proved as a consequence of properties of limits, using the limit definition of the derivative.

instead say simply "differentiate f." This phrasing is tied to the notion of having a derivative to begin with: if the derivative exists at a point, we say "f is differentiable," which is tied to the fact that f can be differentiated.

As we work more and more with the algebraic structure of functions, it is important to strive to develop a big picture view of what we are doing. Here, we can note several general observations based on the rules we have so far. One is that the derivative of any polynomial function will be another polynomial function, and that the degree of the derivative is one less than the degree of the original function. For instance, if $p(t) = 7t^5 - 4t^3 + 8t$, p is a degree 5 polynomial, and its derivative, $p'(t) = 35t^4 - 12t^2 + 8$, is a degree 4 polynomial. Additionally, the derivative of any exponential function is another exponential function: for example, if $g(z) = 7 \cdot 2^z$, then $g'(z) = 7 \cdot 2^z \ln(2)$, which is also exponential.

Furthermore, while our current emphasis is on learning shortcut rules for finding derivatives without directly using the limit definition, we should be certain not to lose sight of the fact that all of the meaning of the derivative still holds that we developed in Chapter 1. That is, anytime we compute a derivative, that derivative measures the instantaneous rate of change of the original function, as well as the slope of the tangent line at any selected point on the curve. The following activity asks you to combine the just-developed derivative rules with some key perspectives that we studied in Chapter 1.

> **Activity 2.1.4.** Each of the following questions asks you to use derivatives to answer key questions about functions. Be sure to think carefully about each question and to use proper notation in your responses.
>
> a. Find the slope of the tangent line to $h(z) = \sqrt{z} + \frac{1}{z}$ at the point where $z = 4$.
>
> b. A population of cells is growing in such a way that its total number in millions is given by the function $P(t) = 2(1.37)^t + 32$, where t is measured in days.
>
>   i. Determine the instantaneous rate at which the population is growing on day 4, and include units on your answer.
>
>   ii. Is the population growing at an increasing rate or growing at a decreasing rate on day 4? Explain.
>
> c. Find an equation for the tangent line to the curve $p(a) = 3a^4 - 2a^3 + 7a^2 - a + 12$ at the point where $a = -1$.
>
> d. What is the difference between being asked to find the *slope* of the tangent line (asked in (a)) and the *equation* of the tangent line (asked in (c))?

Summary

- Given a differentiable function $y = f(x)$, we can express the derivative of f in several different notations: $f'(x)$, $\frac{df}{dx}$, $\frac{dy}{dx}$, and $\frac{d}{dx}[f(x)]$.

- The limit definition of the derivative leads to patterns among certain families of functions that enable us to compute derivative formulas without resorting directly to the limit definition. For example, if f is a power function of the form $f(x) = x^n$, then $f'(x) = nx^{n-1}$ for any real number n other than 0. This is called the Rule for Power Functions.

- We have stated a rule for derivatives of exponential functions in the same spirit as the rule for power functions: for any positive real number a, if $f(x) = a^x$, then $f'(x) = a^x \ln(a)$.

- If we are given a constant multiple of a function whose derivative we know, or a sum of functions whose derivatives we know, the Constant Multiple and Sum Rules make it straightforward to compute the derivative of the overall function. More formally, if $f(x)$ and $g(x)$ are differentiable with derivatives $f'(x)$ and $g'(x)$ and a and b are constants, then

$$\frac{d}{dx}\left[af(x) + bg(x)\right] = af'(x) + bg'(x).$$

Exercises

1. Find the derivative of $y = x^{15/16}$.

$\frac{dy}{dx} =$ [   ]

2. Find the derivative of $f(x) = \dfrac{1}{x^{17}}$.

$f'(x) =$ [   ]

3. Find the derivative of

$y = \sqrt[9]{x}$.

$\frac{dy}{dx} =$ [   ]

4. Find the derivative of $f(t) = 2t^2 - 2t + 15$.

$f'(t) =$ [   ]

5. Find the derivative of $y = 4t^{12} - 7\sqrt{t} + \frac{4}{t}$.

$\frac{dy}{dt} =$ [   ]

6. Find the derivative of $y = \sqrt{x}(x^2 + 4)$.

$\frac{dy}{dx} =$ [   ]

7. Find the derivative of $y = \dfrac{x^5 + 2}{x}$.

$\frac{dy}{dx} =$ [   ]

8. Find an equation for the line tangent to the graph of f at $(2, 27)$, where f is given by $f(x) = 4x^3 - 4x^2 + 11$.

$y = $ [_____]

9. If $f(x) = x^3 + 3x^2 - 189x + 1$, find analytically all values of x for which $f'(x) = 0$. (Enter your answer as a comma separated list of numbers, e.g., -1,0,2)

$x = $ [_____]

10. Let f and g be differentiable functions for which the following information is known: $f(2) = 5$, $g(2) = -3$, $f'(2) = -1/2$, $g'(2) = 2$.

    a. Let h be the new function defined by the rule $h(x) = 3f(x) - 4g(x)$. Determine $h(2)$ and $h'(2)$.

    b. Find an equation for the tangent line to $y = h(x)$ at the point $(2, h(2))$.

    c. Let p be the function defined by the rule $p(x) = -2f(x) + \frac{1}{2}g(x)$. Is p increasing, decreasing, or neither at $a = 2$? Why?

    d. Estimate the value of $p(2.03)$ by using the local linearization of p at the point $(2, p(2))$.

11. Let functions p and q be the piecewise linear functions given by their respective graphs in Figure 2.1.1. Use the graphs to answer the following questions.

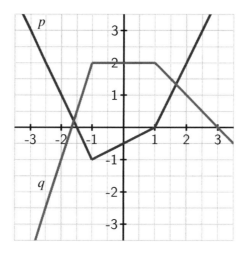

Figure 2.1.1: The graphs of p (in blue) and q (in green).

    a. At what values of x is p not differentiable? At what values of x is q not differentiable? Why?

    b. Let $r(x) = p(x) + 2q(x)$. At what values of x is r not differentiable? Why?

    c. Determine $r'(-2)$ and $r'(0)$.

d. Find an equation for the tangent line to $y = r(x)$ at the point $(2, r(2))$.

12. Consider the functions $r(t) = t^t$ and $s(t) = \arccos(t)$, for which you are given the facts that $r'(t) = t^t(\ln(t)+1)$ and $s'(t) = -\frac{1}{\sqrt{1-t^2}}$. Do not be concerned with where these derivative formulas come from. We restrict our interest in both functions to the domain $0 < t < 1$.

   a. Let $w(t) = 3t^t - 2\arccos(t)$. Determine $w'(t)$.

   b. Find an equation for the tangent line to $y = w(t)$ at the point $(\frac{1}{2}, w(\frac{1}{2}))$.

   c. Let $v(t) = t^t + \arccos(t)$. Is v increasing or decreasing at the instant $t = \frac{1}{2}$? Why?

13. Let $f(x) = a^x$. The goal of this problem is to explore how the value of a affects the derivative of $f(x)$, without assuming we know the rule for $\frac{d}{dx}[a^x]$ that we have stated and used in earlier work in this section.

   a. Use the limit definition of the derivative to show that

$$f'(x) = \lim_{h \to 0} \frac{a^x \cdot a^h - a^x}{h}.$$

   b. Explain why it is also true that

$$f'(x) = a^x \cdot \lim_{h \to 0} \frac{a^h - 1}{h}.$$

   c. Use computing technology and small values of h to estimate the value of

$$L = \lim_{h \to 0} \frac{a^h - 1}{h}$$

   when $a = 2$. Do likewise when $a = 3$.

   d. Note that it would be ideal if the value of the limit L was 1, for then f would be a particularly special function: its derivative would be simply a^x, which would mean that its derivative is itself. By experimenting with different values of a between 2 and 3, try to find a value for a for which

$$L = \lim_{h \to 0} \frac{a^h - 1}{h} = 1.$$

   e. Compute $\ln(2)$ and $\ln(3)$. What does your work in (b) and (c) suggest is true about $\frac{d}{dx}[2^x]$ and $\frac{d}{dx}[3^x]$?

   f. How do your investigations in (d) lead to a particularly important fact about the function $f(x) = e^x$?

2.2 The sine and cosine functions

Motivating Questions

- What is a graphical justification for why $\frac{d}{dx}[a^x] = a^x \ln(a)$?

- What do the graphs of $y = \sin(x)$ and $y = \cos(x)$ suggest as formulas for their respective derivatives?

- Once we know the derivatives of $\sin(x)$ and $\cos(x)$, how do previous derivative rules work when these functions are involved?

Throughout Chapter 2, we will be working to develop shortcut derivative rules that will help us to bypass the limit definition of the derivative in order to quickly determine the formula for $f'(x)$ when we are given a formula for $f(x)$. In Section 2.1, we learned the rule for power functions, that if $f(x) = x^n$, then $f'(x) = nx^{n-1}$, and justified this in part due to results from different n-values when applying the limit definition of the derivative. We also stated the rule for exponential functions, that if a is a positive real number and $f(x) = a^x$, then $f'(x) = a^x \ln(a)$. Later in this present section, we are going to work to conjecture formulas for the sine and cosine functions, primarily through a graphical argument. To help set the stage for doing so, the following preview activity asks you to think about exponential functions and why it is reasonable to think that the derivative of an exponential function is a constant times the exponential function itself.

Preview Activity 2.2.1. Consider the function $g(x) = 2^x$, which is graphed in Figure 2.2.1.

a. At each of $x = -2, -1, 0, 1, 2$, use a straightedge to sketch an accurate tangent line to $y = g(x)$.

b. Use the provided grid to estimate the slope of the tangent line you drew at each point in (a).

c. Use the limit definition of the derivative to estimate $g'(0)$ by using small values of h, and compare the result to your visual estimate for the slope of the tangent line to $y = g(x)$ at $x = 0$ in (b).

d. Based on your work in (a), (b), and (c), sketch an accurate graph of $y = g'(x)$ on the axes adjacent to the graph of $y = g(x)$.

e. Write at least one sentence that explains why it is reasonable to think that $g'(x) = cg(x)$, where c is a constant. In addition, calculate $\ln(2)$, and then discuss how this value, combined with your work above, reasonably suggests that $g'(x) = 2^x \ln(2)$.

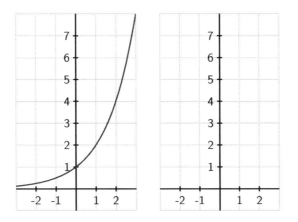

Figure 2.2.1: At left, the graph of $y = g(x) = 2^x$. At right, axes for plotting $y = g'(x)$.

2.2.1 The sine and cosine functions

The sine and cosine functions are among the most important functions in all of mathematics. Sometimes called the *circular* functions due to their genesis in the unit circle , these periodic functions play a key role in modeling repeating phenomena such as the location of a point on a bicycle tire, the behavior of an oscillating mass attached to a spring, tidal elevations, and more. Like polynomial and exponential functions, the sine and cosine functions are considered basic functions, ones that are often used in the building of more complicated functions. As such, we would like to know formulas for $\frac{d}{dx}[\sin(x)]$ and $\frac{d}{dx}[\cos(x)]$, and the next two activities lead us to that end.

Activity 2.2.2. Consider the function $f(x) = \sin(x)$, which is graphed in Figure 2.2.2 below. Note carefully that the grid in the diagram does not have boxes that are 1×1, but rather approximately 1.57×1, as the horizontal scale of the grid is $\pi/2$ units per box.

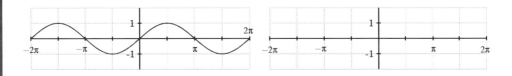

Figure 2.2.2: At left, the graph of $y = f(x) = \sin(x)$.

a. At each of $x = -2\pi, -\frac{3\pi}{2}, -\pi, -\frac{\pi}{2}, 0, \frac{\pi}{2}, \pi, \frac{3\pi}{2}, 2\pi$, use a straightedge to sketch an accurate tangent line to $y = f(x)$.

b. Use the provided grid to estimate the slope of the tangent line you drew at each point. Pay careful attention to the scale of the grid.

c. Use the limit definition of the derivative to estimate $f'(0)$ by using small values of h, and compare the result to your visual estimate for the slope of the tangent line to $y = f(x)$ at $x = 0$ in (b). Using periodicity, what does this result suggest about $f'(2\pi)$? about $f'(-2\pi)$?

d. Based on your work in (a), (b), and (c), sketch an accurate graph of $y = f'(x)$ on the axes adjacent to the graph of $y = f(x)$.

e. What familiar function do you think is the derivative of $f(x) = \sin(x)$?

Activity 2.2.3. Consider the function $g(x) = \cos(x)$, which is graphed in Figure 2.2.3 below. Note carefully that the grid in the diagram does not have boxes that are 1×1, but rather approximately 1.57×1, as the horizontal scale of the grid is $\pi/2$ units per box.

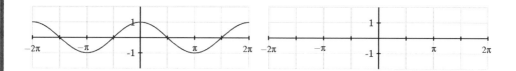

Figure 2.2.3: At left, the graph of $y = g(x) = \cos(x)$.

a. At each of $x = -2\pi, -\frac{3\pi}{2}, -\pi, -\frac{\pi}{2}, 0, \frac{\pi}{2}, \pi, \frac{3\pi}{2}, 2\pi$, use a straightedge to sketch an accurate tangent line to $y = g(x)$.

b. Use the provided grid to estimate the slope of the tangent line you drew at each point. Again, note the scale of the axes and grid.

c. Use the limit definition of the derivative to estimate $g'(\frac{\pi}{2})$ by using small values of h, and compare the result to your visual estimate for the slope of the tangent line to $y = g(x)$ at $x = \frac{\pi}{2}$ in (b). Using periodicity, what does this result suggest about $g'(-\frac{3\pi}{2})$? can symmetry on the graph help you estimate other slopes easily?

d. Based on your work in (a), (b), and (c), sketch an accurate graph of $y = g'(x)$ on the axes adjacent to the graph of $y = g(x)$.

e. What familiar function do you think is the derivative of $g(x) = \cos(x)$?

The results of the two preceding activities suggest that the sine and cosine functions not only have the beautiful interrelationships that are learned in a course in trigonometry — connections such as the identities $\sin^2(x) + \cos^2(x) = 1$ and $\cos(x - \frac{\pi}{2}) = \sin(x)$ — but that they are even further linked through calculus, as the derivative of each involves the other. The following rules summarize the results of the activities[1].

Sine and Cosine Functions

For all real numbers x,

$$\frac{d}{dx}[\sin(x)] = \cos(x) \text{ and } \frac{d}{dx}[\cos(x)] = -\sin(x).$$

We have now added two additional functions to our library of basic functions whose derivatives we know: power functions, exponential functions, and the sine and cosine functions. The constant multiple and sum rules still hold, of course, and all of the inherent meaning of the derivative persists, regardless of the functions that are used to constitute a given choice of $f(x)$. The following activity puts our new knowledge of the derivatives of $\sin(x)$ and $\cos(x)$ to work.

> **Activity 2.2.4.** Answer each of the following questions. Where a derivative is requested, be sure to label the derivative function with its name using proper notation.
>
> a. Determine the derivative of $h(t) = 3\cos(t) - 4\sin(t)$.
>
> b. Find the exact slope of the tangent line to $y = f(x) = 2x + \frac{\sin(x)}{2}$ at the point where $x = \frac{\pi}{6}$.
>
> c. Find the equation of the tangent line to $y = g(x) = x^2 + 2\cos(x)$ at the point where $x = \frac{\pi}{2}$.
>
> d. Determine the derivative of $p(z) = z^4 + 4^z + 4\cos(z) - \sin(\frac{\pi}{2})$.
>
> e. The function $P(t) = 24 + 8\sin(t)$ represents a population of a particular kind of animal that lives on a small island, where P is measured in hundreds and t is measured in decades since January 1, 2010. What is the instantaneous rate of change of P on January 1, 2030? What are the units of this quantity? Write a sentence in everyday language that explains how the population is behaving at this point in time.

Summary

- If we consider the graph of an exponential function $f(x) = a^x$ ($a > 1$), the graph of $f'(x)$ behaves similarly, appearing exponential and as a possibly scaled version of the original function a^x. For $f(x) = 2^x$, careful analysis of the graph and its slopes suggests that $\frac{d}{dx}[2^x] = 2^x \ln(2)$, which is a special case of the rule we stated in Section 2.1.

- By carefully analyzing the graphs of $y = \sin(x)$ and $y = \cos(x)$, plus using the limit definition of the derivative at select points, we found that $\frac{d}{dx}[\sin(x)] = \cos(x)$ and

---

[1]These two rules may be formally proved using the limit definition of the derivative and the expansion identities for $\sin(x + h)$ and $\cos(x + h)$.

$\frac{d}{dx}[\cos(x)] = -\sin(x)$.

- We note that all previously encountered derivative rules still hold, but now may also be applied to functions involving the sine and cosine, plus all of the established meaning of the derivative applies to these trigonometric functions as well.

Exercises

1. Suppose that $V(t) = 24 \cdot 1.07^t + 6\sin(t)$ represents the value of a person's investment portfolio in thousands of dollars in year t, where $t = 0$ corresponds to January 1, 2010.

a. At what instantaneous rate is the portfolio's value changing on January 1, 2012? Include units on your answer.

b. Determine the value of $V''(2)$. What are the units on this quantity and what does it tell you about how the portfolio's value is changing?

c. On the interval $0 \le t \le 20$, graph the function $V(t) = 24 \cdot 1.07^t + 6\sin(t)$ and describe its behavior in the context of the problem. Then, compare the graphs of the functions $A(t) = 24 \cdot 1.07^t$ and $V(t) = 24 \cdot 1.07^t + 6\sin(t)$, as well as the graphs of their derivatives $A'(t)$ and $V'(t)$. What is the impact of the term $6\sin(t)$ on the behavior of the function $V(t)$?

2. Let $f(x) = 3\cos(x) - 2\sin(x) + 6$.

a. Determine the exact slope of the tangent line to $y = f(x)$ at the point where $a = \frac{\pi}{4}$.

b. Determine the tangent line approximation to $y = f(x)$ at the point where $a = \pi$.

c. At the point where $a = \frac{\pi}{2}$, is f increasing, decreasing, or neither?

d. At the point where $a = \frac{3\pi}{2}$, does the tangent line to $y = f(x)$ lie above the curve, below the curve, or neither? How can you answer this question without even graphing the function or the tangent line?

3. In this exercise, we explore how the limit definition of the derivative more formally shows that $\frac{d}{dx}[\sin(x)] = \cos(x)$. Letting $f(x) = \sin(x)$, note that the limit definition of the derivative tells us that

$$f'(x) = \lim_{h \to 0} \frac{\sin(x+h) - \sin(x)}{h}.$$

a. Recall the trigonometric identity for the sine of a sum of angles α and β: $\sin(\alpha + \beta) = \sin(\alpha)\cos(\beta) + \cos(\alpha)\sin(\beta)$. Use this identity and some algebra to show that

$$f'(x) = \lim_{h \to 0} \frac{\sin(x)(\cos(h) - 1) + \cos(x)\sin(h)}{h}.$$

b. Next, note that as h changes, x remains constant. Explain why it therefore makes sense to say that

$$f'(x) = \sin(x) \cdot \lim_{h \to 0} \frac{\cos(h) - 1}{h} + \cos(x) \cdot \lim_{h \to 0} \frac{\sin(h)}{h}.$$

c. Finally, use small values of h to estimate the values of the two limits in (c):

$$\lim_{h \to 0} \frac{\cos(h) - 1}{h} \quad \text{and} \quad \lim_{h \to 0} \frac{\sin(h)}{h}.$$

d. What do your results in (c) thus tell you about $f'(x)$?

e. By emulating the steps taken above, use the limit definition of the derivative to argue convincingly that $\frac{d}{dx}[\cos(x)] = -\sin(x)$.

2.3 The product and quotient rules

Motivating Questions

- How does the algebraic structure of a function direct us in computing its derivative using shortcut rules?

- How do we compute the derivative of a product of two basic functions in terms of the derivatives of the basic functions?

- How do we compute the derivative of a quotient of two basic functions in terms of the derivatives of the basic functions?

- How do the product and quotient rules combine with the sum and constant multiple rules to expand the library of functions we can quickly differentiate?

So far, the basic functions we know how to differentiate include power functions (x^n), exponential functions (a^x), and the two fundamental trigonometric functions ($\sin(x)$ and $\cos(x)$). With the sum rule and constant multiple rules, we can also compute the derivative of combined functions such as

$$f(x) = 7x^{11} - 4 \cdot 9^x + \pi \sin(x) - \sqrt{3}\cos(x),$$

because the function f is fundamentally a sum of basic functions. Indeed, we can now quickly say that $f'(x) = 77x^{10} - 4 \cdot 9^x \ln(9) + \pi \cos(x) + \sqrt{3}\sin(x)$.

But we can of course combine basic functions in ways other than multiplying them by constants and taking sums and differences. For example, we could consider the function that results from a product of two basic functions, such as

$$p(z) = z^3 \cos(z),$$

or another that is generated by the quotient of two basic functions, one like

$$q(t) = \frac{\sin(t)}{2^t}.$$

While the derivative of a sum is the sum of the derivatives, it turns out that the rules for computing derivatives of products and quotients are more complicated. In what follows we explore why this is the case, what the product and quotient rules actually say, and work to expand our repertoire of functions we can easily differentiate. To start, Preview Activity 2.3.1 asks you to investigate the derivative of a product and quotient of two polynomials.

Preview Activity 2.3.1. Let f and g be the functions defined by $f(t) = 2t^2$ and $g(t) = t^3 + 4t$.

    a. Determine $f'(t)$ and $g'(t)$.

b. Let $p(t) = 2t^2(t^3 + 4t)$ and observe that $p(t) = f(t) \cdot g(t)$. Rewrite the formula for p by distributing the $2t^2$ term. Then, compute $p'(t)$ using the sum and constant multiple rules.

c. True or false: $p'(t) = f'(t) \cdot g'(t)$.

d. Let $q(t) = \frac{t^3 + 4t}{2t^2}$ and observe that $q(t) = \frac{g(t)}{f(t)}$. Rewrite the formula for q by dividing each term in the numerator by the denominator and simplify to write q as a sum of constant multiples of powers of t. Then, compute $q'(t)$ using the sum and constant multiple rules.

e. True or false: $q'(t) = \frac{g'(t)}{f'(t)}$.

2.3.1 The product rule

As part (b) of Preview Activity 2.3.1 shows, it is not true in general that the derivative of a product of two functions is the product of the derivatives of those functions. Indeed, the rule for differentiating a function of the form $p(x) = f(x) \cdot g(x)$ in terms of the derivatives of f and g is more complicated than simply taking the product of the derivatives of f and g. To see further why this is the case, as well as to begin to understand how the product rule actually works, we consider an example involving meaningful functions.

Say that an investor is regularly purchasing stock in a particular company. Let $N(t)$ be a function that represents the number of shares owned on day t, where $t = 0$ represents the first day on which shares were purchased. Further, let $S(t)$ be a function that gives the value of one share of the stock on day t; note that the units on $S(t)$ are dollars per share. Moreover, to compute the total value on day t of the stock held by the investor, we use the function $V(t) = N(t) \cdot S(t)$. By taking the product

$$V(t) = N(t)\,\text{shares} \cdot S(t)\,\text{dollars per share,}$$

we have the total value in dollars of the shares held. Observe that over time, both the number of shares and the value of a given share will vary. The derivative $N'(t)$ measures the rate at which the number of shares held is changing, while $S'(t)$ measures the rate at which the value per share is changing. The big question we'd like to answer is: how do these respective rates of change affect the rate of change of the total value function?

To help better understand the relationship among changes in N, S, and V, let's consider some specific data. Suppose that on day 100, the investor owns 520 shares of stock and the stock's current value is $27.50 per share. This tells us that $N(100) = 520$ and $S(100) = 27.50$. In addition, say that on day 100, the investor purchases an additional 12 shares (so the number of shares held is rising at a rate of 12 shares per day), and that on that same day the price of the stock is rising at a rate of 0.75 dollars per share per day. Viewed in calculus notation, this tells us that $N'(100) = 12$ (shares per day) and $S'(100) = 0.75$ (dollars per share per day). At what rate is the value of the investor's total holdings changing on day 100?

Observe that the increase in total value comes from two sources: the growing number of shares, and the rising value of each share. If only the number of shares is rising (and the value of each share is constant), the rate at which which total value would rise is found by computing the product of the current value of the shares with the rate at which the number of shares is changing. That is, the rate at which total value would change is given by

$$S(100) \cdot N'(100) = 27.50 \frac{\text{dollars}}{\text{share}} \cdot 12 \frac{\text{shares}}{\text{day}} = 330 \frac{\text{dollars}}{\text{day}}.$$

Note particularly how the units make sense and explain that we are finding the rate at which the total value V is changing, measured in dollars per day. If instead the number of shares is constant, but the value of each share is rising, then the rate at which the total value would rise is found similarly by taking the product of the number of shares with the rate of change of share value. In particular, the rate total value is rising is

$$N(100) \cdot S'(100) = 520 \,\text{shares} \cdot 0.75 \frac{\text{dollars per share}}{\text{day}} = 390 \frac{\text{dollars}}{\text{day}}.$$

Of course, when both the number of shares is changing and the value of each share is changing, we have to include both of these sources, and hence the rate at which the total value is rising is

$$V'(100) = S(100) \cdot N'(100) + N(100) \cdot S'(100) = 330 + 390 = 720 \frac{\text{dollars}}{\text{day}}.$$

This tells us that we expect the total value of the investor's holdings to rise by about $720 on the 100th day.[1]

Next, we expand our perspective from the specific example above to the more general and abstract setting of a product p of two differentiable functions, f and g. If we have $P(x) = f(x) \cdot g(x)$, our work above suggests that $P'(x) = f(x)g'(x) + g(x)f'(x)$. Indeed, a formal proof using the limit definition of the derivative can be given to show that the following rule, called the *product rule*, holds in general.

Product Rule

If f and g are differentiable functions, then their product $P(x) = f(x) \cdot g(x)$ is also a differentiable function, and

$$P'(x) = f(x)g'(x) + g(x)f'(x).$$

In light of the earlier example involving shares of stock, the product rule also makes sense intuitively: the rate of change of P should take into account both how fast f and g are changing, as well as how large f and g are at the point of interest. Furthermore, we note in words

---

[1]While this example highlights why the product rule is true, there are some subtle issues to recognize. For one, if the stock's value really does rise exactly $0.75 on day 100, and the number of shares really rises by 12 on day 100, then we'd expect that $V(101) = N(101) \cdot S(101) = 532 \cdot 28.25 = 15029$. If, as noted above, we expect the total value to rise by $720, then with $V(100) = N(100) \cdot S(100) = 520 \cdot 27.50 = 14300$, then it seems like we should find that $V(101) = V(100) + 720 = 15020$. Why do the two results differ by 9? One way to understand why this difference occurs is to recognize that $N'(100) = 12$ represents an *instantaneous* rate of change, while our (informal) discussion has also thought of this number as the total change in the number of shares over the course of a single day. The formal proof of the product rule reconciles this issue by taking the limit as the change in the input tends to zero.

what the product rule says: if P is the product of two functions f (the first function) and g (the second), then "the derivative of P is the first times the derivative of the second, plus the second times the derivative of the first." It is often a helpful mental exercise to say this phrasing aloud when executing the product rule.

For example, if $P(z) = z^3 \cdot \cos(z)$, we can now use the product rule to differentiate P. The first function is z^3 and the second function is $\cos(z)$. By the product rule, P' will be given by the first, z^3, times the derivative of the second, $-\sin(z)$, plus the second, $\cos(z)$, times the derivative of the first, $3z^2$. That is,

$$P'(z) = z^3(-\sin(z)) + \cos(z)3z^2 = -z^3\sin(z) + 3z^2\cos(z).$$

The following activity further explores the use of the product rule.

Activity 2.3.2. Use the product rule to answer each of the questions below. Throughout, be sure to carefully label any derivative you find by name. It is not necessary to algebraically simplify any of the derivatives you compute.

    a. Let $m(w) = 3w^{17}4^w$. Find $m'(w)$.

    b. Let $h(t) = (\sin(t) + \cos(t))t^4$. Find $h'(t)$.

    c. Determine the slope of the tangent line to the curve $y = f(x)$ at the point where $a = 1$ if f is given by the rule $f(x) = e^x\sin(x)$.

    d. Find the tangent line approximation $L(x)$ to the function $y = g(x)$ at the point where $a = -1$ if g is given by the rule $g(x) = (x^2 + x)2^x$.

2.3.2 The quotient rule

Because quotients and products are closely linked, we can use the product rule to understand how to take the derivative of a quotient. In particular, let $Q(x)$ be defined by $Q(x) = f(x)/g(x)$, where f and g are both differentiable functions. We desire a formula for Q' in terms of f, g, f', and g'. It turns out that Q is differentiable everywhere that $g(x) \neq 0$. Moreover, taking the formula $Q = f/g$ and multiplying both sides by g, we can observe that

$$f(x) = Q(x) \cdot g(x).$$

Thus, we can use the product rule to differentiate f. Doing so,

$$f'(x) = Q(x)g'(x) + g(x)Q'(x).$$

Since we want to know a formula for Q', we work to solve this most recent equation for $Q'(x)$, finding first that

$$Q'(x)g(x) = f'(x) - Q(x)g'(x).$$

Dividing both sides by $g(x)$, we have

$$Q'(x) = \frac{f'(x) - Q(x)g'(x)}{g(x)}.$$

Finally, we also recall that $Q(x) = \frac{f(x)}{g(x)}$. Using this expression in the preceding equation and simplifying, we have

$$Q'(x) = \frac{f'(x) - \frac{f(x)}{g(x)}g'(x)}{g(x)}$$

$$= \frac{f'(x) - \frac{f(x)}{g(x)}g'(x)}{g(x)} \cdot \frac{g(x)}{g(x)}$$

$$= \frac{g(x)f'(x) - f(x)g'(x)}{g(x)^2}.$$

This shows the fundamental argument for why the *quotient rule* holds.

Quotient Rule

If f and g are differentiable functions, then their quotient $Q(x) = \frac{f(x)}{g(x)}$ is also a differentiable function for all x where $g'(x) \neq 0$ and

$$Q'(x) = \frac{g(x)f'(x) - f(x)g'(x)}{g(x)^2}.$$

Like the product rule, it can be helpful to think of the quotient rule verbally. If a function Q is the quotient of a top function f and a bottom function g, then Q' is given by "the bottom times the derivative of the top, minus the top times the derivative of the bottom, all over the bottom squared." For example, if $Q(t) = \sin(t)/2^t$, then we can identify the top function as $\sin(t)$ and the bottom function as 2^t. By the quotient rule, we then have that Q' will be given by the bottom, 2^t, times the derivative of the top, $\cos(t)$, minus the top, $\sin(t)$, times the derivative of the bottom, $2^t \ln(2)$, all over the bottom squared, $(2^t)^2$. That is,

$$Q'(t) = \frac{2^t \cos(t) - \sin(t)2^t \ln(2)}{(2^t)^2}.$$

In this particular example, it is possible to simplify $Q'(t)$ by removing a factor of 2^t from both the numerator and denominator, hence finding that

$$Q'(t) = \frac{\cos(t) - \sin(t)\ln(2)}{2^t}.$$

In general, we must be careful in doing any such simplification, as we don't want to correctly execute the quotient rule but then find an incorrect overall derivative due to an algebra error. As such, we will often place more emphasis on correctly using derivative rules than we will on simplifying the result that follows. The next activity further explores the use of the quotient rule.

Activity 2.3.3. Use the quotient rule to answer each of the questions below. Throughout, be sure to carefully label any derivative you find by name. That is, if you're given a formula for $f(x)$, clearly label the formula you find for $f'(x)$. It is not necessary to algebraically simplify any of the derivatives you compute.

a. Let $r(z) = \frac{3^z}{z^4+1}$. Find $r'(z)$.

b. Let $v(t) = \frac{\sin(t)}{\cos(t)+t^2}$. Find $v'(t)$.

c. Determine the slope of the tangent line to the curve $R(x) = \dfrac{x^2 - 2x - 8}{x^2 - 9}$ at the point where $x = 0$.

d. When a camera flashes, the intensity I of light seen by the eye is given by the function

$$I(t) = \frac{100t}{e^t},$$

where I is measured in candles and t is measured in milliseconds. Compute $I'(0.5)$, $I'(2)$, and $I'(5)$; include appropriate units on each value; and discuss the meaning of each.

2.3.3 Combining rules

One of the challenges to learning to apply various derivative shortcut rules correctly and effectively is recognizing the fundamental structure of a function. For instance, consider the function given by

$$f(x) = x \sin(x) + \frac{x^2}{\cos(x) + 2}.$$

How do we decide which rules to apply? Our first task is to recognize the overall structure of the given function. Observe that the function f is fundamentally a sum of two slightly less complicated functions, so we can apply the sum rule[2] and get

$$f'(x) = \frac{d}{dx}\left[x \sin(x) + \frac{x^2}{\cos(x) + 2}\right]$$
$$= \frac{d}{dx}[x \sin(x)] + \frac{d}{dx}\left[\frac{x^2}{\cos(x) + 2}\right]$$

Now, the left-hand term above is a product, so the product rule is needed there, while the right-hand term is a quotient, so the quotient rule is required. Applying these rules respectively, we find that

$$f'(x) = (x \cos(x) + \sin(x)) + \frac{(\cos(x) + 2)2x - x^2(-\sin(x))}{(\cos(x) + 2)^2}$$

---

[2]When taking a derivative that involves the use of multiple derivative rules, it is often helpful to use the notation $\frac{d}{dx}[\ ]$ to wait to apply subsequent rules. This is demonstrated in each of the two examples presented here.

$$= x \cos(x) + \sin(x) + \frac{2x \cos(x) + 4x^2 + x^2 \sin(x)}{(\cos(x) + 2)^2}.$$

We next consider how the situation changes with the function defined by

$$s(y) = \frac{y \cdot 7^y}{y^2 + 1}.$$

Overall, s is a quotient of two simpler function, so the quotient rule will be needed. Here, we execute the quotient rule and use the notation $\frac{d}{dy}$ to defer the computation of the derivative of the numerator and derivative of the denominator. Thus,

$$s'(y) = \frac{(y^2 + 1) \cdot \frac{d}{dy} \left[y \cdot 7^y \right] - y \cdot 7^y \cdot \frac{d}{dy} \left[y^2 + 1 \right]}{(y^2 + 1)^2}.$$

Now, there remain two derivatives to calculate. The first one, $\frac{d}{dy} \left[y \cdot 7^y \right]$ calls for use of the product rule, while the second, $\frac{d}{dy} \left[y^2 + 1 \right]$ takes only an elementary application of the sum rule. Applying these rules, we now have

$$s'(y) = \frac{(y^2 + 1)[y \cdot 7^y \ln(7) + 7^y \cdot 1] - y \cdot 7^y [2y]}{(y^2 + 1)^2}.$$

While some minor simplification is possible, we are content to leave $s'(y)$ in its current form, having found the desired derivative of s. In summary, to compute the derivative of s, we applied the quotient rule. In so doing, when it was time to compute the derivative of the top function, we used the product rule; at the point where we found the derivative of the bottom function, we used the sum rule.

In general, one of the main keys to success in applying derivative rules is to recognize the structure of the function, followed by the careful and diligent application of relevant derivative rules. The best way to get good at this process is by doing a large number of exercises, and the next activity provides some practice and exploration to that end.

> **Activity 2.3.4.** Use relevant derivative rules to answer each of the questions below. Throughout, be sure to use proper notation and carefully label any derivative you find by name.
>
> a. Let $f(r) = (5r^3 + \sin(r))(4^r - 2 \cos(r))$. Find $f'(r)$.
>
> b. Let $p(t) = \frac{\cos(t)}{t^6 \cdot 6^t}$. Find $p'(t)$.
>
> c. Let $g(z) = 3z^7 e^z - 2z^2 \sin(z) + \frac{z}{z^2+1}$. Find $g'(z)$.
>
> d. A moving particle has its position in feet at time t in seconds given by the function $s(t) = \frac{3 \cos(t) - \sin(t)}{e^t}$. Find the particle's instantaneous velocity at the moment $t = 1$.
>
> e. Suppose that $f(x)$ and $g(x)$ are differentiable functions and it is known that $f(3) =$

-2, $f'(3) = 7$, $g(3) = 4$, and $g'(3) = -1$. If $p(x) = f(x) \cdot g(x)$ and $q(x) = \dfrac{f(x)}{g(x)}$, calculate $p'(3)$ and $q'(3)$.

As the algebraic complexity of the functions we are able to differentiate continues to increase, it is important to remember that all of the derivative's meaning continues to hold. Regardless of the structure of the function f, the value of $f'(a)$ tells us the instantaneous rate of change of f with respect to x at the moment $x = a$, as well as the slope of the tangent line to $y = f(x)$ at the point $(a, f(a))$.

Summary

- If a function is a sum, product, or quotient of simpler functions, then we can use the sum, product, or quotient rules to differentiate the overall function in terms of the simpler functions and their derivatives.

- The product rule tells us that if P is a product of differentiable functions f and g according to the rule $P(x) = f(x)g(x)$, then

$$P'(x) = f(x)g'(x) + g(x)f'(x).$$

- The quotient rule tells us that if Q is a quotient of differentiable functions f and g according to the rule $Q(x) = \dfrac{f(x)}{g(x)}$, then

$$Q'(x) = \frac{g(x)f'(x) - f(x)g'(x)}{g(x)^2}.$$

- The product and quotient rules now complement the constant multiple and sum rules and enable us to compute the derivative of any function that consists of sums, constant multiples, products, and quotients of basic functions we already know how to differentiate. For instance, if F has the form

$$F(x) = \frac{2a(x) - 5b(x)}{c(x) \cdot d(x)},$$

then F is fundamentally a quotient, and the numerator is a sum of constant multiples and the denominator is a product. Hence the derivative of F can be found by applying the quotient rule and then using the sum and constant multiple rules to differentiate the numerator and the product rule to differentiate the denominator.

Exercises

1. Find the derivative of the function $f(x)$, below. It may be to your advantage to simplify first.

$f(x) = x \cdot 12^x$

$f'(x) = $

2. Find the derivative of the function $f(x)$, below. It may be to your advantage to simplify first.

$$f(x) = (x^8 - \sqrt[7]{x})8^x$$

$f'(x) = $

3. Find the derivative of the function z, below. It may be to your advantage to simplify first.

$$z = \frac{9t + 4}{4t + 5}$$

$\dfrac{dz}{dt} = $

4. Find the derivative of the function $h(r)$, below. It may be to your advantage to simplify first.

$$h(r) = \frac{r^2}{15r + 11}$$

$h'(r) = $

5. Find the derivative of $s(q) = 4 \cos q \sin q$.

$s'(q) = $

6. Find the derivative of $f(x) = x^4 \cos x$

$f'(x) = $

7. Find the derivative of $h(t) = t \sin t + \cos t$

$h'(t) = $

8. Let $h(x) = f(x) \cdot g(x)$, and $k(x) = f(x)/g(x)$. Use the figures below to find the *exact* values of the indicated derivatives.

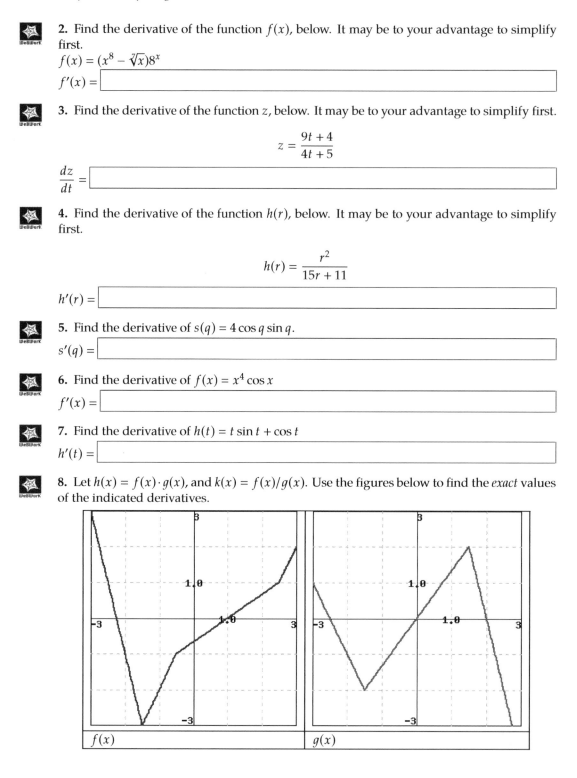

$f(x)$	$g(x)$

A. $h'(0) =$ [   ]

B. $k'(-2) =$ [   ]

(Enter dne *for any answer where the derivative does not exist.)*

9. Let $F(4) = 2, F'(4) = 4, H(4) = 3, H'(4) = 1.$

A. If $G(z) = F(z) \cdot H(z)$, then $G'(4) =$ [   ]

B. If $G(w) = F(w)/H(w)$, then $G'(4) =$ [   ]

10. Let f and g be differentiable functions for which the following information is known: $f(2) = 5, g(2) = -3, f'(2) = -1/2, g'(2) = 2.$

a. Let h be the new function defined by the rule $h(x) = g(x) \cdot f(x)$. Determine $h(2)$ and $h'(2)$.

b. Find an equation for the tangent line to $y = h(x)$ at the point $(2, h(2))$ (where h is the function defined in (a)).

c. Let r be the function defined by the rule $r(x) = \frac{g(x)}{f(x)}$. Is r increasing, decreasing, or neither at $a = 2$? Why?

d. Estimate the value of $r(2.06)$ (where r is the function defined in (c)) by using the local linearization of r at the point $(2, r(2))$.

11. Consider the functions $r(t) = t^t$ and $s(t) = \arccos(t)$, for which you are given the facts that $r'(t) = t^t(\ln(t)+1)$ and $s'(t) = -\frac{1}{\sqrt{1-t^2}}$. Do not be concerned with where these derivative formulas come from. We restrict our interest in both functions to the domain $0 < t < 1$.

a. Let $w(t) = t^t \arccos(t)$. Determine $w'(t)$.

b. Find an equation for the tangent line to $y = w(t)$ at the point $(\frac{1}{2}, w(\frac{1}{2}))$.

c. Let $v(t) = \frac{t^t}{\arccos(t)}$. Is v increasing or decreasing at the instant $t = \frac{1}{2}$? Why?

12. Let functions p and q be the piecewise linear functions given by their respective graphs in Figure 2.3.1. Use the graphs to answer the following questions.

a. Let $r(x) = p(x) \cdot q(x)$. Determine $r'(-2)$ and $r'(0)$.

b. Are there values of x for which $r'(x)$ does not exist? If so, which values, and why?

c. Find an equation for the tangent line to $y = r(x)$ at the point $(2, r(2))$.

d. Let $z(x) = \frac{q(x)}{p(x)}$. Determine $z'(0)$ and $z'(2)$.

e. Are there values of x for which $z'(x)$ does not exist? If so, which values, and why?

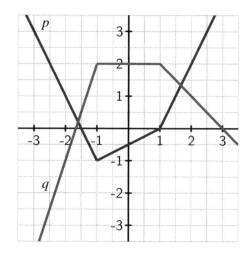

Figure 2.3.1: The graphs of p (in blue) and q (in green).

13. A farmer with large land holdings has historically grown a wide variety of crops. With the price of ethanol fuel rising, he decides that it would be prudent to devote more and more of his acreage to producing corn. As he grows more and more corn, he learns efficiencies that increase his yield per acre. In the present year, he used 7000 acres of his land to grow corn, and that land had an average yield of 170 bushels per acre. At the current time, he plans to increase his number of acres devoted to growing corn at a rate of 600 acres/year, and he expects that right now his average yield is increasing at a rate of 8 bushels per acre per year. Use this information to answer the following questions.

a. Say that the present year is $t = 0$, that $A(t)$ denotes the number of acres the farmer devotes to growing corn in year t, $Y(t)$ represents the average yield in year t (measured in bushels per acre), and $C(t)$ is the total number of bushels of corn the farmer produces. What is the formula for $C(t)$ in terms of $A(t)$ and $Y(t)$? Why?

b. What is the value of $C(0)$? What does it measure?

c. Write an expression for $C'(t)$ in terms of $A(t)$, $A'(t)$, $Y(t)$, and $Y'(t)$. Explain your thinking.

d. What is the value of $C'(0)$? What does it measure?

e. Based on the given information and your work above, estimate the value of $C(1)$.

14. Let $f(v)$ be the gas consumption (in liters/km) of a car going at velocity v (in km/hour). In other words, $f(v)$ tells you how many liters of gas the car uses to go one kilometer if it is traveling at v kilometers per hour. In addition, suppose that $f(80) = 0.05$ and $f'(80) = 0.0004$.

a. Let $g(v)$ be the distance the same car goes on one liter of gas at velocity v. What is the relationship between $f(v)$ and $g(v)$? Hence find $g(80)$ and $g'(80)$.

b. Let $h(v)$ be the gas consumption in liters per hour of a car going at velocity v. In other words, $h(v)$ tells you how many liters of gas the car uses in one hour if it is going at velocity v. What is the algebraic relationship between $h(v)$ and $f(v)$? Hence find $h(80)$ and $h'(80)$.

c. How would you explain the practical meaning of these function and derivative values to a driver who knows no calculus? Include units on each of the function and derivative values you discuss in your response.

2.4 Derivatives of other trigonometric functions

Motivating Questions

- What are the derivatives of the tangent, cotangent, secant, and cosecant functions?

- How do the derivatives of $\tan(x)$, $\cot(x)$, $\sec(x)$, and $\csc(x)$ combine with other derivative rules we have developed to expand the library of functions we can quickly differentiate?

One of the powerful themes in trigonometry is that the entire subject emanates from a very simple idea: locating a point on the unit circle.

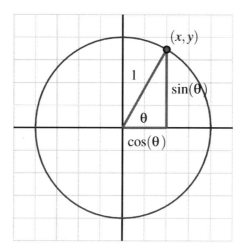

Figure 2.4.1: The unit circle and the definition of the sine and cosine functions.

Because each angle θ corresponds to one and only one point (x, y) on the unit circle, the x- and y-coordinates of this point are each functions of θ. Indeed, this is the very definition of $\cos(\theta)$ and $\sin(\theta)$: $\cos(\theta)$ is the x-coordinate of the point on the unit circle corresponding to the angle θ, and $\sin(\theta)$ is the y-coordinate. From this simple definition, all of trigonometry is founded. For instance, the fundamental trigonometric identity,

$$\sin^2(\theta) + \cos^2(\theta) = 1,$$

is a restatement of the Pythagorean Theorem, applied to the right triangle shown in Figure 2.4.1.

We recall as well that there are four other trigonometric functions, each defined in terms of the sine and/or cosine functions. These six trigonometric functions together offer us a wide range of flexibility in problems involving right triangles. The tangent function is defined by $\tan(\theta) = \frac{\sin(\theta)}{\cos(\theta)}$, while the cotangent function is its reciprocal: $\cot(\theta) = \frac{\cos(\theta)}{\sin(\theta)}$. The secant function is the reciprocal of the cosine function, $\sec(\theta) = \frac{1}{\cos(\theta)}$, and the cosecant function is the reciprocal of the sine function, $\csc(\theta) = \frac{1}{\sin(\theta)}$.

Because we know the derivatives of the sine and cosine function, and the other four trigonometric functions are defined in terms of these familiar functions, we can now develop shortcut differentiation rules for the tangent, cotangent, secant, and cosecant functions. In this section's preview activity, we work through the steps to find the derivative of $y = \tan(x)$.

Preview Activity 2.4.1. Consider the function $f(x) = \tan(x)$, and remember that $\tan(x) = \frac{\sin(x)}{\cos(x)}$.

    a. What is the domain of f?

    b. Use the quotient rule to show that one expression for $f'(x)$ is

$$f'(x) = \frac{\cos(x)\cos(x) + \sin(x)\sin(x)}{\cos^2(x)}.$$

    c. What is the Fundamental Trigonometric Identity? How can this identity be used to find a simpler form for $f'(x)$?

    d. Recall that $\sec(x) = \frac{1}{\cos(x)}$. How can we express $f'(x)$ in terms of the secant function?

    e. For what values of x is $f'(x)$ defined? How does this set compare to the domain of f?

2.4.1 Derivatives of the cotangent, secant, and cosecant functions

In Preview Activity 2.4.1, we found that the derivative of the tangent function can be expressed in several ways, but most simply in terms of the secant function. Next, we develop the derivative of the cotangent function.

Let $g(x) = \cot(x)$. To find $g'(x)$, we observe that $g(x) = \frac{\cos(x)}{\sin(x)}$ and apply the quotient rule. Hence

$$g'(x) = \frac{\sin(x)(-\sin(x)) - \cos(x)\cos(x)}{\sin^2(x)}$$
$$= -\frac{\sin^2(x) + \cos^2(x)}{\sin^2(x)}$$

By the Fundamental Trigonometric Identity, we see that $g'(x) = -\frac{1}{\sin^2(x)}$; recalling that

$\csc(x) = \frac{1}{\sin(x)}$, it follows that we can most simply express g' by the rule

$$g'(x) = -\csc^2(x).$$

Note that neither g nor g' is defined when $\sin(x) = 0$, which occurs at every integer multiple of π. Hence we have the following rule.

Cotangent Function

For all real numbers x such that $x \neq k\pi$, where $k = 0, \pm 1, \pm 2, \ldots$,

$$\frac{d}{dx}[\cot(x)] = -\csc^2(x).$$

Observe that the shortcut rule for the cotangent function is very similar to the rule we discovered in Preview Activity 2.4.1 for the tangent function.

Tangent Function

For all real numbers x such that $x \neq \frac{(2k+1)\pi}{2}$, where $k = \pm 1, \pm 2, \ldots$,

$$\frac{d}{dx}[\tan(x)] = \sec^2(x).$$

In the next two activities, we develop the rules for differentiating the secant and cosecant functions.

Activity 2.4.2. Let $h(x) = \sec(x)$ and recall that $\sec(x) = \frac{1}{\cos(x)}$.

 a. What is the domain of h?

 b. Use the quotient rule to develop a formula for $h'(x)$ that is expressed completely in terms of $\sin(x)$ and $\cos(x)$.

 c. How can you use other relationships among trigonometric functions to write $h'(x)$ only in terms of $\tan(x)$ and $\sec(x)$?

 d. What is the domain of h'? How does this compare to the domain of h?

Activity 2.4.3. Let $p(x) = \csc(x)$ and recall that $\csc(x) = \frac{1}{\sin(x)}$.

 a. What is the domain of p?

 b. Use the quotient rule to develop a formula for $p'(x)$ that is expressed completely in terms of $\sin(x)$ and $\cos(x)$.

 c. How can you use other relationships among trigonometric functions to write $p'(x)$ only in terms of $\cot(x)$ and $\csc(x)$?

d. What is the domain of p'? How does this compare to the domain of p?

The quotient rule has thus enabled us to determine the derivatives of the tangent, cotangent, secant, and cosecant functions, expanding our overall library of basic functions we can differentiate. Moreover, we observe that just as the derivative of any polynomial function is a polynomial, and the derivative of any exponential function is another exponential function, so it is that the derivative of any basic trigonometric function is another function that consists of basic trigonometric functions. This makes sense because all trigonometric functions are periodic, and hence their derivatives will be periodic, too.

As has been and will continue to be the case throughout our work in Chapter 2, the derivative retains all of its fundamental meaning as an instantaneous rate of change and as the slope of the tangent line to the function under consideration. Our present work primarily expands the list of functions for which we can quickly determine a formula for the derivative. Moreover, with the addition of $\tan(x)$, $\cot(x)$, $\sec(x)$, and $\csc(x)$ to our library of basic functions, there are many more functions we can differentiate through the sum, constant multiple, product, and quotient rules.

Activity 2.4.4. Answer each of the following questions. Where a derivative is requested, be sure to label the derivative function with its name using proper notation.

a. Let $f(x) = 5\sec(x) - 2\csc(x)$. Find the slope of the tangent line to f at the point where $x = \frac{\pi}{3}$.

b. Let $p(z) = z^2 \sec(z) - z\cot(z)$. Find the instantaneous rate of change of p at the point where $z = \frac{\pi}{4}$.

c. Let $h(t) = \dfrac{\tan(t)}{t^2 + 1} - 2e^t \cos(t)$. Find $h'(t)$.

d. Let $g(r) = \dfrac{r\sec(r)}{5^r}$. Find $g'(r)$.

e. When a mass hangs from a spring and is set in motion, the object's position oscillates in a way that the size of the oscillations decrease. This is usually called a *damped oscillation*. Suppose that for a particular object, its displacement from equilibrium (where the object sits at rest) is modeled by the function

$$s(t) = \frac{15\sin(t)}{e^t}.$$

Assume that s is measured in inches and t in seconds. Sketch a graph of this function for $t \geq 0$ to see how it represents the situation described. Then compute ds/dt, state the units on this function, and explain what it tells you about the object's motion. Finally, compute and interpret $s'(2)$.

Summary

- The derivatives of the other four trigonometric functions are

$$\frac{d}{dx}[\tan(x)] = \sec^2(x), \quad \frac{d}{dx}[\cot(x)] = -\csc^2(x),$$

$$\frac{d}{dx}[\sec(x)] = \sec(x)\tan(x), \text{ and } \frac{d}{dx}[\csc(x)] = -\csc(x)\cot(x).$$

  Each derivative exists and is defined on the same domain as the original function. For example, both the tangent function and its derivative are defined for all real numbers x such that $x \neq \frac{k\pi}{2}$, where $k = \pm1, \pm2, \ldots$.

- The above four rules for the derivatives of the tangent, cotangent, secant, and cosecant can be used along with the rules for power functions, exponential functions, and the sine and cosine, as well as the sum, constant multiple, product, and quotient rules, to quickly differentiate a wide range of different functions.

Exercises

1. Find the derivative of $h(t) = t \tan t + \sin t$

$h'(t) = $ [   ]

2. Let $f(x) = \dfrac{5\tan(x)}{x}$. Find the following:

    1. $f'(x)$ =

    2. $f'(4)$ =

3. Let $f(x) = \dfrac{\tan(x) - 5}{\sec(x)}$. Find the following:

    1. $f'(x)$ =

    2. $f'(1)$ =

4. Let $f(x) = \dfrac{2x^2 \tan(x)}{\sec(x)}$. Find the following:

    1. $f'(x)$ =

    2. $f'(3)$ =

5. Find the equation of the tangent line to the curve $y = 2\tan x$ at the point $(\pi/4, 2)$. The equation of this tangent line can be written in the form $y = mx + b$ where m is: [   ] and where b is: [   ]

6. An object moving vertically has its height at time t (measured in feet, with time in seconds) given by the function $h(t) = 3 + \frac{2\cos(t)}{1.2^t}$.

   a. What is the object's instantaneous velocity when $t = 2$?

   b. What is the object's acceleration at the instant $t = 2$?

   c. Describe in everyday language the behavior of the object at the instant $t = 2$.

7. Let $f(x) = \sin(x)\cot(x)$.

   a. Use the product rule to find $f'(x)$.

   b. True or false: for all real numbers x, $f(x) = \cos(x)$.

   c. Explain why the function that you found in (a) is almost the opposite of the sine function, but not quite. (Hint: convert all of the trigonometric functions in (a) to sines and cosines, and work to simplify. Think carefully about the domain of f and the domain of f'.)

8. Let $p(z)$ be given by the rule

$$p(z) = \frac{z\tan(z)}{z^2 \sec(z) + 1} + 3e^z + 1.$$

   a. Determine $p'(z)$.

   b. Find an equation for the tangent line to p at the point where $z = 0$.

   c. At $z = 0$, is p increasing, decreasing, or neither? Why?

2.5 The chain rule

Motivating Questions

- What is a composite function and how do we recognize its structure algebraically?

- Given a composite function $C(x) = f(g(x))$ that is built from differentiable functions f and g, how do we compute $C'(x)$ in terms of f, g, f', and g'? What is the statement of the Chain Rule?

In addition to learning how to differentiate a variety of basic functions, we have also been developing our ability to understand how to use rules to differentiate certain algebraic combinations of them. For example, we not only know how to take the derivative of $f(x) = \sin(x)$ and $g(x) = x^2$, but now we can quickly find the derivative of each of the following combinations of f and g:

$$s(x) = 3x^2 - 5\sin(x),$$

$$p(x) = x^2 \sin(x), \text{and}$$

$$q(x) = \frac{\sin(x)}{x^2}.$$

Finding s' uses the sum and constant multiple rules, determining p' requires the product rule, and q' can be attained with the quotient rule. Again, we note the importance of recognizing the algebraic structure of a given function in order to find its derivative: $s(x) = 3g(x) - 5f(x)$, $p(x) = g(x) \cdot f(x)$, and $q(x) = \frac{f(x)}{g(x)}$.

There is one more natural way to algebraically combine basic functions, and that is by *composing* them. For instance, let's consider the function

$$C(x) = \sin(x^2),$$

and observe that any input x passes through a *chain* of functions. In particular, in the process that defines the function $C(x)$, x is first squared, and then the sine of the result is taken. Using an arrow diagram,

$$x \longrightarrow x^2 \longrightarrow \sin(x^2).$$

In terms of the elementary functions f and g, we observe that x is first input in the function g, and then the result is used as the input in f. Said differently, we can write

$$C(x) = f(g(x)) = \sin(x^2)$$

and say that C is the *composition* of f and g. We will refer to g, the function that is first applied to x, as the *inner* function, while f, the function that is applied to the result, is the *outer* function.

The main question that we answer in the present section is: given a composite function $C(x) = f(g(x))$ that is built from differentiable functions f and g, how do we compute $C'(x)$

in terms of f, g, f', and g'? In the same way that the rate of change of a product of two functions, $p(x) = f(x) \cdot g(x)$, depends on the behavior of both f and g, it makes sense intuitively that the rate of change of a composite function $C(x) = f(g(x))$ will also depend on some combination of f and g and their derivatives. The rule that describes how to compute C' in terms of f and g and their derivatives will be called the *chain rule*.

But before we can learn what the chain rule says and why it works, we first need to be comfortable decomposing composite functions so that we can correctly identify the inner and outer functions, as we did in the example above with $C(x) = \sin(x^2)$.

Preview Activity 2.5.1. For each function given below, identify its fundamental algebraic structure. In particular, is the given function a sum, product, quotient, or composition of basic functions? If the function is a composition of basic functions, state a formula for the inner function g and the outer function f so that the overall composite function can be written in the form $f(g(x))$. If the function is a sum, product, or quotient of basic functions, use the appropriate rule to determine its derivative.

a. $h(x) = \tan(2^x)$

b. $p(x) = 2^x \tan(x)$

c. $r(x) = (\tan(x))^2$

d. $m(x) = e^{\tan(x)}$

e. $w(x) = \sqrt{x} + \tan(x)$

f. $z(x) = \sqrt{\tan(x)}$

2.5.1 The chain rule

One of the challenges of differentiating a composite function is that it often cannot be written in an alternate algebraic form. For instance, the function $C(x) = \sin(x^2)$ cannot be expanded or otherwise rewritten, so it presents no alternate approaches to taking the derivative. But other composite functions can be expanded or simplified, and these present a way to begin to explore how the chain rule might have to work. To that end, we consider two examples of composite functions that present alternate means of finding the derivative.

Example 2.5.1. Let $f(x) = -4x + 7$ and $g(x) = 3x - 5$. Determine a formula for $C(x) = f(g(x))$ and compute $C'(x)$. How is C' related to f and g and their derivatives?

Solution. By the rules given for f and g,

$$\begin{aligned}
C(x) &= f(g(x)) \\
&= f(3x - 5) \\
&= -4(3x - 5) + 7 \\
&= -12x + 20 + 7 \\
&= -12x + 27.
\end{aligned}$$

Thus, $C'(x) = -12$. Noting that $f'(x) = -4$ and $g'(x) = 3$, we observe that C' appears to be the product of f' and g'.

From one perspective, Example 2.5.1 may be too elementary. Linear functions are the simplest of all functions, and perhaps composing linear functions (which yields another linear function) does not exemplify the true complexity that is involved with differentiating a composition of more complicated functions. At the same time, we should remember the perspective that any differentiable function is *locally* linear, so any function with a derivative behaves like a line when viewed up close. From this point of view, the fact that the derivatives of f and g are multiplied to find the derivative of their composition turns out to be a key insight.

We now consider a second example involving a nonlinear function to gain further understanding of how differentiating a composite function involves the basic functions that combine to form it.

Example 2.5.2. Let $C(x) = \sin(2x)$. Use the double angle identity to rewrite C as a product of basic functions, and use the product rule to find C'. Rewrite C' in the simplest form possible.

Solution.   By the double angle identity for the sine function,

$$C(x) = \sin(2x) = 2\sin(x)\cos(x).$$

Applying the product rule and simplifying,

$$C'(x) = 2\sin(x)(-\sin(x)) + \cos(x)(2\cos(x)) = 2(\cos^2(x) - \sin^2(x)).$$

Next, we recall that one of the double angle identities for the cosine function tells us that

$$\cos(2x) = \cos^2(x) - \sin^2(x).$$

Substituting this result in our expression for $C'(x)$, we now have that

$$C'(x) = 2\cos(2x).$$

So from Example 2.5.2, we see that if $C(x) = \sin(2x)$, then $C'(x) = 2\cos(2x)$. Letting $g(x) = 2x$ and $f(x) = \sin(x)$, we observe that $C(x) = f(g(x))$. Moreover, with $g'(x) = 2$ and $f'(x) = \cos(x)$, it follows that we can view the structure of $C'(x)$ as

$$C'(x) = 2\cos(2x) = g'(x)f'(g(x)).$$

In this example, we see that for the composite function $C(x) = f(g(x))$, the derivative C' is (as in the example involving linear functions) constituted by multiplying the derivatives of f and g, but with the special condition that f' is evaluated at $g(x)$, rather than at x.

It makes sense intuitively that these two quantities are involved in understanding the rate of change of a composite function: if we are considering $C(x) = f(g(x))$ and asking how fast C is changing at a given x value as x changes, it clearly matters how fast g is changing at x, as well as how fast f is changing at the value of $g(x)$. It turns out that this structure holds not only for the functions in Examples 2.5.1 and Example 2.5.2, but indeed for all differentiable functions[1] as is stated in the Chain Rule.

---

[1]Like other differentiation rules, the Chain Rule can be proved formally using the limit definition of the derivative.

Chain Rule

If g is differentiable at x and f is differentiable at $g(x)$, then the composite function C defined by $C(x) = f(g(x))$ is differentiable at x and

$$C'(x) = f'(g(x))g'(x).$$

As with the product and quotient rules, it is often helpful to think verbally about what the chain rule says: "If C is a composite function defined by an outer function f and an inner function g, then C' is given by the derivative of the outer function, evaluated at the inner function, times the derivative of the inner function."

At least initially in working particular examples requiring the chain rule, it can also be helpful to clearly identify the inner function g and outer function f, compute their derivatives individually, and then put all of the pieces together to generate the derivative of the overall composite function. To see what we mean by this, consider the function

$$r(x) = (\tan(x))^2.$$

The function r is composite, with inner function $g(x) = \tan(x)$ and outer function $f(x) = x^2$. Organizing the key information involving f, g, and their derivatives, we have

$$
\begin{array}{ll}
f(x) = x^2 & g(x) = \tan(x) \\
f'(x) = 2x & g'(x) = \sec^2(x) \\
f'(g(x)) = 2\tan(x) &
\end{array}
$$

Applying the chain rule, which tells us that $r'(x) = f'(g(x))g'(x)$, we find that for $r(x) = (\tan(x))^2$, its derivative is

$$r'(x) = 2\tan(x)\sec^2(x).$$

As a side note, we remark that another way to write $r(x)$ is $r(x) = \tan^2(x)$. Observe that in this format, the composite nature of the function is more implicit, but this is common notation for powers of trigonometric functions: $\cos^4(x)$, $\sin^5(x)$, and $\sec^2(x)$ are all composite functions, with the outer function a power function and the inner function a trigonometric one.

The chain rule now substantially expands the library of functions we can differentiate, as the following activity demonstrates.

Activity 2.5.2. For each function given below, identify an inner function g and outer function f to write the function in the form $f(g(x))$. Determine $f'(x)$, $g'(x)$, and $f'(g(x))$, and then apply the chain rule to determine the derivative of the given function.

a. $h(x) = \cos(x^4)$

b. $p(x) = \sqrt{\tan(x)}$

c. $s(x) = 2^{\sin(x)}$

d. $z(x) = \cot^5(x)$

e. $m(x) = (\sec(x) + e^x)^9$

2.5.2 Using multiple rules simultaneously

The chain rule now joins the sum, constant multiple, product, and quotient rules in our collection of the different techniques for finding the derivative of a function through understanding its algebraic structure and the basic functions that constitute it. It takes substantial practice to get comfortable with navigating multiple rules in a single problem; using proper notation and taking a few extra steps can be particularly helpful as well. We demonstrate with an example and then provide further opportunity for practice in the following activity.

Example 2.5.3. Find a formula for the derivative of $h(t) = 3^{t^2+2t} \sec^4(t)$.

Solution. We first observe that the most basic structure of h is that it is the product of two functions: $h(t) = a(t) \cdot b(t)$ where $a(t) = 3^{t^2+2t}$ and $b(t) = \sec^4(t)$. Therefore, we see that we will need to use the product rule to differentiate h. When it comes time to differentiate a and b in their roles in the product rule, we observe that since each is a composite function, the chain rule will be needed. We therefore begin by working separately to compute $a'(t)$ and $b'(t)$.

Writing $a(t) = f(g(t)) = 3^{t^2+2t}$, and finding the derivatives of f and g, we have

$$
\begin{aligned}
f(t) &= 3^t & g(t) &= t^2 + 2t \\
f'(t) &= 3^t \ln(3) & g'(t) &= 2t + 2 \\
f'(g(t)) &= 3^{t^2+2t} \ln(3) &
\end{aligned}
$$

Thus, by the chain rule, it follows that $a'(t) = f'(g(t))g'(t) = 3^{t^2+2t} \ln(3)(2t + 2)$.

Turning next to b, we write $b(t) = r(s(t)) = \sec^4(t)$ and find the derivatives of r and g. Doing so,

$$
\begin{aligned}
r(t) &= t^4 & s(t) &= \sec(t) \\
r'(t) &= 4t^3 & s'(t) &= \sec(t)\tan(t) \\
r'(s(t)) &= 4\sec^3(t) &
\end{aligned}
$$

By the chain rule, we now know that

$$
b'(t) = r'(s(t))s'(t) = 4\sec^3(t)\sec(t)\tan(t) = 4\sec^4(t)\tan(t).
$$

Now we are finally ready to compute the derivative of the overall function h. Recalling that $h(t) = 3^{t^2+2t} \sec^4(t)$, by the product rule we have

$$
h'(t) = 3^{t^2+2t} \frac{d}{dt}[\sec^4(t)] + \sec^4(t)\frac{d}{dt}[3^{t^2+2t}].
$$

From our work above with a and b, we know the derivatives of 3^{t^2+2t} and $\sec^4(t)$, and therefore

$$
h'(t) = 3^{t^2+2t} 4\sec^4(t)\tan(t) + \sec^4(t)3^{t^2+2t}\ln(3)(2t + 2).
$$

Activity 2.5.3. For each of the following functions, find the function's derivative. State the rule(s) you use, label relevant derivatives appropriately, and be sure to clearly identify your overall answer.

a. $p(r) = 4\sqrt{r^6 + 2e^r}$

b. $m(v) = \sin(v^2)\cos(v^3)$

c. $h(y) = \frac{\cos(10y)}{e^{4y}+1}$

d. $s(z) = 2^{z^2 \sec(z)}$

e. $c(x) = \sin(e^{x^2})$

The chain rule now adds substantially to our ability to do different familiar problems that involve derivatives. Whether finding the equation of the tangent line to a curve, the instantaneous velocity of a moving particle, or the instantaneous rate of change of a certain quantity, if the function under consideration involves a composition of other functions, the chain rule is indispensable.

Activity 2.5.4. Use known derivative rules, including the chain rule, as needed to answer each of the following questions.

a. Find an equation for the tangent line to the curve $y = \sqrt{e^x + 3}$ at the point where $x = 0$.

b. If $s(t) = \dfrac{1}{(t^2 + 1)^3}$ represents the position function of a particle moving horizontally along an axis at time t (where s is measured in inches and t in seconds), find the particle's instantaneous velocity at $t = 1$. Is the particle moving to the left or right at that instant?

c. At sea level, air pressure is 30 inches of mercury. At an altitude of h feet above sea level, the air pressure, P, in inches of mercury, is given by the function $P = 30e^{-0.0000323h}$. Compute dP/dh and explain what this derivative function tells you about air pressure, including a discussion of the units on dP/dh. In addition, determine how fast the air pressure is changing for a pilot of a small plane passing through an altitude of 1000 feet.

d. Suppose that $f(x)$ and $g(x)$ are differentiable functions and that the following information about them is known:

x	$f(x)$	$f'(x)$	$g(x)$	$g'(x)$
-1	2	-5	-3	4
2	-3	4	-1	2

Table 2.5.4: Data for functions f and g.

If $C(x)$ is a function given by the formula $f(g(x))$, determine $C'(2)$. In addition, if $D(x)$ is the function $f(f(x))$, find $D'(-1)$.

2.5.3 The composite version of basic function rules

As we gain more experience with differentiating complicated functions, we will become more comfortable in the process of simply writing down the derivative without taking multiple steps. We demonstrate part of this perspective here by showing how we can find a composite rule that corresponds to two of our basic functions. For instance, we know that $\frac{d}{dx}[\sin(x)] = \cos(x)$. If we instead want to know

$$\frac{d}{dx}[\sin(u(x))],$$

where u is a differentiable function of x, then this requires the chain rule with the sine function as the outer function. Applying the chain rule,

$$\frac{d}{dx}[\sin(u(x))] = \cos(u(x)) \cdot u'(x).$$

Similarly, since $\frac{d}{dx}[a^x] = a^x \ln(a)$, it follows by the chain rule that

$$\frac{d}{dx}[a^{u(x)}] = a^{u(x)} \ln(a) \cdot u'(x).$$

In the process of getting comfortable with derivative rules, an excellent exercise is to write down a list of all basic functions whose derivatives are known, list those derivatives, and then write the corresponding chain rule for the composite version with the inner function being an unknown function $u(x)$ and the outer function being the known basic function. These versions of the chain rule are particularly simple when the inner function is linear, since the derivative of a linear function is a constant. For instance,

$$\frac{d}{dx}\left[(5x + 7)^{10}\right] = 10(5x + 7)^9 \cdot 5,$$

$$\frac{d}{dx}[\tan(17x)] = 17\sec^2(17x), \text{ and}$$

$$\frac{d}{dx}\left[e^{-3x}\right] = -3e^{-3x}.$$

Summary

- A composite function is one where the input variable x first passes through one function, and then the resulting output passes through another. For example, the function $h(x) = 2^{\sin(x)}$ is composite since $x \longrightarrow \sin(x) \longrightarrow 2^{\sin(x)}$.

- Given a composite function $C(x) = f(g(x))$ that is built from differentiable functions f and g, the chain rule tells us that we compute $C'(x)$ in terms of f, g, f', and g' according to the formula

$$C'(x) = f'(g(x))g'(x).$$

Exercises

1. Find the derivative of
$f(x) = e^{4x}(x^2 + 3^x)$
$f'(x) =$ ⬚

2. Find the derivative of
$v(t) = t^5 e^{-ct}$
Assume that c is a constant.
$v'(t) =$ ⬚

3. Find the derivative of
$y = \sqrt{e^{-5t^2} + 5}$
$\frac{dy}{dt} =$ ⬚

4. Find the derivative of
$f(x) = axe^{-bx+10}$
Assume that a and b are constants.
$f'(x) =$ ⬚

5. Use the graph below to find exact values of the indicated derivatives, or state that they do not exist. If a derivative does not exist, enter *dne* in the answer blank. The graph of $f(x)$ is black and has a sharp corner at $x = 2$. The graph of $g(x)$ is blue.

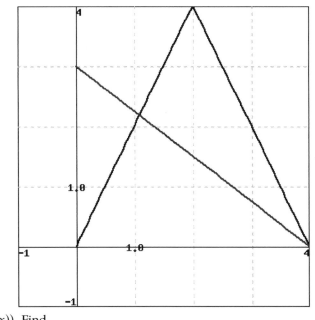

Let $h(x) = f(g(x))$. Find

A. $h'(1) =$

B. $h'(2) =$

C. $h'(3) =$

(Enter dne *for any derivative that does not exist.)*

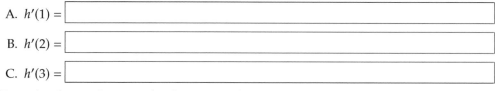

6. Given $F(2) = 3, F'(2) = 4, F(4) = 1, F'(4) = 5$ and $G(1) = 3, G'(1) = 4, G(4) = 2, G'(4) = 7$, find each of the following. (Enter *dne* for any derivative that cannot be computed from this information alone.)

A. $H(4)$ if $H(x) = F(G(x))$

B. $H'(4)$ if $H(x) = F(G(x))$

C. $H(4)$ if $H(x) = G(F(x))$

D. $H'(4)$ if $H(x) = G(F(x))$

E. $H'(4)$ if $H(x) = F(x)/G(x)$

7. Find the derivative of $f(x) = 9x \sin(2x)$

$f'(x) =$

8. Consider the basic functions $f(x) = x^3$ and $g(x) = \sin(x)$.

a. Let $h(x) = f(g(x))$. Find the exact instantaneous rate of change of h at the point where $x = \frac{\pi}{4}$.

b. Which function is changing most rapidly at $x = 0.25$: $h(x) = f(g(x))$ or $r(x) = g(f(x))$? Why?

c. Let $h(x) = f(g(x))$ and $r(x) = g(f(x))$. Which of these functions has a derivative that is periodic? Why?

9. Let $u(x)$ be a differentiable function. For each of the following functions, determine the derivative. Each response will involve u and/or u'.

a. $p(x) = e^{u(x)}$

b. $q(x) = u(e^x)$

c. $r(x) = \cot(u(x))$

d. $s(x) = u(\cot(x))$

e. $a(x) = u(x^4)$

f. $b(x) = u^4(x)$

10. Let functions p and q be the piecewise linear functions given by their respective graphs in Figure 2.5.5. Use the graphs to answer the following questions.

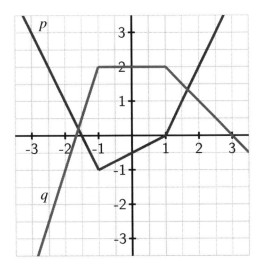

Figure 2.5.5: The graphs of p (in blue) and q (in green).

a. Let $C(x) = p(q(x))$. Determine $C'(0)$ and $C'(3)$.

b. Find a value of x for which $C'(x)$ does not exist. Explain your thinking.

c. Let $Y(x) = q(q(x))$ and $Z(x) = q(p(x))$. Determine $Y'(-2)$ and $Z'(0)$.

11. If a spherical tank of radius 4 feet has h feet of water present in the tank, then the volume of water in the tank is given by the formula

$$V = \frac{\pi}{3}h^2(12 - h).$$

a. At what instantaneous rate is the volume of water in the tank changing with respect to the *height* of the water at the instant $h = 1$? What are the units on this quantity?

b. Now suppose that the height of water in the tank is being regulated by an inflow and outflow (e.g., a faucet and a drain) so that the height of the water at time t is given by the rule $h(t) = \sin(\pi t) + 1$, where t is measured in hours (and h is still measured in feet). At what rate is the height of the water changing with respect to time at the instant $t = 2$?

c. Continuing under the assumptions in (b), at what instantaneous rate is the volume of water in the tank changing with respect to *time* at the instant $t = 2$?

d. What are the main differences between the rates found in (a) and (c)? Include a discussion of the relevant units.

2.6 Derivatives of Inverse Functions

Motivating Questions

- What is the derivative of the natural logarithm function?

- What are the derivatives of the inverse trigonometric functions $\arcsin(x)$ and $\arctan(x)$?

- If g is the inverse of a differentiable function f, how is g' computed in terms of f, f', and g?

Much of mathematics centers on the notion of function. Indeed, throughout our study of calculus, we are investigating the behavior of functions, often doing so with particular emphasis on how fast the output of the function changes in response to changes in the input. Because each function represents a process, a natural question to ask is whether or not the particular process can be reversed. That is, if we know the output that results from the function, can we determine the input that led to it? Connected to this question, we now also ask: if we know how fast a particular process is changing, can we determine how fast the inverse process is changing?

As we have noted, one of the most important functions in all of mathematics is the natural exponential function $f(x) = e^x$. Because the natural logarithm, $g(x) = \ln(x)$, is the inverse of the natural exponential function, the natural logarithm is similarly important. One of our goals in this section is to learn how to differentiate the logarithm function, and thus expand our library of basic functions with known derivative formulas. First, we investigate a more familiar setting to refresh some of the basic concepts surrounding functions and their inverses.

Preview Activity 2.6.1. The equation $y = \frac{5}{9}(x - 32)$ relates a temperature given in x degrees Fahrenheit to the corresponding temperature y measured in degrees Celcius.

- a. Solve the equation $y = \frac{5}{9}(x-32)$ for x to write x (Fahrenheit temperature) in terms of y (Celcius temperature).

- b. Let $C(x) = \frac{5}{9}(x - 32)$ be the function that takes a Fahrenheit temperature as input and produces the Celcius temperature as output. In addition, let $F(y)$ be the function that converts a temperature given in y degrees Celcius to the temperature $F(y)$ measured in degrees Fahrenheit. Use your work in (a) to write a formula for $F(y)$.

- c. Next consider the new function defined by $p(x) = F(C(x))$. Use the formulas for F and C to determine an expression for $p(x)$ and simplify this expression as much as possible. What do you observe?

- d. Now, let $r(y) = C(F(y))$. Use the formulas for F and C to determine an expression for $r(y)$ and simplify this expression as much as possible. What do you observe?

e. What is the value of $C'(x)$? of $F'(y)$? How do these values appear to be related?

2.6.1 Basic facts about inverse functions

A function $f : A \rightarrow B$ is a rule that associates each element in the set A to one and only one element in the set B. We call A the *domain* of f and B the *codomain* of f. If there exists a function $g : B \rightarrow A$ such that $g(f(a)) = a$ for every possible choice of a in the set A and $f(g(b)) = b$ for every b in the set B, then we say that g is the *inverse* of f. We often use the notation f^{-1} (read "f-inverse") to denote the inverse of f. Perhaps the most essential thing to observe about the inverse function is that it undoes the work of f. Indeed, if $y = f(x)$, then

$$f^{-1}(y) = f^{-1}(f(x)) = x,$$

and this leads us to another key observation: writing $y = f(x)$ and $x = f^{-1}(y)$ say the exact same thing. The only difference between the two equations is one of perspective — one is solved for x, while the other is solved for y.

Here we briefly remind ourselves of some key facts about inverse functions.

Note 2.6.1. For a function $f : A \rightarrow B$,

- f has an inverse if and only if f is one-to-one[1] and onto[2];

- provided f^{-1} exists, the domain of f^{-1} is the codomain of f, and the codomain of f^{-1} is the domain of f;

- $f^{-1}(f(x)) = x$ for every x in the domain of f and $f(f^{-1}(y)) = y$ for every y in the codomain of f;

- $y = f(x)$ if and only if $x = f^{-1}(y)$.

The last stated fact reveals a special relationship between the graphs of f and f^{-1}. In particular, if we consider $y = f(x)$ and a point (x, y) that lies on the graph of f, then it is also true that $x = f^{-1}(y)$, which means that the point (y, x) lies on the graph of f^{-1}. This shows us that the graphs of f and f^{-1} are the reflections of one another across the line $y = x$, since reflecting across $y = x$ is precisely the geometric action that swaps the coordinates in an ordered pair. In Figure 2.6.2, we see this exemplified for the function $y = f(x) = 2^x$ and its inverse, with the points $(-1, \frac{1}{2})$ and $(\frac{1}{2}, -1)$ highlighting the reflection of the curves across $y = x$.

---

[1]A function f is *one-to-one* provided that no two distinct inputs lead to the same output.
[2]A function f is *onto* provided that every possible element of the codomain can be realized as an output of the function for some choice of input from the domain.

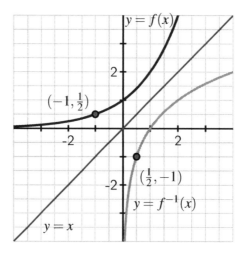

Figure 2.6.2: A graph of a function $y = f(x)$ along with its inverse, $y = f^{-1}(x)$.

To close our review of important facts about inverses, we recall that the natural exponential function $y = f(x) = e^x$ has an inverse function, and its inverse is the natural logarithm, $x = f^{-1}(y) = \ln(y)$. Indeed, writing $y = e^x$ is interchangeable with $x = \ln(y)$, plus $\ln(e^x) = x$ for every real number x and $e^{\ln(y)} = y$ for every positive real number y.

2.6.2 The derivative of the natural logarithm function

In what follows, we determine a formula for the derivative of $g(x) = \ln(x)$. To do so, we take advantage of the fact that we know the derivative of the natural exponential function, which is the inverse of g. In particular, we know that writing $g(x) = \ln(x)$ is equivalent to writing $e^{g(x)} = x$. Now we differentiate both sides of this most recent equation. We observe that

$$\frac{d}{dx}\left[e^{g(x)}\right] = \frac{d}{dx}[x].$$

The righthand side is simply 1; applying the chain rule to the left side, we find that

$$e^{g(x)}g'(x) = 1.$$

Since our goal is to determine $g'(x)$, we solve for $g'(x)$, so

$$g'(x) = \frac{1}{e^{g(x)}}.$$

Finally, we recall that since $g(x) = \ln(x)$, $e^{g(x)} = e^{\ln(x)} = x$, and thus

$$g'(x) = \frac{1}{x}.$$

Natural Logarithm

For all positive real numbers x, $\frac{d}{dx}[\ln(x)] = \frac{1}{x}$.

This rule for the natural logarithm function now joins our list of other basic derivative rules that we have already established. There are two particularly interesting things to note about the fact that $\frac{d}{dx}[\ln(x)] = \frac{1}{x}$. One is that this rule is restricted to only apply to positive values of x, as these are the only values for which the original function is defined. The other is that for the first time in our work, differentiating a basic function of a particular type has led to a function of a very different nature: the derivative of the natural logarithm is not another logarithm, nor even an exponential function, but rather a rational one.

Derivatives of logarithms may now be computed in concert with all of the rules known to date. For instance, if $f(t) = \ln(t^2 + 1)$, then by the chain rule, $f'(t) = \frac{1}{t^2+1} \cdot 2t$.

In addition to the important rule we have derived for the derivative of the natural log functions, there are additional interesting connections to note between the graphs of $f(x) = e^x$ and $f^{-1}(x) = \ln(x)$.

In Figure 2.6.3, we are reminded that since the natural exponential function has the property that its derivative is itself, the slope of the tangent to $y = e^x$ is equal to the height of the curve at that point. For instance, at the point $A = (\ln(0.5), 0.5)$, the slope of the tangent line is $m_A = 0.5$, and at $B = (\ln(5), 5)$, the tangent line's slope is $m_B = 5$. At the corresponding points A' and B' on the graph of the natural logarithm function (which come from reflecting across the line $y = x$), we know that the slope of the tangent line is the reciprocal of the x-coordinate of the point (since $\frac{d}{dx}[\ln(x)] = \frac{1}{x}$). Thus, with $A' = (0.5, \ln(0.5))$, we have $m_{A'} = \frac{1}{0.5} = 2$, and at $B' = (5, \ln(5))$, $m_{B'} = \frac{1}{5}$.

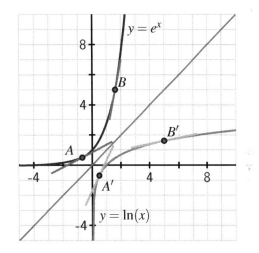

Figure 2.6.3: A graph of the function $y = e^x$ along with its inverse, $y = \ln(x)$, where both functions are viewed using the input variable x.

In particular, we observe that $m_{A'} = \frac{1}{m_A}$ and $m_{B'} = \frac{1}{m_B}$. This is not a coincidence, but in fact holds for any curve $y = f(x)$ and its inverse, provided the inverse exists. One rationale for why this is the case is due to the reflection across $y = x$: in so doing, we essentially change the roles of x and y, thus reversing the rise and run, which leads to the slope of the inverse function at the reflected point being the reciprocal of the slope of the original function. At the close of this section, we will also look at how the chain rule provides us with an algebraic formulation of this general phenomenon.

> **Activity 2.6.2.** For each function given below, find its derivative.
>
> a. $h(x) = x^2 \ln(x)$
>
> b. $p(t) = \frac{\ln(t)}{e^t + 1}$
>
> c. $s(y) = \ln(\cos(y) + 2)$
>
> d. $z(x) = \tan(\ln(x))$
>
> e. $m(z) = \ln(\ln(z))$

2.6.3 Inverse trigonometric functions and their derivatives

Trigonometric functions are periodic, so they fail to be one-to-one, and thus do not have inverses. However, if we restrict the domain of each trigonometric function, we can force the function to be one-to-one. For instance, consider the sine function on the domain $[-\frac{\pi}{2}, \frac{\pi}{2}]$.

Because no output of the sine function is repeated on this interval, the function is one-to-one and thus has an inverse. In particular, if we view $f(x) = \sin(x)$ as having domain $[-\frac{\pi}{2}, \frac{\pi}{2}]$ and codomain $[-1, 1]$, then there exists an inverse function f^{-1} such that

$$f^{-1} : [-1, 1] \to [-\frac{\pi}{2}, \frac{\pi}{2}].$$

We call f^{-1} the *arcsine* (or inverse sine) function and write $f^{-1}(y) = \arcsin(y)$. It is especially important to remember that writing

$$y = \sin(x) \text{ and } x = \arcsin(y)$$

say the exact same thing. We often read "the arcsine of y" as "the angle whose sine is y." For example, we say that $\frac{\pi}{6}$ is the angle whose sine is $\frac{1}{2}$, which can be written more concisely as $\arcsin(\frac{1}{2}) = \frac{\pi}{6}$, which is equivalent to writing $\sin(\frac{\pi}{6}) = \frac{1}{2}$.

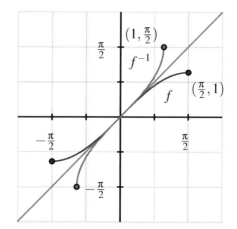

Figure 2.6.4: A graph of $f(x) = \sin(x)$ (in blue), restricted to the domain $[-\frac{\pi}{2}, \frac{\pi}{2}]$, along with its inverse, $f^{-1}(x) = \arcsin(x)$ (in magenta).

Next, we determine the derivative of the arcsine function. Letting $h(x) = \arcsin(x)$, our goal is to find $h'(x)$. Since $h(x)$ is the angle whose sine is x, it is equivalent to write

$$\sin(h(x)) = x.$$

Differentiating both sides of the previous equation, we have

$$\frac{d}{dx}[\sin(h(x))] = \frac{d}{dx}[x],$$

and by the fact that the righthand side is simply 1 and by the chain rule applied to the left side,

$$\cos(h(x))h'(x) = 1.$$

Solving for $h'(x)$, it follows that

$$h'(x) = \frac{1}{\cos(h(x))}.$$

Finally, we recall that $h(x) = \arcsin(x)$, so the denominator of $h'(x)$ is the function $\cos(\arcsin(x))$, or in other words, "the cosine of the angle whose sine is x." A bit of right triangle trigonometry allows us to simplify this expression considerably.

Let's say that $\theta = \arcsin(x)$, so that θ is the angle whose sine is x. From this, it follows that we can picture θ as an angle in a right triangle with hypotenuse 1 and a vertical leg of length x, as shown in Figure 2.6.5. The horizontal leg must be $\sqrt{1 - x^2}$, by the Pythagorean Theorem. Now, note particularly that $\theta = \arcsin(x)$ since $\sin(\theta) = x$, and recall that we want to know a different expression for $\cos(\arcsin(x))$. From the figure, $\cos(\arcsin(x)) = \cos(\theta) = \sqrt{1 - x^2}$.

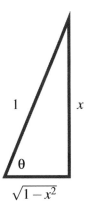

Figure 2.6.5: The right triangle that corresponds to the angle $\theta = \arcsin(x)$.

Thus, returning to our earlier work where we established that if $h(x) = \arcsin(x)$, then $h'(x) = \frac{1}{\cos(\arcsin(x))}$, we have now shown that

$$h'(x) = \frac{1}{\sqrt{1 - x^2}}.$$

Inverse sine

For all real numbers x such that $-1 < x < 1$,

$$\frac{d}{dx}[\arcsin(x)] = \frac{1}{\sqrt{1-x^2}}.$$

Activity 2.6.3. The following prompts in this activity will lead you to develop the derivative of the inverse tangent function.

a. Let $r(x) = \arctan(x)$. Use the relationship between the arctangent and tangent functions to rewrite this equation using only the tangent function.

b. Differentiate both sides of the equation you found in (a). Solve the resulting equation for $r'(x)$, writing $r'(x)$ as simply as possible in terms of a trigonometric function evaluated at $r(x)$.

c. Recall that $r(x) = \arctan(x)$. Update your expression for $r'(x)$ so that it only involves trigonometric functions and the independent variable x.

d. Introduce a right triangle with angle θ so that $\theta = \arctan(x)$. What are the three sides of the triangle?

e. In terms of only x and 1, what is the value of $\cos(\arctan(x))$?

f. Use the results of your work above to find an expression involving only 1 and x for $r'(x)$.

While derivatives for other inverse trigonometric functions can be established similarly, we primarily limit ourselves to the arcsine and arctangent functions. With these rules added to our library of derivatives of basic functions, we can differentiate even more functions using derivative shortcuts. In Activity 2.6.4, we see each of these rules at work.

Activity 2.6.4. Determine the derivative of each of the following functions.

a. $f(x) = x^3 \arctan(x) + e^x \ln(x)$

b. $p(t) = 2^{t \arcsin(t)}$

c. $h(z) = (\arcsin(5z) + \arctan(4 - z))^{27}$

d. $s(y) = \cot(\arctan(y))$

e. $m(v) = \ln(\sin^2(v) + 1)$

f. $g(w) = \arctan\left(\dfrac{\ln(w)}{1 + w^2}\right)$

2.6.4 The link between the derivative of a function and the derivative of its inverse

In Figure 2.6.3, we saw an interesting relationship between the slopes of tangent lines to the natural exponential and natural logarithm functions at points that corresponded to reflection across the line $y = x$. In particular, we observed that for a point such as $(\ln(2), 2)$ on the graph of $f(x) = e^x$, the slope of the tangent line at this point is $f'(\ln(2)) = 2$, while at the corresponding point $(2, \ln(2))$ on the graph of $f^{-1}(x) = \ln(x)$, the slope of the tangent line at this point is $(f^{-1})'(2) = \frac{1}{2}$, which is the reciprocal of $f'(\ln(2))$.

That the two corresponding tangent lines having slopes that are reciprocals of one another is not a coincidence. If we consider the general setting of a differentiable function f with differentiable inverse g such that $y = f(x)$ if and only if $x = g(y)$, then we know that $f(g(x)) = x$ for every x in the domain of f^{-1}. Differentiating both sides of this equation with respect to x, we have

$$\frac{d}{dx}[f(g(x))] = \frac{d}{dx}[x],$$

and by the chain rule,

$$f'(g(x))g'(x) = 1.$$

Solving for $g'(x)$, we have $g'(x) = \frac{1}{f'(g(x))}$. Here we see that the slope of the tangent line to the inverse function g at the point $(x, g(x))$ is precisely the reciprocal of the slope of the tangent line to the original function f at the point $(g(x), f(g(x))) = (g(x), x)$.

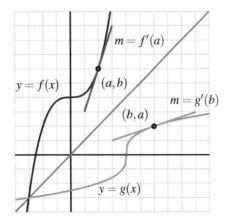

Figure 2.6.6: A graph of function $y = f(x)$ along with its inverse, $y = g(x) = f^{-1}(x)$. Observe that the slopes of the two tangent lines are reciprocals of one another.

To see this more clearly, consider the graph of the function $y = f(x)$ shown in Figure 2.6.6, along with its inverse $y = g(x)$. Given a point (a, b) that lies on the graph of f, we know that (b, a) lies on the graph of g; said differently, $f(a) = b$ and $g(b) = a$. Now, applying the rule that $g'(x) = 1/f'(g(x))$ to the value $x = b$, we have

$$g'(b) = \frac{1}{f'(g(b))} = \frac{1}{f'(a)},$$

which is precisely what we see in the figure: the slope of the tangent line to g at (b, a) is the reciprocal of the slope of the tangent line to f at (a, b), since these two lines are reflections of one another across the line $y = x$.

Derivative of an inverse function

Suppose that f is a differentiable function with inverse g and that (a, b) is a point that lies on the graph of f at which $f'(a) \neq 0$. Then

$$g'(b) = \frac{1}{f'(a)}.$$

More generally, for any x in the domain of g', we have $g'(x) = 1/f'(g(x))$.

The rules we derived for $\ln(x)$, $\arcsin(x)$, and $\arctan(x)$ are all just specific examples of this general property of the derivative of an inverse function. For example, with $g(x) = \ln(x)$ and $f(x) = e^x$, it follows that

$$g'(x) = \frac{1}{f'(g(x))} = \frac{1}{e^{\ln(x)}} = \frac{1}{x}.$$

Summary

- For all positive real numbers x, $\frac{d}{dx}[\ln(x)] = \frac{1}{x}$.

- For all real numbers x such that $-1 < x < 1$, $\frac{d}{dx}[\arcsin(x)] = \frac{1}{\sqrt{1-x^2}}$. In addition, for all real numbers x, $\frac{d}{dx}[\arctan(x)] = \frac{1}{1+x^2}$.

- If g is the inverse of a differentiable function f, then for any point x in the domain of g', $g'(x) = \frac{1}{f'(g(x))}$.

Exercises

1. Find the derivative of the function $f(t)$, below.
$f(t) = \ln(t^2 + 7)$
$f'(t) = $

2. Find the derivative of the function $g(t)$, below. It may be to your advantage to simplify before differentiating.
$g(t) = \cos(\ln(t))$
$g'(t) = $

3. Find the derivative of the function $h(w)$, below. It may be to your advantage to simplify before differentiating.
$h(w) = 5w \arcsin w$

$h'(w) =$ [   ]

4. For $x > 0$, find *and simplify* the derivative of $f(x) = \arctan x + \arctan(1/x)$.

$f'(x) =$ [   ]

(What does your result tell you about f)?

5. Let $(x_0, y_0) = (2, 4)$ and $(x_1, y_1) = (2.5, 4.6)$. Use the following graph of the function f to find the indicated derivatives.

If $h(x) = (f(x))^5$, then

$h'(2) =$ [   ]

If $g(x) = f^{-1}(x)$, then

$g'(4) =$ [   ]

6. Let

$$f(x) = 6\sin^{-1}(x^4)$$

$f'(x) =$ [   ]

NOTE: The webwork system will accept arcsin(x) or $\sin^{-1}(x)$ as the inverse of sin(x).

7. If $f(x) = 7x^4 \arctan(7x^2)$, find $f'(x)$.

$f'(x) =$ [   ]

8. Let $f(x) = 8\sin(x)\sin^{-1}(x)$. Find $f'(x)$.

$f'(x) =$ [   ]

9. Determine the derivative of each of the following functions. Use proper notation and clearly identify the derivative rules you use.

a. $f(x) = \ln(2\arctan(x) + 3\arcsin(x) + 5)$

b. $r(z) = \arctan(\ln(\arcsin(z)))$

c. $q(t) = \arctan^2(3t)\arcsin^4(7t)$

d. $g(v) = \ln\left(\frac{\arctan(v)}{\arcsin(v)+v^2}\right)$

10. Consider the graph of $y = f(x)$ provided in Figure 2.6.7 and use it to answer the following questions.

a. Use the provided graph to estimate the value of $f'(1)$.

b. Sketch an approximate graph of $y = f^{-1}(x)$. Label at least three distinct points on the graph that correspond to three points on the graph of f.

c. Based on your work in (a), what is the value of $(f^{-1})'(-1)$? Why?

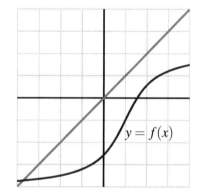

Figure 2.6.7: A function $y = f(x)$

11. Let $f(x) = \frac{1}{4}x^3 + 4$.

a. Sketch a graph of $y = f(x)$ and explain why f is an invertible function.

b. Let g be the inverse of f and determine a formula for g.

c. Compute $f'(x)$, $g'(x)$, $f'(2)$, and $g'(6)$. What is the special relationship between $f'(2)$ and $g'(6)$? Why?

12. Let $h(x) = x + \sin(x)$.

a. Sketch a graph of $y = h(x)$ and explain why h must be invertible.

b. Explain why it does not appear to be algebraically possible to determine a formula for h^{-1}.

c. Observe that the point $(\frac{\pi}{2}, \frac{\pi}{2} + 1)$ lies on the graph of $y = h(x)$. Determine the value of $(h^{-1})'(\frac{\pi}{2} + 1)$.

2.7 Derivatives of Functions Given Implicitly

Motivating Questions

- What does it mean to say that a curve is an implicit function of x, rather than an explicit function of x?

- How does implicit differentiation enable us to find a formula for $\frac{dy}{dx}$ when y is an implicit function of x?

- In the context of an implicit curve, how can we use $\frac{dy}{dx}$ to answer important questions about the tangent line to the curve?

In all of our studies with derivatives to date, we have worked in a setting where we can express a formula for the function of interest explicitly in terms of x. But there are many interesting curves that are determined by an equation involving x and y for which it is impossible to solve for y in terms of x.

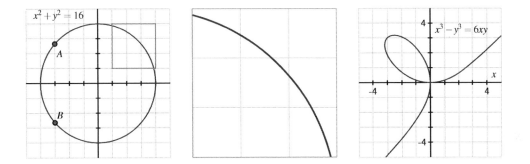

Figure 2.7.1: At left, the circle given by $x^2 + y^2 = 16$. In the middle, the portion of the circle $x^2 + y^2 = 16$ that has been highlighted in the box at left. And at right, the lemniscate given by $x^3 - y^3 = 6xy$.

Perhaps the simplest and most natural of all such curves are circles. Because of the circle's symmetry, for each x value strictly between the endpoints of the horizontal diameter, there are two corresponding y-values. For instance, in Figure 2.7.1, we have labeled $A = (-3, \sqrt{7})$ and $B = (-3, -\sqrt{7})$, and these points demonstrate that the circle fails the vertical line test. Hence, it is impossible to represent the circle through a single function of the form $y = f(x)$. At the same time, portions of the circle can be represented explicitly as a function of x, such as the highlighted arc that is magnified in the center of Figure 2.7.1. Moreover, it is evident that the circle is locally linear, so we ought to be able to find a tangent line to the curve at every point; thus, it makes sense to wonder if we can compute $\frac{dy}{dx}$ at any point on the circle, even though we cannot write y explicitly as a function of x. Finally, we note that the righthand curve in Figure 2.7.1 is called a *lemniscate* and is just one of many fascinating possibilities for implicitly given curves.

In working with implicit functions, we will often be interested in finding an equation for $\frac{dy}{dx}$ that tells us the slope of the tangent line to the curve at a point (x, y). To do so, it will be necessary for us to work with y while thinking of y as a function of x, but without being able to write an explicit formula for y in terms of x. The following preview activity reminds us of some ways we can compute derivatives of functions in settings where the function's formula is not known. For instance, recall the earlier example $\frac{d}{dx}[e^{u(x)}] = e^{u(x)}u'(x)$.

> **Preview Activity 2.7.1.** Let f be a differentiable function of x (whose formula is not known) and recall that $\frac{d}{dx}[f(x)]$ and $f'(x)$ are interchangeable notations. Determine each of the following derivatives of combinations of explicit functions of x, the unknown function f, and an arbitrary constant c.
>
> a. $\frac{d}{dx}\left[x^2 + f(x)\right]$ d. $\frac{d}{dx}\left[f(x^2)\right]$
>
> b. $\frac{d}{dx}\left[x^2 f(x)\right]$ e. $\frac{d}{dx}\left[xf(x) + f(cx) + cf(x)\right]$
>
> c. $\frac{d}{dx}\left[c + x + f(x)^2\right]$

2.7.1 Implicit Differentiation

Because a circle is perhaps the simplest of all curves that cannot be represented explicitly as a single function of x, we begin our exploration of implicit differentiation with the example of the circle given by $x^2 + y^2 = 16$. It is visually apparent that this curve is locally linear, so it makes sense for us to want to find the slope of the tangent line to the curve at any point, and moreover to think that the curve is differentiable. The big question is: how do we find a formula for $\frac{dy}{dx}$, the slope of the tangent line to the circle at a given point on the circle? By viewing y as an *implicit*[1] function of x, we essentially think of y as some function whose formula $f(x)$ is unknown, but which we can differentiate. Just as y represents an unknown formula, so too its derivative with respect to x, $\frac{dy}{dx}$, will be (at least temporarily) unknown.

Consider the equation $x^2 + y^2 = 16$ and view y as an unknown differentiable function of x. Differentiating both sides of the equation with respect to x, we have

$$\frac{d}{dx}\left[x^2 + y^2\right] = \frac{d}{dx}[16].$$

On the right, the derivative of the constant 16 is 0, and on the left we can apply the sum rule, so it follows that

$$\frac{d}{dx}\left[x^2\right] + \frac{d}{dx}\left[y^2\right] = 0.$$

---

[1]Essentially the idea of an implicit function is that it can be broken into pieces where each piece can be viewed as an explicit function of x, and the combination of those pieces constitutes the full implicit function. For the circle, we could choose to take the top half as one explicit function of x, and the bottom half as another.

Next, it is essential that we recognize the different roles being played by x and y. Since x is the independent variable, it is the variable with respect to which we are differentiating, and thus $\frac{d}{dx}\left[x^2\right] = 2x$. But y is the dependent variable and y is an implicit function of x. Thus, when we want to compute $\frac{d}{dx}[y^2]$ it is identical to the situation in Preview Activity 2.7.1 where we computed $\frac{d}{dx}[f(x)^2]$. In both situations, we have an unknown function being squared, and we seek the derivative of the result. This requires the chain rule, by which we find that $\frac{d}{dx}[y^2] = 2y^1\frac{dy}{dx}$. Therefore, continuing our work in differentiating both sides of $x^2 + y^2 = 16$, we now have that

$$2x + 2y\frac{dy}{dx} = 0.$$

Since our goal is to find an expression for $\frac{dy}{dx}$, we solve this most recent equation for $\frac{dy}{dx}$. Subtracting $2x$ from both sides and dividing by $2y$,

$$\frac{dy}{dx} = -\frac{2x}{2y} = -\frac{x}{y}.$$

There are several important things to observe about the result that $\frac{dy}{dx} = -\frac{x}{y}$. First, this expression for the derivative involves both x and y. It makes sense that this should be the case, since for each value of x between -4 and 4, there are two corresponding points on the circle, and the slope of the tangent line is different at each of these points. Second, this formula is entirely consistent with our understanding of circles. If we consider the radius from the origin to the point (a, b), the slope of this line segment is $m_r = \frac{b}{a}$. The tangent line to the circle at (a, b) will be perpendicular to the radius, and thus have slope $m_t = -\frac{a}{b}$, as shown in Figure 2.7.2. Finally, the slope of the tangent line is zero at $(0, 4)$ and $(0, -4)$, and is undefined at $(-4, 0)$ and $(4, 0)$; all of these values are consistent with the formula $\frac{dy}{dx} = -\frac{x}{y}$.

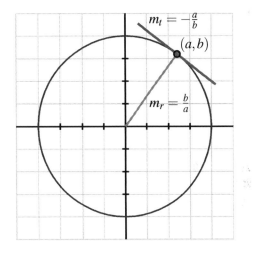

Figure 2.7.2: The circle given by $x^2 + y^2 = 16$ with point (a, b) on the circle and the tangent line at that point, with labeled slopes of the radial line, m_r, and tangent line, m_t.

We consider the following more complicated example to investigate and demonstrate some additional algebraic issues that arise in problems involving implicit differentiation.

Example 2.7.3. For the curve given implicitly by $x^3 + y^2 - 2xy = 2$, shown in Figure 2.7.4, find the slope of the tangent line at $(-1, 1)$.

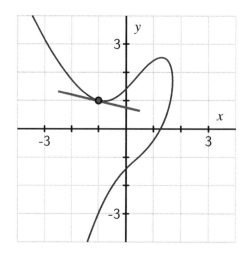

Figure 2.7.4: The curve $x^3 + y^2 - 2xy = 2$.

Solution.   We begin by differentiating the curve's equation implicitly. Taking the derivative of each side with respect to x,

$$\frac{d}{dx}[x^3 + y^2 - 2xy] = \frac{d}{dx}[2],$$

by the sum rule and the fact that the derivative of a constant is zero, we have

$$\frac{d}{dx}[x^3] + \frac{d}{dx}[y^2] - \frac{d}{dx}[2xy] = 0.$$

For the three derivatives we now must execute, the first uses the simple power rule, the second requires the chain rule (since y is an implicit function of x), and the third necessitates the product rule (again since y is a function of x). Applying these rules, we now find that

$$3x^2 + 2y\frac{dy}{dx} - [2x\frac{dy}{dx} + 2y] = 0.$$

Remembering that our goal is to find an expression for $\frac{dy}{dx}$ so that we can determine the slope of a particular tangent line, we want to solve the preceding equation for $\frac{dy}{dx}$. To do so, we get all of the terms involving $\frac{dy}{dx}$ on one side of the equation and then factor. Expanding and then subtracting $3x^2 - 2y$ from both sides, it follows that

$$2y\frac{dy}{dx} - 2x\frac{dy}{dx} = 2y - 3x^2.$$

Factoring the left side to isolate $\frac{dy}{dx}$, we have

$$\frac{dy}{dx}(2y - 2x) = 2y - 3x^2.$$

Finally, we divide both sides by $(2y - 2x)$ and conclude that

$$\frac{dy}{dx} = \frac{2y - 3x^2}{2y - 2x}.$$

Here again, the expression for $\frac{dy}{dx}$ depends on both x and y. To find the slope of the tangent line at $(-1, 1)$, we substitute this point in the formula for $\frac{dy}{dx}$, using the notation

$$\frac{dy}{dx}\bigg|_{(-1,1)} = \frac{2(1) - 3(-1)^2}{2(1) - 2(-1)} = -\frac{1}{4}.$$

This value matches our visual estimate of the slope of the tangent line shown in Figure 2.7.4.

Example 2.7.3 shows that it is possible when differentiating implicitly to have multiple terms involving $\frac{dy}{dx}$. Regardless of the particular curve involved, our approach will be similar each time. After differentiating, we expand so that each side of the equation is a sum of terms, some of which involve $\frac{dy}{dx}$. Next, addition and subtraction are used to get all terms involving $\frac{dy}{dx}$ on one side of the equation, with all remaining terms on the other. Finally, we factor to get a single instance of $\frac{dy}{dx}$, and then divide to solve for $\frac{dy}{dx}$.

Note, too, that since $\frac{dy}{dx}$ is often a function of both x and y, we use the notation

$$\frac{dy}{dx}\bigg|_{(a,b)}$$

to denote the evaluation of $\frac{dy}{dx}$ at the point (a, b). This is analogous to writing $f'(a)$ when f' depends on a single variable.

Finally, there is a big difference between writing $\frac{d}{dx}$ and $\frac{dy}{dx}$. For example,

$$\frac{d}{dx}[x^2 + y^2]$$

gives an instruction to take the derivative with respect to x of the quantity $x^2 + y^2$, presumably where y is a function of x. On the other hand,

$$\frac{dy}{dx}(x^2 + y^2)$$

means the product of the derivative of y with respect to x with the quantity $x^2 + y^2$. Understanding this notational subtlety is essential.

The following activities present opportunities to explore several different problems involving implicit differentiation.

Activity 2.7.2. Consider the curve defined by the equation $x = y^5 - 5y^3 + 4y$, whose graph is pictured in Figure 2.7.5.

a. Explain why it is not possible to express y as an explicit function of x.

b. Use implicit differentiation to find a formula for dy/dx.

c. Use your result from part (b) to find an equation of the line tangent to the graph of $x = y^5 - 5y^3 + 4y$ at the point $(0, 1)$.

d. Use your result from part (b) to determine all of the points at which the graph of $x = y^5 - 5y^3 + 4y$ has a vertical tangent line.

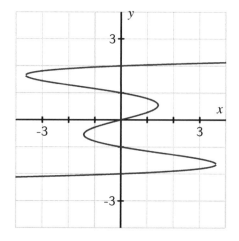

Figure 2.7.5: The curve $x = y^5 - 5y^3 + 4y$.

Two natural questions to ask about any curve involve where the tangent line can be vertical or horizontal. To be horizontal, the slope of the tangent line must be zero, while to be vertical, the slope must be undefined. It is typically the case when differentiating implicitly that the formula for $\frac{dy}{dx}$ is expressed as a quotient of functions of x and y, say

$$\frac{dy}{dx} = \frac{p(x, y)}{q(x, y)}.$$

Thus, we observe that the tangent line will be horizontal precisely when the numerator is zero and the denominator is nonzero, making the slope of the tangent line zero. Similarly, the tangent line will be vertical whenever $q(x, y) = 0$ and $p(x, y) \neq 0$, making the slope undefined. If both x and y are involved in an equation such as $p(x, y) = 0$, we try to solve for one of them in terms of the other, and then use the resulting condition in the original equation that defines the curve to find an equation in a single variable that we can solve to determine the point(s) that lie on the curve at which the condition holds. It is not always possible to execute the desired algebra due to the possibly complicated combinations of functions that often arise.

Activity 2.7.3. Consider the curve defined by the equation $y(y^2 - 1)(y - 2) = x(x - 1)(x - 2)$, whose graph is pictured in Figure 2.7.6. Through implicit differentiation, it can be shown that

$$\frac{dy}{dx} = \frac{(x-1)(x-2) + x(x-2) + x(x-1)}{(y^2-1)(y-2) + 2y^2(y-2) + y(y^2-1)}.$$

Use this fact to answer each of the following questions.

a. Determine all points (x, y) at which the tangent line to the curve is horizontal. (Use technology appropriately to find the needed zeros of the relevant polynomial function.)

b. Determine all points (x, y) at which the tangent line is vertical. (Use technology appropriately to find the needed zeros of the relevant polynomial function.)

c. Find the equation of the tangent line to the curve at one of the points where $x = 1$.

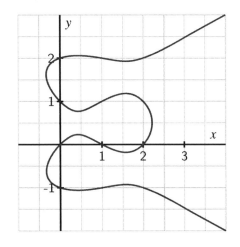

Figure 2.7.6: $y(y^2-1)(y-2) = x(x-1)(x-2)$.

Activity 2.7.4. For each of the following curves, use implicit differentiation to find dy/dx and determine the equation of the tangent line at the given point.

a. $x^3 - y^3 = 6xy$, $(-3,3)$

b. $\sin(y) + y = x^3 + x$, $(0,0)$

c. $3xe^{-xy} = y^2$, $(0.619061, 1)$

Summary

- When we have an equation involving x and y where y cannot be solved for explicitly in terms of x, but where portions of the curve can be thought of as being generated by explicit functions of x, we say that y is an implicit function of x. A good example of such a curve is the unit circle.

- In the process of implicit differentiation, we take the equation that generates an implicitly given curve and differentiate both sides with respect to x while treating y as a function of x. In so doing, the chain rule leads $\frac{dy}{dx}$ to arise, and then we may subsequently solve for $\frac{dy}{dx}$ using algebra.

- While $\frac{dy}{dx}$ may now involve both the variables x and y, $\frac{dy}{dx}$ still measures the slope of the tangent line to the curve, and thus this derivative may be used to decide when the tangent line is horizontal ($\frac{dy}{dx} = 0$) or vertical ($\frac{dy}{dx}$ is undefined), or to find the equation of the tangent line at a particular point on the curve.

Exercises

1. Find dy/dx in terms of x and y if $x^5y - x - 9y - 8 = 0$.

$$\frac{dy}{dx} = \boxed{}$$

2. Find $\dfrac{dy}{dx}$ in terms of x and y if $x \ln y + y^2 = 6 \ln x$.

$$\frac{dy}{dx} = \boxed{}$$

3. Find dy/dx in terms of x and y if $\arcsin(x^3y) = xy^3$.

$$\frac{dy}{dx} = \boxed{}$$

4. Find the slope of the tangent to the curve $x^3 + 2xy + y^2 = 64$ at $(1, 7)$.

The slope is $\boxed{}$.

(Enter undef *if the slope is not defined at this point.)*

5. Use implicit differentiation to find an equation of the tangent line to the curve $xy^3 + 4xy = 40$ at the point $(8, 1)$.

The equation $\boxed{}$ defines the tangent line to the curve at the point $(8, 1)$.

6. Consider the curve given by the equation $2y^3 + y^2 - y^5 = x^4 - 2x^3 + x^2$. Find all points at which the tangent line to the curve is horizontal or vertical. Be sure to use a graphing utility to plot this implicit curve and to visually check the results of algebraic reasoning that you use to determine where the tangent lines are horizontal and vertical.

7. For the curve given by the equation $\sin(x + y) + \cos(x - y) = 1$, find the equation of the tangent line to the curve at the point $(\frac{\pi}{2}, \frac{\pi}{2})$.

8. Implicit differentiation enables us a different perspective from which to see why the rule $\frac{d}{dx}[a^x] = a^x \ln(a)$ holds, if we assume that $\frac{d}{dx}[\ln(x)] = \frac{1}{x}$. This exercise leads you through the key steps to do so.

- a. Let $y = a^x$. Rewrite this equation using the natural logarithm function to write x in terms of y (and the constant a).

- b. Differentiate both sides of the equation you found in (a) with respect to x, keeping in mind that y is implicitly a function of x.

- c. Solve the equation you found in (b) for $\frac{dy}{dx}$, and then use the definition of y to write $\frac{dy}{dx}$ solely in terms of x. What have you found?

2.8 Using Derivatives to Evaluate Limits

Motivating Questions

- How can derivatives be used to help us evaluate indeterminate limits of the form $\frac{0}{0}$?

- What does it mean to say that $\lim_{x \to \infty} f(x) = L$ and $\lim_{x \to a} f(x) = \infty$?

- How can derivatives assist us in evaluating indeterminate limits of the form $\frac{\infty}{\infty}$?

Because differential calculus is based on the definition of the derivative, and the definition of the derivative involves a limit, there is a sense in which all of calculus rests on limits. In addition, the limit involved in the limit definition of the derivative is one that always generates an indeterminate form of $\frac{0}{0}$. If f is a differentiable function for which $f'(x)$ exists, then when we consider

$$f'(x) = \lim_{h \to 0} \frac{f(x+h) - f(x)}{h},$$

it follows that not only does $h \to 0$ in the denominator, but also $(f(x+h) - f(x)) \to 0$ in the numerator, since f is continuous. Thus, the fundamental form of the limit involved in the definition of $f'(x)$ is $\frac{0}{0}$. Remember, saying a limit has an indeterminate form only means that we don't yet know its value and have more work to do: indeed, limits of the form $\frac{0}{0}$ can take on any value, as is evidenced by evaluating $f'(x)$ for varying values of x for a function such as $f'(x) = x^2$.

Of course, we have learned many different techniques for evaluating the limits that result from the derivative definition, and including a large number of shortcut rules that enable us to evaluate these limits quickly and easily. In this section, we turn the situation upside-down: rather than using limits to evaluate derivatives, we explore how to use derivatives to evaluate certain limits. This topic will combine several different ideas, including limits, derivative shortcuts, local linearity, and the tangent line approximation.

Preview Activity 2.8.1. Let h be the function given by $h(x) = \frac{x^5 + x - 2}{x^2 - 1}$.

a. What is the domain of h?

b. Explain why $\lim_{x \to 1} \dfrac{x^5 + x - 2}{x^2 - 1}$ results in an indeterminate form.

c. Next we will investigate the behavior of both the numerator and denominator of h near the point where $x = 1$. Let $f(x) = x^5 + x - 2$ and $g(x) = x^2 - 1$. Find the local linearizations of f and g at $a = 1$, and call these functions $L_f(x)$ and $L_g(x)$, respectively.

d. Explain why $h(x) \approx \dfrac{L_f(x)}{L_g(x)}$ for x near $a = 1$.

e. Using your work from (c) and (d), evaluate

$$\lim_{x \to 1} \frac{L_f(x)}{L_g(x)}.$$

What do you think your result tells us about $\lim_{x \to 1} h(x)$?

f. Investigate the function $h(x)$ graphically and numerically near $x = 1$. What do you think is the value of $\lim_{x \to 1} h(x)$?

2.8.1 Using derivatives to evaluate indeterminate limits of the form $\frac{0}{0}$.

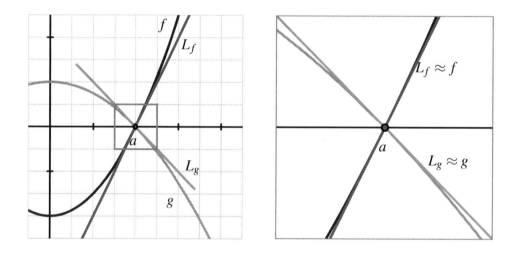

Figure 2.8.1: At left, the graphs of f and g near the value a, along with their tangent line approximations L_f and L_g at $x = a$. At right, zooming in on the point a and the four graphs.

The fundamental idea of Preview Activity 2.8.1 — that we can evaluate an indeterminate limit of the form $\frac{0}{0}$ by replacing each of the numerator and denominator with their local linearizations at the point of interest — can be generalized in a way that enables us to easily evaluate a wide range of limits. We begin by assuming that we have a function $h(x)$ that can be written in the form $h(x) = \frac{f(x)}{g(x)}$ where f and g are both differentiable at $x = a$ and for which $f(a) = g(a) = 0$. We are interested in finding a way to evaluate the indeterminate limit given by $\lim_{x \to a} h(x)$. In Figure 2.8.1, we see a visual representation of the situation involving such functions f and g. In particular, we see that both f and g have an x-intercept at the point where $x = a$. In addition, since each function is differentiable, each is locally linear, and we can find their respective tangent line approximations L_f and L_g at $x = a$, which are also shown in the figure. Since we are interested in the limit of $\frac{f(x)}{g(x)}$ as $x \to a$, the individual behaviors of $f(x)$ and $g(x)$ as $x \to a$ are key to understand. Here, we take advantage of the fact that each function and its tangent line approximation become indistinguishable as $x \to a$.

First, let's reall that $L_f(x) = f'(a)(x - a) + f(a)$ and $L_g(x) = g'(a)(x - a) + g(a)$. The critical observation we make is that when taking the limit, because x is getting arbitrarily close to a, we can replace f with L_f and replace g with L_g, and thus we observe that

$$\lim_{x \to a} \frac{f(x)}{g(x)} = \lim_{x \to a} \frac{L_f(x)}{L_g(x)}$$
$$= \lim_{x \to a} \frac{f'(a)(x - a) + f(a)}{g'(a)(x - a) + g(a)}.$$

Next, we remember a key fundamental assumption: that both $f(a) = 0$ and $g(a) = 0$, as this is precisely what makes the original limit indeterminate. Substituting these values for $f(a)$ and $g(a)$ in the limit above, we now have

$$\lim_{x \to a} \frac{f(x)}{g(x)} = \lim_{x \to a} \frac{f'(a)(x - a)}{g'(a)(x - a)}$$
$$= \lim_{x \to a} \frac{f'(a)}{g'(a)},$$

where the latter equality holds since x is approaching (but not equal to) a, so $\frac{x-a}{x-a} = 1$. Finally, we note that $\frac{f'(a)}{g'(a)}$ is constant with respect to x, and thus

$$\lim_{x \to a} \frac{f(x)}{g(x)} = \frac{f'(a)}{g'(a)}.$$

We have, of course, implicitly made the assumption that $g'(a) \neq 0$, which is essential to the overall limit having the value $\frac{f'(a)}{g'(a)}$. We summarize our work above with the statement of L'Hôpital's Rule, which is the formal name of the result we have shown.

L'Hôpital's Rule

Let f and g be differentiable at $x = a$, and suppose that $f(a) = g(a) = 0$ and that $g'(a) \neq 0$. Then $\lim_{x \to a} \frac{f(x)}{g(x)} = \frac{f'(a)}{g'(a)}$.

In practice, we typically work with a slightly more general version of L'Hôpital's Rule, which states that (under the identical assumptions as the boxed rule above and the extra assumption that g' is continuous at $x = a$)

$$\lim_{x \to a} \frac{f(x)}{g(x)} = \lim_{x \to a} \frac{f'(x)}{g'(x)},$$

provided the righthand limit exists. This form reflects the fundamental benefit of L'Hôpital's Rule: if $\frac{f(x)}{g(x)}$ produces an indeterminate limit of form $\frac{0}{0}$ as $x \to a$, it is equivalent to consider the limit of the quotient of the two functions' derivatives, $\frac{f'(x)}{g'(x)}$. For example, if we consider the limit from Preview Activity 2.8.1,

$$\lim_{x \to 1} \frac{x^5 + x - 2}{x^2 - 1},$$

by L'Hôpital's Rule we have that

$$\lim_{x \to 1} \frac{x^5 + x - 2}{x^2 - 1} = \lim_{x \to 1} \frac{5x^4 + 1}{2x} = \frac{6}{2} = 3.$$

By being able to replace the numerator and denominator with their respective derivatives, we often move from an indeterminate limit to one whose value we can easily determine.

> **Activity 2.8.2.** Evaluate each of the following limits. If you use L'Hôpital's Rule, indicate where it was used, and be certain its hypotheses are met before you apply it.
>
> a. $\lim_{x \to 0} \frac{\ln(1+x)}{x}$
>
> b. $\lim_{x \to \pi} \frac{\cos(x)}{x}$
>
> c. $\lim_{x \to 1} \frac{2\ln(x)}{1 - e^{x-1}}$
>
> d. $\lim_{x \to 0} \frac{\sin(x) - x}{\cos(2x) - 1}$

While L'Hôpital's Rule can be applied in an entirely algebraic way,

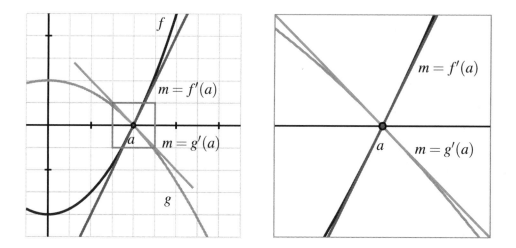

Figure 2.8.2: Two functions f and g that satisfy L'Hôpital's Rule.

it is important to remember that the genesis of the rule is graphical: the main idea is that the slopes of the tangent lines to f and g at $x = a$ determine the value of the limit of $\frac{f(x)}{g(x)}$ as

$x \to a$. We see this in Figure 2.8.2, which is a modified version of Figure 2.8.1, where we can see from the grid that $f'(a) = 2$ and $g'(a) = -1$, hence by L'Hôpital's Rule,

$$\lim_{x \to a} \frac{f(x)}{g(x)} = \frac{f'(a)}{g'(a)} = \frac{2}{-1} = -2.$$

Indeed, what we observe is that it's not the fact that f and g both approach zero that matters most, but rather the *rate* at which each approaches zero that determines the value of the limit. This is a good way to remember what L'Hôpital's Rule says: if $f(a) = g(a) = 0$, the the limit of $\frac{f(x)}{g(x)}$ as $x \to a$ is given by the ratio of the slopes of f and g at $x = a$.

Activity 2.8.3. In this activity, we reason graphically from the following figure to evaluate limits of ratios of functions about which some information is known.

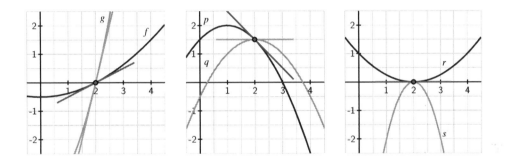

Figure 2.8.3: Three graphs referenced in the questions of Activity 2.8.3.

a. Use the left-hand graph to determine the values of $f(2)$, $f'(2)$, $g(2)$, and $g'(2)$. Then, evaluate $\lim\limits_{x \to 2} \frac{f(x)}{g(x)}$.

b. Use the middle graph to find $p(2)$, $p'(2)$, $q(2)$, and $q'(2)$. Then, determine the value of $\lim\limits_{x \to 2} \frac{p(x)}{q(x)}$.

c. Use the right-hand graph to compute $r(2)$, $r'(2)$, $s(2)$, $s'(2)$. Explain why you cannot determine the exact value of $\lim\limits_{x \to 2} \frac{r(x)}{s(x)}$ without further information being provided, but that you can determine the sign of $\lim\limits_{x \to 2} \frac{r(x)}{s(x)}$. In addition, state what the sign of the limit will be, with justification.

2.8.2 Limits involving ∞

The concept of infinity, denoted ∞, arises naturally in calculus, like it does in much of mathematics. It is important to note from the outset that ∞ is a concept, but not a number itself. Indeed, the notion of ∞ naturally invokes the idea of limits. Consider, for example, the function $f(x) = \frac{1}{x}$, whose graph is pictured in Figure 2.8.4.

We note that $x = 0$ is not in the domain of f, so we may naturally wonder what happens as $x \to 0$. As $x \to 0^+$, we observe that $f(x)$ *increases without bound*. That is, we can make the value of $f(x)$ as large as we like by taking x closer and closer (but not equal) to 0, while keeping $x > 0$. This is a good way to think about what infinity represents: a quantity is tending to infinity if there is no single number that the quantity is always less than.

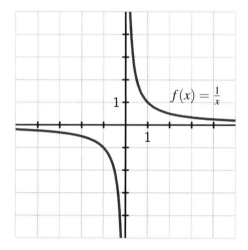

Figure 2.8.4: The graph of $f(x) = \frac{1}{x}$.

Recall that when we write $\lim_{x \to a} f(x) = L$, this means that can make $f(x)$ as close to L as we'd like by taking x sufficiently close (but not equal) to a. We thus expand this notation and language to include the possibility that either L or a can be ∞. For instance, for $f(x) = \frac{1}{x}$, we now write

$$\lim_{x \to 0^+} \frac{1}{x} = \infty,$$

by which we mean that we can make $\frac{1}{x}$ as large as we like by taking x sufficiently close (but not equal) to 0. In a similar way, we naturally write

$$\lim_{x \to \infty} \frac{1}{x} = 0,$$

since we can make $\frac{1}{x}$ as close to 0 as we'd like by taking x sufficiently large (i.e., by letting x increase without bound).

In general, we understand the notation $\lim_{x \to a} f(x) = \infty$ to mean that we can make $f(x)$ as large as we'd like by taking x sufficiently close (but not equal) to a, and the notation $\lim_{x \to \infty} f(x) = L$ to mean that we can make $f(x)$ as close to L as we'd like by taking x sufficiently large. This notation applies to left- and right-hand limits, plus we can also use limits involving $-\infty$. For example, returning to Figure 2.8.4 and $f(x) = \frac{1}{x}$, we can say that

$$\lim_{x \to 0^-} \frac{1}{x} = -\infty \quad \text{and} \quad \lim_{x \to -\infty} \frac{1}{x} = 0.$$

Finally, we write

$$\lim_{x \to \infty} f(x) = \infty$$

when we can make the value of $f(x)$ as large as we'd like by taking x sufficiently large. For example,

$$\lim_{x \to \infty} x^2 = \infty.$$

Note particularly that limits involving infinity identify *vertical* and *horizontal asymptotes* of a function. If $\lim_{x \to a} f(x) = \infty$, then $x = a$ is a vertical asymptote of f, while if $\lim_{x \to \infty} f(x) = L$, then $y = L$ is a horizontal asymptote of f. Similar statements can be made using $-\infty$, as well as with left- and right-hand limits as $x \to a^-$ or $x \to a^+$.

In precalculus classes, it is common to study the *end behavior* of certain families of functions, by which we mean the behavior of a function as $x \to \infty$ and as $x \to -\infty$. Here we briefly examine a library of some familiar functions and note the values of several limits involving ∞.

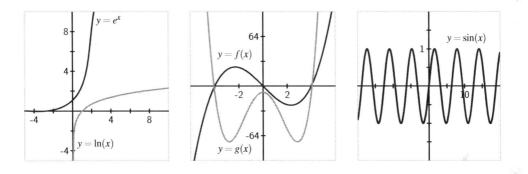

Figure 2.8.5: Graphs of some familiar functions whose end behavior as $x \to \pm\infty$ is known. In the middle graph, $f(x) = x^3 - 16x$ and $g(x) = x^4 - 16x^2 - 8$.

For the natural exponential function e^x, we note that $\lim_{x \to \infty} e^x = \infty$ and $\lim_{x \to -\infty} e^x = 0$, while for the related exponential decay function e^{-x}, observe that these limits are reversed, with $\lim_{x \to \infty} e^{-x} = 0$ and $\lim_{x \to -\infty} e^{-x} = \infty$. Turning to the natural logarithm function, we have $\lim_{x \to 0^+} \ln(x) = -\infty$ and $\lim_{x \to \infty} \ln(x) = \infty$. While both e^x and $\ln(x)$ grow without bound as $x \to \infty$, the exponential function does so much more quickly than the logarithm function does. We'll soon use limits to quantify what we mean by "quickly."

For polynomial functions of the form $p(x) = a_n x^n + a_{n-1} x^{n-1} + \cdots a_1 x + a_0$, the end behavior depends on the sign of a_n and whether the highest power n is even or odd. If n is even and a_n is positive, then $\lim_{x \to \infty} p(x) = \infty$ and $\lim_{x \to -\infty} p(x) = \infty$, as in the plot of g in Figure 2.8.5. If instead a_n is negative, then $\lim_{x \to \infty} p(x) = -\infty$ and $\lim_{x \to -\infty} p(x) = -\infty$. In the situation where n is odd, then either $\lim_{x \to \infty} p(x) = \infty$ and $\lim_{x \to -\infty} p(x) = -\infty$ (which occurs when a_n is positive, as in the graph of f in Figure 2.8.5), or $\lim_{x \to \infty} p(x) = -\infty$ and $\lim_{x \to -\infty} p(x) = \infty$ (when a_n is negative).

A function can fail to have a limit as $x \to \infty$. For example, consider the plot of the sine function at right in Figure 2.8.5. Because the function continues oscillating between -1 and 1 as $x \to \infty$, we say that $\lim_{x \to \infty} \sin(x)$ does not exist.

Finally, it is straightforward to analyze the behavior of any rational function as $x \to \infty$. Consider, for example, the function

$$q(x) = \frac{3x^2 - 4x + 5}{7x^2 + 9x - 10}.$$

Note that both $(3x^2 - 4x + 5) \to \infty$ as $x \to \infty$ and $(7x^2 + 9x - 10) \to \infty$ as $x \to \infty$. Here we say that $\lim_{x \to \infty} q(x)$ has indeterminate form $\frac{\infty}{\infty}$, much like we did when we encountered limits of the form $\frac{0}{0}$. We can determine the value of this limit through a standard algebraic approach. Multiplying the numerator and denominator each by $\frac{1}{x^2}$, we find that

$$\lim_{x \to \infty} q(x) = \lim_{x \to \infty} \frac{3x^2 - 4x + 5}{7x^2 + 9x - 10} \cdot \frac{\frac{1}{x^2}}{\frac{1}{x^2}}$$

$$= \lim_{x \to \infty} \frac{3 - 4\frac{1}{x} + 5\frac{1}{x^2}}{7 + 9\frac{1}{x} - 10\frac{1}{x^2}} = \frac{3}{7}$$

since $\frac{1}{x^2} \to 0$ and $\frac{1}{x} \to 0$ as $x \to \infty$. This shows that the rational function q has a horizontal asymptote at $y = \frac{3}{7}$. A similar approach can be used to determine the limit of any rational function as $x \to \infty$.

But how should we handle a limit such as

$$\lim_{x \to \infty} \frac{x^2}{e^x}?$$

Here, both $x^2 \to \infty$ and $e^x \to \infty$, but there is not an obvious algebraic approach that enables us to find the limit's value. Fortunately, it turns out that L'Hôpital's Rule extends to cases involving infinity.

L'Hôpital's Rule (∞)

If f and g are differentiable and both approach zero or both approach $\pm\infty$ as $x \to a$ (where a is allowed to be ∞) , then

$$\lim_{x \to a} \frac{f(x)}{g(x)} = \lim_{x \to a} \frac{f'(x)}{g'(x)}.$$

(To be technically correct, we need to the additional hypothesis that $g'(x) \neq 0$ on an open interval that contains a or in every neighborhood of infinity if a is ∞; this is almost always met in practice.)

To evaluate $\lim_{x \to \infty} \frac{x^2}{e^x}$, we observe that we can apply L'Hôpital's Rule, since both $x^2 \to \infty$ and $e^x \to \infty$. Doing so, it follows that

$$\lim_{x \to \infty} \frac{x^2}{e^x} = \lim_{x \to \infty} \frac{2x}{e^x}.$$

This updated limit is still indeterminate and of the form $\frac{\infty}{\infty}$, but it is simpler since $2x$ has replaced x^2. Hence, we can apply L'Hôpital's Rule again, by which we find that

$$\lim_{x \to \infty} \frac{x^2}{e^x} = \lim_{x \to \infty} \frac{2x}{e^x} = \lim_{x \to \infty} \frac{2}{e^x}.$$

Now, since 2 is constant and $e^x \to \infty$ as $x \to \infty$, it follows that $\frac{2}{e^x} \to 0$ as $x \to \infty$, which shows that

$$\lim_{x \to \infty} \frac{x^2}{e^x} = 0.$$

> **Activity 2.8.4.** Evaluate each of the following limits. If you use L'Hôpital's Rule, indicate where it was used, and be certain its hypotheses are met before you apply it.
>
> a. $\lim_{x \to \infty} \frac{x}{\ln(x)}$
>
> b. $\lim_{x \to \infty} \frac{e^x + x}{2e^x + x^2}$
>
> c. $\lim_{x \to 0^+} \frac{\ln(x)}{\frac{1}{x}}$
>
> d. $\lim_{x \to \frac{\pi}{2}^-} \frac{\tan(x)}{x - \frac{\pi}{2}}$
>
> e. $\lim_{x \to \infty} xe^{-x}$

When we are considering the limit of a quotient of two functions $\frac{f(x)}{g(x)}$ that results in an indeterminate form of $\frac{\infty}{\infty}$, in essence we are asking which function is growing faster without bound. We say that the function g *dominates* the function f as $x \to \infty$ provided that

$$\lim_{x \to \infty} \frac{f(x)}{g(x)} = 0,$$

whereas f dominates g provided that $\lim_{x \to \infty} \frac{f(x)}{g(x)} = \infty$. Finally, if the value of $\lim_{x \to \infty} \frac{f(x)}{g(x)}$ is finite and nonzero, we say that f and g *grow at the same rate*. For example, from earlier work we know that $\lim_{x \to \infty} \frac{x^2}{e^x} = 0$, so e^x dominates x^2, while $\lim_{x \to \infty} \frac{3x^2 - 4x + 5}{7x^2 + 9x - 10} = \frac{3}{7}$, so $f(x) = 3x^2 - 4x + 5$ and $g(x) = 7x^2 + 9x - 10$ grow at the same rate.

Summary

- Derivatives be used to help us evaluate indeterminate limits of the form $\frac{0}{0}$ through L'Hôpital's Rule, which is developed by replacing the functions in the numerator and denominator with their tangent line approximations. In particular, if $f(a) = g(a) = 0$ and f and g are differentiable at a, L'Hôpital's Rule tells us that

  $$\lim_{x \to a} \frac{f(x)}{g(x)} = \lim_{x \to a} \frac{f'(x)}{g'(x)}.$$

- When we write $x \to \infty$, this means that x is increasing without bound. We thus use ∞ along with limit notation to write $\lim_{x \to \infty} f(x) = L$, which means we can make $f(x)$ as close to L as we like by choosing x to be sufficiently large, and similarly $\lim_{x \to a} f(x) = \infty$, which means we can make $f(x)$ as large as we like by choosing x sufficiently close to a.

- A version of L'Hôpital's Rule also allows us to use derivatives to assist us in evaluating indeterminate limits of the form $\frac{\infty}{\infty}$. In particular, If f and g are differentiable and both approach zero or both approach $\pm\infty$ as $x \to a$ (where a is allowed to be ∞), then

$$\lim_{x \to a} \frac{f(x)}{g(x)} = \lim_{x \to a} \frac{f'(x)}{g'(x)}.$$

Exercises

1. For the figures below, determine the nature of $\lim\limits_{x \to a} \dfrac{f(x)}{g(x)}$, if $f(x)$ is shown as the blue curve and $g(x)$ as the black curve.

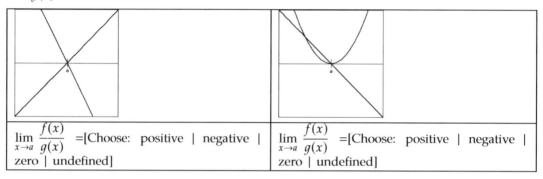

$\lim\limits_{x \to a} \dfrac{f(x)}{g(x)}$ =[Choose: positive | negative | zero | undefined]	$\lim\limits_{x \to a} \dfrac{f(x)}{g(x)}$ =[Choose: positive | negative | zero | undefined]

2. Find the limit: $\lim\limits_{x \to 4} \dfrac{\ln(x/4)}{x^2 - 16}$ = [   ]
(Enter undefined *if the limit does not exist.)*

3. Compute the following limits using l'Hôpital's rule if appropriate. Use INF to denote ∞ and MINF to denote $-\infty$.

$\lim\limits_{x \to 0} \dfrac{1 - \cos(7x)}{1 - \cos(6x)}$ = [   ]

$\lim\limits_{x \to 1} \dfrac{7^x - 6^x - 1}{x^2 - 1}$ = [   ]

4. Evaluate the limit using L'Hopital's rule.

$\lim\limits_{x \to \infty} \dfrac{13x^2}{e^{8x}}$ = [   ] help (limits)

5. Let f and g be differentiable functions about which the following information is known: $f(3) = g(3) = 0$, $f'(3) = g'(3) = 0$, $f''(3) = -2$, and $g''(3) = 1$. Let a new function h be given by the rule $h(x) = \frac{f(x)}{g(x)}$. On the same set of axes, sketch possible graphs of f and g near $x = 3$, and use the provided information to determine the value of

$$\lim_{x \to 3} h(x).$$

Provide explanation to support your conclusion.

6. Find all vertical and horizontal asymptotes of the function

$$R(x) = \frac{3(x - a)(x - b)}{5(x - a)(x - c)},$$

where a, b, and c are distinct, arbitrary constants. In addition, state all values of x for which R is not continuous. Sketch a possible graph of R, clearly labeling the values of a, b, and c.

7. Consider the function $g(x) = x^{2x}$, which is defined for all $x > 0$. Observe that $\lim_{x \to 0^+} g(x)$ is indeterminate due to its form of 0^0. (Think about how we know that $0^k = 0$ for all $k > 0$, while $b^0 = 1$ for all $b \neq 0$, but that neither rule can apply to 0^0.)

 a. Let $h(x) = \ln(g(x))$. Explain why $h(x) = 2x \ln(x)$.

 b. Next, explain why it is equivalent to write $h(x) = \frac{2\ln(x)}{\frac{1}{x}}$.

 c. Use L'Hôpital's Rule and your work in (b) to compute $\lim_{x \to 0^+} h(x)$.

 d. Based on the value of $\lim_{x \to 0^+} h(x)$, determine $\lim_{x \to 0^+} g(x)$.

8. Recall we say that function g **dominates** function f provided that $\lim_{x \to \infty} f(x) = \infty$, $\lim_{x \to \infty} g(x) = \infty$, and $\lim_{x \to \infty} \frac{f(x)}{g(x)} = 0$.

 a. Which function dominates the other: $\ln(x)$ or $\sqrt{x}$?

 b. Which function dominates the other: $\ln(x)$ or $\sqrt[n]{x}$? (n can be any positive integer)

 c. Explain why e^x will dominate any polynomial function.

 d. Explain why x^n will dominate $\ln(x)$ for any positive integer n.

 e. Give any example of two nonlinear functions such that neither dominates the other.

CHAPTER 3

Using Derivatives

3.1 Using derivatives to identify extreme values

Motivating Questions

- What are the critical numbers of a function f and how are they connected to identifying the most extreme values the function achieves?

- How does the first derivative of a function reveal important information about the behavior of the function, including the function's extreme values?

- How can the second derivative of a function be used to help identify extreme values of the function?

In many different settings, we are interested in knowing where a function achieves its least and greatest values. These can be important in applications — say to identify a point at which maximum profit or minimum cost occurs — or in theory to understand how to characterize the behavior of a function or a family of related functions. Consider the simple and familiar example of a parabolic function such as $s(t) = -16t^2 + 32t + 48$ (shown at left in Figure 3.1.1) that represents the height of an object tossed vertically: its maximum value occurs at the vertex of the parabola and represents the highest value that the object reaches. Moreover, this maximum value identifies an especially important point on the graph, the point at which the curve changes from increasing to decreasing.

More generally, for any function we consider, we can investigate where its lowest and highest points occur in comparison to points nearby or to all possible points on the graph. Given a function f, we say that $f(c)$ is a *global* or *absolute maximum* provided that $f(c) \geq f(x)$ for all x in the domain of f, and similarly call $f(c)$ a *global* or *absolute minimum* whenever $f(c) \leq f(x)$ for all x in the domain of f. For instance, for the function g given at right in Figure 3.1.1, g has a global maximum of $g(c)$, but g does not appear to have a global minimum, as the graph of g seems to decrease without bound. We note that the point $(c, g(c))$ marks a fundamental change in the behavior of g, where g changes from increasing to decreasing; similar things happen at both $(a, g(a))$ and $(b, g(b))$, although these points are not global mins or maxes.

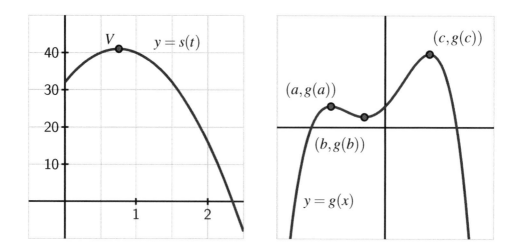

Figure 3.1.1: At left, $s(t) = -16t^2 + 24t + 32$ whose vertex is $(\frac{3}{4}, 41)$; at right, a function g that demonstrates several high and low points.

For any function f, we say that $f(c)$ is a *local maximum* or *relative maximum* provided that $f(c) \geq f(x)$ for all x near c, while $f(c)$ is called a *local* or *relative minimum* whenever $f(c) \leq f(x)$ for all x near c. Any maximum or minimum may be called an *extreme value* of f. For example, in Figure 3.1.1, g has a relative minimum of $g(b)$ at the point $(b, g(b))$ and a relative maximum of $g(a)$ at $(a, g(a))$. We have already identified the global maximum of g as $g(c)$; this global maximum can also be considered a relative maximum.

We would like to use fundamental calculus ideas to help us identify and classify key function behavior, including the location of relative extremes. Of course, if we are given a graph of a function, it is often straightforward to locate these important behaviors visually. We investigate this situation in the following preview activity.

Preview Activity 3.1.1. Consider the function h given by the graph in Figure 3.1.2. Use the graph to answer each of the following questions.

a. Identify all of the values of c for which $h(c)$ is a local maximum of h.

b. Identify all of the values of c for which $h(c)$ is a local minimum of h.

c. Does h have a global maximum on the interval $[-3, 3]$? If so, what is the value of this global maximum?

d. Does h have a global minimum on the interval $[-3, 3]$? If so, what is its value?

e. Identify all values of c for which $h'(c) = 0$.

f. Identify all values of c for which $h'(c)$ does not exist.

g. True or false: every relative maximum and minimum of h occurs at a point where $h'(c)$ is either zero or does not exist.

h. True or false: at every point where $h'(c)$ is zero or does not exist, h has a relative maximum or minimum.

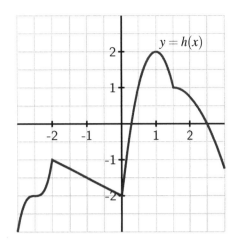

Figure 3.1.2: The graph of a function h on the interval $[-3, 3]$.

3.1.1 Critical numbers and the first derivative test

If a function has a relative extreme value at a point $(c, f(c))$, the function must change its behavior at c regarding whether it is increasing or decreasing before or after the point.

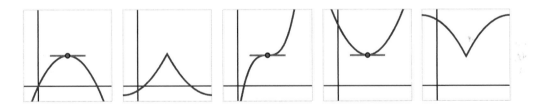

Figure 3.1.3: From left to right, a function with a relative maximum where its derivative is zero; a function with a relative maximum where its derivative is undefined; a function with neither a maximum nor a minimum at a point where its derivative is zero; a function with a relative minimum where its derivative is zero; and a function with a relative minimum where its derivative is undefined.

For example, if a continuous function has a relative maximum at c, such as those pictured in the two leftmost functions in Figure 3.1.3, then it is both necessary and sufficient that the function change from being increasing just before c to decreasing just after c. In the same way, a continuous function has a relative minimum at c if and only if the function changes from decreasing to increasing at c. See, for instance, the two functions pictured at right in Figure 3.1.3. There are only two possible ways for these changes in behavior to occur: either $f'(c) = 0$ or $f'(c)$ is undefined.

Because these values of c are so important, we call them *critical numbers*. More specifically, we say that a function f has a *critical number* at $x = c$ provided that c is in the domain of f, and $f'(c) = 0$ or $f'(c)$ is undefined. Critical numbers provide us with the only possible locations where the function f may have relative extremes. Note that not every critical number produces a maximum or minimum; in the middle graph of Figure 3.1.3, the function pictured there has a horizontal tangent line at the noted point, but the function is increasing before and increasing after, so the critical number does not yield a location where the function is greater than every value nearby, nor less than every value nearby.

We also sometimes use the terminology that, when c is a critical number, that $(c, f(c))$ is a *critical point* of the function, or that $f(c)$ is a *critical value* .

The *first derivative test* summarizes how sign changes in the first derivative indicate the presence of a local maximum or minimum for a given function.

First Derivative Test

If p is a critical number of a continuous function f that is differentiable near p (except possibly at $x = p$), then f has a relative maximum at p if and only if f' changes sign from positive to negative at p, and f has a relative minimum at p if and only if f' changes sign from negative to positive at p.

We consider an example to show one way the first derivative test can be used to identify the relative extreme values of a function.

Example 3.1.4. Let f be a function whose derivative is given by the formula $f'(x) = e^{-2x}(3 - x)(x + 1)^2$. Determine all critical numbers of f and decide whether a relative maximum, relative minimum, or neither occurs at each.

Solution.   Since we already have $f'(x)$ written in factored form, it is straightforward to find the critical numbers of f. Since $f'(x)$ is defined for all values of x, we need only determine where $f'(x) = 0$. From the equation

$$e^{-2x}(3 - x)(x + 1)^2 = 0$$

and the zero product property, it follows that $x = 3$ and $x = -1$ are critical numbers of f. (Note particularly that there is no value of x that makes $e^{-2x} = 0$.)

Next, to apply the first derivative test, we'd like to know the sign of $f'(x)$ at inputs near the critical numbers. Because the critical numbers are the only locations at which f' can change sign, it follows that the sign of the derivative is the same on each of the intervals created by the critical numbers: for instance, the sign of f' must be the same for every $x < -1$. We create a first derivative sign chart to summarize the sign of f' on the relevant intervals along with the corresponding behavior of f.

$$f'(x) = e^{-2x}(3-x)(x+1)^2$$

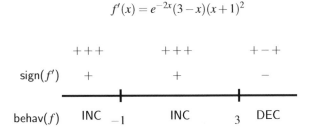

Figure 3.1.5: The first derivative sign chart for a function f whose derivative is given by the formula $f'(x) = e^{-2x}(3-x)(x+1)^2$.

The first derivative sign chart in Figure 3.1.5 comes from thinking about the sign of each of the terms in the factored form of $f'(x)$ at one selected point in the interval under consideration. For instance, for $x < -1$, we could consider $x = -2$ and determine the sign of e^{-2x}, $(3-x)$, and $(x+1)^2$ at the value $x = -2$. We note that both e^{-2x} and $(x+1)^2$ are positive regardless of the value of x, while $(3-x)$ is also positive at $x = -2$. Hence, each of the three terms in f' is positive, which we indicate by writing "$+++$." Taking the product of three positive terms obviously results in a value that is positive, which we denote by the "$+$" in the interval to the left of $x = -1$ indicating the overall sign of f'. And, since f' is positive on that interval, we further know that f is increasing, which we summarize by writing "INC" to represent the corresponding behavior of f. In a similar way, we find that f' is positive and f is increasing on $-1 < x < 3$, and f' is negative and f is decreasing for $x > 3$.

Now, by the first derivative test, to find relative extremes of f we look for critical numbers at which f' changes sign. In this example, f' only changes sign at $x = 3$, where f' changes from positive to negative, and thus f has a relative maximum at $x = 3$. While f has a critical number at $x = -1$, since f is increasing both before and after $x = -1$, f has neither a minimum nor a maximum at $x = -1$.

Activity 3.1.2. Suppose that $g(x)$ is a function continuous for every value of $x \neq 2$ whose first derivative is $g'(x) = \frac{(x+4)(x-1)^2}{x-2}$. Further, assume that it is known that g has a vertical asymptote at $x = 2$.

a. Determine all critical numbers of g.

b. By developing a carefully labeled first derivative sign chart, decide whether g has as a local maximum, local minimum, or neither at each critical number.

c. Does g have a global maximum? global minimum? Justify your claims.

d. What is the value of $\lim_{x\to\infty} g'(x)$? What does the value of this limit tell you about the long-term behavior of g?

e. Sketch a possible graph of $y = g(x)$.

3.1.2 The second derivative test

Recall that the second derivative of a function tells us several important things about the behavior of the function itself. For instance, if f'' is positive on an interval, then we know that f' is increasing on that interval and, consequently, that f is concave up, which also tells us that throughout the interval the tangent line to $y = f(x)$ lies below the curve at every point. In this situation where we know that $f'(p) = 0$, it turns out that the sign of the second derivative determines whether f has a local minimum or local maximum at the critical number p.

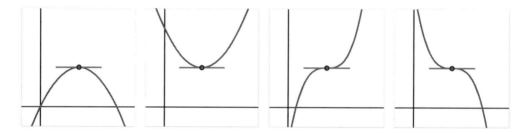

Figure 3.1.6: Four possible graphs of a function f with a horizontal tangent line at a critical point.

In Figure 3.1.6, we see the four possibilities for a function f that has a critical number p at which $f'(p) = 0$, provided $f''(p)$ is not zero on an interval including p (except possibly at p). On either side of the critical number, f'' can be either positive or negative, and hence f can be either concave up or concave down. In the first two graphs, f does not change concavity at p, and in those situations, f has either a local minimum or local maximum. In particular, if $f'(p) = 0$ and $f''(p) < 0$, then we know f is concave down at p with a horizontal tangent line, and this guarantees f has a local maximum there. This fact, along with the corresponding statement for when $f''(p)$ is positive, is stated in the *second derivative test*.

> **Second Derivative Test**
>
> If p is a critical number of a continuous function f such that $f'(p) = 0$ and $f''(p) \neq 0$, then f has a relative maximum at p if and only if $f''(p) < 0$, and f has a relative minimum at p if and only if $f''(p) > 0$.

In the event that $f''(p) = 0$, the second derivative test is inconclusive. That is, the test doesn't provide us any information. This is because if $f''(p) = 0$, it is possible that f has a local minimum, local maximum, or neither.[1]

---

[1]Consider the functions $f(x) = x^4$, $g(x) = -x^4$, and $h(x) = x^3$ at the critical point $p = 0$.

Just as a first derivative sign chart reveals all of the increasing and decreasing behavior of a function, we can construct a second derivative sign chart that demonstrates all of the important information involving concavity.

Example 3.1.7. Let $f(x)$ be a function whose first derivative is $f'(x) = 3x^4 - 9x^2$. Construct both first and second derivative sign charts for f, fully discuss where f is increasing and decreasing and concave up and concave down, identify all relative extreme values, and sketch a possible graph of f.

Solution. Since we know $f'(x) = 3x^4 - 9x^2$, we can find the critical numbers of f by solving $3x^4 - 9x^2 = 0$. Factoring, we observe that

$$0 = 3x^2(x^2 - 3) = 3x^2(x + \sqrt{3})(x - \sqrt{3}),$$

so that $x = 0, \pm\sqrt{3}$ are the three critical numbers of f. It then follows that the first derivative sign chart for f is given in Figure 3.1.8.

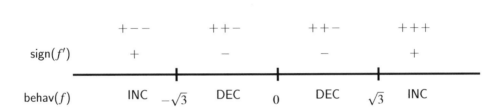

Figure 3.1.8: The first derivative sign chart for f when $f'(x) = 3x^4 - 9x^2 = 3x^2(x^2 - 3)$.

Thus, f is increasing on the intervals $(-\infty, -\sqrt{3})$ and $(\sqrt{3}, \infty)$, while f is decreasing on $(-\sqrt{3}, 0)$ and $(0, \sqrt{3})$. Note particularly that by the first derivative test, this information tells us that f has a local maximum at $x = -\sqrt{3}$ and a local minimum at $x = \sqrt{3}$. While f also has a critical number at $x = 0$, neither a maximum nor minimum occurs there since f' does not change sign at $x = 0$.

Next, we move on to investigate concavity. Differentiating $f'(x) = 3x^4 - 9x^2$, we see that $f''(x) = 12x^3 - 18x$. Since we are interested in knowing the intervals on which f'' is positive and negative, we first find where $f''(x) = 0$. Observe that

$$0 = 12x^3 - 18x = 12x\left(x^2 - \frac{3}{2}\right) = 12x\left(x + \sqrt{\frac{3}{2}}\right)\left(x - \sqrt{\frac{3}{2}}\right),$$

which implies that $x = 0, \pm\sqrt{\frac{3}{2}}$. Building a sign chart for f'' in the exact same way we do for f', we see the result shown in Figure 3.1.9.

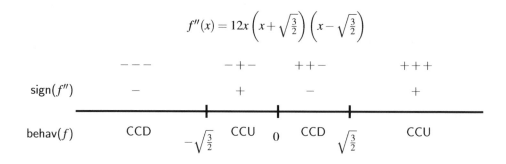

$$f''(x) = 12x\left(x + \sqrt{\tfrac{3}{2}}\right)\left(x - \sqrt{\tfrac{3}{2}}\right)$$

Figure 3.1.9: The second derivative sign chart for f when $f''(x) = 12x^3 - 18x = 12x^2\left(x^2 - \sqrt{\tfrac{3}{2}}\right)$.

Therefore, f is concave down on the intervals $(-\infty, -\sqrt{\tfrac{3}{2}})$ and $(0, \sqrt{\tfrac{3}{2}})$, and concave up on $(-\sqrt{\tfrac{3}{2}}, 0)$ and $(\sqrt{\tfrac{3}{2}}, \infty)$.

Putting all of the above information together, we now see a complete and accurate possible graph of f in Figure 3.1.10.

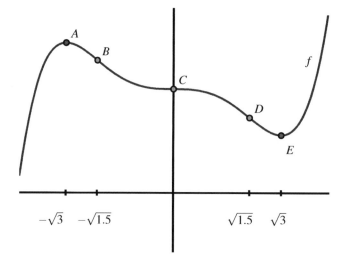

Figure 3.1.10: A possible graph of the function f in Example 3.1.7.

The point $A = (-\sqrt{3}, f(-\sqrt{3}))$ is a local maximum, as f is increasing prior to A and decreasing after; similarly, the point $E = (\sqrt{3}, f(\sqrt{3}))$ is a local minimum. Note, too, that f is concave down at A and concave up at B, which is consistent both with our second derivative sign chart and the second derivative test. At points B and D, concavity changes, as we saw in the results of the second derivative sign chart in Figure 3.1.9. Finally, at point C, f has a critical point with a horizontal tangent line, but neither a maximum nor a minimum occurs there since f is decreasing both before and after C. It is also the case that concavity changes at C.

While we completely understand where f is increasing and decreasing, where f is concave up and concave down, and where f has relative extremes, we do not know any specific information about the y-coordinates of points on the curve. For instance, while we know that f has a local maximum at $x = -\sqrt{3}$, we don't know the value of that maximum because we do not know $f(-\sqrt{3})$. Any vertical translation of our sketch of f in Figure 3.1.10 would satisfy the given criteria for f.

Points B, C, and D in Figure 3.1.10 are locations at which the concavity of f changes. We give a special name to any such point: if p is a value in the domain of a continuous function f at which f changes concavity, then we say that $(p, f(p))$ is an *inflection point* of f. Just as we look for locations where f changes from increasing to decreasing at points where $f'(p) = 0$ or $f'(p)$ is undefined, so too we find where $f''(p) = 0$ or $f''(p)$ is undefined to see if there are points of inflection at these locations.

It is important at this point in our study to remind ourselves of the big picture that derivatives help to paint: the sign of the first derivative f' tells us *whether* the function f is increasing or decreasing, while the sign of the second derivative f'' tells us *how* the function f is increasing or decreasing.

> **Activity 3.1.3.** Suppose that g is a function whose second derivative, g'', is given by the following graph.
>
> a. Find the x-coordinates of all points of inflection of g.
>
> b. Fully describe the concavity of g by making an appropriate sign chart.
>
> c. Suppose you are given that $g'(-1.67857351) = 0$. Is there is a local maximum, local minimum, or neither (for the function g) at this critical number of g, or is it impossible to say? Why?
>
>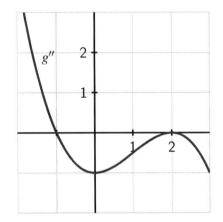
>
> **Figure 3.1.11:** The graph of $y = g''(x)$.
>
> d. Assuming that $g''(x)$ is a polynomial (and that all important behavior of g'' is seen

in the graph above), what degree polynomial do you think $g(x)$ is? Why?

As we will see in more detail in the following section, derivatives also help us to understand families of functions that differ only by changing one or more parameters. For instance, we might be interested in understanding the behavior of all functions of the form $f(x) = a(x - h)^2 + k$ where a, h, and k are numbers that may vary. In the following activity, we investigate a particular example where the value of a single parameter has considerable impact on how the graph appears.

Activity 3.1.4. Consider the family of functions given by $h(x) = x^2 + \cos(kx)$, where k is an arbitrary positive real number.

a. Use a graphing utility to sketch the graph of h for several different k-values, including $k = 1, 3, 5, 10$. Plot $h(x) = x^2 + \cos(3x)$ on the axes provided. What is the smallest value of k at which you think you can see (just by looking at the graph) at least one inflection point on the graph of h?

b. Explain why the graph of h has no inflection points if $k \leq \sqrt{2}$, but infinitely many inflection points if $k > \sqrt{2}$.

c. Explain why, no matter the value of k, h can only have finitely many critical numbers.

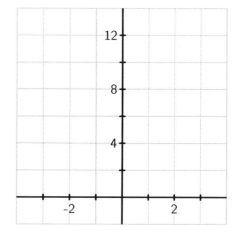

Figure 3.1.12: Axes for plotting $y = h(x)$.

Summary

- The critical numbers of a continuous function f are the values of p for which $f'(p) = 0$ or $f'(p)$ does not exist. These values are important because they identify horizontal tangent lines or corner points on the graph, which are the only possible locations at which a local maximum or local minimum can occur.

- Given a differentiable function f, whenever f' is positive, f is increasing; whenever f' is negative, f is decreasing. The first derivative test tells us that at any point where f changes from increasing to decreasing, f has a local maximum, while conversely at any point where f changes from decreasing to increasing f has a local minimum.

- Given a twice differentiable function f, if we have a horizontal tangent line at $x = p$ and $f''(p)$ is nonzero, then the fact that f'' tells us the concavity of f will determine whether f has a maximum or minimum at $x = p$. In particular, if $f'(p) = 0$ and $f''(p) < 0$, then f is concave down at p and f has a local maximum there, while if $f'(p) = 0$ and $f''(p) > 0$, then f has a local minimum at p. If $f'(p) = 0$ and $f''(p) = 0$, then the second derivative does not tell us whether f has a local extreme at p or not.

Exercises

1. Use a graph below of $f(x) = 3e^{-9x^2}$ to estimate the x-values of any critical points and inflection points of $f(x)$.

critical points (enter as a comma-separated list): $x =$ [   ]

inflection points (enter as a comma-separated list): $x =$ [   ]
Next, use derivatives to find the x-values of any critical points and inflection points exactly.

critical points (enter as a comma-separated list): $x =$ [   ]

inflection points (enter as a comma-separated list): $x =$ [   ]

2. Find the inflection points of $f(x) = 2x^4 + 27x^3 - 21x^2 + 15$. (Give your answers as a comma separated list, e.g., 3,-2.)

inflection points = [   ]

3. The following shows graphs of three functions, A (in black), B (in blue), and C (in green). If these are the graphs of three functions f, f', and f'', identify which is which.

(For each enter A, B or C).

$f =$ [   ]; $f' =$ [   ]; $f'' =$ [   ]

4. This problem concerns a function about which the following information is known:

- f is a differentiable function defined at every real number x
- $f(0) = -1/2$

- $y = f'(x)$ has its graph given at center in Figure 3.1.13

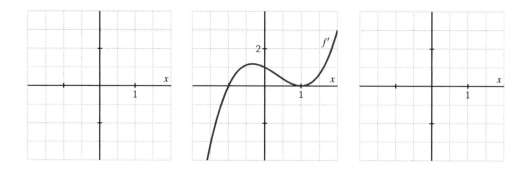

Figure 3.1.13: At center, a graph of $y = f'(x)$; at left, axes for plotting $y = f(x)$; at right, axes for plotting $y = f''(x)$.

a. Construct a first derivative sign chart for f. Clearly identify all critical numbers of f, where f is increasing and decreasing, and where f has local extrema.

b. On the right-hand axes, sketch an approximate graph of $y = f''(x)$.

c. Construct a second derivative sign chart for f. Clearly identify where f is concave up and concave down, as well as all inflection points.

d. On the left-hand axes, sketch a possible graph of $y = f(x)$.

5. Suppose that g is a differentiable function and $g'(2) = 0$. In addition, suppose that on $1 < x < 2$ and $2 < x < 3$ it is known that $g'(x)$ is positive.

a. Does g have a local maximum, local minimum, or neither at $x = 2$? Why?

b. Suppose that $g''(x)$ exists for every x such that $1 < x < 3$. Reasoning graphically, describe the behavior of $g''(x)$ for x-values near 2.

c. Besides being a critical number of g, what is special about the value $x = 2$ in terms of the behavior of the graph of g?

6. Suppose that h is a differentiable function whose first derivative is given by the graph in Figure 3.1.14.

a. How many real number solutions can the equation $h(x) = 0$ have? Why?

b. If $h(x) = 0$ has two distinct real solutions, what can you say about the signs of the two solutions? Why?

c. Assume that $\lim_{x\to\infty} h'(x) = 3$, as appears to be indicated in Figure 3.1.14. How will the graph of $y = h(x)$ appear as $x \to \infty$? Why?

d. Describe the concavity of $y = h(x)$ as fully as you can from the provided information.

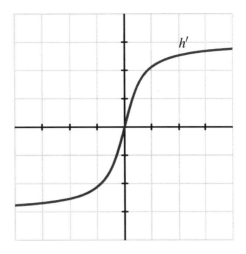

Figure 3.1.14: The graph of $y = h'(x)$.

7. Let p be a function whose second derivative is $p''(x) = (x + 1)(x - 2)e^{-x}$.

a. Construct a second derivative sign chart for p and determine all inflection points of p.

b. Suppose you also know that $x = \frac{\sqrt{5}-1}{2}$ is a critical number of p. Does p have a local minimum, local maximum, or neither at $x = \frac{\sqrt{5}-1}{2}$? Why?

c. If the point $(2, \frac{12}{e^2})$ lies on the graph of $y = p(x)$ and $p'(2) = -\frac{5}{e^2}$, find the equation of the tangent line to $y = p(x)$ at the point where $x = 2$. Does the tangent line lie above the curve, below the curve, or neither at this value? Why?

3.2 Using derivatives to describe families of functions

Motivating Questions

- Given a family of functions that depends on one or more parameters, how does the shape of the graph of a typical function in the family depend on the value of the parameters?

- How can we construct first and second derivative sign charts of functions that depend on one or more parameters while allowing those parameters to remain arbitrary constants?

Mathematicians are often interested in making general observations, say by describing patterns that hold in a large number of cases. For example, think about the Pythagorean Theorem: it doesn't tell us something about a single right triangle, but rather a fact about *every* right triangle, thus providing key information about every member of the right triangle family. In the next part of our studies, we would like to use calculus to help us make general observations about families of functions that depend on one or more parameters. People who use applied mathematics, such as engineers and economists, often encounter the same types of functions in various settings where only small changes to certain constants occur. These constants are called *parameters*.

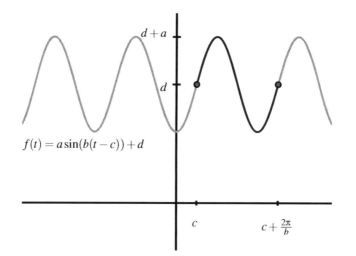

Figure 3.2.1: The graph of $f(t) = a \sin(b(t - c)) + d$ based on parameters a, b, c, and d.

We are already familiar with certain families of functions. For example, $f(t) = a \sin(b(t - c)) + d$ is a stretched and shifted version of the sine function with amplitude a, period $\frac{2\pi}{b}$, phase shift c, and vertical shift d. We understand from experience with trigonometric functions that a affects the size of the oscillation, b the rapidity of oscillation, and c where the oscillation starts, as shown in Figure 3.2.1, while d affects the vertical positioning of the graph.

In addition, there are several basic situations that we already understand completely. For instance, every function of the form $y = mx + b$ is a line with slope m and y-intercept $(0, b)$. Note that the form $y = mx + b$ allows us to consider every possible line by using two parameters (except for vertical lines which are of the form $x = a$). Further, we understand that the value of m affects the line's steepness and whether the line rises or falls from left to right, while the value of b situates the line vertically on the coordinate axes.

For other less familiar families of functions, we would like to use calculus to understand and classify where key behavior occurs: where members of the family are increasing or decreasing, concave up or concave down, where relative extremes occur, and more, all in terms of the parameters involved. To get started, we revisit a common collection of functions to see how calculus confirms things we already know.

> **Preview Activity 3.2.1.** Let a, h, and k be arbitrary real numbers with $a \neq 0$, and let f be the function given by the rule $f(x) = a(x - h)^2 + k$.
>
> a. What familiar type of function is f? What information do you know about f just by looking at its form? (Think about the roles of a, h, and k.)
>
> b. Next we use some calculus to develop familiar ideas from a different perspective. To start, treat a, h, and k as constants and compute $f'(x)$.
>
> c. Find all critical numbers of f. (These will depend on at least one of a, h, and k.)
>
> d. Assume that $a < 0$. Construct a first derivative sign chart for f.
>
> e. Based on the information you've found above, classify the critical values of f as maxima or minima.

3.2.1 Describing families of functions in terms of parameters

Given a family of functions that depends on one or more parameters, our goal is to describe the key characteristics of the overall behavior of each member of the familiy in terms of those parameters. By finding the first and second derivatives and constructing first and second derivative sign charts (each of which may depend on one or more of the parameters), we can often make broad conclusions about how each member of the family will appear. The fundamental steps for this analysis are essentially identical to the work we did in Section 3.1, as we demonstrate through the following example.

Example 3.2.2. Consider the two-parameter family of functions given by $g(x) = axe^{-bx}$, where a and b are positive real numbers. Fully describe the behavior of a typical member of the family in terms of a and b, including the location of all critical numbers, where g is increasing, decreasing, concave up, and concave down, and the long term behavior of g.

Solution. We begin by computing $g'(x)$. By the product rule,

$$g'(x) = ax\frac{d}{dx}\left[e^{-bx}\right] + e^{-bx}\frac{d}{dx}[ax],$$

and thus by applying the chain rule and constant multiple rule, we find that

$$g'(x) = axe^{-bx}(-b) + e^{-bx}(a).$$

To find the critical numbers of g, we solve the equation $g'(x) = 0$. Here, it is especially helpful to factor $g'(x)$. We thus observe that setting the derivative equal to zero implies

$$0 = ae^{-bx}(-bx + 1).$$

Since we are given that $a \neq 0$ and we know that $e^{-bx} \neq 0$ for all values of x, the only way the preceding equation can hold is when $-bx + 1 = 0$. Solving for x, we find that $x = \frac{1}{b}$, and this is therefore the only critical number of g.

Now, recall that we have shown $g'(x) = ae^{-bx}(1 - bx)$ and that the only critical number of g is $x = \frac{1}{b}$. This enables us to construct the first derivative sign chart for g that is shown in Figure 3.2.3.

$$g'(x) = ae^{-bx}(1 - bx)$$

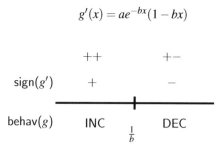

Figure 3.2.3: The first derivative sign chart for $g(x) = axe^{-bx}$.

Note particularly that in $g'(x) = ae^{-bx}(1 - bx)$, the term ae^{-bx} is always positive, so the sign depends on the linear term $(1 - bx)$, which is zero when $x = \frac{1}{b}$. Note that this line has negative slope $(-b)$, so $(1 - bx)$ is positive for $x < \frac{1}{b}$ and negative for $x > \frac{1}{b}$. Hence we can not only conclude that g is always increasing for $x < \frac{1}{b}$ and decreasing for $x > \frac{1}{b}$, but also that g has a global maximum at $(\frac{1}{b}, g(\frac{1}{b}))$ and no local minimum.

We turn next to analyzing the concavity of g. With $g'(x) = -abxe^{-bx} + ae^{-bx}$, we differentiate to find that

$$g''(x) = -abxe^{-bx}(-b) + e^{-bx}(-ab) + ae^{-bx}(-b).$$

Combining like terms and factoring, we now have

$$g''(x) = ab^2xe^{-bx} - 2abe^{-bx} = abe^{-bx}(bx - 2).$$

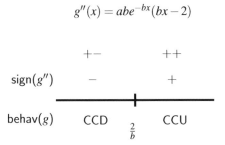

$$g''(x) = abe^{-bx}(bx - 2)$$

Figure 3.2.4: The second derivative sign chart for $g(x) = axe^{-bx}$.

Similar to our work with the first derivative, we observe that abe^{-bx} is always positive, and thus the sign of g'' depends on the sign of $(bx - 2)$, which is zero when $x = \frac{2}{b}$. Since $(bx - 2)$ represents a line with positive slope (b), the value of $(bx - 2)$ is negative for $x < \frac{2}{b}$ and positive for $x > \frac{2}{b}$, and thus the sign chart for g'' is given by the one shown in Figure 3.2.4. Thus, g is concave down for all $x < \frac{2}{b}$ and concave up for all $x > \frac{2}{b}$.

Finally, we analyze the long term behavior of g by considering two limits. First, we note that

$$\lim_{x \to \infty} g(x) = \lim_{x \to \infty} axe^{-bx} = \lim_{x \to \infty} \frac{ax}{e^{bx}}.$$

Since this limit has indeterminate form $\frac{\infty}{\infty}$, we can apply L'Hôpital's Rule and thus find that $\lim_{x \to \infty} g(x) = 0$. In the other direction,

$$\lim_{x \to -\infty} g(x) = \lim_{x \to -\infty} axe^{-bx} = -\infty,$$

since $ax \to -\infty$ and $e^{-bx} \to \infty$ as $x \to -\infty$. Hence, as we move left on its graph, g decreases without bound, while as we move to the right, $g(x) \to 0$.

All of the above information now allows us to produce the graph of a typical member of this family of functions without using a graphing utility (and without choosing particular values for a and b), as shown in Figure 3.2.5.

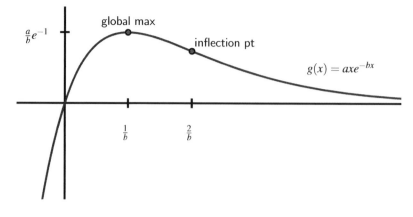

Figure 3.2.5: The graph of $g(x) = axe^{-bx}$.

We note that the value of b controls the horizontal location of the global maximum and the inflection point, as neither depends on a. The value of a affects the vertical stretch of the graph. For example, the global maximum occurs at the point $(\frac{1}{b}, g(\frac{1}{b})) = (\frac{1}{b}, \frac{a}{b}e^{-1})$, so the larger the value of a, the greater the value of the global maximum.

The kind of work we've completed in Example 3.2.2 can often be replicated for other families of functions that depend on parameters. Normally we are most interested in determining all critical numbers, a first derivative sign chart, a second derivative sign chart, and some analysis of the limit of the function as $x \to \infty$. Throughout, we strive to work with the parameters as arbitrary constants. If stuck, it is always possible to experiment with some particular values of the parameters present to reduce the algebraic complexity of our work. The following sequence of activities offers several key examples where we see that the values of different parameters substantially affect the behavior of individual functions within a given family.

> **Activity 3.2.2.** Consider the family of functions defined by $p(x) = x^3 - ax$, where $a \neq 0$ is an arbitrary constant.
>
> a. Find $p'(x)$ and determine the critical numbers of p. How many critical numbers does p have?
>
> b. Construct a first derivative sign chart for p. What can you say about the overall behavior of p if the constant a is positive? Why? What if the constant a is negative? In each case, describe the relative extremes of p.
>
> c. Find $p''(x)$ and construct a second derivative sign chart for p. What does this tell you about the concavity of p? What role does a play in determining the concavity of p?
>
> d. Without using a graphing utility, sketch and label typical graphs of $p(x)$ for the

cases where $a > 0$ and $a < 0$. Label all inflection points and local extrema.

e. Finally, use a graphing utility to test your observations above by entering and plotting the function $p(x) = x^3 - ax$ for at least four different values of a. Write several sentences to describe your overall conclusions about how the behavior of p depends on a.

Activity 3.2.3. Consider the two-parameter family of functions of the form $h(x) = a(1 - e^{-bx})$, where a and b are positive real numbers.

a. Find the first derivative and the critical numbers of h. Use these to construct a first derivative sign chart and determine for which values of x the function h is increasing and decreasing.

b. Find the second derivative and build a second derivative sign chart. For which values of x is a function in this family concave up? concave down?

c. What is the value of $\lim_{x\to\infty} a(1 - e^{-bx})$? $\lim_{x\to-\infty} a(1 - e^{-bx})$?

d. How does changing the value of b affect the shape of the curve?

e. Without using a graphing utility, sketch the graph of a typical member of this family. Write several sentences to describe the overall behavior of a typical function h and how this behavior depends on a and b.

Activity 3.2.4. Let $L(t) = \frac{A}{1+ce^{-kt}}$, where A, c, and k are all positive real numbers.

a. Observe that we can equivalently write $L(t) = A(1+ce^{-kt})^{-1}$. Find $L'(t)$ and explain why L has no critical numbers. Is L always increasing or always decreasing? Why?

b. Given the fact that

$$L''(t) = Ack^2 e^{-kt} \frac{ce^{-kt} - 1}{(1 + ce^{-kt})^3},$$

find all values of t such that $L''(t) = 0$ and hence construct a second derivative sign chart. For which values of t is a function in this family concave up? concave down?

c. What is the value of $\lim_{t\to\infty} \frac{A}{1+ce^{-kt}}$? $\lim_{t\to-\infty} \frac{A}{1+ce^{-kt}}$?

d. Find the value of $L(x)$ at the inflection point found in (b).

e. Without using a graphing utility, sketch the graph of a typical member of this family. Write several sentences to describe the overall behavior of a typical function L and how this behavior depends on A, c, and k number.

f. Explain why it is reasonable to think that the function $L(t)$ models the growth of a population over time in a setting where the largest possible population the surrounding environment can support is A.

Summary

- Given a family of functions that depends on one or more parameters, by investigating how critical numbers and locations where the second derivative is zero depend on the values of these parameters, we can often accurately describe the shape of the function in terms of the parameters.

- In particular, just as we can created first and second derivative sign charts for a single function, we often can do so for entire families of functions where critical numbers and possible inflection points depend on arbitrary constants. These sign charts then reveal where members of the family are increasing or decreasing, concave up or concave down, and help us to identify relative extremes and inflection points.

Exercises

1. For some positive constant C, a patient's temperature change, T, due to a dose, D, of a drug is given by $T = \left(\frac{C}{2} - \frac{D}{3}\right)D^2$.
What dosage maximizes the temperature change?

$D = $ ⬚

The sensitivity of the body to the drug is defined as dT/dD. What dosage maximizes sensitivity?

$D = $ ⬚

2. The figure below gives the behavior of the derivative of $g(x)$ on $-2 \le x \le 2$.

Graph of $g'(x)$ (not $g(x)$)
Sketch a graph of $g(x)$ and use your sketch to answer the following questions.
A. Where does the graph of $g(x)$ have inflection points?

$x = $ ⬚

Enter your answer as a comma-separated list of values, or enter none *if there are none.*
B. Where are the global maxima and minima of g on $[-2, 2]$?

minimum at $x = $ ⬚

maximum at $x = $ ⬚

C. If $g(-2) = -8$, what are possible values for $g(0)$?

$g(0)$ is in ⬚

(Enter your answer as an interval, or union of intervals, giving the possible values. Thus if you know $-1 < g(0) \le 3$, *enter (-1,3]. Enter* infinity *for* ∞, *the interval* [2,2] *to indicate a single point).*
How is the value of $g(2)$ related to the value of $g(0)$?

$g(2)$ [   ] $g(0)$

(Enter the appropriate mathematical equality or inequality, =, <, >, etc.)

3. Consider the one-parameter family of functions given by $p(x) = x^3 - ax^2$, where $a > 0$.

a. Sketch a plot of a typical member of the family, using the fact that each is a cubic polynomial with a repeated zero at $x = 0$ and another zero at $x = a$.

b. Find all critical numbers of p.

c. Compute p'' and find all values for which $p''(x) = 0$. Hence construct a second derivative sign chart for p.

d. Describe how the location of the critical numbers and the inflection point of p change as a changes. That is, if the value of a is increased, what happens to the critical numbers and inflection point?

4. Let $q(x) = \frac{e^{-x}}{x-c}$ be a one-parameter family of functions where $c > 0$.

a. Explain why q has a vertical asymptote at $x = c$.

b. Determine $\lim_{x \to \infty} q(x)$ and $\lim_{x \to -\infty} q(x)$.

c. Compute $q'(x)$ and find all critical numbers of q.

d. Construct a first derivative sign chart for q and determine whether each critical number leads to a local minimum, local maximum, or neither for the function q.

e. Sketch a typical member of this family of functions with important behaviors clearly labeled.

5. Let $E(x) = e^{-\frac{(x-m)^2}{2s^2}}$, where m is any real number and s is a positive real number.

a. Compute $E'(x)$ and hence find all critical numbers of E.

b. Construct a first derivative sign chart for E and classify each critical number of the function as a local minimum, local maximum, or neither.

c. It can be shown that $E''(x)$ is given by the formula

$$E''(x) = e^{-\frac{(x-m)^2}{2s^2}} \left(\frac{(x-m)^2 - s^2}{s^4} \right).$$

Find all values of x for which $E''(x) = 0$.

d. Determine $\lim_{x \to \infty} E(x)$ and $\lim_{x \to -\infty} E(x)$.

e. Construct a labeled graph of a typical function E that clearly shows how important points on the graph of $y = E(x)$ depend on m and s.

3.3 Global Optimization

Motivating Questions

- What are the differences between finding relative extreme values and global extreme values of a function?

- How is the process of finding the global maximum or minimum of a function over the function's entire domain different from determining the global maximum or minimum on a restricted domain?

- For a function that is guaranteed to have both a global maximum and global minimum on a closed, bounded interval, what are the possible points at which these extreme values occur?

We have seen that we can use the first derivative of a function to determine where the function is increasing or decreasing, and the second derivative to know where the function is concave up or concave down. Each of these approaches provides us with key information that helps us determine the overall shape and behavior of the graph, as well as whether the function has a relative minimum or relative maximum at a given critical number. Remember that the difference between a relative maximum and a global maximum is that there is a relative maximum of f at $x = p$ if $f(p) \geq f(x)$ for all x near p, while there is a global maximum at p if $f(p) \geq f(x)$ for all x in the domain of f.

For instance, in Figure 3.3.1, we see a function f that has a global maximum at $x = c$ and a relative maximum at $x = a$, since $f(c)$ is greater than $f(x)$ for every value of x, while $f(a)$ is only greater than the value of $f(x)$ for x near a. Since the function appears to decrease without bound, f has no global minimum, though clearly f has a relative minimum at $x = b$.

Our emphasis in this section is on finding the global extreme values of a function (if they exist). In so doing, we will either be interested in the behavior of the function over its entire domain or on some restricted portion. The former situation is familiar and similar to work that we did in the two preceding sections of the text. We explore this through a particular example in the following preview activity.

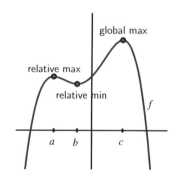

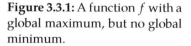

Figure 3.3.1: A function f with a global maximum, but no global minimum.

Preview Activity 3.3.1. Let $f(x) = 2 + \frac{3}{1+(x+1)^2}$.

a. Determine all of the critical numbers of f.

b. Construct a first derivative sign chart for f and thus determine all intervals on which f is increasing or decreasing.

c. Does f have a global maximum? If so, why, and what is its value and where is the maximum attained? If not, explain why.

d. Determine $\lim_{x \to \infty} f(x)$ and $\lim_{x \to -\infty} f(x)$.

e. Explain why $f(x) > 2$ for every value of x.

f. Does f have a global minimum? If so, why, and what is its value and where is the minimum attained? If not, explain why.

3.3.1 Global Optimization

For the functions in Figure 3.3.1 and Preview Activity 3.3.1, we were interested in finding the global minimum and global maximum on the entire domain, which turned out to be $(-\infty, \infty)$ for each. At other times, our perspective on a function might be more focused due to some restriction on its domain. For example, rather than considering $f(x) = 2 + \frac{3}{1+(x+1)^2}$ for every value of x, perhaps instead we are only interested in those x for which $0 \le x \le 4$, and we would like to know which values of x in the interval $[0, 4]$ produce the largest possible and smallest possible values of f. We are accustomed to critical numbers playing a key role in determining the location of extreme values of a function; now, by restricting the domain to an interval, it makes sense that the endpoints of the interval will also be important to consider, as we see in the following activity. When limiting ourselves to a particular interval, we will often refer to the *absolute* maximum or minimum value, rather than the *global* maximum or minimum.

Activity 3.3.2. Let $g(x) = \frac{1}{3}x^3 - 2x + 2$.

a. Find all critical numbers of g that lie in the interval $-2 \le x \le 3$.

b. Use a graphing utility to construct the graph of g on the interval $-2 \le x \le 3$.

c. From the graph, determine the x-values at which the absolute minimum and absolute maximum of g occur on the interval $[-2, 3]$.

d. How do your answers change if we instead consider the interval $-2 \le x \le 2$?

e. What if we instead consider the interval $-2 \le x \le 1$?

In Activity 3.3.2, we saw how the absolute maximum and absolute minimum of a function on a closed, bounded interval $[a, b]$, depend not only on the critical numbers of the function, but also on the selected values of a and b. These observations demonstrate several important facts that hold much more generally. First, we state an important result called the Extreme

Value Theorem.

> ### The Extreme Value Theorem
>
> If f is a continuous function on a closed interval $[a, b]$, then f attains both an absolute minimum and absolute maximum on $[a, b]$. That is, for some value x_m such that $a \leq x_m \leq b$, it follows that $f(x_m) \leq f(x)$ for all x in $[a, b]$. Similarly, there is a value x_M in $[a, b]$ such that $f(x_M) \geq f(x)$ for all x in $[a, b]$. Letting $m = f(x_m)$ and $M = f(x_M)$, it follows that $m \leq f(x) \leq M$ for all x in $[a, b]$.

The Extreme Value Theorem tells us that provided a function is continuous, on any closed interval $[a, b]$ the function has to achieve both an absolute minimum and an absolute maximum. Note, however, that this result does not tell us where these extreme values occur, but rather only that they must exist. As seen in the examples of Activity 3.3.2, it is apparent that the only possible locations for relative extremes are either the endpoints of the interval or at a critical number (the latter being where a relative minimum or maximum could occur, which is a potential location for an absolute extreme).

Note 3.3.2. Thus, we have the following approach to finding the absolute maximum and minimum of a continuous function f on the interval $[a, b]$:

- find all critical numbers of f that lie in the interval;

- evaluate the function f at each critical number in the interval and at each endpoint of the interval;

- from among the noted function values, the smallest is the absolute minimum of f on the interval, while the largest is the absolute maximum.

> **Activity 3.3.3.** Find the *exact* absolute maximum and minimum of each function on the stated interval.
>
> a. $h(x) = xe^{-x}$, $[0, 3]$
>
> b. $p(t) = \sin(t) + \cos(t)$, $[-\frac{\pi}{2}, \frac{\pi}{2}]$
>
> c. $q(x) = \frac{x^2}{x-2}$, $[3, 7]$
>
> d. $f(x) = 4 - e^{-(x-2)^2}$, $(-\infty, \infty)$
>
> e. $h(x) = xe^{-ax}$, $[0, \frac{2}{a}]$ $(a > 0)$
>
> f. $f(x) = b - e^{-(x-a)^2}$, $(-\infty, \infty)$, $a, b > 0$

One of the big lessons in finding absolute extreme values is the realization that the interval we choose has nearly the same impact on the problem as the function under consideration. Consider, for instance, the function pictured in Figure 3.3.3.

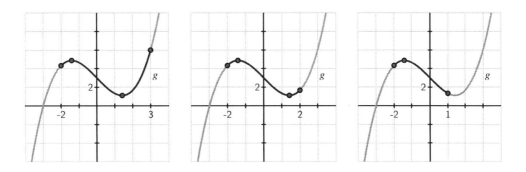

Figure 3.3.3: A function g considered on three different intervals.

In sequence, from left to right, as we see the interval under consideration change from $[-2, 3]$ to $[-2, 2]$ to $[-2, 1]$, we move from having two critical numbers in the interval with the absolute minimum at one critical number and the absolute maximum at the right endpoint, to still having both critical numbers in the interval but then with the absolute minimum and maximum at the two critical numbers, to finally having just one critical number in the interval with the absolute maximum at one critical number and the absolute minimum at one endpoint. It is particularly essential to always remember to only consider the critical numbers that lie within the interval.

3.3.2 Moving toward applications

In Section 3.4, we will focus almost exclusively on applied optimization problems: problems where we seek to find the absolute maximum or minimum value of a function that represents some physical situation. We conclude this current section with an example of one such problem because it highlights the role that a closed, bounded domain can play in finding absolute extrema. In addition, these problems often involve considerable preliminary work to develop the function which is to be optimized, and this example demonstrates that process.

Example 3.3.4. A 20 cm piece of wire is cut into two pieces. One piece is used to form a square and the other an equilateral triangle. How should the wire be cut to maximize the total area enclosed by the square and triangle? to minimize the area?

Solution. We begin by constructing a picture that exemplifies the given situation. The primary variable in the problem is where we decide to cut the wire. We thus label that point x, and note that the remaining portion of the wire then has length $20 - x$

As shown in Figure 3.3.5, we see that the x cm of the wire that are used to form the equilateral triangle result in a triangle with three sides of length $\frac{x}{3}$. For the remaining $20-x$ cm of wire, the square that results will have each side of length $\frac{20-x}{4}$.

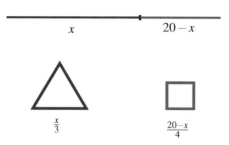

Figure 3.3.5: A 20 cm piece of wire cut into two pieces, one of which forms an equilateral triangle, the other which yields a square.

At this point, we note that there are obvious restrictions on x: in particular, $0 \leq x \leq 20$. In the extreme cases, all of the wire is being used to make just one figure. For instance, if $x = 0$, then all 20 cm of wire are used to make a square that is 5×5.

Now, our overall goal is to find the absolute minimum and absolute maximum areas that can be enclosed. We note that the area of the triangle is $A_\triangle = \frac{1}{2}bh = \frac{1}{2} \cdot \frac{x}{3} \cdot \frac{x\sqrt{3}}{6}$, since the height of an equilateral triangle is $\sqrt{3}$ times half the length of the base. Further, the area of the square is $A_\square = s^2 = \left(\frac{20-x}{4}\right)^2$. Therefore, the total area function is

$$A(x) = \frac{\sqrt{3}x^2}{36} + \left(\frac{20-x}{4}\right)^2.$$

Again, note that we are only considering this function on the restricted domain $[0, 20]$ and we seek its absolute minimum and absolute maximum.

Differentiating $A(x)$, we have

$$A'(x) = \frac{\sqrt{3}x}{18} + 2\left(\frac{20-x}{4}\right)\left(-\frac{1}{4}\right) = \frac{\sqrt{3}}{18}x + \frac{1}{8}x - \frac{5}{2}.$$

Setting $A'(x) = 0$, it follows that $x = \frac{180}{4\sqrt{3}+9} \approx 11.3007$ is the only critical number of A, and we note that this lies within the interval $[0, 20]$.

Evaluating A at the critical number and endpoints, we see that

- $A\left(\frac{180}{4\sqrt{3}+9}\right) = \frac{\sqrt{3}(\frac{180}{4\sqrt{3}+9})^2}{4} + \left(\frac{20-\frac{180}{4\sqrt{3}+9}}{4}\right)^2 \approx 10.8741$

- $A(0) = 25$

- $A(20) = \frac{\sqrt{3}}{36}(400) = \frac{100}{9}\sqrt{3} \approx 19.2450$

Thus, the absolute minimum occurs when $x \approx 11.3007$ and results in the minimum area of approximately 10.8741 square centimeters, while the absolute maximum occurs when we invest all of the wire in the square (and none in the triangle), resulting in 25 square centimeters of area. These results are confirmed by a plot of $y = A(x)$ on the interval $[0, 20]$, as shown in Figure 3.3.6.

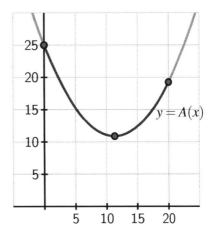

Figure 3.3.6: A plot of the area function from Example 3.3.4.

Activity 3.3.4. A piece of cardboard that is 10×15 (each measured in inches) is being made into a box without a top. To do so, squares are cut from each corner of the box and the remaining sides are folded up. If the box needs to be at least 1 inch deep and no more than 3 inches deep, what is the maximum possible volume of the box? what is the minimum volume? Justify your answers using calculus.

a. Draw a labeled diagram that shows the given information. What variable should we introduce to represent the choice we make in creating the box? Label the diagram appropriately with the variable, and write a sentence to state what the variable represents.

b. Determine a formula for the function V (that depends on the variable in (a)) that tells us the volume of the box.

c. What is the domain of the function V? That is, what values of x make sense for input? Are there additional restrictions provided in the problem?

d. Determine all critical numbers of the function V.

e. Evaluate V at each of the endpoints of the domain and at any critical numbers that lie in the domain.

f. What is the maximum possible volume of the box? the minimum?

The approaches shown in Example 3.3.4 and experienced in Activity 3.3.4 include standard steps that we undertake in almost every applied optimization problem: we draw a picture to demonstrate the situation, introduce one or more variables to represent quantities that are changing, work to find a function that models the quantity to be optimized, and then decide an appropriate domain for that function. Once that work is done, we are in the familiar situation of finding the absolute minimum and maximum of a function over a particular domain, at which time we apply the calculus ideas that we have been studying to this point in Chapter 3.

Summary

- To find relative extreme values of a function, we normally use a first derivative sign chart and classify all of the function's critical numbers. If instead we are interested in absolute extreme values, we first decide whether we are considering the entire domain of the function or a particular interval.

- In the case of finding global extremes over the function's entire domain, we again use a first or second derivative sign chart in an effort to make overall conclusions about whether or not the function can have a absolute maximum or minimum. If we are working to find absolute extremes on a restricted interval, then we first identify all critical numbers of the function that lie in the interval.

- For a continuous function on a closed, bounded interval, the only possible points at which absolute extreme values occur are the critical numbers and the endpoints. Thus, to find said absolute extremes, we simply evaluate the function at each endpoint and each critical number in the interval, and then we compare the results to decide which is largest (the absolute maximum) and which is smallest (the absolute minimum).

Exercises

1. Based on the given information about each function, decide whether the function has global maximum, a global minimum, neither, both, or that it is not possible to say without more information. Assume that each function is twice differentiable and defined for all real numbers, unless noted otherwise. In each case, write one sentence to explain your conclusion.

a. f is a function such that $f''(x) < 0$ for every x.

b. g is a function with two critical numbers a and b (where $a < b$), and $g'(x) < 0$ for $x < a$, $g'(x) < 0$ for $a < x < b$, and $g'(x) > 0$ for $x > b$.

c. h is a function with two critical numbers a and b (where $a < b$), and $h'(x) < 0$ for $x < a$, $h'(x) > 0$ for $a < x < b$, and $h'(x) < 0$ for $x > b$. In addition, $\lim_{x \to \infty} h(x) = 0$ and $\lim_{x \to -\infty} h(x) = 0$.

d. p is a function differentiable everywhere except at $x = a$ and $p''(x) > 0$ for $x < a$ and $p''(x) < 0$ for $x > a$.

2. For each family of functions that depends on one or more parameters, determine the function's absolute maximum and absolute minimum on the given interval.

a. $p(x) = x^3 - a^2x$, $[0, a]$ $(a > 0)$

b. $r(x) = axe^{-bx}$, $[\frac{1}{2b}, b]$ $(a, b > 0)$

c. $w(x) = a(1 - e^{-bx})$, $[b, 3b]$ $(a, b > 0)$

d. $s(x) = \sin(kx)$, $[\frac{\pi}{3k}, \frac{5\pi}{6k}]$

3. For each of the functions described below (each continuous on $[a, b]$), state the location of the function's absolute maximum and absolute minimum on the interval $[a, b]$, or say there is not enough information provided to make a conclusion. Assume that any critical numbers mentioned in the problem statement represent all of the critical numbers the function has in $[a, b]$. In each case, write one sentence to explain your answer.

a. $f'(x) \leq 0$ for all x in $[a, b]$

b. g has a critical number at c such that $a < c < b$ and $g'(x) > 0$ for $x < c$ and $g'(x) < 0$ for $x > c$

c. $h(a) = h(b)$ and $h''(x) < 0$ for all x in $[a, b]$

d. $p(a) > 0$, $p(b) < 0$, and for the critical number c such that $a < c < b$, $p'(x) < 0$ for $x < c$ and $p'(x) > 0$ for $x > c$

4. Let $s(t) = 3\sin(2(t - \frac{\pi}{6})) + 5$. Find the exact absolute maximum and minimum of s on the provided intervals by testing the endpoints and finding and evaluating all relevant critical numbers of s.

a. $[\frac{\pi}{6}, \frac{7\pi}{6}]$

c. $[0, 2\pi]$

b. $[0, \frac{\pi}{2}]$

d. $[\frac{\pi}{3}, \frac{5\pi}{6}]$

3.4 Applied Optimization

Motivating Questions

- In a setting where a situation is described for which optimal parameters are sought, how do we develop a function that models the situation and use calculus to find the desired maximum or minimum?

Near the conclusion of Section 3.3, we considered two examples of optimization problems where determining the function to be optimized was part of a broader question. In Example 3.3.4, we sought to use a single piece of wire to build two geometric figures (an equilateral triangle and square) and to understand how various choices for how to cut the wire led to different values of the area enclosed. One of our conclusions was that in order to maximize the total combined area enclosed by the triangle and square, all of the wire must be used to make a square. In the subsequent Activity 3.3.4, we investigated how the volume of a box constructed from a piece of cardboard by removing squares from each corner and folding up the sides depends on the size of the squares removed.

Both of these problems exemplify situations where there is not a function explicitly provided to optimize. Rather, we first worked to understand the given information in the problem, drawing a figure and introducing variables, and then sought to develop a formula for a function that models the quantity (area or volume, in the two examples, respectively) to be optimized. Once the function was established, we then considered what domain was appropriate on which to pursue the desired absolute minimum or maximum (or both). At this point in the problem, we are finally ready to apply the ideas of calculus to determine and justify the absolute minimum or maximum. Thus, what is primarily different about problems of this type is that the problem-solver must do considerable work to introduce variables and develop the correct function and domain to represent the described situation.

Throughout what follows in the current section, the primary emphasis is on the reader solving problems. Initially, some substantial guidance is provided, with the problems progressing to require greater independence as we move along.

Preview Activity 3.4.1. According to U.S. postal regulations, the girth plus the length of a parcel sent by mail may not exceed 108 inches, where by "girth" we mean the perimeter of the smallest end. What is the largest possible volume of a rectangular parcel with a square end that can be sent by mail? What are the dimensions of the package of largest volume?

    a. Let x represent the length of one side of the square end and y the length of the longer side. Label these quantities appropriately on the image shown in Figure 3.4.1.

    b. What is the quantity to be optimized in this problem? Find a formula for this quantity in terms of x and y.

c. The problem statement tells us that the parcel's girth plus length may not exceed 108 inches. In order to maximize volume, we assume that we will actually need the girth plus length to equal 108 inches. What equation does this produce involving x and y?

d. Solve the equation you found in (c) for one of x or y (whichever is easier).

e. Now use your work in (b) and (d) to determine a formula for the volume of the parcel so that this formula is a function of a single variable.

f. Over what domain should we consider this function? Note that both x and y must be positive; how does the constraint that girth plus length is 108 inches produce intervals of possible values for x and y?

g. Find the absolute maximum of the volume of the parcel on the domain you established in (f) and hence also determine the dimensions of the box of greatest volume. Justify that you've found the maximum using calculus.

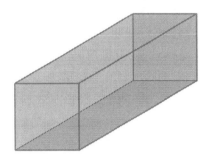

Figure 3.4.1: A rectangular parcel with a square end.

3.4.1 More applied optimization problems

Many of the steps in Preview Activity 3.4.1 are ones that we will execute in any applied optimization problem. We briefly summarize those here to provide an overview of our approach in subsequent questions.

Note 3.4.2.

- Draw a picture and introduce variables. It is essential to first understand what quantities are allowed to vary in the problem and then to represent those values with variables. Constructing a figure with the variables labeled is almost always an essential first step. Sometimes drawing several diagrams can be especially helpful to get a sense of the situation. A nice example of this can be seen at http://gvsu.edu/s/99, where the choice of where to bend a piece of wire into the shape of a rectangle determines both the rectangle's shape and area.

- Identify the quantity to be optimized as well as any key relationships among the variable quantities. Essentially this step involves writing equations that involve the variables that have been introduced: one to represent the quantity whose minimum or maximum is sought, and possibly others that show how multiple variables in the problem may be interrelated.

- Determine a function of a single variable that models the quantity to be optimized; this may involve using other relationships among variables to eliminate one or more variables in the function formula. For example, in Preview Activity 3.4.1, we initially found that $V = x^2y$, but then the additional relationship that $4x + y = 108$ (girth plus length equals 108 inches) allows us to relate x and y and thus observe equivalently that $y = 108 - 4x$. Substituting for y in the volume equation yields $V(x) = x^2(108 - 4x)$, and thus we have written the volume as a function of the single variable x.

- Decide the domain on which to consider the function being optimized. Often the physical constraints of the problem will limit the possible values that the independent variable can take on. Thinking back to the diagram describing the overall situation and any relationships among variables in the problem often helps identify the smallest and largest values of the input variable.

- Use calculus to identify the absolute maximum and/or minimum of the quantity being optimized. This always involves finding the critical numbers of the function first. Then, depending on the domain, we either construct a first derivative sign chart (for an open or unbounded interval) or evaluate the function at the endpoints and critical numbers (for a closed, bounded interval), using ideas we've studied so far in Chapter 3.

- Finally, we make certain we have answered the question: does the question seek the absolute maximum of a quantity, or the values of the variables that produce the maximum? That is, finding the absolute maximum volume of a parcel is different from finding the dimensions of the parcel that produce the maximum.

Activity 3.4.2. A soup can in the shape of a right circular cylinder is to be made from two materials. The material for the side of the can costs $0.015 per square inch and the material for the lids costs $0.027 per square inch. Suppose that we desire to construct a can that has a volume of 16 cubic inches. What dimensions minimize the cost of the can?

   a. Draw a picture of the can and label its dimensions with appropriate variables.

   b. Use your variables to determine expressions for the volume, surface area, and cost of the can.

   c. Determine the total cost function as a function of a single variable. What is the domain on which you should consider this function?

   d. Find the absolute minimum cost and the dimensions that produce this value.

Familiarity with common geometric formulas is particularly helpful in problems like the one in Activity 3.4.2. Sometimes those involve perimeter, area, volume, or surface area. At other times, the constraints of a problem introduce right triangles (where the Pythagorean Theorem applies) or other functions whose formulas provide relationships among variables present.

> **Activity 3.4.3.** A hiker starting at a point P on a straight road walks east towards point Q, which is on the road and 3 kilometers from point P.
>
> Two kilometers due north of point Q is a cabin. The hiker will walk down the road for a while, at a pace of 8 kilometers per hour. At some point Z between P and Q, the hiker leaves the road and makes a straight line towards the cabin through the woods, hiking at a pace of 3 kph, as pictured in Figure 3.4.3. In order to minimize the time to go from P to Z to the cabin, where should the hiker turn into the forest?

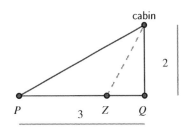

Figure 3.4.3: A hiker walks from P to Z to the cabin, as pictured.

In more geometric problems, we often use curves or functions to provide natural constraints. For instance, we could investigate which isosceles triangle that circumscribes a unit circle has the smallest area, which you can explore for yourself at http://gvsu.edu/s/9b. Or similarly, for a region bounded by a parabola, we might seek the rectangle of largest area that fits beneath the curve, as shown at http://gvsu.edu/s/9c. The next activity is similar to the latter problem.

> **Activity 3.4.4.** Consider the region in the x-y plane that is bounded by the x-axis and the function $f(x) = 25 - x^2$. Construct a rectangle whose base lies on the x-axis and is centered at the origin, and whose sides extend vertically until they intersect the curve $y = 25 - x^2$. Which such rectangle has the maximum possible area? Which such rectangle has the greatest perimeter? Which has the greatest combined perimeter and area? (Challenge: answer the same questions in terms of positive parameters a and b for the function $f(x) = b - ax^2$.)

> **Activity 3.4.5.** A trough is being constructed by bending a 4×24 (measured in feet) rectangular piece of sheet metal.

Two symmetric folds 2 feet apart will be made parallel to the longest side of the rectangle so that the trough has cross-sections in the shape of a trapezoid, as pictured in Figure 3.4.4. At what angle should the folds be made to produce the trough of maximum volume?

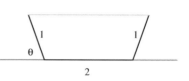

Figure 3.4.4: A cross-section of the trough formed by folding to an angle of θ.

Summary

- While there is no single algorithm that works in every situation where optimization is used, in most of the problems we consider, the following steps are helpful: draw a picture and introduce variables; identify the quantity to be optimized and find relationships among the variables; determine a function of a single variable that models the quantity to be optimized; decide the domain on which to consider the function being optimized; use calculus to identify the absolute maximum and/or minimum of the quantity being optimized.

Exercises

1. An open box is to be made out of a 10-inch by 14-inch piece of cardboard by cutting out squares of equal size from the four corners and bending up the sides. Find the dimensions of the resulting box that has the largest volume.

Dimensions of the bottom of the box: [＿＿＿＿] x [＿＿＿＿]

Height of the box: [＿＿＿＿]

2. A rectangular storage container with an open top is to have a volume of 22 cubic meters. The length of its base is twice the width. Material for the base costs 14 dollars per square meter. Material for the sides costs 8 dollars per square meter. Find the cost of materials for the cheapest such container.

Total cost = [＿＿＿＿] (Round to the nearest penny and include monetary units. For example, if your answer is 1.095, enter $1.10 including the dollar sign and second decimal place.)

3. A cattle rancher wants to enclose a rectangular area and then divide it into four pens with fencing parallel to one side of the rectangle (see the figure below). There are 520 feet of fencing available to complete the job. What is the largest possible total area of the four pens?

Largest area = [   ] (include units)

4. The top and bottom margins of a poster are 8 cm and the side margins are each 2 cm. If the area of printed material on the poster is fixed at 386 square centimeters, find the dimensions of the poster with the smallest area.

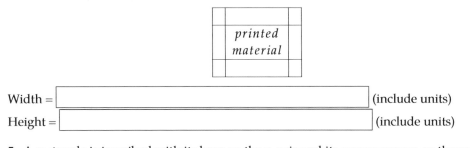

Width = [   ] (include units)

Height = [   ] (include units)

5. A rectangle is inscribed with its base on the *x*-axis and its upper corners on the parabola $y = 12 - x^2$. What are the dimensions of such a rectangle with the greatest possible area?

Width = [   ]

Height = [   ]

6. A rectangular box with a square bottom and closed top is to be made from two materials. The material for the side costs $1.50 per square foot and the material for the bottom costs $3.00 per square foot. If you are willing to spend $15 on the box, what is the largest volume it can contain? Justify your answer completely using calculus.

7. A farmer wants to start raising cows, horses, goats, and sheep, and desires to have a rectangular pasture for the animals to graze in. However, no two different kinds of animals can graze together. In order to minimize the amount of fencing she will need, she has decided to enclose a large rectangular area and then divide it into four equally sized pens by adding three segments of fence inside the large rectangle that are parallel to two existing sides. She has decided to purchase 7500 ft of fencing. What is the maximum possible area that each of the four pens will enclose?

8. Two vertical poles of heights 60 ft and 80 ft stand on level ground, with their bases 100 ft apart. A cable that is stretched from the top of one pole to some point on the ground between the poles, and then to the top of the other pole. What is the minimum possible length of cable required? Justify your answer completely using calculus.

9. A company is designing propane tanks that are cylindrical with hemispherical ends. Assume that the company wants tanks that will hold 1000 cubic feet of gas, and that the ends are more expensive to make, costing $5 per square foot, while the cylindrical barrel between the ends costs $2 per square foot. Use calculus to determine the minimum cost to construct such a tank.

3.5 Related Rates

Motivating Questions

- If two quantities that are related, such as the radius and volume of a spherical balloon, are both changing as implicit functions of time, how are their rates of change related? That is, how does the relationship between the values of the quantities affect the relationship between their respective derivatives with respect to time?

In most of our applications of the derivative so far, we have worked in settings where one quantity (often called y) depends explicitly on another (say x), and in some way we have been interested in the instantaneous rate at which y changes with respect to x, leading us to compute $\frac{dy}{dx}$. These settings emphasize how the derivative enables us to quantify how the quantity y is changing as x changes at a given x-value.

We are next going to consider situations where multiple quantities are related to one another and changing, but where each quantity can be considered an implicit function of the variable t, which represents time. Through knowing how the quantities are related, we will be interested in determining how their respective rates of change with respect to time are related. For example, suppose that air is being pumped into a spherical balloon in such a way that its volume increases at a constant rate of 20 cubic inches per second. It makes sense that since the balloon's volume and radius are related, by knowing how fast the volume is changing, we ought to be able to relate this rate to how fast the radius is changing. More specifically, can we find how fast the radius of the balloon is increasing at the moment the balloon's diameter is 12 inches?

The following preview activity leads you through the steps to answer this question.

> **Preview Activity 3.5.1.** A spherical balloon is being inflated at a constant rate of 20 cubic inches per second. How fast is the radius of the balloon changing at the instant the balloon's diameter is 12 inches? Is the radius changing more rapidly when $d = 12$ or when $d = 16$? Why?
>
>   a. Draw several spheres with different radii, and observe that as volume changes, the radius, diameter, and surface area of the balloon also change.
>
>   b. Recall that the volume of a sphere of radius r is $V = \frac{4}{3}\pi r^3$. Note well that in the setting of this problem, *both V and r* are changing as time t changes, and thus both V and r may be viewed as implicit functions of t, with respective derivatives $\frac{dV}{dt}$ and $\frac{dr}{dt}$. Differentiate both sides of the equation $V = \frac{4}{3}\pi r^3$ with respect to t (using the chain rule on the right) to find a formula for $\frac{dV}{dt}$ that depends on both r and $\frac{dr}{dt}$.
>
>   c. At this point in the problem, by differentiating we have "related the rates" of change of V and r. Recall that we are given in the problem that the balloon is

being inflated at a constant *rate* of 20 cubic inches per second. Is this rate the value of $\frac{dr}{dt}$ or $\frac{dV}{dt}$? Why?

d. From part (c), we know the value of $\frac{dV}{dt}$ at every value of t. Next, observe that when the diameter of the balloon is 12, we know the value of the radius. In the equation $\frac{dV}{dt} = 4\pi r^2 \frac{dr}{dt}$, substitute these values for the relevant quantities and solve for the remaining unknown quantity, which is $\frac{dr}{dt}$. How fast is the radius changing at the instant $d = 12$?

e. How is the situation different when $d = 16$? When is the radius changing more rapidly, when $d = 12$ or when $d = 16$?

3.5.1 Related Rates Problems

In problems where two or more quantities can be related to one another, and all of the variables involved can be viewed as implicit functions of time, t, we are often interested in how the rates of change of the individual quantities with respect to time are themselves related; we call these *related rates* problems. Often these problems involve identifying one or more key underlying geometric relationships to relate the variables involved. Once we have an equation establishing the fundamental relationship among variables, we differentiate implicitly with respect to time to find connections among the rates of change.

For example, consider the situation where sand is being dumped by a conveyor belt on a pile so that the sand forms a right circular cone, as pictured in Figure 3.5.1.

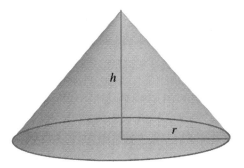

Figure 3.5.1: A conical pile of sand.

As sand falls from the conveyor belt onto the top of the pile, obviously several features of the sand pile will change: the volume of the pile will grow, the height will increase, and the radius will get bigger, too. All of these quantities are related to one another, and the rate at which each is changing is related to the rate at which sand falls from the conveyor.

The first key steps in any related rates problem involve identifying which variables are changing and how they are related. In the current problem involving a conical pile of sand, we observe that the radius and height of the pile are related to the volume of the pile by the standard equation for the volume of a cone,

$$V = \frac{1}{3}\pi r^2 h.$$

Viewing each of V, r, and h as functions of t, we can differentiate implicitly to determine an equation that relates their respective rates of change. Taking the derivative of each side of the equation with respect to t,

$$\frac{d}{dt}[V] = \frac{d}{dt}\left[\frac{1}{3}\pi r^2 h\right].$$

On the left, $\frac{d}{dt}[V]$ is simply $\frac{dV}{dt}$. On the right, the situation is more complicated, as both r and h are implicit functions of t, hence we have to use the product and chain rules. Doing so, we find that

$$\frac{dV}{dt} = \frac{d}{dt}\left[\frac{1}{3}\pi r^2 h\right]$$
$$= \frac{1}{3}\pi r^2 \frac{d}{dt}[h] + \frac{1}{3}\pi h \frac{d}{dt}[r^2]$$
$$= \frac{1}{3}\pi r^2 \frac{dh}{dt} + \frac{1}{3}\pi h 2r \frac{dr}{dt}$$

Note particularly how we are using ideas from Section 2.7 on implicit differentiation. There we found that when y is an implicit function of x, $\frac{d}{dx}[y^2] = 2y\frac{dy}{dx}$. The exact same thing is occurring here when we compute $\frac{d}{dt}[r^2] = 2r\frac{dr}{dt}$.

With our arrival at the equation

$$\frac{dV}{dt} = \frac{1}{3}\pi r^2 \frac{dh}{dt} + \frac{2}{3}\pi rh \frac{dr}{dt},$$

we have now related the rates of change of V, h, and r. If we are given sufficient information, we may then find the value of one or more of these rates of change at one or more points in time. Say, for instance, that we know the following: (a) sand falls from the conveyor in such a way that the height of the pile is always half the radius, and (b) sand falls from the conveyor belt at a constant rate of 10 cubic feet per minute. With this information given, we can answer questions such as: how fast is the height of the sandpile changing at the moment the radius is 4 feet?

The information that the height is always half the radius tells us that for all values of t, $h = \frac{1}{2}r$. Differentiating with respect to t, it follows that $\frac{dh}{dt} = \frac{1}{2}\frac{dr}{dt}$. These relationships enable us to relate $\frac{dV}{dt}$ exclusively to just one of r or h. Substituting the expressions involving r and $\frac{dr}{dt}$ for h and $\frac{dh}{dt}$, we now have that

$$\frac{dV}{dt} = \frac{1}{3}\pi r^2 \cdot \frac{1}{2}\frac{dr}{dt} + \frac{2}{3}\pi r \cdot \frac{1}{2}r \cdot \frac{dr}{dt}. \tag{3.5.1}$$

Since sand falls from the conveyor at the constant rate of 10 cubic feet per minute, this tells us the value of $\frac{dV}{dt}$, the rate at which the volume of the sand pile changes. In particular,

$\frac{dV}{dt} = 10$ ft^3/min. Furthermore, since we are interested in how fast the height of the pile is changing at the instant $r = 4$, we use the value $r = 4$ along with $\frac{dV}{dt} = 10$ in Equation (3.5.1), and hence find that

$$10 = \frac{1}{3}\pi 4^2 \cdot \frac{1}{2}\frac{dr}{dt}\Big|_{r=4} + \frac{2}{3}\pi 4 \cdot \frac{1}{2}4 \cdot \frac{dr}{dt}\Big|_{r=4} = \frac{8}{3}\pi \frac{dr}{dt}\Big|_{r=4} + \frac{16}{3}\pi \frac{dr}{dt}\Big|_{r=4}.$$

With only the value of $\frac{dr}{dt}\big|_{r=4}$ remaining unknown, we solve for $\frac{dr}{dt}\big|_{r=4}$ and find that $10 = 8\pi \frac{dr}{dt}\big|_{r=4}$, so that

$$\frac{dr}{dt}\Big|_{r=4} = \frac{10}{8\pi} \approx 0.39789$$

feet per second. Because we were interested in how fast the height of the pile was changing at this instant, we want to know $\frac{dh}{dt}$ when $r = 4$. Since $\frac{dh}{dt} = \frac{1}{2}\frac{dr}{dt}$ for all values of t, it follows

$$\frac{dh}{dt}\Big|_{r=4} = \frac{5}{8\pi} \approx 0.19894 \text{ ft/min.}$$

Note particularly how we distinguish between the notations $\frac{dr}{dt}$ and $\frac{dr}{dt}\big|_{r=4}$. The former represents the rate of change of r with respect to t at an arbitrary value of t, while the latter is the rate of change of r with respect to t at a particular moment, in fact the moment $r = 4$. While we don't know the exact value of t, because information is provided about the value of r, it is important to distinguish that we are using this more specific data.

The relationship between h and r, with $h = \frac{1}{2}r$ for all values of t, enables us to transition easily between questions involving r and h. Indeed, had we known this information at the problem's outset, we could have immediately simplified our work. Using $h = \frac{1}{2}r$, it follows that since $V = \frac{1}{3}\pi r^2 h$, we can write V solely in terms of r to have

$$V = \frac{1}{3}\pi r^2 \left(\frac{1}{2}h\right) = \frac{1}{6}\pi r^3.$$

From this last equation, differentiating with respect to t implies

$$\frac{dV}{dt} = \frac{1}{2}\pi r^2 \frac{dr}{dt},$$

from which the same conclusions made earlier about $\frac{dr}{dt}$ and $\frac{dh}{dt}$ can be made.

Our work with the sandpile problem above is similar in many ways to our approach in Preview Activity 3.5.1, and these steps are typical of most related rates problems. In certain ways, they also resemble work we do in applied optimization problems, and here we summarize the main approach for consideration in subsequent problems.

Note 3.5.2.

- Identify the quantities in the problem that are changing and choose clearly defined variable names for them. Draw one or more figures that clearly represent the situation.

- Determine all rates of change that are known or given and identify the rate(s) of change to be found.

- Find an equation that relates the variables whose rates of change are known to those variables whose rates of change are to be found.

- Differentiate implicitly with respect to t to relate the rates of change of the involved quantities.

- Evaluate the derivatives and variables at the information relevant to the instant at which a certain rate of change is sought. Use proper notation to identify when a derivative is being evaluated at a particular instant, such as $\frac{dr}{dt}\big|_{r=4}$.

In the first step of identifying changing quantities and drawing a picture, it is important to think about the dynamic ways in which the involved quantities change. Sometimes a sequence of pictures can be helpful; for some already-drawn pictures that can be easily modified as applets built in Geogebra, see the following links[1] which represent

- how a circular oil slick's area grows as its radius increases http://gvsu.edu/s/9n;

- how the location of the base of a ladder and its height along a wall change as the ladder slides http://gvsu.edu/s/9o;

- how the water level changes in a conical tank as it fills with water at a constant rate http://gvsu.edu/s/9p (compare the problem in Activity 3.5.2);

- how a skateboarder's shadow changes as he moves past a lamppost http://gvsu.edu/s/9q.

Drawing well-labeled diagrams and envisioning how different parts of the figure change is a key part of understanding related rates problems and being successful at solving them.

Activity 3.5.2. A water tank has the shape of an inverted circular cone (point down) with a base of radius 6 feet and a depth of 8 feet. Suppose that water is being pumped into the tank at a constant instantaneous rate of 4 cubic feet per minute.

a. Draw a picture of the conical tank, including a sketch of the water level at a point in time when the tank is not yet full. Introduce variables that measure the radius of the water's surface and the water's depth in the tank, and label them on your figure.

b. Say that r is the radius and h the depth of the water at a given time, t. What equation relates the radius and height of the water, and why?

c. Determine an equation that relates the volume of water in the tank at time t to the depth h of the water at that time.

d. Through differentiation, find an equation that relates the instantaneous rate of change of water volume with respect to time to the instantaneous rate of change

---

[1]We again refer to the work of Prof. Marc Renault of Shippensburg University, found at http://gvsu.edu/s/5p.

of water depth at time t.

   e. Find the instantaneous rate at which the water level is rising when the water in the tank is 3 feet deep.

   f. When is the water rising most rapidly: at $h = 3$, $h = 4$, or $h = 5$?

Recognizing familiar geometric configurations is one way that we relate the changing quantities in a given problem. For instance, while the problem in Activity 3.5.2 is centered on a conical tank, one of the most important observations is that there are two key right triangles present. In another setting, a right triangle might be indicative of an opportunity to take advantage of the Pythagorean Theorem to relate the legs of the triangle. But in the conical tank, the fact that the water at any time fills a portion of the tank in such a way that the ratio of radius to depth is constant turns out to be the most important relationship with which to work. That enables us to write r in terms of h and reduce the overall problem to one that involves only one variable, where the volume of water depends simply on h, and hence to subsequently relate $\frac{dV}{dt}$ and $\frac{dh}{dt}$. In other situations where a changing angle is involved, a right triangle may offer the opportunity to find relationships among various parts of the triangle using trigonometric functions.

Activity 3.5.3. A television camera is positioned 4000 feet from the base of a rocket launching pad. The angle of elevation of the camera has to change at the correct rate in order to keep the rocket in sight. In addition, the auto-focus of the camera has to take into account the increasing distance between the camera and the rocket. We assume that the rocket rises vertically. (A similar problem is discussed and pictured dynamically at http://gvsu.edu/s/9t. Exploring the applet at the link will be helpful to you in answering the questions that follow.)

   a. Draw a figure that summarizes the given situation. What parts of the picture are changing? What parts are constant? Introduce appropriate variables to represent the quantities that are changing.

   b. Find an equation that relates the camera's angle of elevation to the height of the rocket, and then find an equation that relates the instantaneous rate of change of the camera's elevation angle to the instantaneous rate of change of the rocket's height (where all rates of change are with respect to time).

   c. Find an equation that relates the distance from the camera to the rocket to the rocket's height, as well as an equation that relates the instantaneous rate of change of distance from the camera to the rocket to the instantaneous rate of change of the rocket's height (where all rates of change are with respect to time).

   d. Suppose that the rocket's speed is 600 ft/sec at the instant it has risen 3000 feet. How fast is the distance from the television camera to the rocket changing at that moment? If the camera is following the rocket, how fast is the camera's angle of elevation changing at that same moment?

   e. If from an elevation of 3000 feet onward the rocket continues to rise at 600 feet/sec, will the rate of change of distance with respect to time be greater when the elevation is 4000 feet than it was at 3000 feet, or less? Why?

In addition to being able to find instantaneous rates of change at particular points in time, we are often able to make more general observations about how particular rates themselves will change over time. For instance, when a conical tank (point down) is filling with water at a constant rate, we naturally intuit that the depth of the water should increase more slowly over time. Note how carefully we need to speak: we mean to say that while the depth, h, of the water is increasing, its rate of change $\frac{dh}{dt}$ is decreasing (both as a function of t and as a function of h). These observations may often be made by taking the general equation that relates the various rates and solving for one of them, and doing this without substituting any particular values for known variables or rates. For instance, in the conical tank problem in Activity 3.5.2, we established that

$$\frac{dV}{dt} = \frac{1}{16}\pi h^2 \frac{dh}{dt},$$

and hence

$$\frac{dh}{dt} = \frac{16}{\pi h^2}\frac{dV}{dt}.$$

Provided that $\frac{dV}{dt}$ is constant, it is immediately apparent that as h gets larger, $\frac{dh}{dt}$ will get smaller, while always remaining positive. Hence, the depth of the water is increasing at a decreasing rate.

Activity 3.5.4. As pictured in the applet at http://gvsu.edu/s/9q, a skateboarder who is 6 feet tall rides under a 15 foot tall lamppost at a constant rate of 3 feet per second. We are interested in understanding how fast his shadow is changing at various points in time.

a. Draw an appropriate right triangle that represents a snapshot in time of the skateboarder, lamppost, and his shadow. Let x denote the horizontal distance from the base of the lamppost to the skateboarder and s represent the length of his shadow. Label these quantities, as well as the skateboarder's height and the lamppost's height on the diagram.

b. Observe that the skateboarder and the lamppost represent parallel line segments in the diagram, and thus similar triangles are present. Use similar triangles to establish an equation that relates x and s.

c. Use your work in (b) to find an equation that relates $\frac{dx}{dt}$ and $\frac{ds}{dt}$.

d. At what rate is the length of the skateboarder's shadow increasing at the instant the skateboarder is 8 feet from the lamppost?

e. As the skateboarder's distance from the lamppost increases, is his shadow's length increasing at an increasing rate, increasing at a decreasing rate, or increasing at a constant rate?

      f. Which is moving more rapidly: the skateboarder or the tip of his shadow? Explain, and justify your answer.

As we progress further into related rates problems, less direction will be provided. In the first three activities of this section, we have been provided with guided instruction to build a solution in a step by step way. For the closing activity and the following exercises, most of the detailed work is left to the reader.

> **Activity 3.5.5.** A baseball diamond is 90′ square. A batter hits a ball along the third base line and runs to first base. At what rate is the distance between the ball and first base changing when the ball is halfway to third base, if at that instant the ball is traveling 100 feet/sec? At what rate is the distance between the ball and the runner changing at the same instant, if at the same instant the runner is 1/8 of the way to first base running at 30 feet/sec?

Summary

- When two or more related quantities are changing as implicit functions of time, their rates of change can be related by implicitly differentiating the equation that relates the quantities themselves. For instance, if the sides of a right triangle are all changing as functions of time, say having lengths x, y, and z, then these quantities are related by the Pythagorean Theorem: $x^2 + y^2 = z^2$. It follows by implicitly differentiating with respect to t that their rates are related by the equation

$$2x\frac{dx}{dt} + 2y\frac{dy}{dt} = 2z\frac{dz}{dt},$$

so that if we know the values of x, y, and z at a particular time, as well as two of the three rates, we can deduce the value of the third.

Exercises

1. Gravel is being dumped from a conveyor belt at a rate of 10 cubic feet per minute. It forms a pile in the shape of a right circular cone whose base diameter and height are always the same. How fast is the height of the pile increasing when the pile is 15 feet high? Recall that the volume of a right circular cone with height h and radius of the base r is given by $V = \frac{1}{3}\pi r^2 h$.

When the pile is 15 feet high, its height is increasing at ⬚ feet per minute.

2. A street light is at the top of a 12 foot tall pole. A 6 foot tall woman walks away from the pole with a speed of 8 ft/sec along a straight path. How fast is the tip of her shadow moving when she is 50 feet from the base of the pole?

The tip of the shadow is moving at ⬚ ft/sec.

3. Water is leaking out of an inverted conical tank at a rate of 8600.0 cm^3/min at the same time that water is being pumped into the tank at a constant rate. The tank has height 12.0 m and the the diameter at the top is 4.0 m. If the water level is rising at a rate of 24.0 cm/min when the height of the water is 5.0 m, find the rate at which water is being pumped into the tank in cubic centimeters per minute.

Answer: [   ] cm^3/min

4. A sailboat is sitting at rest near its dock. A rope attached to the bow of the boat is drawn in over a pulley that stands on a post on the end of the dock that is 5 feet higher than the bow. If the rope is being pulled in at a rate of 2 feet per second, how fast is the boat approaching the dock when the length of rope from bow to pulley is 13 feet?

5. A swimming pool is 60 feet long and 25 feet wide. Its depth varies uniformly from 3 feet at the shallow end to 15 feet at the deep end, as shown in the Figure 3.5.3.

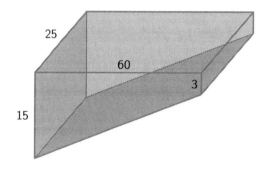

Figure 3.5.3: The swimming pool.

Suppose the pool has been emptied and is now being filled with water at a rate of 800 cubic feet per minute. At what rate is the depth of water (measured at the deepest point of the pool) increasing when it is 5 feet deep at that end? Over time, describe how the depth of the water will increase: at an increasing rate, at a decreasing rate, or at a constant rate. Explain.

6. A baseball diamond is a square with sides 90 feet long. Suppose a baseball player is advancing from second to third base at the rate of 24 feet per second, and an umpire is standing on home plate. Let θ be the angle between the third baseline and the line of sight from the umpire to the runner. How fast is θ changing when the runner is 30 feet from third base?

7. Sand is being dumped off a conveyor belt onto a pile in such a way that the pile forms in the shape of a cone whose radius is always equal to its height. Assuming that the sand is being dumped at a rate of 10 cubic feet per minute, how fast is the height of the pile changing when there are 1000 cubic feet on the pile?

CHAPTER 4

The Definite Integral

4.1 Determining distance traveled from velocity

Motivating Questions

- If we know the velocity of a moving body at every point in a given interval, can we determine the distance the object has traveled on the time interval?

- How is the problem of finding distance traveled related to finding the area under a certain curve?

- What does it mean to antidifferentiate a function and why is this process relevant to finding distance traveled?

- If velocity is negative, how does this impact the problem of finding distance traveled?

In the very first section of the text, we considered a situation where a moving object had a known position at time t. In particular, we stipulated that a tennis ball tossed into the air had its height s (in feet) at time t (in seconds) given by $s(t) = 64 - 16(t - 1)^2$. From this starting point, we investigated the average velocity of the ball on a given interval $[a, b]$, computed by the difference quotient $\frac{s(b)-s(a)}{b-a}$, and eventually found that we could determine the exact instantaneous velocity of the ball at time t by taking the derivative of the position function,

$$s'(t) = \lim_{h \to 0} \frac{s(t + h) - s(t)}{h}.$$

Thus, given a differentiable position function, we are able to know the exact velocity of the moving object at any point in time.

Moreover, from this foundational problem involving position and velocity we have learned a great deal. Given a differentiable function f, we are now able to find its derivative and use this derivative to determine the function's instantaneous rate of change at any point in the domain, as well as to find where the function is increasing or decreasing, is concave up or concave down, and has relative extremes. The vast majority of the problems and applications we have considered have involved the situation where a particular function is known and we seek information that relies on knowing the function's instantaneous rate of change. That is, we have typically proceeded from a function f to its derivative, f', and then used the meaning of the derivative to help us answer important questions.

In a much smaller number of situations so far, we have encountered the reverse situation where we instead know the derivative, f', and have tried to deduce information about f. It is this particular problem that will be the focus of our attention in most of Chapter 4: if we know the instantaneous rate of change of a function, are we able to determine the function itself? To begin, we start with a more focused question: if we know the instantaneous velocity of an object moving along a straight line path, can we determine its corresponding position function?

Preview Activity 4.1.1. Suppose that a person is taking a walk along a long straight path and walks at a constant rate of 3 miles per hour.

a. On the left-hand axes provided in Figure 4.1.1, sketch a labeled graph of the velocity function $v(t) = 3$.

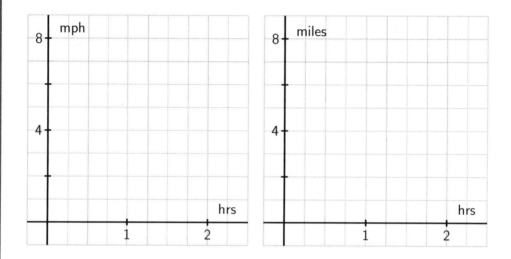

Figure 4.1.1: At left, axes for plotting $y = v(t)$; at right, for plotting $y = s(t)$.

Note that while the scale on the two sets of axes is the same, the units on the right-hand axes differ from those on the left. The right-hand axes will be used in question (d).

b. How far did the person travel during the two hours? How is this distance related to the area of a certain region under the graph of $y = v(t)$?

c. Find an algebraic formula, $s(t)$, for the position of the person at time t, assuming that $s(0) = 0$. Explain your thinking.

d. On the right-hand axes provided in Figure 4.1.1, sketch a labeled graph of the position function $y = s(t)$.

e. For what values of t is the position function s increasing? Explain why this is the case using relevant information about the velocity function v.

4.1.1 Area under the graph of the velocity function

In Preview Activity 4.1.1, we encountered a fundamental fact: when a moving object's velocity is constant (and positive), the area under the velocity curve over a given interval tells us the distance the object traveled.

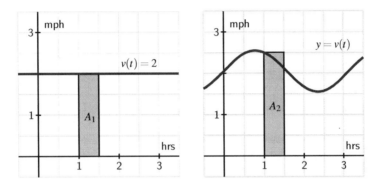

Figure 4.1.2: At left, a constant velocity function; at right, a non-constant velocity function.

As seen at left in Figure 4.1.2, if we consider an object moving at 2 miles per hour over the time interval $[1, 1.5]$, then the area A_1 of the shaded region under $y = v(t)$ on $[1, 1.5]$ is

$$A_1 = 2 \, \frac{\text{miles}}{\text{hour}} \cdot \frac{1}{2} \, \text{hours} = 1 \, \text{mile}.$$

This principle holds in general simply due to the fact that distance equals rate times time, provided the rate is constant. Thus, if $v(t)$ is constant on the interval $[a, b]$, then the distance traveled on $[a, b]$ is the area A that is given by

$$A = v(a)(b - a) = v(a)\Delta t,$$

where Δt is the change in t over the interval. Note, too, that we could use any value of $v(t)$ on the interval $[a, b]$, since the velocity is constant; we simply chose $v(a)$, the value at the interval's left endpoint. For several examples where the velocity function is piecewise constant, see http://gvsu.edu/s/9T.[1]

The situation is obviously more complicated when the velocity function is not constant. At the same time, on relatively small intervals on which $v(t)$ does not vary much, the area

---

[1]Marc Renault, calculus applets.

principle allows us to estimate the distance the moving object travels on that time interval. For instance, for the non-constant velocity function shown at right in Figure 4.1.2, we see that on the interval $[1, 1.5]$, velocity varies from $v(1) = 2.5$ down to $v(1.5) \approx 2.1$. Hence, one estimate for distance traveled is the area of the pictured rectangle,

$$A_2 = v(1)\Delta t = 2.5 \, \frac{\text{miles}}{\text{hour}} \cdot \frac{1}{2} \, \text{hours} = 1.25 \, \text{miles}.$$

Because v is decreasing on $[1, 1.5]$ and the rectangle lies above the curve, clearly $A_2 = 1.25$ is an over-estimate of the actual distance traveled.

If we want to estimate the area under the non-constant velocity function on a wider interval, say $[0, 3]$, it becomes apparent that one rectangle probably will not give a good approximation. Instead, we could use the six rectangles pictured in Figure 4.1.3, find the area of each rectangle, and add up the total. Obviously there are choices to make and issues to understand: how many rectangles should we use? where should we evaluate the function to decide the rectangle's height? what happens if velocity is sometimes negative? can we attain the exact area under any non-constant curve?

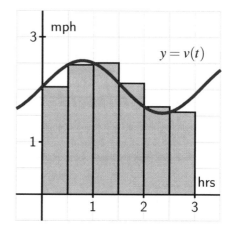

Figure 4.1.3: Using six rectangles to estimate the area under $y = v(t)$ on $[0, 3]$.

These questions and more are ones we will study in what follows; for now it suffices to realize that the simple idea of the area of a rectangle gives us a powerful tool for estimating both distance traveled from a velocity function as well as the area under an arbitrary curve. To explore the setting of multiple rectangles to approximate area under a non-constant velocity function, see the applet found at http://gvsu.edu/s/9U.[2]

> **Activity 4.1.2.** Suppose that a person is walking in such a way that her velocity varies slightly according to the information given in Table 4.1.4 and graph given in Figure 4.1.5.
>
> | t | 0.00 | 0.25 | 0.50 | 0.75 | 1.00 | 1.25 | 1.50 | 1.75 | 2.00 |
> |------|------|------|------|------|------|------|------|------|------|
> | $v(t)$ | 1.500 | 1.789 | 1.938 | 1.992 | 2.000 | 2.008 | 2.063 | 2.211 | 2.500 |
>
> **Table 4.1.4:** Velocity data for the person walking.

[2]Marc Renault, calculus applets.

a. Using the grid, graph, and given data appropriately, estimate the distance traveled by the walker during the two hour interval from $t = 0$ to $t = 2$. You should use time intervals of width $\Delta t = 0.5$, choosing a way to use the function consistently to determine the height of each rectangle in order to approximate distance traveled.

b. How could you get a better approximation of the distance traveled on $[0, 2]$? Explain, and then find this new estimate.

c. Now suppose that you know that v is given by $v(t) = 0.5t^3 - 1.5t^2 + 1.5t + 1.5$. Remember that v is the derivative of the walker's position function, s. Find a formula for s so that $s' = v$.

d. Based on your work in (c), what is the value of $s(2) - s(0)$? What is the meaning of this quantity?

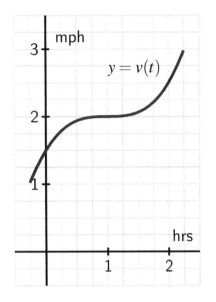

Figure 4.1.5: The graph of $y = v(t)$.

4.1.2 Two approaches: area and antidifferentiation

When the velocity of a moving object is positive, the object's position is always increasing. While we will soon consider situations where velocity is negative and think about the ramifications of this condition on distance traveled, for now we continue to assume that we are working with a positive velocity function. In that setting, we have established that whenever v is actually constant, the exact distance traveled on an interval is the area under the velocity curve; furthermore, we have observed that when v is not constant, we can estimate the total distance traveled by finding the areas of rectangles that help to approximate the area under the velocity curve on the given interval. Hence, we see the importance of the problem of finding the area between a curve and the horizontal axis: besides being an interesting geometric question, in the setting of the curve being the (positive) velocity of a moving object, the area under the curve over an interval tells us the exact distance traveled on the interval. We can estimate this area any time we have a graph of the velocity function or a table of data that tells us some relevant values of the function.

In Activity 4.1.2, we also encountered an alternate approach to finding the distance traveled. In particular, if we know a formula for the instantaneous velocity, $y = v(t)$, of the moving body at time t, then we realize that v must be the derivative of some corresponding position function s. If we can find a formula for s from the formula for v, it follows that we know the position of the object at time t. In addition, under the assumption that velocity is positive, the change in position over a given interval then tells us the distance traveled on that interval.

For a simple example, consider the situation from Preview Activity 4.1.1, where a person is walking along a straight line and has velocity function $v(t) = 3$ mph.

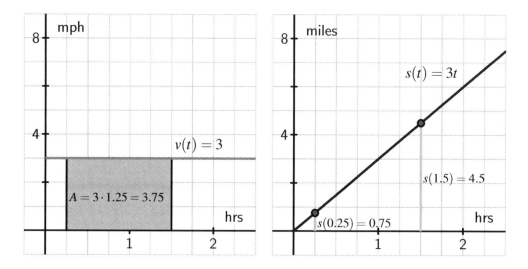

Figure 4.1.6: The velocity function $v(t) = 3$ and corresponding position function $s(t) = 3t$.

As pictured in Figure 4.1.6, we see the already noted relationship between area and distance traveled on the left-hand graph of the velocity function. In addition, because the velocity is constant at 3, we know that if[3] $s(t) = 3t$, then $s'(t) = 3$, so $s(t) = 3t$ is a function whose derivative is $v(t)$. Furthermore, we now observe that $s(1.5) = 4.5$ and $s(0.25) = 0.75$, which are the respective locations of the person at times $t = 0.25$ and $t = 1.5$, and therefore

$$s(1.5) - s(0.25) = 4.5 - 0.75 = 3.75 \text{ miles.}$$

This is not only the change in position on $[0.25, 1.5]$, but also precisely the distance traveled on $[0.25, 1.5]$, which can also be computed by finding the area under the velocity curve over the same interval. There are profound ideas and connections present in this example that we will spend much of the remainder of Chapter 4 studying and exploring.

For now, it is most important to observe that if we are given a formula for a velocity function v, it can be very helpful to find a function s that satisfies $s' = v$. In this context, we say that

---

[3]Here we are making the implicit assumption that $s(0) = 0$; we will further discuss the different possibilities for values of $s(0)$ in subsequent study.

s is an *antiderivative* of v. More generally, just as we say that f' is the derivative of f for a given function f, if we are given a function g and G is a function such that $G' = g$, we say that G is an *antiderivative* of g. For example, if $g(x) = 3x^2 + 2x$, an antiderivative of g is $G(x) = x^3 + x^2$, since $G'(x) = g(x)$. Note that we say "an" antiderivative of g rather than "the" antiderivative of g because $H(x) = x^3 + x^2 + 5$ is also a function whose derivative is g, and thus H is another antiderivative of g.

Activity 4.1.3. A ball is tossed vertically in such a way that its velocity function is given by $v(t) = 32 - 32t$, where t is measured in seconds and v in feet per second. Assume that this function is valid for $0 \le t \le 2$.

a. For what values of t is the velocity of the ball positive? What does this tell you about the motion of the ball on this interval of time values?

b. Find an antiderivative, s, of v that satisfies $s(0) = 0$.

c. Compute the value of $s(1) - s(\frac{1}{2})$. What is the meaning of the value you find?

d. Using the graph of $y = v(t)$ provided in Figure 4.1.7, find the exact area of the region under the velocity curve between $t = \frac{1}{2}$ and $t = 1$. What is the meaning of the value you find?

e. Answer the same questions as in (c) and (d) but instead using the interval $[0, 1]$.

f. What is the value of $s(2) - s(0)$? What does this result tell you about the flight of the ball? How is this value connected to the provided graph of $y = v(t)$? Explain.

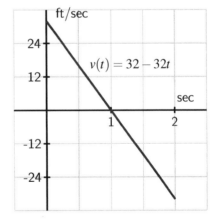

Figure 4.1.7: The graph of $y = v(t)$.

4.1.3 When velocity is negative

Most of our work in this section has occurred under the assumption that velocity is positive. This hypothesis guarantees that the movement of the object under consideration is always in a single direction, and hence ensures that the moving body's change in position is the same as the distance it travels on a given interval. As we saw in Activity 4.1.3, there are natural settings in which a moving object's velocity is negative; we would like to understand this scenario fully as well.

Consider a simple example where a person goes for a walk on a beach along a stretch of very straight shoreline that runs east-west. We can naturally assume that their initial position is $s(0) = 0$, and further stipulate that their position function increases as they move east from their starting location. For instance, a position of $s = 1$ mile represents being one mile east of the start location, while $s = -1$ tells us the person is one mile west of where they began walking on the beach. Now suppose the person walks in the following manner. From the outset at $t = 0$, the person walks due east at a constant rate of 3 mph for 1.5 hours. After 1.5 hours, the person stops abruptly and begins walking due west at the constant rate of 4 mph and does so for 0.5 hours. Then, after another abrupt stop and start, the person resumes walking at a constant rate of 3 mph to the east for one more hour. What is the total distance the person traveled on the time interval $t = 0$ to $t = 3$? What is the person's total change in position over that time?

On one hand, these are elementary questions to answer because the velocity involved is constant on each interval. From $t = 0$ to $t = 1.5$, the person traveled

$$D_{[0,1.5]} = 3 \text{ miles per hour} \cdot 1.5 \text{ hours} = 4.5 \text{ miles.}$$

Similarly, on $t = 1.5$ to $t = 2$, having a different rate, the distance traveled is

$$D_{[1.5,2]} = 4 \text{ miles per hour} \cdot 0.5 \text{ hours} = 2 \text{ miles.}$$

Finally, similar calculations reveal that in the final hour, the person walked

$$D_{[2,3]} = 3 \text{ miles per hour} \cdot 1 \text{ hours} = 3 \text{ miles,}$$

so the total distance traveled is

$$D = D_{[0,1.5]} + D_{[1.5,2]} + D_{[2,3]} = 4.5 + 2 + 3 = 9.5 \text{ miles.}$$

Since the velocity on $1.5 < t < 2$ is actually $v = -4$, being negative to indicate motion in the westward direction, this tells us that the person first walked 4.5 miles east, then 2 miles west, followed by 3 more miles east. Thus, the walker's total change in position is

$$\text{change in position} = 4.5 - 2 + 3 = 5.5 \text{ miles.}$$

While we have been able to answer these questions fairly easily, it is also important to think about this problem graphically in order that we can generalize our solution to the more complicated setting when velocity is not constant, as well as to note the particular impact that negative velocity has.

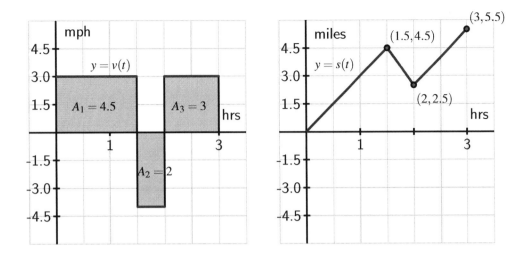

Figure 4.1.8: At left, the velocity function of the person walking; at right, the corresponding position function.

In Figure 4.1.8, we see how the distances we computed above can be viewed as areas: $A_1 = 4.5$ comes from taking rate times time $(3 \cdot 1.5)$, as do A_2 and A_3 for the second and third rectangles. The big new issue is that while A_2 is an area (and is therefore positive), because this area involves an interval on which the velocity function is negative, its area has a negative sign associated with it. This helps us to distinguish between distance traveled and change in position.

The distance traveled is the sum of the areas,

$$D = A_1 + A_2 + A_3 = 4.5 + 2 + 3 = 9.5 \text{ miles.}$$

But the change in position has to account for the sign associated with the area, where those above the t-axis are considered positive while those below the t-axis are viewed as negative, so that

$$s(3) - s(0) = (+4.5) + (-2) + (+3) = 5.5 \text{ miles,}$$

assigning the "-2" to the area in the interval $[1.5, 2]$ because there velocity is negative and the person is walking in the "negative" direction. In other words, the person walks 4.5 miles in the positive direction, followed by two miles in the negative direction, and then 3 more miles in the positive direction. This affect of velocity being negative is also seen in the graph of the function $y = s(t)$, which has a negative slope (specifically, its slope is -4) on the interval $1.5 < t < 2$ since the velocity is -4 on that interval, which shows the person's position function is decreasing due to the fact that she is walking east, rather than west. On the intervals where she is walking west, the velocity function is positive and the slope of the position function s is therefore also positive.

To summarize, we see that if velocity is sometimes negative, this makes the moving object's change in position different from its distance traveled. By viewing the intervals on which velocity is positive and negative separately, we may compute the distance traveled on each such interval, and then depending on whether we desire total distance traveled or total change in position, we may account for negative velocities that account for negative change in position, while still contributing positively to total distance traveled. We close this section with one additional activity that further explores the effects of negative velocity on the problem of finding change in position and total distance traveled.

Activity 4.1.4. Suppose that an object moving along a straight line path has its velocity v (in meters per second) at time t (in seconds) given by the piecewise linear function whose graph is pictured at left in Figure 4.1.9. We view movement to the right as being in the positive direction (with positive velocity), while movement to the left is in the negative direction.

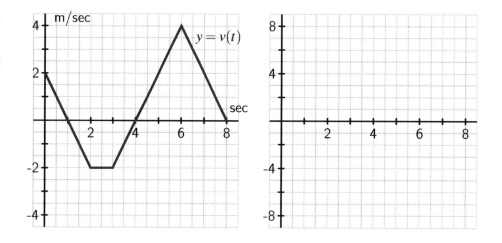

Figure 4.1.9: The velocity function of a moving object.

Suppose further that the object's initial position at time $t = 0$ is $s(0) = 1$.

a. Determine the total distance traveled and the total change in position on the time interval $0 \le t \le 2$. What is the object's position at $t = 2$?

b. On what time intervals is the moving object's position function increasing? Why? On what intervals is the object's position decreasing? Why?

c. What is the object's position at $t = 8$? How many total meters has it traveled to get to this point (including distance in both directions)? Is this different from the object's total change in position on $t = 0$ to $t = 8$?

d. Find the exact position of the object at $t = 1, 2, 3, \ldots, 8$ and use this data to sketch an accurate graph of $y = s(t)$ on the axes provided at right in Figure 4.1.9. How can you use the provided information about $y = v(t)$ to determine the concavity of s on each relevant interval?

Summary

- If we know the velocity of a moving body at every point in a given interval and the velocity is positive throughout, we can estimate the object's distance traveled and in some circumstances determine this value exactly.

- In particular, when velocity is positive on an interval, we can find the total distance traveled by finding the area under the velocity curve and above the t-axis on the given time interval. We may only be able to estimate this area, depending on the shape of the velocity curve.

- An antiderivative of a function f is a new function F whose derivative is f. That is, F is an antiderivative of f provided that $F' = f$. In the context of velocity and position, if we know a velocity function v, an antiderivative of v is a position function s that satisfies $s' = v$. If v is positive on a given interval, say $[a, b]$, then the change in position, $s(b) - s(a)$, measures the distance the moving object traveled on $[a, b]$.

- In the setting where velocity is sometimes negative, this means that the object is sometimes traveling in the opposite direction (depending on whether velocity is positive or negative), and thus involves the object backtracking. To determine distance traveled, we have to think about the problem separately on intervals where velocity is positive and negative and account for the change in position on each such interval.

Exercises

1. A car comes to a stop six seconds after the driver applies the brakes. While the brakes are on, the following velocities are recorded:

Time since brakes applied (sec)	0	2	4	6
Velocity (ft/s)	88	45	16	0

Give lower and upper estimates (using all of the available data) for the distance the car traveled after the brakes were applied.

lower: [   ]

upper: [   ]
(for each, include units)
On a sketch of velocity against time, show the lower and upper estimates you found above..

2. The velocity of a car is $f(t) = 9t$ meters/second. Use a graph of $f(t)$ to find the exact distance traveled by the car, in meters, from $t = 0$ to $t = 10$ seconds.

distance = [   ] (include units)

3. The velocity of a particle moving along the x-axis is given by $f(t) = 12 - 4t$ cm/sec. Use a graph of $f(t)$ to find the exact change in position of the particle from time $t = 0$ to $t = 4$ seconds.

change in position = [   ] (include units)

4. Two cars start at the same time and travel in the same direction along a straight road. The figure below gives the velocity, v (in km/hr), of each car as a function of time (in hr).

The velocity of car A is given by the solid, blue curve, and the velocity of car B by dashed, red curve.

(a)

Which car attains the larger maximum velocity?

(b)

Which stops first?

(c)

Which travels farther?

5. Suppose that an accelerating car goes from 0 mph to 66.8 mph in five seconds. Its velocity is given in the following table, converted from miles per hour to feet per second, so that all time measurements are in seconds. (Note: 1 mph is 22/15 feet per sec = 22/15 ft/s.) Find the average acceleration of the car over each of the first two seconds.

t	0	1	2	3	4	5
$v(t)$	0.00	33.41	57.91	75.73	89.09	98.00

average acceleration over the first second = [   ]
(include units)

average aceleration over the second second = [   ]
(include units)

6. The velocity function is $v(t) = t^2 - 6t + 8$ for a particle moving along a line. Find the displacement (net distance covered) of the particle during the time interval $[-2, 5]$.

displacement = [   ]

7. Along the eastern shore of Lake Michigan from Lake Macatawa (near Holland) to Grand Haven, there is a bike bath that runs almost directly north-south. For the purposes of this problem, assume the road is completely straight, and that the function $s(t)$ tracks the position of the biker along this path in miles north of Pigeon Lake, which lies roughly halfway between the ends of the bike path.

Suppose that the biker's velocity function is given by the graph in Figure 4.1.10 on the time interval $0 \le t \le 4$ (where t is measured in hours), and that $s(0) = 1$.

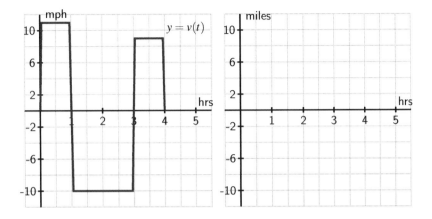

Figure 4.1.10: The graph of the biker's velocity, $y = v(t)$, at left. At right, axes to plot an approximate sketch of $y = s(t)$.

a. Approximately how far north of Pigeon Lake was the cyclist when she was the greatest distance away from Pigeon Lake? At what time did this occur?

b. What is the cyclist's total change in position on the time interval $0 \le t \le 2$? At $t = 2$, was she north or south of Pigeon Lake?

c. What is the total distance the biker traveled on $0 \le t \le 4$? At the end of the ride, how close was she to the point at which she started?

d. Sketch an approximate graph of $y = s(t)$, the position function of the cyclist, on the interval $0 \le t \le 4$. Label at least four important points on the graph of s.

8. A toy rocket is launched vertically from the ground on a day with no wind. The rocket's vertical velocity at time t (in seconds) is given by $v(t) = 500 - 32t$ feet/sec.

a. At what time after the rocket is launched does the rocket's velocity equal zero? Call this time value a. What happens to the rocket at $t = a$?

b. Find the value of the total area enclosed by $y = v(t)$ and the t-axis on the interval $0 \le t \le a$. What does this area represent in terms of the physical setting of the problem?

c. Find an antiderivative s of the function v. That is, find a function s such that $s'(t) = v(t)$.

d. Compute the value of $s(a) - s(0)$. What does this number represent in terms of the physical setting of the problem?

e. Compute $s(5) - s(1)$. What does this number tell you about the rocket's flight?

9. An object moving along a horizontal axis has its instantaneous velocity at time t in seconds given by the function v pictured in Figure 4.1.11, where v is measured in feet/sec.

Assume that the curves that make up the parts of the graph of $y = v(t)$ are either portions of straight lines or portions of circles.

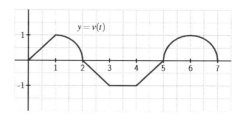

Figure 4.1.11: The graph of $y = v(t)$, the velocity function of a moving object.

a. Determine the exact total distance the object traveled on $0 \le t \le 2$.

b. What is the value and meaning of $s(5) - s(2)$, where $y = s(t)$ is the position function of the moving object?

c. On which time interval did the object travel the greatest distance: $[0, 2]$, $[2, 4]$, or $[5, 7]$?

d. On which time interval(s) is the position function s increasing? At which point(s) does s achieve a relative maximum?

10. Filters at a water treatment plant become dirtier over time and thus become less effective; they are replaced every 30 days. During one 30-day period, the rate at which pollution passes through the filters into a nearby lake (in units of particulate matter per day) is measured every 6 days and is given in the following table. The time t is measured in days since the filters were replaced.

Day, t	0	6	12	18	24	30
Rate of pollution in units per day, $p(t)$	7	8	10	13	18	35

Table 4.1.12: Pollution data for the water filters.

a. Plot the given data on a set of axes with time on the horizontal axis and the rate of pollution on the vertical axis.

b. Explain why the amount of pollution that entered the lake during this 30-day period would be given exactly by the area bounded by $y = p(t)$ and the t-axis on the time interval $[0, 30]$.

c. Estimate the total amount of pollution entering the lake during this 30-day period. Carefully explain how you determined your estimate.

4.2 Riemann Sums

Motivating Questions

- How can we use a Riemann sum to estimate the area between a given curve and the horizontal axis over a particular interval?

- What are the differences among left, right, middle, and random Riemann sums?

- How can we write Riemann sums in an abbreviated form?

In Section 4.1, we learned that if we have a moving object with velocity function v, whenever $v(t)$ is positive, the area between $y = v(t)$ and the t-axis over a given time interval tells us the distance traveled by the object over that time period; in addition, if $v(t)$ is sometimes negative and we view the area of any region below the t-axis as having an associated negative sign, then the sum of these signed areas over a given interval tells us the moving object's change in position over the time interval.

For instance, for the velocity function given in Figure 4.2.1, if the areas of shaded regions are A_1, A_2, and A_3 as labeled, then the total distance D traveled by the moving object on $[a, b]$ is

$$D = A_1 + A_2 + A_3,$$

while the total change in the object's position on $[a, b]$ is

$$s(b) - s(a) = A_1 - A_2 + A_3.$$

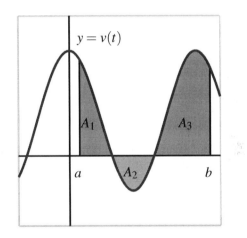

Figure 4.2.1: A velocity function that is sometimes negative.

Because the motion is in the negative direction on the interval where $v(t) < 0$, we subtract A_2 when determining the object's total change in position.

Of course, finding D and $s(b) - s(a)$ for the situation given in Figure 4.2.1 presumes that we can actually find the areas represented by A_1, A_2, and A_3. In most of our work in Section 4.1, such as in Activities 4.1.3 and Activity 4.1.4, we worked with velocity functions that were either constant or linear, so that by finding the areas of rectangles and triangles, we could find the area bounded by the velocity function and the horizontal axis exactly. But when the curve that bounds a region is not one for which we have a known formula for area, we are unable to find this area exactly. Indeed, this is one of our biggest goals in Chapter 4: to learn how to find the exact area bounded between a curve and the horizontal axis for as many different types of functions as possible.

To begin, we expand on the ideas in Activity 4.1.2, where we encountered a nonlinear velocity function and approximated the area under the curve using four and eight rectangles, respectively. In the following preview activity, we focus on three different options for deciding how to find the heights of the rectangles we will use.

Preview Activity 4.2.1. A person walking along a straight path has her velocity in miles per hour at time t given by the function $v(t) = 0.25t^3 - 1.5t^2 + 3t + 0.25$, for times in the interval $0 \le t \le 2$. The graph of this function is also given in each of the three diagrams in Figure 4.2.2.

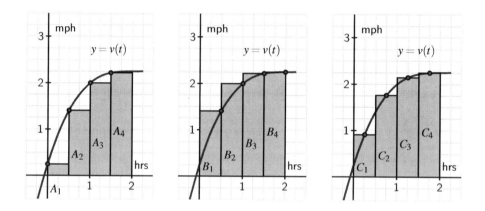

Figure 4.2.2: Three approaches to estimating the area under $y = v(t)$ on the interval $[0, 2]$.

Note that in each diagram, we use four rectangles to estimate the area under $y = v(t)$ on the interval $[0, 2]$, but the method by which the four rectangles' respective heights are decided varies among the three individual graphs.

   a. How are the heights of rectangles in the left-most diagram being chosen? Explain, and hence determine the value of

$$S = A_1 + A_2 + A_3 + A_4$$

by evaluating the function $y = v(t)$ at appropriately chosen values and observing the width of each rectangle. Note, for example, that

$$A_3 = v(1) \cdot \frac{1}{2} = 2 \cdot \frac{1}{2} = 1.$$

b. Explain how the heights of rectangles are being chosen in the middle diagram and find the value of

$$T = B_1 + B_2 + B_3 + B_4.$$

c. Likewise, determine the pattern of how heights of rectangles are chosen in the right-most diagram and determine

$$U = C_1 + C_2 + C_3 + C_4.$$

d. Of the estimates S, T, and U, which do you think is the best approximation of D, the total distance the person traveled on $[0, 2]$? Why?

4.2.1 Sigma Notation

It is apparent from several different problems we have considered that sums of areas of rectangles is one of the main ways to approximate the area under a curve over a given interval. Intuitively, we expect that using a larger number of thinner rectangles will provide a way to improve the estimates we are computing. As such, we anticipate dealing with sums with a large number of terms. To do so, we introduce the use of so-called *sigma notation*, named for the Greek letter Σ, which is the capital letter S in the Greek alphabet.

For example, say we are interested in the sum

$$1 + 2 + 3 + \cdots + 100,$$

which is the sum of the first 100 natural numbers. Sigma notation provides a shorthand notation that recognizes the general pattern in the terms of the sum. It is equivalent to write

$$\sum_{k=1}^{100} k = 1 + 2 + 3 + \cdots + 100.$$

We read the symbol $\sum_{k=1}^{100} k$ as "the sum from k equals 1 to 100 of k." The variable k is usually called the index of summation, and the letter that is used for this variable is immaterial. Each sum in sigma notation involves a function of the index; for example,

$$\sum_{k=1}^{10} (k^2 + 2k) = (1^2 + 2 \cdot 1) + (2^2 + 2 \cdot 2) + (3^2 + 2 \cdot 3) + \cdots + (10^2 + 2 \cdot 10),$$

and more generally,

$$\sum_{k=1}^{n} f(k) = f(1) + f(2) + \cdots + f(n).$$

Sigma notation allows us the flexibility to easily vary the function being used to track the pattern in the sum, as well as to adjust the number of terms in the sum simply by changing the value of n. We test our understanding of this new notation in the following activity.

Activity 4.2.2. For each sum written in sigma notation, write the sum long-hand and evaluate the sum to find its value. For each sum written in expanded form, write the sum in sigma notation.

a. $\sum_{k=1}^{5}(k^2 + 2)$

b. $\sum_{i=3}^{6}(2i - 1)$

c. $3 + 7 + 11 + 15 + \cdots + 27$

d. $4 + 8 + 16 + 32 + \cdots + 256$

e. $\sum_{i=1}^{6} \frac{1}{2^i}$

4.2.2 Riemann Sums

When a moving body has a positive velocity function $y = v(t)$ on a given interval $[a, b]$, we know that the area under the curve over the interval is the total distance the body travels on $[a, b]$. While this is the fundamental motivating force behind our interest in the area bounded by a function, we are also interested more generally in being able to find the exact area bounded by $y = f(x)$ on an interval $[a, b]$, regardless of the meaning or context of the function f. For now, we continue to focus on determining an accurate estimate of this area through the use of a sum of the areas of rectangles, doing so in the setting where $f(x) \geq 0$ on $[a, b]$. Throughout, unless otherwise indicated, we also assume that f is continuous on $[a, b]$.

The first choice we make in any such approximation is the number of rectangles.

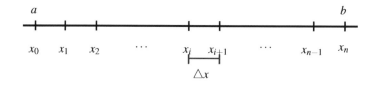

Figure 4.2.3: Subdividing the interval $[a, b]$ into n subintervals of equal length Δx.

If we say that the total number of rectangles is n, and we desire n rectangles of equal width to subdivide the interval $[a, b]$, then each rectangle must have width $\Delta x = \frac{b-a}{n}$. We observe further that $x_1 = x_0 + \Delta x$, $x_2 = x_0 + 2\Delta x$, and thus in general $x_i = a + i\Delta x$, as pictured in Figure 4.2.3.

We use each subinterval $[x_i, x_{i+1}]$ as the base of a rectangle, and next must choose how to decide the height of the rectangle that will be used to approximate the area under $y = f(x)$ on the subinterval. There are three standard choices: use the left endpoint of each subinterval, the right endpoint of each subinterval, or the midpoint of each. These are precisely the options encountered in Preview Activity 4.2.1 and seen in Figure 4.2.2. We next explore how these choices can be reflected in sigma notation.

If we now consider an arbitrary positive function f on $[a, b]$ with the interval subdivided as shown in Figure 4.2.3, and choose to use left endpoints, then on each interval of the form

$[x_i, x_{i+1}]$, the area of the rectangle formed is given by

$$A_{i+1} = f(x_i) \cdot \Delta x,$$

as seen in Figure 4.2.4.

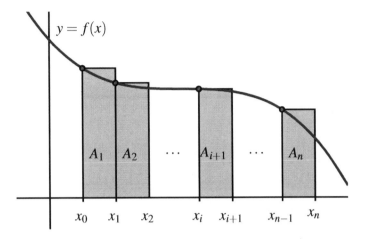

Figure 4.2.4: Subdividing the interval $[a, b]$ into n subintervals of equal length Δx and approximating the area under $y = f(x)$ over $[a, b]$ using left rectangles.

If we let L_n denote the sum of the areas of rectangles whose heights are given by the function value at each respective left endpoint, then we see that

$$L_n = A_1 + A_2 + \cdots + A_{i+1} + \cdots + A_n$$
$$= f(x_0) \cdot \Delta x + f(x_1) \cdot \Delta x + \cdots + f(x_i) \cdot \Delta x + \cdots + f(x_{n-1}) \cdot \Delta x.$$

In the more compact sigma notation, we have

$$L_n = \sum_{i=0}^{n-1} f(x_i)\Delta x.$$

Note particularly that since the index of summation begins at 0 and ends at $n - 1$, there are indeed n terms in this sum. We call L_n the *left Riemann sum* for the function f on the interval $[a, b]$.

There are now two fundamental issues to explore: the number of rectangles we choose to use and the selection of the pattern by which we identify the height of each rectangle. It is best to explore these choices dynamically, and the applet[1] found at http://gvsu.edu/s/a9 is a particularly useful one. There we see the image shown in Figure 4.2.5, but with the opportunity to adjust the slider bars for the left endpoint and the number of subintervals.

[1]Marc Renault, Geogebra Calculus Applets.

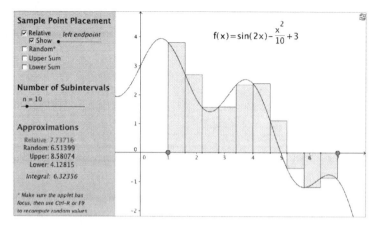

Figure 4.2.5: A snapshot of the applet found at http://gvsu.edu/s/a9.

By moving the sliders, we can see how the heights of the rectangles change as we consider left endpoints, midpoints, and right endpoints, as well as the impact that a larger number of narrower rectangles has on the approximation of the exact area bounded by the function and the horizontal axis.

To see how the Riemann sums for right endpoints and midpoints are constructed, we consider Figure 4.2.6.

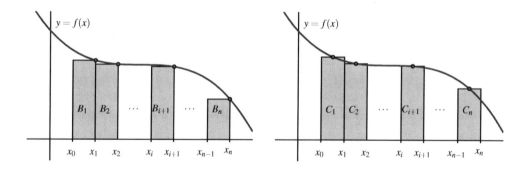

Figure 4.2.6: Riemann sums using right endpoints and midpoints.

For the sum with right endpoints, we see that the area of the rectangle on an arbitrary interval $[x_i, x_{i+1}]$ is given by $B_{i+1} = f(x_{i+1}) \cdot \Delta x$, so that the sum of all such areas of rectangles is given by

$$R_n = B_1 + B_2 + \cdots + B_{i+1} + \cdots + B_n$$
$$= f(x_1) \cdot \Delta x + f(x_2) \cdot \Delta x + \cdots + f(x_{i+1}) \cdot \Delta x + \cdots + f(x_n) \cdot \Delta x$$
$$= \sum_{i=1}^{n} f(x_i)\Delta x.$$

We call R_n the *right Riemann sum* for the function f on the interval $[a, b]$. For the sum that uses midpoints, we introduce the notation

$$\overline{x}_{i+1} = \frac{x_i + x_{i+1}}{2}$$

so that $\overline{x}_{i+1}$ is the midpoint of the interval $[x_i, x_{i+1}]$. For instance, for the rectangle with area C_1 in Figure 4.2.6, we now have

$$C_1 = f(\overline{x}_1) \cdot \Delta x.$$

Hence, the sum of all the areas of rectangles that use midpoints is

$$
\begin{aligned}
M_n &= C_1 + C_2 + \cdots + C_{i+1} + \cdots + C_n \\
&= f(\overline{x}_1) \cdot \Delta x + f(\overline{x}_2) \cdot \Delta x + \cdots + f(\overline{x}_{i+1}) \cdot \Delta x + \cdots + f(\overline{x}_n) \cdot \Delta x \\
&= \sum_{i=1}^{n} f(\overline{x}_i) \Delta x,
\end{aligned}
$$

and we say that M_n is the *middle Riemann sum* for f on $[a, b]$.

When $f(x) \geq 0$ on $[a, b]$, each of the Riemann sums L_n, R_n, and M_n provides an estimate of the area under the curve $y = f(x)$ over the interval $[a, b]$; momentarily, we will discuss the meaning of Riemann sums in the setting when f is sometimes negative. We also recall that in the context of a nonnegative velocity function $y = v(t)$, the corresponding Riemann sums are approximating the distance traveled on $[a, b]$ by the moving object with velocity function v.

There is a more general way to think of Riemann sums, and that is to not restrict the choice of where the function is evaluated to determine the respective rectangle heights. That is, rather than saying we'll always choose left endpoints, or always choose midpoints, we simply say that a point x_{i+1}^* will be selected at random in the interval $[x_i, x_{i+1}]$ (so that $x_i \leq x_{i+1}^* \leq x_{i+1}$), which makes the Riemann sum given by

$$f(x_1^*) \cdot \Delta x + f(x_2^*) \cdot \Delta x + \cdots + f(x_{i+1}^*) \cdot \Delta x + \cdots + f(x_n^*) \cdot \Delta x = \sum_{i=1}^{n} f(x_i^*) \Delta x.$$

At http://gvsu.edu/s/a9, the applet noted earlier and referenced in Figure 4.2.5, by unchecking the "relative" box at the top left, and instead checking "random," we can easily explore the effect of using random point locations in subintervals on a given Riemann sum. In computational practice, we most often use L_n, R_n, or M_n, while the random Riemann sum is useful in theoretical discussions. In the following activity, we investigate several different Riemann sums for a particular velocity function.

Activity 4.2.3. Suppose that an object moving along a straight line path has its velocity in feet per second at time t in seconds given by $v(t) = \frac{2}{9}(t - 3)^2 + 2$.

a. Carefully sketch the region whose exact area will tell you the value of the distance the object traveled on the time interval $2 \leq t \leq 5$.

    b. Estimate the distance traveled on $[2, 5]$ by computing L_4, R_4, and M_4.

    c. Does averaging L_4 and R_4 result in the same value as M_4? If not, what do you think the average of L_4 and R_4 measures?

    d. For this question, think about an arbitrary function f, rather than the particular function v given above. If f is positive and increasing on $[a, b]$, will L_n over-estimate or under-estimate the exact area under f on $[a, b]$? Will R_n over- or under-estimate the exact area under f on $[a, b]$? Explain.

4.2.3 When the function is sometimes negative

For a Riemann sum such as

$$L_n = \sum_{i=0}^{n-1} f(x_i)\Delta x,$$

we can of course compute the sum even when f takes on negative values. We know that when f is positive on $[a, b]$, the corresponding left Riemann sum L_n estimates the area bounded by f and the horizontal axis over the interval.

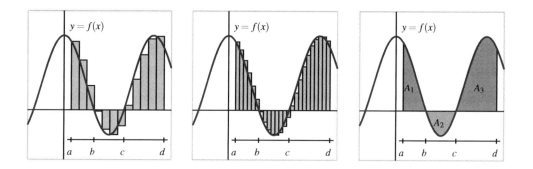

Figure 4.2.7: At left and center, two left Riemann sums for a function f that is sometimes negative; at right, the areas bounded by f on the interval $[a, d]$.

For a function such as the one pictured in Figure 4.2.7, where in the first figure a left Riemann sum is being taken with 12 subintervals over $[a, d]$, we observe that the function is negative on the interval $b \leq x \leq c$, and so for the four left endpoints that fall in $[b, c]$, the terms $f(x_i)\Delta x$ have negative function values. This means that those four terms in the Riemann sum produce an estimate of the *opposite* of the area bounded by $y = f(x)$ and the x-axis on $[b, c]$.

In Figure 4.2.7, we also see evidence that by increasing the number of rectangles used in a Riemann sum, it appears that the approximation of the area (or the opposite of the area) bounded by a curve appears to improve. For instance, in the middle graph, we use 24 left rectangles, and from the shaded areas, it appears that we have decreased the error from the approximation that uses 12. When we proceed to Section 4.3, we will discuss the natural idea of letting the number of rectangles in the sum increase without bound.

For now, it is most important for us to observe that, in general, any Riemann sum of a continuous function f on an interval $[a, b]$ approximates the difference between the area that lies above the horizontal axis on $[a, b]$ and under f and the area that lies below the horizontal axis on $[a, b]$ and above f. In the notation of Figure 4.2.7, we may say that

$$L_{24} \approx A_1 - A_2 + A_3,$$

where L_{24} is the left Riemann sum using 24 subintervals shown in the middle graph, and A_1 and A_3 are the areas of the regions where f is positive on the interval of interest, while A_2 is the area of the region where f is negative. We will also call the quantity $A_1 - A_2 + A_3$ the *net signed area* bounded by f over the interval $[a, d]$, where by the phrase "signed area" we indicate that we are attaching a minus sign to the areas of regions that fall below the horizontal axis.

Finally, we recall from the introduction to this present section that in the context where the function f represents the velocity of a moving object, the total sum of the areas bounded by the curve tells us the total distance traveled over the relevant time interval, while the total net signed area bounded by the curve computes the object's change in position on the interval.

Activity 4.2.4. Suppose that an object moving along a straight line path has its velocity v (in feet per second) at time t (in seconds) given by

$$v(t) = \frac{1}{2}t^2 - 3t + \frac{7}{2}.$$

a. Compute M_5, the middle Riemann sum, for v on the time interval $[1, 5]$. Be sure to clearly identify the value of Δt as well as the locations of $t_0, t_1, \cdots, t_5$. In addition, provide a careful sketch of the function and the corresponding rectangles that are being used in the sum.

b. Building on your work in (a), estimate the total change in position of the object on the interval $[1, 5]$.

c. Building on your work in (a) and (b), estimate the total distance traveled by the object on $[1, 5]$.

d. Use appropriate computing technology[a] to compute M_{10} and M_{20}. What exact value do you think the middle sum eventually approaches as n increases without bound? What does that number represent in the physical context of the overall problem?

[a]For instance, consider the applet at http://gvsu.edu/s/a9 and change the function and adjust the locations of the blue points that represent the interval endpoints a and b.

Summary

- A Riemann sum is simply a sum of products of the form $f(x_i^*)\Delta x$ that estimates the area between a positive function and the horizontal axis over a given interval. If the function is sometimes negative on the interval, the Riemann sum estimates the difference between the areas that lie above the horizontal axis and those that lie below the axis.

- The three most common types of Riemann sums are left, right, and middle sums, plus we can also work with a more general, random Riemann sum. The only difference among these sums is the location of the point at which the function is evaluated to determine the height of the rectangle whose area is being computed in the sum. For a left Riemann sum, we evaluate the function at the left endpoint of each subinterval, while for right and middle sums, we use right endpoints and midpoints, respectively.

- The left, right, and middle Riemann sums are denoted L_n, R_n, and M_n, with formulas

$$L_n = f(x_0)\Delta x + f(x_1)\Delta x + \cdots + f(x_{n-1})\Delta x = \sum_{i=0}^{n-1} f(x_i)\Delta x,$$

$$R_n = f(x_1)\Delta x + f(x_2)\Delta x + \cdots + f(x_n)\Delta x = \sum_{i=1}^{n} f(x_i)\Delta x,$$

$$M_n = f(\overline{x}_1)\Delta x + f(\overline{x}_2)\Delta x + \cdots + f(\overline{x}_n)\Delta x = \sum_{i=1}^{n} f(\overline{x}_i)\Delta x,$$

where $x_0 = a$, $x_i = a + i\Delta x$, and $x_n = b$, using $\Delta x = \frac{b-a}{n}$. For the midpoint sum, $\overline{x}_i = (x_{i-1} + x_i)/2$.

Exercises

1. The rectangles in the graph below illustrate a left endpoint Riemann sum for $f(x) = \frac{-x^2}{4} + 2x$ on the interval $[2, 6]$.

The value of this left endpoint Riemann sum is [   ], and this Riemann sum is [Choose: [select an answer] | an overestimate of | equal to | an underestimate of | there is ambiguity] the area of the region enclosed by $y = f(x)$, the x-axis, and the vertical lines x = 2 and x = 6.

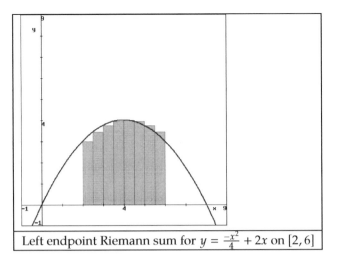

Left endpoint Riemann sum for $y = \frac{-x^2}{4} + 2x$ on $[2, 6]$

The rectangles in the graph below illustrate a right endpoint Riemann sum for $f(x) = \frac{-x^2}{4} + 2x$ on the interval $[2, 6]$.

The value of this right endpoint Riemann sum is [   ], and this Riemann sum is [Choose: [select an answer] | an overestimate of | equal to | an underestimate of | there is ambiguity] the area of the region enclosed by $y = f(x)$, the x-axis, and the vertical lines x = 2 and x = 6.

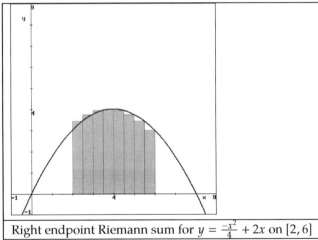

Right endpoint Riemann sum for $y = \frac{-x^2}{4} + 2x$ on $[2, 6]$

2. Your task is to estimate how far an object traveled during the time interval $0 \le t \le 8$, but you only have the following data about the velocity of the object.

time (sec)	0	1	2	3	4	5	6	7	8
velocity (feet/sec)	-3	-2	-3	-1	-2	-1	4	1	2

To get an idea of what the velocity function might look like, you pick up a black pen, plot the data points, and connect them by curves. Your sketch looks something like the black curve in the graph below.

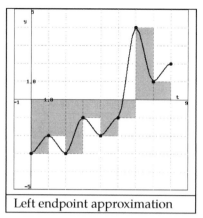

Left endpoint approximation

You decide to use a left endpoint Riemann sum to estimate the total displacement. So, you pick up a blue pen and draw rectangles whose height is determined by the velocity measurement at the left endpoint of each one-second interval. By using the left endpoint Riemann sum as an approximation, you are assuming that the actual velocity is approximately constant on each one-second interval (or, equivalently, that the actual acceleration is approximately zero on each one-second interval), and that the velocity and acceleration have discontinuous jumps every second. This assumption is probably incorrect because it is likely that the velocity and acceleration change continuously over time. However, you decide to use this approximation anyway since it seems like a reasonable approximation to the actual velocity given the limited amount of data.

(A) Using the left endpoint Riemann sum, find approximately how far the object traveled. Your answers must include the correct units.

Total displacement =

Total distance traveled =

Using the same data, you also decide to estimate how far the object traveled using a right endpoint Riemann sum. So, you sketch the curve again with a black pen, and draw rectangles whose height is determined by the velocity measurement at the right endpoint of each one-second interval.

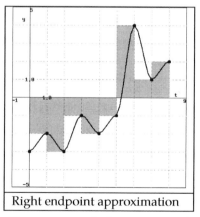

Right endpoint approximation

(B) Using the right endpoint Riemann sum, find approximately how far the object traveled.

Your answers must include the correct units.

Total displacement = [   ]

Total distance traveled = [   ]

3. On a sketch of $y = e^x$, represent the left Riemann sum with $n = 2$ approximating $\int_2^3 e^x \, dx$. Write out the terms of the sum, but do not evaluate it:

Sum = [   ] + [   ]

On another sketch, represent the right Riemann sum with $n = 2$ approximating $\int_2^3 e^x \, dx$. Write out the terms of the sum, but do not evaluate it:

Sum = [   ] + [   ]

Which sum is an overestimate?

Which sum is an underestimate?

4. Consider the function $f(x) = 3x + 4$.

a. Compute M_4 for $y = f(x)$ on the interval $[2, 5]$. Be sure to clearly identify the value of Δx, as well as the locations of $x_0, x_1, \ldots, x_4$. Include a careful sketch of the function and the corresponding rectangles being used in the sum.

b. Use a familiar geometric formula to determine the exact value of the area of the region bounded by $y = f(x)$ and the x-axis on $[2, 5]$.

c. Explain why the values you computed in (a) and (b) turn out to be the same. Will this be true if we use a number different than $n = 4$ and compute M_n? Will L_4 or R_4 have the same value as the exact area of the region found in (b)?

d. Describe the collection of functions g for which it will always be the case that M_n, regardless of the value of n, gives the exact net signed area bounded between the function g and the x-axis on the interval $[a, b]$.

5. Let S be the sum given by

$$S = ((1.4)^2 + 1) \cdot 0.4 + ((1.8)^2 + 1) \cdot 0.4 + ((2.2)^2 + 1) \cdot 0.4 + ((2.6)^2 + 1) \cdot 0.4 + ((3.0)^2 + 1) \cdot 0.4.$$

a. Assume that S is a right Riemann sum. For what function f and what interval $[a, b]$ is S an approximation of the area under f and above the x-axis on $[a, b]$? Why?

b. How does your answer to (a) change if S is a left Riemann sum? a middle Riemann sum?

c. Suppose that S really is a right Riemann sum. What is geometric quantity does S approximate?

d. Use sigma notation to write a new sum R that is the right Riemann sum for the same function, but that uses twice as many subintervals as S.

6. A car traveling along a straight road is braking and its velocity is measured at several different points in time, as given in the following table.

seconds, t	0	0.3	0.6	0.9	1.2	1.5	1.8
Velocity in ft/sec, $v(t)$	100	88	74	59	40	19	0

Table 4.2.8: Data for the braking car.

a. Plot the given data on a set of axes with time on the horizontal axis and the velocity on the vertical axis.

b. Estimate the total distance traveled during the car the time brakes using a middle Riemann sum with 3 subintervals.

c. Estimate the total distance traveled on $[0, 1.8]$ by computing L_6, R_6, and $\frac{1}{2}(L_6 + R_6)$.

d. Assuming that $v(t)$ is always decreasing on $[0, 1.8]$, what is the maximum possible distance the car traveled before it stopped? Why?

7. The rate at which pollution escapes a scrubbing process at a manufacturing plant increases over time as filters and other technologies become less effective. For this particular example, assume that the rate of pollution (in tons per week) is given by the function r that is pictured in Figure 4.2.9.

a. Use the graph to estimate the value of M_4 on the interval $[0, 4]$.

b. What is the meaning of M_4 in terms of the pollution discharged by the plant?

c. Suppose that $r(t) = 0.5e^{0.5t}$. Use this formula for r to compute L_5 on $[0, 4]$.

d. Determine an upper bound on the total amount of pollution that can escape the plant during the pictured four week time period that is accurate within an error of at most one ton of pollution.

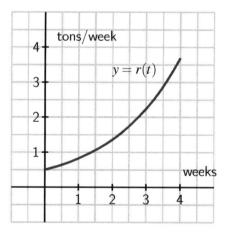

Figure 4.2.9: The rate, $r(t)$, of pollution in tons per week.

4.3 The Definite Integral

Motivating Questions

- How does increasing the number of subintervals affect the accuracy of the approximation generated by a Riemann sum?

- What is the definition of the definite integral of a function f over the interval $[a, b]$?

- What does the definite integral measure exactly, and what are some of the key properties of the definite integral?

In Figure 4.2.7, which is repeated below as Figure 4.3.1, we see visual evidence that increasing the number of rectangles in a Riemann sum improves the accuracy of the approximation of the net signed area that is bounded by the given function on the interval under consideration.

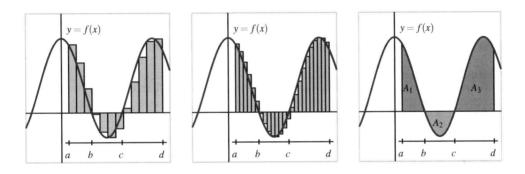

Figure 4.3.1: At left and center, two left Riemann sums for a function f that is sometimes negative; at right, the exact areas bounded by f on the interval $[a, d]$.

We thus explore the natural idea of allowing the number of rectangles to increase without bound in an effort to compute the exact net signed area bounded by a function on an interval. In addition, it is important to think about the differences among left, right, and middle Riemann sums and the different results they generate as the value of n increases. As we have done throughout our investigations with area, we begin with functions that are exclusively positive on the interval under consideration.

> **Preview Activity 4.3.1.** Consider the applet found at http://gvsu.edu/s/a9[a]. There, you will initially see the situation shown in Figure 4.3.2.

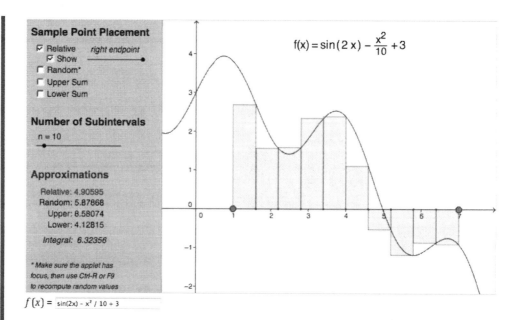

Sample Point Placement
- ☑ Relative   *right endpoint*
- ☑ Show ————————●
- ☐ Random*
- ☐ Upper Sum
- ☐ Lower Sum

Number of Subintervals

n = 10

●————————

Approximations

Relative: 4.90595
Random: 5.87868
Upper: 8.58074
Lower: 4.12815

Integral: 6.32356

* *Make sure the applet has focus, then use Ctrl-R or F9 to recompute random values*

$f(x) = $ sin(2x) - x² / 10 + 3

$$f(x) = \sin(2x) - \frac{x^2}{10} + 3$$

Figure 4.3.2: A right Riemann sum with 10 subintervals for the function $f(x) = \sin(2x) - \frac{x^2}{10} + 3$ on the interval $[1, 7]$. The value of the sum is $R_{10} = 4.90595$.

Note that the value of the chosen Riemann sum is displayed next to the word "relative," and that you can change the type of Riemann sum being computed by dragging the point on the slider bar below the phrase "sample point placement."

Explore to see how you can change the window in which the function is viewed, as well as the function itself. You can set the minimum and maximum values of x by clicking and dragging on the blue points that set the endpoints; you can change the function by typing a new formula in the "f(x)" window at the bottom; and you can adjust the overall window by "panning and zooming" by using the Shift key and the scrolling feature of your mouse. More information on how to pan and zoom can be found at http://gvsu.edu/s/Fl.

Work accordingly to adjust the applet so that it uses a left Riemann sum with $n = 5$ subintervals for the function is $f(x) = 2x + 1$. You should see the updated figure shown in Figure 4.3.3. Then, answer the following questions.

a. Update the applet (and view window, as needed) so that the function being considered is $f(x) = 2x + 1$ on $[1, 4]$, as directed above. For this function on this interval, compute L_n, M_n, R_n for $n = 5$, $n = 25$, and $n = 100$. What appears to be the exact area bounded by $f(x) = 2x + 1$ and the x-axis on $[1, 4]$?

b. Use basic geometry to determine the exact area bounded by $f(x) = 2x + 1$ and the *x*-axis on $[1, 4]$.

c. Based on your work in (a) and (b), what do you observe occurs when we increase the number of subintervals used in the Riemann sum?

d. Update the applet to consider the function $f(x) = x^2 + 1$ on the interval $[1, 4]$ (note that you need to enter "x ^ 2 + 1" for the function formula). Use the applet to compute L_n, M_n, R_n for $n = 5$, $n = 25$, and $n = 100$. What do you conjecture is the exact area bounded by $f(x) = x^2 + 1$ and the *x*-axis on $[1, 4]$?

e. Why can we not compute the exact value of the area bounded by $f(x) = x^2 + 1$ and the *x*-axis on $[1, 4]$ using a formula like we did in (b)?

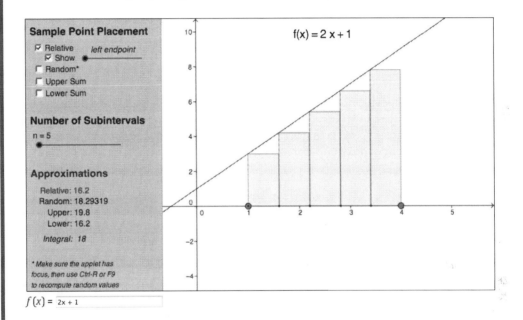

Figure 4.3.3: A left Riemann sum with 5 subintervals for the function $f(x) = 2x + 1$ on the interval $[1, 4]$. The value of the sum is $L_5 = 16.2$.

---

[a]Marc Renault, Shippensburg University, Geogebra Applets for Calclulus, http://gvsu.edu/s/5p.

4.3.1 The definition of the definite integral

In both examples in Preview Activity 4.3.1, we saw that as the number of rectangles got larger and larger, the values of L_n, M_n, and R_n all grew closer and closer to the same value. It turns out that this occurs for any continuous function on an interval $[a, b]$, and even more generally for a Riemann sum using any point x^*_{i+1} in the interval $[x_i, x_{i+1}]$. Said differently, as we let $n \to \infty$, it doesn't really matter where we choose to evaluate the function within a given subinterval, because

$$\lim_{n\to\infty} L_n = \lim_{n\to\infty} R_n = \lim_{n\to\infty} M_n = \lim_{n\to\infty} \sum_{i=1}^{n} f(x_i^*)\Delta x.$$

That these limits always exist (and share the same value) for a continuous[1] function f allows us to make the following definition.

Definition 4.3.4. The **definite integral** of a continuous function f on the interval $[a, b]$, denoted $\int_a^b f(x)\,dx$, is the real number given by

$$\int_a^b f(x)\,dx = \lim_{n\to\infty} \sum_{i=1}^{n} f(x_i^*)\Delta x,$$

where $\Delta x = \frac{b-a}{n}$, $x_i = a + i\Delta x$ (for $i = 0, \ldots, n$), and x_i^* satisfies $x_{i-1} \le x_i^* \le x_i$ (for $i = 1, \ldots, n$).

We call the symbol $\int$ the *integral sign*, the values a and b the *limits of integration*, and the function f the *integrand*. The process of determining the real number $\int_a^b f(x)\,dx$ is called *evaluating the definite integral*. While we will come to understand that there are several different interpretations of the value of the definite integral, for now the most important is that $\int_a^b f(x)\,dx$ measures the net signed area bounded by $y = f(x)$ and the x-axis on the interval $[a, b]$.

For example, in the notation of the definite integral, if f is the function pictured in Figure 4.3.5 and A_1, A_2, and A_3 are the exact areas bounded by f and the x-axis on the respective intervals $[a, b]$, $[b, c]$, and $[c, d]$, then

$$\int_a^b f(x)\,dx = A_1, \quad \int_b^c f(x)\,dx = -A_2,$$

$$\int_c^d f(x)\,dx = A_3,$$

and $\int_a^d f(x)\,dx = A_1 - A_2 + A_3.$

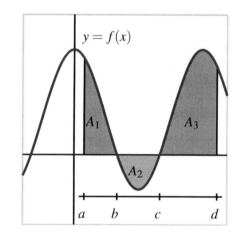

Figure 4.3.5: A continuous function f on the interval $[a, d]$.

---

[1]It turns out that a function need not be continuous in order to have a definite integral. For our purposes, we assume that the functions we consider are continuous on the interval(s) of interest. It is straightforward to see that any function that is piecewise continuous on an interval of interest will also have a well-defined definite integral.

We can also use definite integrals to express the change in position and distance traveled by a moving object. In the setting of a velocity function v on an interval $[a, b]$, it follows from our work above and in preceding sections that the change in position, $s(b) - s(a)$, is given by

$$s(b) - s(a) = \int_a^b v(t) \, dt.$$

If the velocity function is nonnegative on $[a, b]$, then $\int_a^b v(t) \, dt$ tells us the distance the object traveled. When velocity is sometimes negative on $[a, b]$, the areas bounded by the function on intervals where v does not change sign can be found using integrals, and the sum of these values will tell us the distance the object traveled.

If we wish to compute the value of a definite integral using the definition, we have to take the limit of a sum. While this is possible to do in select circumstances, it is also tedious and time-consuming; moreover, computing these limits does not offer much additional insight into the meaning or interpretation of the definite integral. Instead, in Section 4.4, we will learn the Fundamental Theorem of Calculus, a result that provides a shortcut for evaluating a large class of definite integrals. This will enable us to determine the exact net signed area bounded by a continuous function and the x-axis in many circumstances, including examples such as $\int_1^4 (x^2 + 1) \, dx$, which we approximated by Riemann sums in Preview Activity 4.3.1.

For now, our goal is to understand the meaning and properties of the definite integral, rather than how to actually compute its value using ideas in calculus. Thus, we temporarily rely on the net signed area interpretation of the definite integral and observe that if a given curve produces regions whose areas we can compute exactly through known area formulas, we can thus compute the exact value of the integral.

For instance, if we wish to evaluate the definite integral $\int_1^4 (2x + 1) \, dx$, we can observe that the region bounded by this function and the x-axis is the trapezoid shown in Figure 4.3.6, and by the known formula for the area of a trapezoid, its area is $A = \frac{1}{2}(3 + 9) \cdot 3 = 18$, so

$$\int_1^4 (2x + 1) \, dx = 18.$$

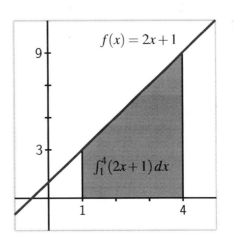

Figure 4.3.6: The area bounded by $f(x) = 2x + 1$ and the x-axis on the interval $[1, 4]$.

Activity 4.3.2. Use known geometric formulas and the net signed area interpretation of the definite integral to evaluate each of the definite integrals below.

a. $\int_0^1 3x\,dx$

b. $\int_{-1}^4 (2 - 2x)\,dx$

c. $\int_{-1}^1 \sqrt{1 - x^2}\,dx$

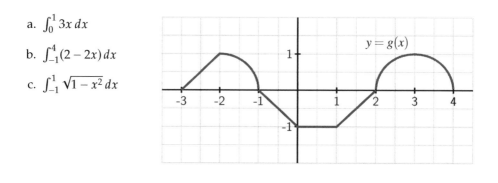

Figure 4.3.7: A function g that is piecewise defined; each piece of the function is part of a circle or part of a line.

d. $\int_{-3}^4 g(x)\,dx$, where g is the function pictured in Figure 4.3.7. Assume that each portion of g is either part of a line or part of a circle.

4.3.2 Some properties of the definite integral

With the perspective that the definite integral of a function f over an interval $[a, b]$ measures the net signed area bounded by f and the x-axis over the interval, we naturally arrive at several different standard properties of the definite integral. In addition, it is helpful to remember that the definite integral is defined in terms of Riemann sums that fundamentally consist of the areas of rectangles.

If we consider the definite integral $\int_a^a f(x)\,dx$ for any real number a, it is evident that no area is being bounded because the interval begins and ends with the same point. Hence,

If f is a continuous function and a is a real number, then $\int_a^a f(x)\,dx = 0$.

Next, we consider the results of subdividing a given interval. In Figure 4.3.8, we see that

$$\int_a^b f(x)\,dx = A_1, \quad \int_b^c f(x)\,dx = A_2,$$

$$\text{and } \int_a^c f(x)\,dx = A_1 + A_2,$$

which is indicative of the following general rule.

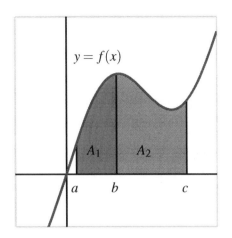

Figure 4.3.8: The area bounded by $y = f(x)$ on the interval $[a, c]$.

If f is a continuous function and a, b, and c are real numbers, then

$$\int_a^c f(x)\,dx = \int_a^b f(x)\,dx + \int_b^c f(x)\,dx.$$

While this rule is most apparent in the situation where $a < b < c$, it in fact holds in general for any values of a, b, and c. This result is connected to another property of the definite integral, which states that if we reverse the order of the limits of integration, we change the sign of the integral's value.

If f is a continuous function and a and b are real numbers, then

$$\int_b^a f(x)\,dx = -\int_a^b f(x)\,dx.$$

This result makes sense because if we integrate from a to b, then in the defining Riemann sum $\Delta x = \frac{b-a}{n}$, while if we integrate from b to a, $\Delta x = \frac{a-b}{n} = -\frac{b-a}{n}$, and this is the only change in the sum used to define the integral.

There are two additional properties of the definite integral that we need to understand. Recall that when we worked with derivative rules in Chapter 2, we found that both the Constant Multiple Rule and the Sum Rule held. The Constant Multiple Rule tells us that if f is a differentiable function and k is a constant, then

$$\frac{d}{dx}[kf(x)] = kf'(x),$$

and the Sum Rule states that if f and g are differentiable functions, then

$$\frac{d}{dx}[f(x) + g(x)] = f'(x) + g'(x).$$

These rules are useful because they enable us to deal individually with the simplest parts of certain functions and take advantage of the elementary operations of addition and multiplying by a constant. They also tell us that the process of taking the derivative respects addition and multiplying by constants in the simplest possible way.

It turns out that similar rules hold for the definite integral. First, let's consider the situation pictured in Figure 4.3.9,

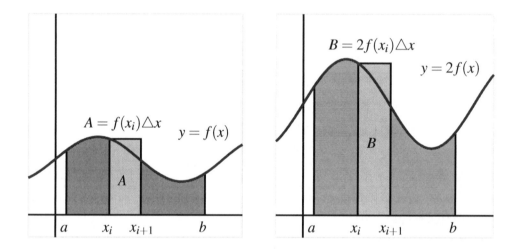

Figure 4.3.9: The areas bounded by $y = f(x)$ and $y = 2f(x)$ on $[a, b]$.

where we examine the effect of multiplying a function by a factor of 2 on the area it bounds with the x-axis. Because multiplying the function by 2 doubles its height at every x-value, we see that if we consider a typical rectangle from a Riemann sum, the difference in area comes from the changed height of the rectangle: $f(x_i)$ for the original function, versus $2f(x_i)$ in the doubled function, in the case of left sum. Hence, in Figure 4.3.9, we see that for the pictured rectangles with areas A and B, it follows $B = 2A$. As this will happen in every such rectangle, regardless of the value of n and the type of sum we use, we see that in the limit, the area of the red region bounded by $y = 2f(x)$ will be twice that of the area of the blue region bounded by $y = f(x)$. As there is nothing special about the value 2 compared to an arbitrary constant k, it turns out that the following general principle holds.

Constant Multiple Rule

If f is a continuous function and k is any real number then

$$\int_a^b k \cdot f(x)\,dx = k \int_a^b f(x)\,dx.$$

Finally, we see a similar situation geometrically with the sum of two functions f and g.

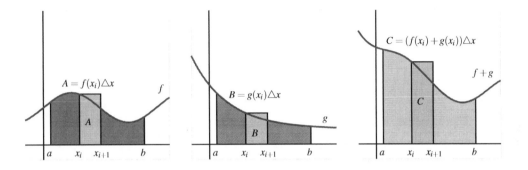

Figure 4.3.10: The areas bounded by $y = f(x)$ and $y = g(x)$ on $[a, b]$, as well as the area bounded by $y = f(x) + g(x)$.

In particular, as shown in Figure 4.3.10, if we take the sum of two functions f and g, at every point in the interval, the height of the function $f + g$ is given by $(f + g)(x_i) = f(x_i) + g(x_i)$, which is the sum of the individual function values of f and g (taken at left endpoints). Hence, for the pictured rectangles with areas A, B, and C, it follows that $C = A + B$, and because this will occur for every such rectangle, in the limit the area of the gray region will be the sum of the areas of the blue and red regions. Stated in terms of definite integrals, we have the following general rule.

Sum Rule

If f and g are continuous functions, then

$$\int_a^b [f(x) + g(x)]\,dx = \int_a^b f(x)\,dx + \int_a^b g(x)\,dx.$$

More generally, the Constant Multiple and Sum Rules can be combined to make the observation that for any continuous functions f and g and any constants c and k,

$$\int_a^b [cf(x) \pm kg(x)]\,dx = c \int_a^b f(x)\,dx \pm k \int_a^b g(x)\,dx.$$

Activity 4.3.3. Suppose that the following information is known about the functions f, g, x^2, and x^3:

- $\int_0^2 f(x)\,dx = -3$; $\int_2^5 f(x)\,dx = 2$

- $\int_0^2 g(x)\,dx = 4$; $\int_2^5 g(x)\,dx = -1$

- $\int_0^2 x^2\,dx = \frac{8}{3}$; $\int_2^5 x^2\,dx = \frac{117}{3}$

- $\int_0^2 x^3\,dx = 4$; $\int_2^5 x^3\,dx = \frac{609}{4}$

Use the provided information and the rules discussed in the preceding section to evaluate each of the following definite integrals.

a. $\int_5^2 f(x)\,dx$

b. $\int_0^5 g(x)\,dx$

c. $\int_0^5 (f(x) + g(x))\,dx$

d. $\int_2^5 (3x^2 - 4x^3)\,dx$

e. $\int_5^0 (2x^3 - 7g(x))\,dx$

4.3.3 How the definite integral is connected to a function's average value

One of the most valuable applications of the definite integral is that it provides a way to meaningfully discuss the average value of a function, even for a function that takes on infinitely many values. Recall that if we wish to take the average of n numbers $y_1, y_2, \ldots, y_n$, we do so by computing

$$AVG = \frac{y_1 + y_2 + \cdots + y_n}{n}.$$

Since integrals arise from Riemann sums in which we add n values of a function, it should not be surprising that evaluating an integral is something like averaging the output values of a function. Consider, for instance, the right Riemann sum R_n of a function f, which is given by

$$R_n = f(x_1)\Delta x + f(x_2)\Delta x + \cdots + f(x_n)\Delta x = (f(x_1) + f(x_2) + \cdots + f(x_n))\Delta x.$$

Since $\Delta x = \frac{b-a}{n}$, we can thus write

$$R_n = (f(x_1) + f(x_2) + \cdots + f(x_n)) \cdot \frac{b-a}{n}$$
$$= (b-a)\frac{f(x_1) + f(x_2) + \cdots + f(x_n)}{n}. \tag{4.3.1}$$

Here, we see that the right Riemann sum with n subintervals is the length of the interval $(b-a)$ times the average of the n function values found at the right endpoints. And just as with our efforts to compute area, we see that the larger the value of n we use, the more accurate our average of the values of f will be. Indeed, we will define the average value of f on $[a, b]$ to be

$$f_{\text{AVG}[a,b]} = \lim_{n \to \infty} \frac{f(x_1) + f(x_2) + \cdots + f(x_n)}{n}.$$

But we also know that for any continuous function f on $[a, b]$, taking the limit of a Riemann sum leads precisely to the definite integral. That is, $\lim_{n \to \infty} R_n = \int_a^b f(x)\,dx$, and thus taking the limit as $n \to \infty$ in Equation (4.3.1), we have that

$$\int_a^b f(x)\,dx = (b - a) \cdot f_{\text{AVG}[a,b]}. \tag{4.3.2}$$

Solving Equation (4.3.2) for $f_{\text{AVG}[a,b]}$, we have the following general principle.

Average value of a function

If f is a continuous function on $[a, b]$, then its average value on $[a, b]$ is given by the formula

$$f_{\text{AVG}[a,b]} = \frac{1}{b - a} \cdot \int_a^b f(x)\,dx.$$

Observe that Equation (4.3.2) tells us another way to interpret the definite integral: the definite integral of a function f from a to b is the length of the interval $(b - a)$ times the average value of the function on the interval. In addition, Equation (4.3.2) has a natural visual interpretation when the function f is nonnegative on $[a, b]$.

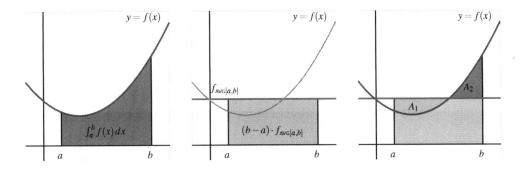

Figure 4.3.11: A function $y = f(x)$, the area it bounds, and its average value on $[a, b]$.

Consider Figure 4.3.11, where we see at left the shaded region whose area is $\int_a^b f(x)\,dx$, at center the shaded rectangle whose dimensions are $(b - a)$ by $f_{\text{AVG}[a,b]}$, and at right these two figures superimposed. Specifically, note that in dark green we show the horizontal line $y = f_{\text{AVG}[a,b]}$. Thus, the area of the green rectangle is given by $(b - a) \cdot f_{\text{AVG}[a,b]}$, which is precisely the value of $\int_a^b f(x)\,dx$. Said differently, the area of the blue region in the left figure is the same as that of the green rectangle in the center figure; this can also be seen by observing that the areas A_1 and A_2 in the rightmost figure appear to be equal. Ultimately, the average value of a function enables us to construct a rectangle whose area is the same as the value of the definite integral of the function on the interval. The java applet[2] at http://gvsu.edu/s/az provides an opportunity to explore how the average value of the function changes as the interval changes, through an image similar to that found in Figure 4.3.11.

Activity 4.3.4. Suppose that $v(t) = \sqrt{4 - (t - 2)^2}$ tells us the instantaneous velocity of a moving object on the interval $0 \le t \le 4$, where t is measured in minutes and v is measured in meters per minute.

   a. Sketch an accurate graph of $y = v(t)$. What kind of curve is $y = \sqrt{4 - (t - 2)^2}$?

   b. Evaluate $\int_0^4 v(t)\,dt$ exactly.

   c. In terms of the physical problem of the moving object with velocity $v(t)$, what is the meaning of $\int_0^4 v(t)\,dt$? Include units on your answer.

   d. Determine the exact average value of $v(t)$ on $[0, 4]$. Include units on your answer.

   e. Sketch a rectangle whose base is the line segment from $t = 0$ to $t = 4$ on the t-axis such that the rectangle's area is equal to the value of $\int_0^4 v(t)\,dt$. What is the rectangle's exact height?

   f. How can you use the average value you found in (d) to compute the total distance traveled by the moving object over $[0, 4]$?

Summary

- Any Riemann sum of a continuous function f on an interval $[a, b]$ provides an estimate of the net signed area bounded by the function and the horizontal axis on the interval. Increasing the number of subintervals in the Riemann sum improves the accuracy of this estimate, and letting the number of subintervals increase without bound results in the values of the corresponding Riemann sums approaching the exact value of the enclosed net signed area.

- When we take the just described limit of Riemann sums, we arrive at what we call the definite integral of f over the interval $[a, b]$. In particular, the symbol $\int_a^b f(x)\,dx$ denotes the definite integral of f over $[a, b]$, and this quantity is defined by the equation

$$\int_a^b f(x)\,dx = \lim_{n \to \infty} \sum_{i=1}^n f(x_i^*)\Delta x,$$

   where $\Delta x = \frac{b-a}{n}$, $x_i = a + i\Delta x$ (for $i = 0, \ldots, n$), and x_i^* satisfies $x_{i-1} \le x_i^* \le x_i$ (for $i = 1, \ldots, n$).

---

[2]David Austin, http://gvsu.edu/s/5r.

- The definite integral $\int_a^b f(x)\,dx$ measures the exact net signed area bounded by f and the horizontal axis on $[a,b]$; in addition, the value of the definite integral is related to what we call the average value of the function on $[a,b]$: $f_{\text{AVG}[a,b]} = \frac{1}{b-a} \cdot \int_a^b f(x)\,dx$. In the setting where we consider the integral of a velocity function v, $\int_a^b v(t)\,dt$ measures the exact change in position of the moving object on $[a,b]$; when v is nonnegative, $\int_a^b v(t)\,dt$ is the object's distance traveled on $[a,b]$.

- The definite integral is a sophisticated sum, and thus has some of the same natural properties that finite sums have. Perhaps most important of these is how the definite integral respects sums and constant multiples of functions, which can be summarized by the rule

$$\int_a^b [c f(x) \pm k g(x)]\,dx = c \int_a^b f(x)\,dx \pm k \int_a^b g(x)\,dx$$

where f and g are continuous functions on $[a,b]$ and c and k are arbitrary constants.

Exercises

1. Use the following figure, which shows a graph of $f(x)$ to find each of the indicated integrals.

Note that the first area (with vertical, red shading) is 55 and the second (with oblique, black shading) is 5.

A. $\int_a^b f(x)dx = $

B. $\int_b^c f(x)dx = $

C. $\int_a^c f(x)dx = $

D. $\int_a^c |f(x)|dx = $

2. Use the graph of $f(x)$ shown below to find the following integrals.

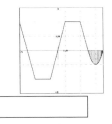

A. $\int_{-4}^0 f(x)dx = $

B. If the vertical red shaded area in the graph has area A, estimate: $\int_{-4}^6 f(x)dx = $
(*Your estimate may be written in terms of A.*)

3. Find the average value of $f(x) = 6x + 5$ over $[2, 6]$

average value = _____

4. The figure below to the left is a graph of $f(x)$, and below to the right is $g(x)$.

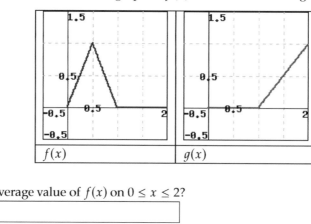

(a)

What is the average value of $f(x)$ on $0 \le x \le 2$?

avg value = _____

(b)

What is the average value of $g(x)$ on $0 \le x \le 2$?

avg value = _____

(c)

What is the average value of $f(x) \cdot g(x)$ on $0 \le x \le 2$?

avg value = _____

(d)

Is the following statement true?

$$\text{Average}(f) \cdot \text{Average}(g) = \text{Average}(f \cdot g)$$

5. Use the figure below, which shows the graph of $y = f(x)$, to answer the following questions.

A. Estimate the integral: $\int_{-3}^{3} f(x)\, dx \approx$ _____

(You will certainly want to use an enlarged version of the graph to obtain your estimate.)

B. Which of the following average values of f is larger?

6. Suppose $\int_{10}^{14.5} f(x)dx = 5$, $\int_{10}^{11.5} f(x)dx = 6$, $\int_{13}^{14.5} f(x)dx = 8$.

$\int_{11.5}^{13} f(x)dx =$ _____

$$\int_{13}^{11.5} (5f(x) - 6)dx = \boxed{}$$

7. The velocity of an object moving along an axis is given by the piecewise linear function v that is pictured in Figure 4.3.12. Assume that the object is moving to the right when its velocity is positive, and moving to the left when its velocity is negative. Assume that the given velocity function is valid for $t = 0$ to $t = 4$.

a. Write an expression involving definite integrals whose value is the total change in position of the object on the interval $[0, 4]$.

b. Use the provided graph of v to determine the value of the total change in position on $[0, 4]$.

c. Write an expression involving definite integrals whose value is the total distance traveled by the object on $[0, 4]$. What is the exact value of the total distance traveled on $[0, 4]$?

d. What is the object's exact average velocity on $[0, 4]$?

e. Find an algebraic formula for the object's position function on $[0, 1.5]$ that satisfies $s(0) = 0$.

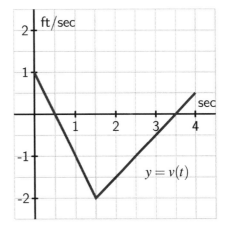

Figure 4.3.12: The velocity function of a moving object.

8. Suppose that the velocity of a moving object is given by $v(t) = t(t-1)(t-3)$, measured in feet per second, and that this function is valid for $0 \le t \le 4$.

a. Write an expression involving definite integrals whose value is the total change in position of the object on the interval $[0, 4]$.

b. Use appropriate technology (such as http://gvsu.edu/s/a9[3]) to compute Riemann sums to estimate the object's total change in position on $[0, 4]$. Work to ensure that your estimate is accurate to two decimal places, and explain how you know this to be the case.

c. Write an expression involving definite integrals whose value is the total distance traveled by the object on $[0, 4]$.

d. Use appropriate technology to compute Riemann sums to estimate the object's total distance travelled on $[0, 4]$. Work to ensure that your estimate is accurate to two decimal places, and explain how you know this to be the case.

---

[3]Marc Renault, Shippensburg University.

e. What is the object's average velocity on $[0, 4]$, accurate to two decimal places?

9. Consider the graphs of two functions f and g that are provided in Figure 4.3.13. Each piece of f and g is either part of a straight line or part of a circle.

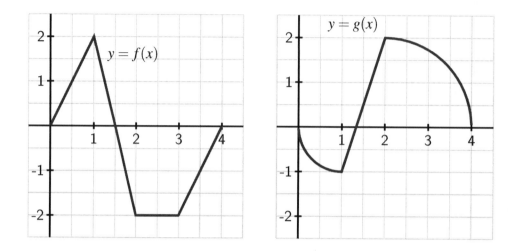

Figure 4.3.13: Two functions f and g.

a. Determine the exact value of $\int_0^1 [f(x) + g(x)] \, dx$.

b. Determine the exact value of $\int_1^4 [2f(x) - 3g(x)] \, dx$.

c. Find the exact average value of $h(x) = g(x) - f(x)$ on $[0, 4]$.

d. For what constant c does the following equation hold?

$$\int_0^4 c \, dx = \int_0^4 [f(x) + g(x)] \, dx$$

10. Let $f(x) = 3 - x^2$ and $g(x) = 2x^2$.

a. On the interval $[-1, 1]$, sketch a labeled graph of $y = f(x)$ and write a definite integral whose value is the exact area bounded by $y = f(x)$ on $[-1, 1]$.

b. On the interval $[-1, 1]$, sketch a labeled graph of $y = g(x)$ and write a definite integral whose value is the exact area bounded by $y = g(x)$ on $[-1, 1]$.

c. Write an expression involving a difference of definite integrals whose value is the exact area that lies between $y = f(x)$ and $y = g(x)$ on $[-1, 1]$.

d. Explain why your expression in (c) has the same value as the single integral $\int_{-1}^1 [f(x) - g(x)] \, dx$.

e. Explain why, in general, if $p(x) \geq q(x)$ for all x in $[a, b]$, the exact area between $y = p(x)$ and $y = q(x)$ is given by

$$\int_a^b [p(x) - q(x)]\, dx.$$

4.4 The Fundamental Theorem of Calculus

Motivating Questions

- How can we find the exact value of a definite integral without taking the limit of a Riemann sum?

- What is the statement of the Fundamental Theorem of Calculus, and how do antiderivatives of functions play a key role in applying the theorem?

- What is the meaning of the definite integral of a rate of change in contexts other than when the rate of change represents velocity?

Much of our work in Chapter 4 has been motivated by the velocity-distance problem: if we know the instantaneous velocity function, $v(t)$, for a moving object on a given time interval $[a, b]$, can we determine its exact distance traveled on $[a, b]$? In the vast majority of our discussion in Sections 4.1- Section 4.3, we have focused on the fact that this distance traveled is connected to the area bounded by $y = v(t)$ and the t-axis on $[a, b]$. In particular, for any nonnegative velocity function $y = v(t)$ on $[a, b]$, we know that the exact area bounded by the velocity curve and the t-axis on the interval tells us the total distance traveled, which is also the value of the definite integral $\int_a^b v(t)\, dt$. In the situation where velocity is sometimes negative, the total area bounded by the velocity function still tells us distance traveled, while the net signed area that the function bounds tells us the object's change in position.

Recall, for instance, the introduction to Section 4.2, where we observed that for the velocity function in Figure 4.4.1, the total distance D traveled by the moving object on $[a, b]$ is

$$D = A_1 + A_2 + A_3,$$

while the total change in the object's position on $[a, b]$ is

$$s(b) - s(a) = A_1 - A_2 + A_3.$$

The areas A_1, A_2, and A_3, which are each given by definite integrals, may be computed through limits of Riemann sums (and in select special circumstances through familiar geometric formulas).

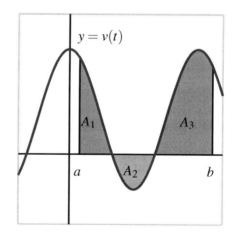

Figure 4.4.1: A velocity function that is sometimes negative.

In the present section we turn our attention to an alternate approach, similar to the one we encountered in Activity 4.1.3. To explore these ideas further, we consider the following preview activity.

Preview Activity 4.4.1. A student with a third floor dormitory window 32 feet off the ground tosses a water balloon straight up in the air with an initial velocity of 16 feet per second. It turns out that the instantaneous velocity of the water balloon is given by $v(t) = -32t + 16$, where v is measured in feet per second and t is measured in seconds.

a. Let $s(t)$ represent the height of the water balloon above ground at time t, and note that s is an antiderivative of v. That is, v is the derivative of s: $s'(t) = v(t)$. Find a formula for $s(t)$ that satisfies the initial condition that the balloon is tossed from 32 feet above ground. In other words, make your formula for s satisfy $s(0) = 32$.

b. When does the water balloon reach its maximum height? When does it land?

c. Compute $s(\frac{1}{2}) - s(0)$, $s(2) - s(\frac{1}{2})$, and $s(2) - s(0)$. What do these represent?

d. What is the total vertical distance traveled by the water balloon from the time it is tossed until the time it lands?

e. Sketch a graph of the velocity function $y = v(t)$ on the time interval $[0, 2]$. What is the total net signed area bounded by $y = v(t)$ and the t-axis on $[0, 2]$? Answer this question in two ways: first by using your work above, and then by using a familiar geometric formula to compute areas of certain relevant regions.

4.4.1 The Fundamental Theorem of Calculus

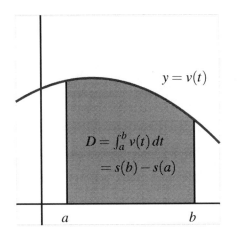

Consider the setting where we know the position function $s(t)$ of an object moving along an axis, as well as its corresponding velocity function $v(t)$, and for the moment let us assume that $v(t)$ is positive on $[a, b]$. Then, as shown in Figure 4.4.2, we know two different perspectives on the distance, D, the object travels: one is that $D = s(b) - s(a)$, which is the object's change in position. The other is that the distance traveled is the area under the velocity curve, which is given by the definite integral, so $D = \int_a^b v(t)\,dt$.

Figure 4.4.2: Finding distance traveled when we know a velocity function v.

Of course, since both of these expressions tell us the distance traveled, it follows that they are equal, so

$$s(b) - s(a) = \int_a^b v(t)\,dt. \qquad (4.4.1)$$

Furthermore, we know that Equation (4.4.1) holds even when velocity is sometimes negative, since $s(b) - s(a)$ is the object's change in position over $[a, b]$, which is simultaneously measured by the total net signed area on $[a, b]$ given by $\int_a^b v(t)\,dt$.

Perhaps the most powerful part of Equation (4.4.1) lies in the fact that we can compute the integral's value if we can find a formula for s. Remember, s and v are related by the fact that v is the derivative of s, or equivalently that s is an antiderivative of v. For example, if we have an object whose velocity is $v(t) = 3t^2 + 40$ feet per second (which is always nonnegative), and wish to know the distance traveled on the interval $[1, 5]$, we have that

$$D = \int_1^5 v(t)\,dt = \int_1^5 (3t^2 + 40)\,dt = s(5) - s(1),$$

where s is an antiderivative of v. We know that the derivative of t^3 is $3t^2$ and that the derivative of $40t$ is 40, so it follows that if $s(t) = t^3 + 40t$, then s is a function whose derivative is $v(t) = s'(t) = 3t^2 + 40$, and thus we have found an antiderivative of v. Therefore,

$$D = \int_1^5 3t^2 + 40\,dt = s(5) - s(1)$$
$$= (5^3 + 40 \cdot 5) - (1^3 + 40 \cdot 1) = 284 \text{ feet.}$$

Note the key lesson of this example: to find the distance traveled, we needed to compute the area under a curve, which is given by the definite integral. But to evaluate the integral, we found an antiderivative, s, of the velocity function, and then computed the total change in s on the interval. In particular, observe that we have found the exact area of the region shown in Figure 4.4.3, and done so without a familiar formula (such as those for the area of a triangle or circle) and without directly computing the limit of a Riemann sum.

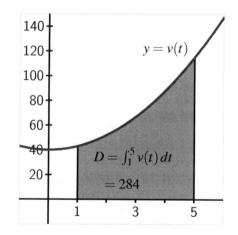

Figure 4.4.3: The exact area of the region enclosed by $v(t) = 3t^2 + 40$ on $[1, 5]$.

As we proceed to thinking about contexts other than just velocity and position, it is advantageous to have a shorthand symbol for a function's antiderivative. In the general setting of a continuous function f, we will often denote an antiderivative of f by F, so that the relationship between F and f is that $F'(x) = f(x)$ for all relevant x. Using the notation V in place of s (so that V is an antiderivative of v) in Equation (4.4.1), we find it is equivalent to write that

$$V(b) - V(a) = \int_a^b v(t)\,dt. \tag{4.4.2}$$

Now, in the general setting of wanting to evaluate the definite integral $\int_a^b f(x)\,dx$ for an arbitrary continuous function f, we could certainly think of f as representing the velocity of some moving object, and x as the variable that represents time. And again, Equations (4.4.1) and (4.4.2) hold for any continuous velocity function, even when v is sometimes negative. This leads us to see that Equation (4.4.2) tells us something even more important than the change in position of a moving object: it offers a shortcut route to evaluating any definite integral, provided that we can find an antiderivative of the integrand. The Fundamental Theorem of Calculus (FTC) summarizes these observations.

Fundamental Theorem of Calculus

If f is a continuous function on $[a, b]$, and F is any antiderivative of f, then $\int_a^b f(x)\,dx = F(b) - F(a)$.

A common alternate notation for $F(b) - F(a)$ is

$$F(b) - F(a) = F(x)\big|_a^b,$$

where we read the righthand side as "the function F evaluated from a to b." In this notation, the FTC says that

$$\int_a^b f(x)\,dx = F(x)\big|_a^b.$$

The FTC opens the door to evaluating exactly a wide range of integrals. In particular, if we are interested in a definite integral for which we can find an antiderivative F for the integrand f, then we can evaluate the integral exactly. For instance since $\frac{d}{dx}[\frac{1}{3}x^3] = x^2$, the FTC tells us that

$$\int_0^1 x^2\,dx = \frac{1}{3}x^3\bigg|_0^1$$
$$= \frac{1}{3}(1)^3 - \frac{1}{3}(0)^3$$
$$= \frac{1}{3}.$$

But finding an antiderivative can be far from simple; in fact, often finding a formula for an antiderivative is very hard or even impossible. While we can differentiate just about any function, even some relatively simple ones don't have an elementary antiderivative. A significant portion of integral calculus (which is the main focus of second semester college calculus) is devoted to understanding the problem of finding antiderivatives.

Activity 4.4.2. Use the Fundamental Theorem of Calculus to evaluate each of the following integrals exactly. For each, sketch a graph of the integrand on the relevant interval and write one sentence that explains the meaning of the value of the integral in terms of the (net signed) area bounded by the curve.

a. $\int_{-1}^{4}(2-2x)\,dx$

b. $\int_{0}^{\frac{\pi}{2}}\sin(x)\,dx$

c. $\int_{0}^{1}e^x\,dx$

d. $\int_{-1}^{1}x^5\,dx$

e. $\int_{0}^{2}(3x^3-2x^2-e^x)\,dx$

4.4.2 Basic antiderivatives

The general problem of finding an antiderivative is difficult. In part, this is due to the fact that we are trying to undo the process of differentiating, and the undoing is much more difficult than the doing. For example, while it is evident that an antiderivative of $f(x) = \sin(x)$ is $F(x) = -\cos(x)$ and that an antiderivative of $g(x) = x^2$ is $G(x) = \frac{1}{3}x^3$, combinations of f and g can be far more complicated. Consider such functions as

$$5\sin(x) - 4x^2, \quad x^2\sin(x), \quad \frac{\sin(x)}{x^2}, \quad \text{and } \sin(x^2).$$

What is involved in trying to find an antiderivative for each? From our experience with derivative rules, we know that while derivatives of sums and constant multiples of basic functions are simple to execute, derivatives involving products, quotients, and composites of familiar functions are much more complicated. Thus, it stands to reason that antidifferentiating products, quotients, and composites of basic functions may be even more challenging. We defer our study of all but the most elementary antiderivatives to later in the text.

We do note that each time we have a function for which we know its derivative, we have a *function-derivative pair*, which also leads us to knowing the antiderivative of a function. For instance, since we know that

$$\frac{d}{dx}[-\cos(x)] = \sin(x),$$

it follows that $F(x) = -\cos(x)$ is an antiderivative of $f(x) = \sin(x)$. It is equivalent to say that $f(x) = \sin(x)$ is the derivative of $F(x) = -\cos(x)$, and thus F and f together form the function-derivative pair. Clearly, every basic derivative rule leads us to such a pair, and thus to a known antiderivative. In Activity 4.4.3, we will construct a list of most of the basic antiderivatives we know at this time. Furthermore, those rules will enable us to antidifferentiate sums and constant multiples of basic functions. For example, if $f(x) = 5\sin(x) - 4x^2$, note that since $-\cos(x)$ is an antiderivative of $\sin(x)$ and $\frac{1}{3}x^3$ is an antiderivative of x^2, it follows that

$$F(x) = -5\cos(x) - \frac{4}{3}x^3$$

is an antiderivative of f, by the sum and constant multiple rules for differentiation.

Finally, before proceeding to build a list of common functions whose antiderivatives we know, we revisit the fact that each function has more than one antiderivative. Because the derivative of any constant is zero, any time we seek an arbitrary antiderivative, we may add a constant of our choice. For instance, if we want to determine an antiderivative of $g(x) = x^2$, we know that $G(x) = \frac{1}{3}x^3$ is one such function. But we could alternately have chosen $G(x) = \frac{1}{3}x^3 + 7$, since in this case as well, $G'(x) = x^2$. In some contexts later on in calculus, it is important to discuss the most general antiderivative of a function. If $g(x) = x^2$, we say that the *general antiderivative* of g is

$$G(x) = \frac{1}{3}x^3 + C,$$

where C represents an arbitrary real number constant. Regardless of the formula for g, including $+C$ in the formula for its antiderivative G results in the most general possible antiderivative.

Our primary current interest in antiderivatives is for use in evaluating definite integrals by the Fundamental Theorem of Calculus. In that situation, the arbitrary constant C is irrelevant, and thus we usually omit it. To see why, consider the definite integral

$$\int_0^1 x^2 \, dx.$$

For the integrand $g(x) = x^2$, suppose we find and use the general antiderivative $G(x) = \frac{1}{3}x^3 + C$. Then, by the FTC,

$$\begin{aligned}
\int_0^1 x^2 \, dx &= \frac{1}{3}x^3 + C \Big|_0^1 \\
&= \left(\frac{1}{3}(1)^3 + C\right) - \left(\frac{1}{3}(0)^3 + C\right) \\
&= \frac{1}{3} + C - 0 - C \\
&= \frac{1}{3}.
\end{aligned}$$

Specifically, we observe that the C-values appear as opposites in the evaluation of the integral and thus do not affect the definite integral's value. In the same way, the potential inclusion of $+C$ with the antiderivative has no bearing on any definite integral, and thus we generally choose to omit this possible constant whenever we evaluate an integral using the Fundamental Theorem of Calculus.

In the following activity, we work to build a list of basic functions whose antiderivatives we already know.

Activity 4.4.3. Use your knowledge of derivatives of basic functions to complete the above table of antiderivatives. For each entry, your task is to find a function F whose derivative is the given function f. When finished, use the FTC and the results in the table to evaluate the three given definite integrals.

given function, $f(x)$	antiderivative, $F(x)$
k, (k is constant)	
x^n, $n \neq -1$	
$\frac{1}{x}$, $x > 0$	
$\sin(x)$	
$\cos(x)$	
$\sec(x)\tan(x)$	
$\csc(x)\cot(x)$	
$\sec^2(x)$	
$\csc^2(x)$	
e^x	
a^x ($a > 1$)	
$\frac{1}{1+x^2}$	
$\frac{1}{\sqrt{1-x^2}}$	

Table 4.4.4: Familiar basic functions and their antiderivatives.

a. $\displaystyle\int_0^1 \left(x^3 - x - e^x + 2\right) dx$

b. $\displaystyle\int_0^{\pi/3} \left(2\sin(t) - 4\cos(t) + \sec^2(t) - \pi\right) dt$

c. $\displaystyle\int_0^1 \left(\sqrt{x} - x^2\right) dx$

4.4.3 The total change theorem

As we use the Fundamental Theorem of Calculus to evaluate definite integrals, it is essential that we remember and understand the meaning of the numbers we find. We briefly summarize three key interpretations to date.

- For a moving object with instantaneous velocity $v(t)$, the object's change in position on the time interval $[a, b]$ is given by $\int_a^b v(t)\, dt$, and whenever $v(t) \geq 0$ on $[a, b]$, $\int_a^b v(t)\, dt$ tells us the total distance traveled by the object on $[a, b]$.

- For any continuous function f, its definite integral $\int_a^b f(x)\, dx$ represents the total net signed area bounded by $y = f(x)$ and the x-axis on $[a, b]$, where regions that lie below the x-axis have a minus sign associated with their area.

• The value of a definite integral is linked to the average value of a function: for a continuous function f on $[a, b]$, its average value $f_{\text{AVG}[a,b]}$ is given by

$$f_{\text{AVG}[a,b]} = \frac{1}{b-a} \int_a^b f(x)\,dx.$$

The Fundamental Theorem of Calculus now enables us to evaluate exactly (without taking a limit of Riemann sums) any definite integral for which we are able to find an antiderivative of the integrand.

A slight change in notational perspective allows us to gain even more insight into the meaning of the definite integral. To begin, recall Equation (4.4.2), where we wrote the Fundamental Theorem of Calculus for a velocity function v with antiderivative V as

$$V(b) - V(a) = \int_a^b v(t)\,dt.$$

If we instead replace V with s (which represents position) and replace v with s' (since velocity is the derivative of position), Equation (4.4.2) equivalently reads

$$s(b) - s(a) = \int_a^b s'(t)\,dt. \tag{4.4.3}$$

In words, this version of the FTC tells us that the total change in the object's position function on a particular interval is given by the definite integral of the position function's derivative over that interval.

Of course, this result is not limited to only the setting of position and velocity. Writing the result in terms of a more general function f, we have the Total Change Theorem.

Total Change Theorem

If f is a continuously differentiable function on $[a, b]$ with derivative f', then $f(b) - f(a) = \int_a^b f'(x)\,dx$. That is, the definite integral of the derivative of a function on $[a, b]$ is the total change of the function itself on $[a, b]$.

The Total Change Theorem tells us more about the relationship between the graph of a function and that of its derivative. Recall Figure 1.4.1, which provided one of the first times we saw that heights on the graph of the derivative function come from slopes on the graph of the function itself. That observation occurred in the context where we knew f and were seeking f'; if now instead we think about knowing f' and seeking information about f, we can instead say the following:

differences in heights on f correspond to net signed areas bounded by f'.

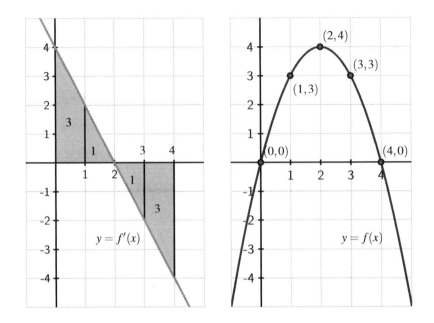

Figure 4.4.5: The graphs of $f'(x) = 4 - 2x$ (at left) and an antiderivative $f(x) = 4x - x^2$ at right. Differences in heights on f correspond to net signed areas bounded by f'.

To see why this is so, say we consider the difference $f(1) - f(0)$. Note that this value is 3, in part because $f(1) = 3$ and $f(0) = 0$, but also because the net signed area bounded by $y = f'(x)$ on $[0, 1]$ is 3. That is, $f(1) - f(0) = \int_0^1 f'(x) \, dx$. A similar pattern holds throughout, including the fact that since the total net signed area bounded by f' on $[0, 4]$ is 0, $\int_0^4 f'(x) \, dx = 0$, so it must be that $f(4) - f(0) = 0$, so $f(4) = f(0)$.

Beyond this general observation about area, the Total Change Theorem enables us to consider interesting and important problems where we know the rate of change, and answer key questions about the function whose rate of change we know.

Example 4.4.6. Suppose that pollutants are leaking out of an underground storage tank at a rate of $r(t)$ gallons/day, where t is measured in days. It is conjectured that $r(t)$ is given by the formula $r(t) = 0.0069t^3 - 0.125t^2 + 11.079$ over a certain 12-day period. The graph of $y = r(t)$ is given in Figure 4.4.7. What is the meaning of $\int_4^{10} r(t) \, dt$ and what is its value? What is the average rate at which pollutants are leaving the tank on the time interval $4 \leq t \leq 10$?

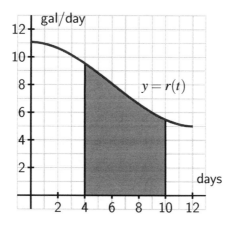

Figure 4.4.7: The rate $r(t)$ of pollution leaking from a tank, measured in gallons per day.

Solution. We know that since $r(t) \geq 0$, the value of $\int_4^{10} r(t)\,dt$ is the area under the curve on the interval $[4, 10]$. If we think about this area from the perspective of a Riemann sum, the rectangles will have heights measured in gallons per day and widths measured in days, thus the area of each rectangle will have units of

$$\frac{\text{gallons}}{\text{day}} \cdot \text{days} = \text{gallons}.$$

Thus, the definite integral tells us the total number of gallons of pollutant that leak from the tank from day 4 to day 10. The Total Change Theorem tells us the same thing: if we let $R(t)$ denote the function that measures the total number of gallons of pollutant that have leaked from the tank up to day t, then $R'(t) = r(t)$, and

$$\int_4^{10} r(t)\,dt = R(10) - R(4),$$

which is the total change in the function that measures total gallons leaked over time, thus the number of gallons that have leaked from day 4 to day 10.

To compute the exact value, we use the Fundamental Theorem of Calculus. Antidifferentiating $r(t) = 0.0069t^3 - 0.125t^2 + 11.079$, we find that

$$\int_4^{10} 0.0069t^3 - 0.125t^2 + 11.079\,dt = \left. 0.0069 \cdot \frac{1}{4}t^4 - 0.125 \cdot \frac{1}{3}t^3 + 11.079t \right|_4^{10}$$

$$\approx 44.282.$$

Thus, approximately 44.282 gallons of pollutant leaked over the six day time period.

To find the average rate at which pollutant leaked from the tank over $4 \leq t \leq 10$, we want to compute the average value of r on $[4, 10]$. Thus,

$$r_{\text{AVG}[4,10]} = \frac{1}{10-4} \int_4^{10} r(t)\, dt \approx \frac{44.282}{6} = 7.380,$$

which has its units measured in gallons per day.

Activity 4.4.4. During a 40-minute workout, a person riding an exercise machine burns calories at a rate of c calories per minute, where the function $y = c(t)$ is given in Figure 4.4.8. On the interval $0 \leq t \leq 10$, the formula for c is $c(t) = -0.05t^2 + t + 10$, while on $30 \leq t \leq 40$, its formula is $c(t) = -0.05t^2 + 3t - 30$.

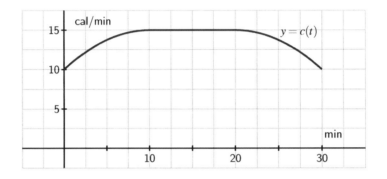

Figure 4.4.8: The rate $c(t)$ at which a person exercising burns calories, measured in calories per minute.

a. What is the exact total number of calories the person burns during the first 10 minutes of her workout?

b. Let $C(t)$ be an antiderivative of $c(t)$. What is the meaning of $C(40) - C(0)$ in the context of the person exercising? Include units on your answer.

c. Determine the exact average rate at which the person burned calories during the 40-minute workout.

d. At what time(s), if any, is the instantaneous rate at which the person is burning calories equal to the average rate at which she burns calories, on the time interval $0 \leq t \leq 40$?

Summary

- We can find the exact value of a definite integral without taking the limit of a Riemann sum or using a familiar area formula by finding the antiderivative of the integrand, and hence applying the Fundamental Theorem of Calculus.

- The Fundamental Theorem of Calculus says that if f is a continuous function on $[a, b]$ and F is an antiderivative of f, then

$$\int_a^b f(x)\,dx = F(b) - F(a).$$

  Hence, if we can find an antiderivative for the integrand f, evaluating the definite integral comes from simply computing the change in F on $[a, b]$.

- A slightly different perspective on the FTC allows us to restate it as the Total Change Theorem, which says that

$$\int_a^b f'(x)\,dx = f(b) - f(a),$$

  for any continuously differentiable function f. This means that the definite integral of the instantaneous rate of change of a function f on an interval $[a, b]$ is equal to the total change in the function f on $[a, b]$.

Exercises

1. Use the following figure, which shows a graph of $f(x)$ to find each of the indicated integrals.

Note that the first area (with vertical, red shading) is 9 and the second (with oblique, black shading) is 3.

A. $\int_a^b f(x)dx = $

B. $\int_b^c f(x)dx = $

C. $\int_a^c f(x)dx = $

D. $\int_a^c |f(x)|dx = $

2. Use the graph of $f(x)$ shown below to find the following integrals.

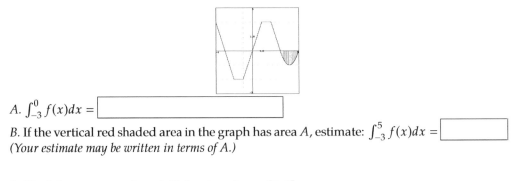

A. $\int_{-3}^{0} f(x)dx =$ [   ]

B. If the vertical red shaded area in the graph has area A, estimate: $\int_{-3}^{5} f(x)dx =$ [   ]
(*Your estimate may be written in terms of A.*)

3. Find the average value of $f(x) = 4x + 4$ over $[4, 9]$

average value = [   ]

4. The figure below to the left is a graph of $f(x)$, and below to the right is $g(x)$.

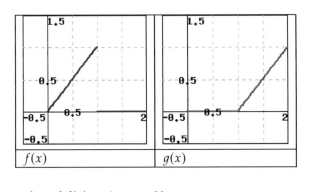

| $f(x)$ | $g(x)$ |

(*a*)
What is the average value of $f(x)$ on $0 \le x \le 2$?

avg value = [   ]
(*b*)
What is the average value of $g(x)$ on $0 \le x \le 2$?

avg value = [   ]
(*c*)
What is the average value of $f(x) \cdot g(x)$ on $0 \le x \le 2$?

avg value = [   ]
(*d*)
Is the following statement true?

$$\text{Average}(f) \cdot \text{Average}(g) = \text{Average}(f \cdot g)$$

5. Use the figure below, which shows the graph of $y = f(x)$, to answer the following questions.

A. Estimate the integral: $\int_{-3}^{3} f(x)\,dx \approx$ [   ]

(You will certainly want to use an enlarged version of the graph to obtain your estimate.)

B. Which of the following average values of f is larger?

6. Suppose $\displaystyle\int_{3}^{7.5} f(x)dx = 4, \quad \int_{3}^{4.5} f(x)dx = 1, \quad \int_{6}^{7.5} f(x)dx = 1.$

$\displaystyle\int_{4.5}^{6} f(x)dx =$ [   ]

$\displaystyle\int_{6}^{4.5} (4f(x) - 1)dx =$ [   ]

7. The instantaneous velocity (in meters per minute) of a moving object is given by the function v as pictured in Figure 4.4.9. Assume that on the interval $0 \le t \le 4$, $v(t)$ is given by $v(t) = -\frac{1}{4}t^3 + \frac{3}{2}t^2 + 1$, and that on every other interval v is piecewise linear, as shown.

a. Determine the exact distance traveled by the object on the time interval $0 \le t \le 4$.

b. What is the object's average velocity on $[12, 24]$?

c. At what time is the object's acceleration greatest?

d. Suppose that the velocity of the object is increased by a constant value c for all values of t. What value of c will make the object's total distance traveled on $[12, 24]$ be 210 meters?

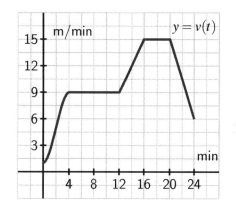

Figure 4.4.9: The velocity function of a moving body.

8. A function f is given piecewise by the formula

$$f(x) = \begin{cases} -x^2 + 2x + 1, & \text{if } 0 \le x < 2 \\ -x + 3, & \text{if } 2 \le x < 3 \\ x^2 - 8x + 15, & \text{if } 3 \le x \le 5 \end{cases}$$

a. Determine the exact value of the net signed area enclosed by f and the x-axis on the interval $[2, 5]$.

b. Compute the exact average value of f on $[0, 5]$.

c. Find a formula for a function g on $5 \leq x \leq 7$ so that if we extend the above definition of f so that $f(x) = g(x)$ if $5 \leq x \leq 7$, it follows that $\int_0^7 f(x)\,dx = 0$.

9. When an aircraft attempts to climb as rapidly as possible, its climb rate (in feet per minute) decreases as altitude increases, because the air is less dense at higher altitudes. Given below is a table showing performance data for a certain single engine aircraft, giving its climb rate at various altitudes, where $c(h)$ denotes the climb rate of the airplane at an altitude h.

h (feet)	0	1000	2000	3000	4000	5000	6000	7000	8000	9000	10,000
c (ft/min)	925	875	830	780	730	685	635	585	535	490	440

Let a new function called $m(h)$ measure the number of minutes required for a plane at altitude h to climb the next foot of altitude.

a. Determine a similar table of values for $m(h)$ and explain how it is related to the table above. Be sure to explain the units.

b. Give a careful interpretation of a function whose derivative is $m(h)$. Describe what the input is and what the output is. Also, explain in plain English what the function tells us.

c. Determine a definite integral whose value tells us exactly the number of minutes required for the airplane to ascend to 10,000 feet of altitude. Clearly explain why the value of this integral has the required meaning.

d. Use the Riemann sum M_5 to estimate the value of the integral you found in (c). Include units on your result.

10. In Chapter 1, we showed that for an object moving along a straight line with position function $s(t)$, the object's "average velocity on the interval $[a, b]$" is given by

$$AV_{[a,b]} = \frac{s(b) - s(a)}{b - a}.$$

More recently in Chapter 4, we found that for an object moving along a straight line with velocity function $v(t)$, the object's "average value of its velocity function on $[a, b]$" is

$$v_{AVG[a,b]} = \frac{1}{b - a} \int_a^b v(t)\,dt.$$

Are the "average velocity on the interval $[a, b]$" and the "average value of the velocity function on $[a, b]$" the same thing? Why or why not? Explain.

CHAPTER 5

Evaluating Integrals

5.1 Constructing Accurate Graphs of Antiderivatives

Motivating Questions

- Given the graph of a function's derivative, how can we construct a completely accurate graph of the original function?

- How many antiderivatives does a given function have? What do those antiderivatives all have in common?

- Given a function f, how does the rule $A(x) = \int_0^x f(t)\,dt$ define a new function A?

A recurring theme in our discussion of differential calculus has been the question "Given information about the derivative of an unknown function f, how much information can we obtain about f itself?" For instance, in Activity 1.8.3, we explored the situation where the graph of $y = f'(x)$ was known (along with the value of f at a single point) and endeavored to sketch a possible graph of f near the known point. In Example 3.1.4 — and indeed throughout Section 3.1 — we investigated how the first derivative test enables us to use information regarding f' to determine where the original function f is increasing and decreasing, as well as where f has relative extreme values. Further, if we know a formula or graph of f', by computing f'' we can find where the original function f is concave up and concave down. Thus, the combination of knowing f' and f'' enables us to fully understand the shape of the graph of f.

We returned to this question in even more detail in Section 4.1; there, we considered the situation where we knew the instantaneous velocity of a moving object and worked from that information to determine as much information as possible about the object's position function. We found key connections between the net-signed area under the velocity function and the corresponding change in position of the function; in Section 4.4, the Total Change Theorem further illuminated these connections between f' and f in a more general setting, such as the one found in Figure 4.4.5, showing that the total change in the value of f over an interval $[a, b]$ is determined by the exact net-signed area bounded by f' and the x-axis on the same interval.

In what follows, we explore these issues still further, with a particular emphasis on the situation where we possess an accurate graph of the derivative function along with a single value of the function f. From that information, we desire to completely determine an accurate graph of f that not only represents correctly where f is increasing, decreasing, concave up, and concave down, but also allows us to find an accurate function value at any point of interest to us.

Preview Activity 5.1.1. Suppose that the following information is known about a function f: the graph of its derivative, $y = f'(x)$, is given in Figure 5.1.1. Further, assume that f' is piecewise linear (as pictured) and that for $x \le 0$ and $x \ge 6$, $f'(x) = 0$. Finally, it is given that $f(0) = 1$.

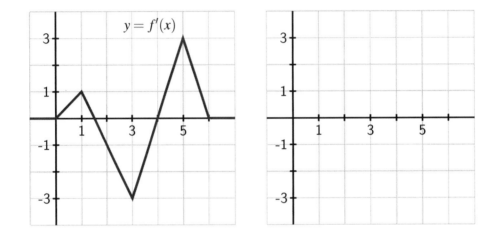

Figure 5.1.1: At left, the graph of $y = f'(x)$; at right, axes for plotting $y = f(x)$.

a. On what interval(s) is f an increasing function? On what intervals is f decreasing?

b. On what interval(s) is f concave up? concave down?

c. At what point(s) does f have a relative minimum? a relative maximum?

d. Recall that the Total Change Theorem tells us that

$$f(1) - f(0) = \int_0^1 f'(x)\,dx.$$

What is the exact value of $f(1)$?

e. Use the given information and similar reasoning to that in (d) to determine the exact value of $f(2)$, $f(3)$, $f(4)$, $f(5)$, and $f(6)$.

f. Based on your responses to all of the preceding questions, sketch a complete and accurate graph of $y = f(x)$ on the axes provided, being sure to indicate the behavior of f for $x < 0$ and $x > 6$.

5.1.1 Constructing the graph of an antiderivative

Preview Activity 5.1.1 demonstrates that when we can find the exact area under a given graph on any given interval, it is possible to construct an accurate graph of the given function's antiderivative: that is, we can find a representation of a function whose derivative is the given one. While we have considered this question at different points throughout our study, it is important to note here that we now can determine not only the overall shape of the antiderivative, but also the actual *height* of the antiderivative at any point of interest.

Indeed, this is one key consequence of the Fundamental Theorem of Calculus: if we know a function f and wish to know information about its antiderivative, F, provided that we have some starting point a for which we know the value of $F(a)$, we can determine the value of $F(b)$ via the definite integral. In particular, since $F(b) - F(a) = \int_a^b f(x)\,dx$, it follows that

$$F(b) = F(a) + \int_a^b f(x)\,dx. \tag{5.1.1}$$

Moreover, in the discussion surrounding Figure 4.4.5, we made the observation that differences in heights of a function correspond to net-signed areas bounded by its derivative. Rephrasing this in terms of a given function f and its antiderivative F, we observe that on an interval $[a, b]$,

> *differences in heights on the antiderivative (such as $F(b) - F(a)$) correspond to the net-signed area bounded by the original function on the interval $[a, b]$ ($\int_a^b f(x)\,dx$).*

For example, say that $f(x) = x^2$ and that we are interested in an antiderivative of f that satisfies $F(1) = 2$. Thinking of $a = 1$ and $b = 2$ in Equation (5.1.1), it follows from the Fundamental Theorem of Calculus that

$$F(2) = F(1) + \int_1^2 x^2\,dx$$

$$= 2 + \frac{1}{3}x^3\Big|_1^2$$

$$= 2 + \left(\frac{8}{3} - \frac{1}{3}\right)$$

$$= \frac{13}{3}.$$

In this way, we see that if we are given a function f for which we can find the exact net-signed area bounded by f on a given interval, along with one value of a corresponding antiderivative F, we can find any other value of F that we seek, and in this way construct a completely accurate graph of F. We have two main options for finding the exact net-signed area: using the Fundamental Theorem of Calculus (which requires us to find an algebraic formula for an antiderivative of the given function f), or, in the case where f has nice geometric properties, finding net-signed areas through the use of known area formulas.

Activity 5.1.2. Suppose that the function $y = f(x)$ is given by the graph shown in Figure 5.1.2, and that the pieces of f are either portions of lines or portions of circles. In addition, let F be an antiderivative of f and say that $F(0) = -1$. Finally, assume that for $x \leq 0$ and $x \geq 7$, $f(x) = 0$.

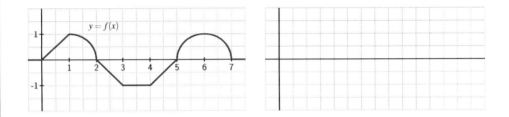

Figure 5.1.2: At left, the graph of $y = f(x)$.

a. On what interval(s) is F an increasing function? On what intervals is F decreasing?

b. On what interval(s) is F concave up? concave down? neither?

c. At what point(s) does F have a relative minimum? a relative maximum?

d. Use the given information to determine the exact value of $F(x)$ for $x = 1, 2, \ldots, 7$. In addition, what are the values of $F(-1)$ and $F(8)$?

e. Based on your responses to all of the preceding questions, sketch a complete and accurate graph of $y = F(x)$ on the axes provided, being sure to indicate the behavior of F for $x < 0$ and $x > 7$. Clearly indicate the scale on the vertical and horizontal axes of your graph.

f. What happens if we change one key piece of information: in particular, say that G is an antiderivative of f and $G(0) = 0$. How (if at all) would your answers to the preceding questions change? Sketch a graph of G on the same axes as the graph of F you constructed in (e).

5.1.2 Multiple antiderivatives of a single function

In the final question of Activity 5.1.2, we encountered a very important idea: a given function f has more than one antiderivative. In addition, any antiderivative of f is determined uniquely by identifying the value of the desired antiderivative at a single point. For example, suppose that f is the function given at left in Figure 5.1.3,

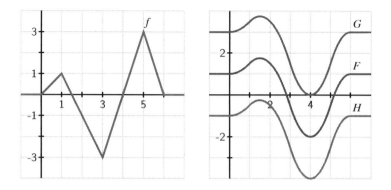

Figure 5.1.3: At left, the graph of $y = f(x)$. At right, three different antiderivatives of f.

and we say that F is an antiderivative of f that satisfies $F(0) = 1$.

Then, using Equation (5.1.1), we can compute $F(1) = 1.5$, $F(2) = 1.5$, $F(3) = -0.5$, $F(4) = -2$, $F(5) = -0.5$, and $F(6) = 1$, plus we can use the fact that $F' = f$ to ascertain where F is increasing and decreasing, concave up and concave down, and has relative extremes and inflection points. Through work similar to what we encountered in Preview Activity 5.1.1 and Activity 5.1.2, we ultimately find that the graph of F is the one given in blue in Figure 5.1.3.

If we instead chose to consider a function G that is an antiderivative of f but has the property that $G(0) = 3$, then G will have the exact same shape as F (since both share the derivative f), but G will be shifted vertically away from the graph of F, as pictured in red in Figure 5.1.3. Note that $G(1) - G(0) = \int_0^1 f(x)\,dx = 0.5$, just as $F(1) - F(0) = 0.5$,, but since $G(0) = 3$, $G(1) = G(0) + 0.5 = 3.5$, whereas $F(1) = F(0) + 0.5 = 1.5$, since $F(0) = 1$. In the same way, if we assigned a different initial value to the antiderivative, say $H(0) = -1$, we would get still another antiderivative, as shown in magenta in Figure 5.1.3.

This example demonstrates an important fact that holds more generally:

> If G and H are both antiderivatives of a function f, then the function $G - H$ must be constant.

To see why this result holds, observe that if G and H are both antiderivatives of f, then $G' = f$ and $H' = f$. Hence,

$$\frac{d}{dx}[G(x) - H(x)] = G'(x) - H'(x) = f(x) - f(x) = 0.$$

Since the only way a function can have derivative zero is by being a constant function, it follows that the function $G - H$ must be constant.

Further, we now see that if a function has a single antiderivative, it must have infinitely many: we can add any constant of our choice to the antiderivative and get another antiderivative.

For this reason, we sometimes refer to the *general antiderivative* of a function f. For example, if $f(x) = x^2$, its general antiderivative is $F(x) = \frac{1}{3}x^3 + C$, where we include the "$+C$" to indicate that F includes *all* of the possible antiderivatives of f. To identify a particular antiderivative of f, we must be provided a single value of the antiderivative F (this value is often called an *initial condition*). In the present example, suppose that condition is $F(2) = 3$; substituting the value of 2 for x in $F(x) = \frac{1}{3}x^3 + C$, we find that

$$3 = \frac{1}{3}(2)^3 + C,$$

and thus $C = 3 - \frac{8}{3} = \frac{1}{3}$. Therefore, the particular antiderivative in this case is $F(x) = \frac{1}{3}x^3 + \frac{1}{3}$.

Activity 5.1.3. For each of the following functions, sketch an accurate graph of the antiderivative that satisfies the given initial condition. In addition, sketch the graph of two additional antiderivatives of the given function, and state the corresponding initial conditions that each of them satisfy. If possible, find an algebraic formula for the antiderivative that satisfies the initial condition.

a. original function: $g(x) = |x| - 1$; initial condition: $G(-1) = 0$; interval for sketch: $[-2, 2]$

b. original function: $h(x) = \sin(x)$; initial condition: $H(0) = 1$; interval for sketch: $[0, 4\pi]$

c. original function: $p(x) = \begin{cases} x^2, & \text{if } 0 < x \le 1 \\ -(x-2)^2, & \text{if } 1 < x < 2; \text{ initial condition: } P(0) = 1; \\ 0 & \text{otherwise} \end{cases}$

   interval for sketch: $[-1, 3]$

5.1.3 Functions defined by integrals

In Equation (5.1.1), we found an important rule that enables us to compute the value of the antiderivative F at a point b, provided that we know $F(a)$ and can evaluate the definite integral from a to b of f. Again, that rule is

$$F(b) = F(a) + \int_a^b f(x)\,dx.$$

In several examples, we have used this formula to compute several different values of $F(b)$ and then plotted the points $(b, F(b))$ to assist us in generating an accurate graph of F. That suggests that we may want to think of b, the upper limit of integration, as a variable itself. To that end, we introduce the idea of an *integral function*, a function whose formula involves a definite integral.

Given a continuous function f, we define the corresponding integral function A according to the rule

$$A(x) = \int_a^x f(t)\,dt. \qquad (5.1.2)$$

Note particularly that because we are using the variable x as the independent variable in the function A, and x determines the other endpoint of the interval over which we integrate (starting from a), we need to use a variable other than x as the variable of integration. A standard choice is t, but any variable other than x is acceptable.

One way to think of the function A is as the "net-signed area from a up to x" function, where we consider the region bounded by $y = f(t)$ on the relevant interval. For example, in Figure 5.1.4, we see a given function f pictured at left, and its corresponding area function (choosing $a = 0$), $A(x) = \int_0^x f(t)\,dt$ shown at right.

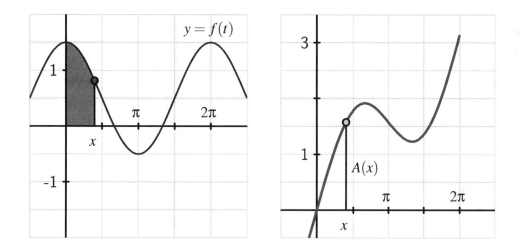

Figure 5.1.4: At left, the graph of the given function f. At right, the area function $A(x) = \int_0^x f(t)\,dt$.

Note particularly that the function A measures the net-signed area from $t = 0$ to $t = x$ bounded by the curve $y = f(t)$; this value is then reported as the corresponding height on the graph of $y = A(x)$. It is even more natural to think of this relationship between f and A dynamically. At http://gvsu.edu/s/cz, we find a java applet[1] that brings the static picture in Figure 5.1.4 to life. There, the user can move the red point on the function f and see how the corresponding height changes at the light blue point on the graph of A.

The choice of a is somewhat arbitrary. In the activity that follows, we explore how the value of a affects the graph of the integral function, as well as some additional related is-

[1]David Austin, Grand Valley State University

sues.

Activity 5.1.4. Suppose that g is given by the graph at left in Figure 5.1.5 and that A is the corresponding integral function defined by $A(x) = \int_1^x g(t)\, dt$.

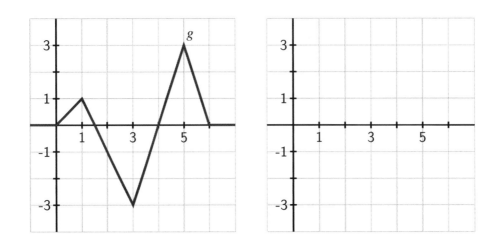

Figure 5.1.5: At left, the graph of $y = g(t)$; at right, axes for plotting $y = A(x)$, where A is defined by the formula $A(x) = \int_1^x g(t)\, dt$.

a. On what interval(s) is A an increasing function? On what intervals is A decreasing? Why?

b. On what interval(s) do you think A is concave up? concave down? Why?

c. At what point(s) does A have a relative minimum? a relative maximum?

d. Use the given information to determine the exact values of $A(0)$, $A(1)$, $A(2)$, $A(3)$, $A(4)$, $A(5)$, and $A(6)$.

e. Based on your responses to all of the preceding questions, sketch a complete and accurate graph of $y = A(x)$ on the axes provided, being sure to indicate the behavior of A for $x < 0$ and $x > 6$.

f. How does the graph of B compare to A if B is instead defined by $B(x) = \int_0^x g(t)\, dt$?

Summary

- Given the graph of a function f, we can construct the graph of its antiderivative F provided that (a) we know a starting value of F, say $F(a)$, and (b) we can evaluate the integral $\int_a^b f(x)\,dx$ exactly for relevant choices of a and b. For instance, if we wish to know $F(3)$, we can compute $F(3) = F(a) + \int_a^3 f(x)\,dx$. When we combine this information about the function values of F together with our understanding of how the behavior of $F' = f$ affects the overall shape of F, we can develop a completely accurate graph of the antiderivative F.

- Because the derivative of a constant is zero, if F is an antiderivative of f, it follows that $G(x) = F(x) + C$ will also be an antiderivative of f. Moreover, any two antiderivatives of a function f differ precisely by a constant. Thus, any function with at least one antiderivative in fact has infinitely many, and the graphs of any two antiderivatives will differ only by a vertical translation.

- Given a function f, the rule $A(x) = \int_a^x f(t)\,dt$ defines a new function A that measures the net-signed area bounded by f on the interval $[a, x]$. We call the function A the integral function corresponding to f.

Exercises

1. Use the graph of $f(x)$ shown below to find the following integrals.

A. $\int_{-5}^0 f(x)dx =$

B. If the vertical red shaded area in the graph has area A, estimate: $\int_{-5}^7 f(x)dx =$
(Your estimate may be written in terms of A.)

2. Consider the graph of the function $f(x)$ shown below.

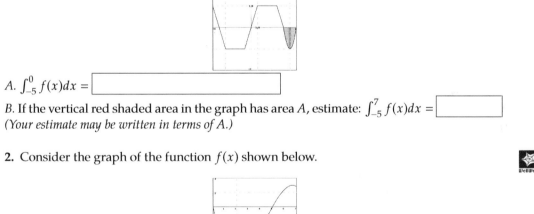

(Click on the graph for a larger version)
A. Estimate the integral
$\int_0^7 f(x)dx \approx$
B. If F is an antiderivative of the same function f and $F(0) = 40$, estimate $F(7)$:
$F(7) \approx$

3. Assume f' is given by the graph below. Suppose f is continuous and that $f(5) = 0$.

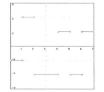

Sketch, on a sheet of work paper, an accurate graph of f, and use it to find each of

$f(0) = $

and

$f(7) = $

Then find the value of the integral:

$\int_0^7 f'(x)\,dx = $

(Note that you can do this in two different ways!)

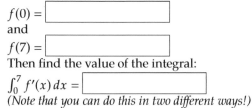

4. The figure below shows f.

If $F' = f$ and $F(0) = 0$, find $F(b)$ for $b = 1, 2, 3, 4, 5, 6$, and fill these values in the following table.

b	1	2	3	4	5	6
$F(b)$						

5. A moving particle has its velocity given by the quadratic function v pictured in Figure 5.1.6. In addition, it is given that $A_1 = \frac{7}{6}$ and $A_2 = \frac{8}{3}$, as well as that for the corresponding position function s, $s(0) = 0.5$.

a. Use the given information to determine $s(1)$, $s(3)$, $s(5)$, and $s(6)$.

b. On what interval(s) is s increasing? On what interval(s) is s decreasing?

c. On what interval(s) is s concave up? On what interval(s) is s concave down?

d. Sketch an accurate, labeled graph of s on the axes at right in Figure 5.1.6.

e. Note that $v(t) = -2 + \frac{1}{2}(t-3)^2$. Find a formula for s.

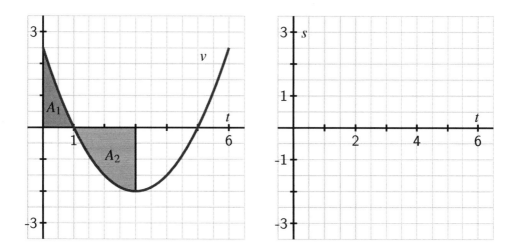

Figure 5.1.6: At left, the given graph of v. At right, axes for plotting s.

6. A person exercising on a treadmill experiences different levels of resistance and thus burns calories at different rates, depending on the treadmill's setting. In a particular workout, the rate at which a person is burning calories is given by the piecewise constant function c pictured in Figure 5.1.7. Note that the units on c are "calories per minute."

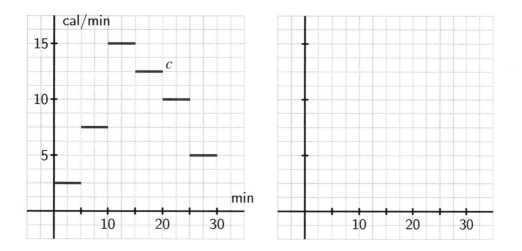

Figure 5.1.7: At left, the given graph of c. At right, axes for plotting C.

a. Let C be an antiderivative of c. What does the function C measure? What are its units?

b. Assume that $C(0) = 0$. Determine the exact value of $C(t)$ at the values $t = 5, 10, 15, 20, 25, 30$.

c. Sketch an accurate graph of C on the axes provided at right in Figure 5.1.7. Be certain to label the scale on the vertical axis.

d. Determine a formula for C that does not involve an integral and is valid for $5 \le t \le 10$.

7. Consider the piecewise linear function f given in Figure 5.1.8. Let the functions A, B, and C be defined by the rules $A(x) = \int_{-1}^{x} f(t)\,dt$, $B(x) = \int_{0}^{x} f(t)\,dt$, and $C(x) = \int_{1}^{x} f(t)\,dt$.

a. For the values $x = -1, 0, 1, \ldots, 6$, make a table that lists corresponding values of $A(x)$, $B(x)$, and $C(x)$.

b. On the axes provided in Figure 5.1.8, sketch the graphs of A, B, and C.

c. How are the graphs of A, B, and C related?

d. How would you best describe the relationship between the function A and the function f?

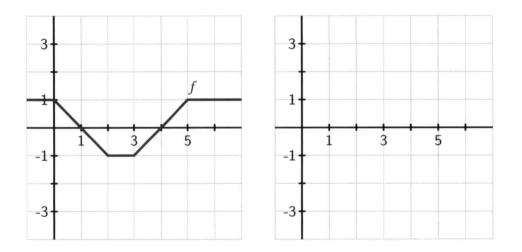

Figure 5.1.8: At left, the given graph of f. At right, axes for plotting A, B, and C.

5.2 The Second Fundamental Theorem of Calculus

Motivating Questions

- How does the integral function $A(x) = \int_1^x f(t)\,dt$ define an antiderivative of f?

- What is the statement of the Second Fundamental Theorem of Calculus?

- How do the First and Second Fundamental Theorems of Calculus enable us to formally see how differentiation and integration are almost inverse processes?

In Section 4.4, we learned the Fundamental Theorem of Calculus (FTC), which from here forward will be referred to as the *First* Fundamental Theorem of Calculus, as in this section we develop a corresponding result that follows it. In particular, recall that the First FTC tells us that if f is a continuous function on $[a, b]$ and F is any antiderivative of f (that is, $F' = f$), then

$$\int_a^b f(x)\,dx = F(b) - F(a).$$

We have typically used this result in two settings: (1) where f is a function whose graph we know and for which we can compute the exact area bounded by f on a certain interval $[a, b]$, we can compute the change in an antiderivative F over the interval; and (2) where f is a function for which it is easy to determine an algebraic formula for an antiderivative, we may evaluate the integral exactly and hence determine the net-signed area bounded by the function on the interval. For the former, see Preview Activity 5.1.1 or Activity 5.1.2. For the latter, we can easily evaluate exactly integrals such as

$$\int_1^4 x^2\,dx,$$

since we know that the function $F(x) = \frac{1}{3}x^3$ is an antiderivative of $f(x) = x^2$. Thus,

$$\begin{aligned}
\int_1^4 x^2\,dx &= \frac{1}{3}x^3 \Big|_1^4 \\
&= \frac{1}{3}(4)^3 - \frac{1}{3}(1)^3 \\
&= 21.
\end{aligned}$$

Here we see that the First FTC can be viewed from at least two perspectives: first, as a tool to find the difference $F(b) - F(a)$ for an antiderivative F of the integrand f. In this situation, we need to be able to determine the value of the integral $\int_a^b f(x)\,dx$ exactly, perhaps through known geometric formulas for area. It is possible that we may not have a formula for F itself. From a second perspective, the First FTC provides a way to find the exact value of a definite integral, and hence a certain net-signed area exactly, by finding an antiderivative of the integrand and evaluating its total change over the interval. In this latter case, we need to know a formula for the antiderivative F, as this enables us to compute net-signed areas exactly through definite integrals, as demonstrated in Figure 5.2.1.

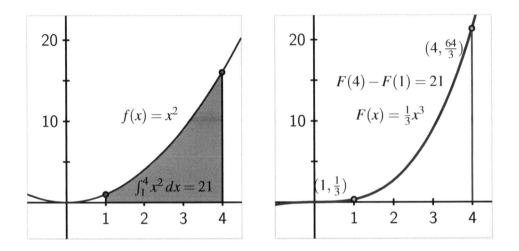

Figure 5.2.1: At left, the graph of $f(x) = x^2$ on the interval $[1, 4]$ and the area it bounds. At right, the antiderivative function $F(x) = \frac{1}{3}x^3$, whose total change on $[1, 4]$ is the value of the definite integral at left.

We recall further that the value of a definite integral may have additional meaning depending on context: change in position when the integrand is a velocity function, total pollutant leaked from a tank when the integrand is the rate at which pollution is leaking, or other total changes that correspond to a given rate function that is the integrand. In addition, the value of the definite integral is always connected to the average value of a continuous function on a given interval: $f_{\text{AVG}[a,b]} = \frac{1}{b-a} \int_a^b f(x)\,dx$.

Next, remember that in the last part of Section 5.1, we studied integral functions of the form $A(x) = \int_c^x f(t)\,dt$. Figure 5.1.4 is a particularly important image to keep in mind as we work with integral functions, and the corresponding java applet at gvsu.edu/s/cz is likewise foundational to our understanding of the function A. In what follows, we use the First FTC to gain additional understanding of the function $A(x) = \int_c^x f(t)\,dt$, where the integrand f is given (either through a graph or a formula), and c is a constant. In particular, we investigate further the special nature of the relationship between the functions A and f.

Preview Activity 5.2.1. Consider the function A defined by the rule

$$A(x) = \int_1^x f(t)\,dt,$$

where $f(t) = 4 - 2t$.

   a. Compute $A(1)$ and $A(2)$ exactly.

   b. Use the First Fundamental Theorem of Calculus to find a formula for $A(x)$ that does not involve integrals. That is, use the first FTC to evaluate $\int_1^x (4 - 2t) \, dt$.

   c. Observe that f is a linear function; what kind of function is A?

   d. Using the formula you found in (b) that does not involve integrals, compute $A'(x)$.

   e. While we have defined f by the rule $f(t) = 4 - 2t$, it is equivalent to say that f is given by the rule $f(x) = 4 - 2x$. What do you observe about the relationship between A and f?

5.2.1 The Second Fundamental Theorem of Calculus

The result of Preview Activity 5.2.1 is not particular to the function $f(t) = 4 - 2t$, nor to the choice of "1" as the lower bound in the integral that defines the function A. For instance, if we let $f(t) = \cos(t) - t$ and set $A(x) = \int_2^x f(t) \, dt$, then we can determine a formula for A without integrals by the First FTC. Specifically,

$$A(x) = \int_2^x (\cos(t) - t) \, dt$$

$$= \sin(t) - \frac{1}{2}t^2 \Big|_2^x$$

$$= \sin(x) - \frac{1}{2}x^2 - (\sin(2) - 2).$$

Differentiating $A(x)$, since $(\sin(2) - 2)$ is constant, it follows that

$$A'(x) = \cos(x) - x,$$

and thus we see that $A'(x) = f(x)$. This tells us that for this particular choice of f, A is an antiderivative of f. More specifically, since $A(2) = \int_2^2 f(t) \, dt = 0$, A is the only antiderivative of f for which $A(2) = 0$.

In general, if f is any continuous function, and we define the function A by the rule

$$A(x) = \int_c^x f(t) \, dt,$$

where c is an arbitrary constant, then we can show that A is an antiderivative of f. To see why, let's demonstrate that $A'(x) = f(x)$ by using the limit definition of the derivative. Doing so, we observe that

$$A'(x) = \lim_{h \to 0} \frac{A(x + h) - A(x)}{h}$$

$$= \lim_{h \to 0} \frac{\int_c^{x+h} f(t) \, dt - \int_c^x f(t) \, dt}{h}$$

$$= \lim_{h \to 0} \frac{\int_x^{x+h} f(t)\,dt}{h}, \qquad\qquad (5.2.1)$$

where Equation (5.2.1) in the preceding chain follows from the fact that $\int_c^x f(t)\,dt + \int_x^{x+h} f(t)\,dt = \int_c^{x+h} f(t)\,dt$. Now, observe that for small values of h,

$$\int_x^{x+h} f(t)\,dt \approx f(x) \cdot h,$$

by a simple left-hand approximation of the integral. Thus, as we take the limit in Equation (5.2.1), it follows that

$$A'(x) = \lim_{h \to 0} \frac{\int_x^{x+h} f(t)\,dt}{h} = \lim_{h \to 0} \frac{f(x) \cdot h}{h} = f(x).$$

Hence, A is indeed an antiderivative of f. In addition, $A(c) = \int_c^c f(t)\,dt = 0$. The preceding argument demonstrates the truth of the Second Fundamental Theorem of Calculus, which we state as follows.

The Second Fundamental Theorem of Calculus

If f is a continuous function and c is any constant, then f has a unique antiderivative A that satisfies $A(c) = 0$, and that antiderivative is given by the rule $A(x) = \int_c^x f(t)\,dt$.

Activity 5.2.2. Suppose that f is the function given in Figure 5.2.2 and that f is a piecewise function whose parts are either portions of lines or portions of circles, as pictured.

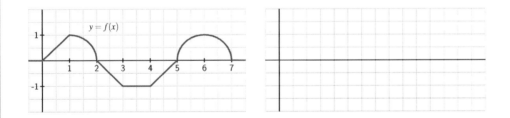

Figure 5.2.2: At left, the graph of $y = f(x)$. At right, axes for sketching $y = A(x)$.

In addition, let A be the function defined by the rule $A(x) = \int_2^x f(t)\,dt$.

a. What does the Second FTC tell us about the relationship between A and f?

b. Compute $A(1)$ and $A(3)$ exactly.

c. Sketch a precise graph of $y = A(x)$ on the axes at right that accurately reflects where A is increasing and decreasing, where A is concave up and concave down, and the exact values of A at $x = 0, 1, \ldots, 7$.

    d. How is A similar to, but different from, the function F that you found in Activity 5.1.2?

    e. With as little additional work as possible, sketch precise graphs of the functions $B(x) = \int_3^x f(t)\,dt$ and $C(x) = \int_1^x f(t)\,dt$. Justify your results with at least one sentence of explanation.

5.2.2 Understanding Integral Functions

The Second FTC provides us with a means to construct an antiderivative of any continuous function. In particular, if we are given a continuous function g and wish to find an antiderivative of G, we can now say that

$$G(x) = \int_c^x g(t)\,dt$$

provides the rule for such an antiderivative, and moreover that $G(c) = 0$. Note especially that we know that $G'(x) = g(x)$. We sometimes want to write this relationship between G and g from a different notational perspective. In particular, observe that

$$\frac{d}{dx}\left[\int_c^x g(t)\,dt\right] = g(x). \tag{5.2.2}$$

This result can be particularly useful when we're given an integral function such as G and wish to understand properties of its graph by recognizing that $G'(x) = g(x)$, while not necessarily being able to exactly evaluate the definite integral $\int_c^x g(t)\,dt$. To see how this is the case, we consider the following example.

Example 5.2.3. Investigate the behavior of the integral function

$$E(x) = \int_0^x e^{-t^2}\,dt.$$

Solution. E is closely related to the well known *error function*[1], a function that is particularly important in probability and statistics. It turns out that the function e^{-t^2} does not have an elementary antiderivative that we can express without integrals. That is, whereas a function such as $f(t) = 4 - 2t$ has elementary antiderivative $F(t) = 4t - t^2$, we are unable to find a simple formula for an antiderivative of e^{-t^2} that does not involve a definite integral. We will learn more about finding (complicated) algebraic formulas for antiderivatives without definite integrals in the chapter on infinite series.

Returning our attention to the function E, while we cannot evaluate E exactly for any value other than $x = 0$, we still can gain a tremendous amount of information about the function

---

[1]The error function is defined by the rule $\operatorname{erf}(x) = \frac{2}{\sqrt{\pi}}\int_0^x e^{-t^2}\,dt$ and has the key property that $0 \le \operatorname{erf}(x) < 1$ for all $x \ge 0$ and moreover that $\lim_{x\to\infty}\operatorname{erf}(x) = 1$.

E. To begin, applying the rule in Equation (5.2.2) to E, it follows that

$$E'(x) = \frac{d}{dx}\left[\int_0^x e^{-t^2}\,dt\right] = e^{-x^2},$$

so we know a formula for the derivative of E. Moreover, we know that $E(0) = 0$. This information is precisely the type we were given in problems such as the one in Activity 3.1.2 and others in Section 3.1, where we were given information about the derivative of a function, but lacked a formula for the function itself.

Here, using the first and second derivatives of E, along with the fact that $E(0) = 0$, we can determine more information about the behavior of E. First, with $E'(x) = e^{-x^2}$, we note that for all real numbers x, $e^{-x^2} > 0$, and thus $E'(x) > 0$ for all x. Thus E is an always increasing function. Further, we note that as $x \to \infty$, $E'(x) = e^{-x^2} \to 0$, hence the slope of the function E tends to zero as $x \to \infty$ (and similarly as $x \to -\infty$). Indeed, it turns out (due to some more sophisticated analysis) that E has horizontal asymptotes as x increases or decreases without bound.

In addition, we can observe that $E''(x) = -2xe^{-x^2}$, and that $E''(0) = 0$, while $E''(x) < 0$ for $x > 0$ and $E''(x) > 0$ for $x < 0$. This information tells us that E is concave up for $x < 0$ and concave down for $x > 0$ with a point of inflection at $x = 0$.

The only thing we lack at this point is a sense of how big E can get as x increases. If we use a midpoint Riemann sum with 10 subintervals to estimate $E(2)$, we see that $E(2) \approx 0.8822$; a similar calculation to estimate $E(3)$ shows little change ($E(3) \approx 0.8862$), so it appears that as x increases without bound, E approaches a value just larger than 0.886, which aligns with the fact that E has horizontal asymptotes. Putting all of this information together (and using the symmetry of $f(t) = e^{-t^2}$), we see the results shown in Figure 5.2.4.

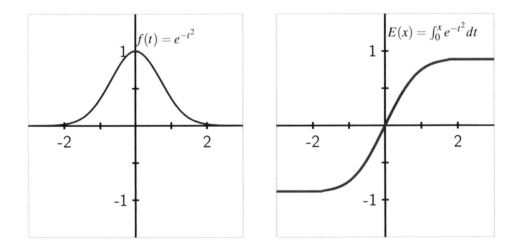

Figure 5.2.4: At left, the graph of $f(t) = e^{-t^2}$. At right, the integral function $E(x) = \int_0^x e^{-t^2}\,dt$, which is the unique antiderivative of f that satisfies $E(0) = 0$.

Again, E is the antiderivative of $f(t) = e^{-t^2}$ that satisfies $E(0) = 0$. Moreover, the values on the graph of $y = E(x)$ represent the net-signed area of the region bounded by $f(t) = e^{-t^2}$ from 0 up to x. We see that the value of E increases rapidly near zero but then levels off as x increases since there is less and less additional accumulated area bounded by $f(t) = e^{-t^2}$ as x increases.

Activity 5.2.3. Suppose that $f(t) = \frac{t}{1+t^2}$ and $F(x) = \int_0^x f(t)\,dt$.

    a. On the axes at left in Figure 5.2.5, plot a graph of $f(t) = \frac{t}{1+t^2}$ on the interval $-10 \leq t \leq 10$. Clearly label the vertical axes with appropriate scale.

    b. What is the key relationship between F and f, according to the Second FTC?

    c. Use the first derivative test to determine the intervals on which F is increasing and decreasing.

    d. Use the second derivative test to determine the intervals on which F is concave up and concave down. Note that $f'(t)$ can be simplified to be written in the form $f'(t) = \frac{1-t^2}{(1+t^2)^2}$.

    e. Using technology appropriately, estimate the values of $F(5)$ and $F(10)$ through appropriate Riemann sums.

    f. Sketch an accurate graph of $y = F(x)$ on the righthand axes provided, and clearly label the vertical axes with appropriate scale.

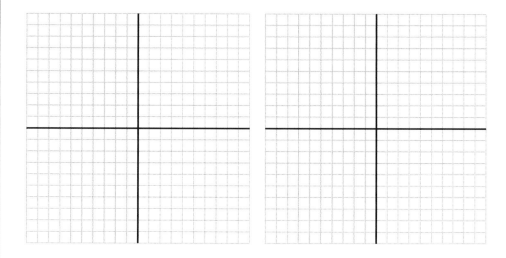

Figure 5.2.5: Axes for plotting f and F.

5.2.3 Differentiating an Integral Function

We have seen that the Second FTC enables us to construct an antiderivative F of any continuous function f by defining F by the corresponding integral function $F(x) = \int_c^x f(t)\,dt$. Said differently, if we have a function of the form $F(x) = \int_c^x f(t)\,dt$, then we know that $F'(x) = \frac{d}{dx}\left[\int_c^x f(t)\,dt\right] = f(x)$. This shows that integral functions, while perhaps having the most complicated formulas of any functions we have encountered, are nonetheless particularly simple to differentiate. For instance, if

$$F(x) = \int_\pi^x \sin(t^2)\,dt,$$

then by the Second FTC, we know immediately that

$$F'(x) = \sin(x^2).$$

Stating this result more generally for an arbitrary function f, we know by the Second FTC that

$$\frac{d}{dx}\left[\int_a^x f(t)\,dt\right] = f(x).$$

In words, the last equation essentially says that "the derivative of the integral function whose integrand is f, is f." In this sense, we see that if we first integrate the function f from $t = a$ to $t = x$, and then differentiate with respect to x, these two processes "undo" one another.

Taking a different approach, say we begin with a function $f(t)$ and differentiate with respect to t. What happens if we follow this by integrating the result from $t = a$ to $t = x$? That is, what can we say about the quantity

$$\int_a^x \frac{d}{dt}\left[f(t)\right]\,dt?$$

Here, we use the First FTC and note that $f(t)$ is an antiderivative of $\frac{d}{dt}\left[f(t)\right]$. Applying this result and evaluating the antiderivative function, we see that

$$\int_a^x \frac{d}{dt}\left[f(t)\right]\,dt = f(t)\Big|_a^x$$
$$= f(x) - f(a).$$

Thus, we see that if we apply the processes of first differentiating f and then integrating the result from a to x, we return to the function f, minus the constant value $f(a)$. So in this situation, the two processes almost undo one another, up to the constant $f(a)$.

The observations made in the preceding two paragraphs demonstrate that differentiating and integrating (where we integrate from a constant up to a variable) are almost inverse processes. In one sense, this should not be surprising: integrating involves antidifferentiating, which reverses the process of differentiating. On the other hand, we see that there is some subtlety involved, as integrating the derivative of a function does not quite produce the function itself. This is connected to a key fact we observed in Section 5.1, which is that any function has an entire family of antiderivatives, and any two of those antiderivatives differ only by a constant.

> **Activity 5.2.4.** Evaluate each of the following derivatives and definite integrals. Clearly cite whether you use the First or Second FTC in so doing.
>
> a. $\frac{d}{dx}\left[\int_4^x e^{t^2}\, dt\right]$
>
> b. $\int_{-2}^x \frac{d}{dt}\left[\frac{t^4}{1+t^4}\right]\, dt$
>
> c. $\frac{d}{dx}\left[\int_x^1 \cos(t^3)\, dt\right]$
>
> d. $\int_3^x \frac{d}{dt}\left[\ln(1+t^2)\right]\, dt$
>
> e. $\frac{d}{dx}\left[\int_4^{x^3} \sin(t^2)\, dt\right]$.

Summary

- For a continuous function f, the integral function $A(x) = \int_1^x f(t)\, dt$ defines an antiderivative of f.

- The Second Fundamental Theorem of Calculus is the formal, more general statement of the preceding fact: if f is a continuous function and c is any constant, then $A(x) = \int_c^x f(t)\, dt$ is the unique antiderivative of f that satisfies $A(c) = 0$.

- Together, the First and Second FTC enable us to formally see how differentiation and integration are almost inverse processes through the observations that

$$\int_c^x \frac{d}{dt}\left[f(t)\right]\, dt = f(x) - f(c)$$

and

$$\frac{d}{dx}\left[\int_c^x f(t)\, dt\right] = f(x).$$

Exercises

1. Let $g(x) = \int_2^x f(t)\, dt$, where $f(t)$ is given in the figure below.

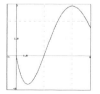

Find each of the following:

A. $g(2) =$ [   ]

B. $g'(3) =$ [   ]

C. The interval (with endpoints given to the nearest 0.25) where g is concave up:

interval = [   ]

(Give your answer as an interval or a list of intervals, e.g., (-infinity,8] or (1,5),(7,10), or enter nonefor no intervals.)

D. The value of x where g takes its maximum on the interval $0 \le x \le 8$.

$x =$ [   ]

2. Find the derivative: $\dfrac{d}{dx} \displaystyle\int_{x}^{a} \cos(\tan(t))\, dt =$ [   ]

3. Find a good numerical approximation to $F(9)$ for the function with the properties that $F'(x) = e^{-x^2/5}$ and $F(0) = 2$.

$F(9) \approx$ [   ]

4. Let g be the function pictured at left in Figure 5.2.6, and let F be defined by $F(x) = \int_{2}^{x} g(t)\, dt$. Assume that the shaded areas have values $A_1 = 4.29$, $A_2 = 12.75$, $A_3 = 0.36$, and $A_4 = 1.79$. Assume further that the portion of A_2 that lies between $x = 0.5$ and $x = 2$ is 6.06.

Sketch a carefully labeled graph of F on the axes provided, and include a written analysis of how you know where F is zero, increasing, decreasing, CCU, and CCD.

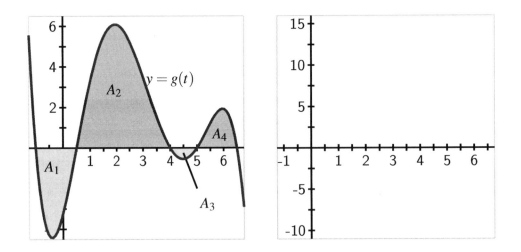

Figure 5.2.6: At left, the graph of g. At right, axes for plotting F.

5. The tide removes sand from the beach at a small ocean park at a rate modeled by the function

$$R(t) = 2 + 5\sin\left(\frac{4\pi t}{25}\right)$$

A pumping station adds sand to the beach at rate modeled by the function

$$S(t) = \frac{15t}{1 + 3t}$$

Both $R(t)$ and $S(t)$ are measured in cubic yards of sand per hour, t is measured in hours, and the valid times are $0 \leq t \leq 6$. At time $t = 0$, the beach holds 2500 cubic yards of sand.

a. What definite integral measures how much sand the tide will remove during the time period $0 \leq t \leq 6$? Why?

b. Write an expression for $Y(x)$, the total number of cubic yards of sand on the beach at time x. Carefully explain your thinking and reasoning.

c. At what instantaneous rate is the total number of cubic yards of sand on the beach at time $t = 4$ changing?

d. Over the time interval $0 \leq t \leq 6$, at what time t is the amount of sand on the beach least? What is this minimum value? Explain and justify your answers fully.

6. When an aircraft attempts to climb as rapidly as possible, its climb rate (in feet per minute) decreases as altitude increases, because the air is less dense at higher altitudes. Given below is a table showing performance data for a certain single engine aircraft, giving its climb rate at various altitudes, where $c(h)$ denotes the climb rate of the airplane at an altitude h.

h (feet)	0	1000	2000	3000	4000	5000	6000	7000	8000	9000	10,000
c (ft/min)	925	875	830	780	730	685	635	585	535	490	440

Table 5.2.7: Data for the climbing aircraft.

Let a new function m, that also depends on h, (say $y = m(h)$) measure the number of minutes required for a plane at altitude h to climb the next foot of altitude.

a. Determine a similar table of values for $m(h)$ and explain how it is related to the table above. Be sure to discuss the units on m.

b. Give a careful interpretation of a function whose derivative is $m(h)$. Describe what the input is and what the output is. Also, explain in plain English what the function tells us.

c. Determine a definite integral whose value tells us exactly the number of minutes required for the airplane to ascend to 10,000 feet of altitude. Clearly explain why the value of this integral has the required meaning.

d. Determine a formula for a function $M(h)$ whose value tells us the exact number of minutes required for the airplane to ascend to h feet of altitude.

e. Estimate the values of $M(6000)$ and $M(10000)$ as accurately as you can. Include units on your results.

5.3 Integration by Substitution

Motivating Questions

- How can we begin to find algebraic formulas for antiderivatives of more complicated algebraic functions?

- What is an indefinite integral and how is its notation used in discussing antiderivatives?

- How does the technique of u-substitution work to help us evaluate certain indefinite integrals, and how does this process rely on identifying function-derivative pairs?

In Section 4.4, we learned the key role that antiderivatives play in the process of evaluating definite integrals exactly. In particular, the Fundamental Theorem of Calculus tells us that if F is any antiderivative of f, then

$$\int_a^b f(x)\,dx = F(b) - F(a).$$

Furthermore, we realized that each elementary derivative rule developed in Chapter 2 leads to a corresponding elementary antiderivative, as summarized in Table 4.4.4. Thus, if we wish to evaluate an integral such as

$$\int_0^1 \left(x^3 - \sqrt{x} + 5^x\right)\,dx,$$

it is straightforward to do so, since we can easily antidifferentiate $f(x) = x^3 - \sqrt{x} + 5^x$. In particular, since a function F whose derivative is f is given by $F(x) = \frac{1}{4}x^4 - \frac{2}{3}x^{3/2} + \frac{1}{\ln(5)}5^x$, the Fundamental Theorem of Calculus tells us that

$$
\begin{aligned}
\int_0^1 \left(x^3 - \sqrt{x} + 5^x\right)\,dx &= \frac{1}{4}x^4 - \frac{2}{3}x^{3/2} + \frac{1}{\ln(5)}5^x \Big|_0^1 \\
&= \left(\frac{1}{4}(1)^4 - \frac{2}{3}(1)^{3/2} + \frac{1}{\ln(5)}5^1\right) - \left(0 - 0 + \frac{1}{\ln(5)}5^0\right) \\
&= -\frac{5}{12} + \frac{4}{\ln(5)}.
\end{aligned}
$$

Because an algebraic formula for an antiderivative of f enables us to evaluate the definite integral $\int_a^b f(x)\,dx$ exactly, we see that we have a natural interest in being able to find such algebraic antiderivatives. Note that we emphasize *algebraic* antiderivatives, as opposed to any antiderivative, since we know by the Second Fundamental Theorem of Calculus that $G(x) = \int_a^x f(t)\,dt$ is indeed an antiderivative of the given function f, but one that still involves a definite integral. One of our main goals in this section and the one following is to develop understanding, in select circumstances, of how to "undo" the process of differentiation in order to find an algebraic antiderivative for a given function.

Preview Activity 5.3.1. In Section 2.5, we learned the Chain Rule and how it can be applied to find the derivative of a composite function. In particular, if u is a differentiable function of x, and f is a differentiable function of $u(x)$, then

$$\frac{d}{dx}\left[f(u(x))\right] = f'(u(x)) \cdot u'(x).$$

In words, we say that the derivative of a composite function $c(x) = f(u(x))$, where f is considered the "outer" function and u the "inner" function, is "the derivative of the outer function, evaluated at the inner function, times the derivative of the inner function."

a. For each of the following functions, use the Chain Rule to find the function's derivative. Be sure to label each derivative by name (e.g., the derivative of $g(x)$ should be labeled $g'(x)$).

    i. $g(x) = e^{3x}$ iii. $p(x) = \arctan(2x)$

    ii. $h(x) = \sin(5x + 1)$ iv. $q(x) = (2 - 7x)^4$

                                                   v. $r(x) = 3^{4-11x}$

b. For each of the following functions, use your work in (a) to help you determine the general antiderivative[a] of the function. Label each antiderivative by name (e.g., the antiderivative of m should be called M). In addition, check your work by computing the derivative of each proposed antiderivative.

    i. $m(x) = e^{3x}$ iv. $v(x) = (2 - 7x)^3$

    ii. $n(x) = \cos(5x + 1)$ v. $w(x) = 3^{4-11x}$

    iii. $s(x) = \frac{1}{1+4x^2}$

c. Based on your experience in parts (a) and (b), conjecture an antiderivative for each of the following functions. Test your conjectures by computing the derivative of each proposed antiderivative.

    i. $a(x) = \cos(\pi x)$ iii. $c(x) = xe^{x^2}$

    ii. $b(x) = (4x + 7)^{11}$

---

[a]Recall that the general antiderivative of a function includes "+C" to reflect the entire family of functions that share the same derivative.

5.3.1 Reversing the Chain Rule: First Steps

In Preview Activity 5.3.1, we saw that it is usually straightforward to antidifferentiate a function of the form

$$h(x) = f(u(x)),$$

whenever f is a familiar function whose antiderivative is known and $u(x)$ is a linear function. For example, if we consider

$$h(x) = (5x - 3)^6,$$

in this context the outer function f is $f(u) = u^6$, while the inner function is $u(x) = 5x - 3$. Since the antiderivative of f is $F(u) = \frac{1}{7}u^7 + C$, we see that the antiderivative of h is

$$H(x) = \frac{1}{7}(5x - 3)^7 \cdot \frac{1}{5} + C = \frac{1}{35}(5x - 3)^7 + C.$$

The inclusion of the constant $\frac{1}{5}$ is essential precisely because the derivative of the inner function is $u'(x) = 5$. Indeed, if we now compute $H'(x)$, we find by the Chain Rule (and Constant Multiple Rule) that

$$H'(x) = \frac{1}{35} \cdot 7(5x - 3)^6 \cdot 5 = (5x - 3)^6 = h(x),$$

and thus H is indeed the general antiderivative of h.

Hence, in the special case where the outer function is familiar and the inner function is linear, we can antidifferentiate composite functions according to the following rule.

> If $h(x) = f(ax + b)$ and F is a known algebraic antiderivative of f, then the general antiderivative of h is given by
>
> $$H(x) = \frac{1}{a}F(ax + b) + C.$$

When discussing antiderivatives, it is often useful to have shorthand notation that indicates the instruction to find an antiderivative. Thus, in a similar way to how the notation

$$\frac{d}{dx}[f(x)]$$

represents the derivative of $f(x)$ with respect to x, we use the notation of the *indefinite integral*,

$$\int f(x)\,dx$$

to represent the general antiderivative of f with respect to x. For instance, returning to the earlier example with $h(x) = (5x - 3)^6$ above, we can rephrase the relationship between h and its antiderivative H through the notation

$$\int (5x - 3)^6\,dx = \frac{1}{35}(5x - 6)^7 + C.$$

When we find an antiderivative, we will often say that we *evaluate an indefinite integral*; said differently, the instruction to evaluate an indefinite integral means to find the general antiderivative. Just as the notation $\frac{d}{dx}[\Box]$ means "find the derivative with respect to x of $\Box$," the notation $\int \Box \, dx$ means "find a function of x whose derivative is $\Box$."

Activity 5.3.2. Evaluate each of the following indefinite integrals. Check each antiderivative that you find by differentiating.

a. $\int \sin(8 - 3x) \, dx$

b. $\int \sec^2(4x) \, dx$

c. $\int \frac{1}{11x - 9} \, dx$

d. $\int \csc(2x + 1) \cot(2x + 1) \, dx$

e. $\int \frac{1}{\sqrt{1 - 16x^2}} \, dx$

f. $\int 5^{-x} \, dx$

5.3.2 Reversing the Chain Rule: u-substitution

Of course, a natural question arises from our recent work: what happens when the inner function is not a linear function? For example, can we find antiderivatives of such functions as

$$g(x) = xe^{x^2} \text{ and } h(x) = e^{x^2}?$$

It is important to explicitly remember that differentiation and antidifferentiation are essentially inverse processes; that they are not quite inverse processes is due to the $+C$ that arises when antidifferentiating. This close relationship enables us to take any known derivative rule and translate it to a corresponding rule for an indefinite integral. For example, since

$$\frac{d}{dx}\left[x^5\right] = 5x^4,$$

we can equivalently write

$$\int 5x^4 \, dx = x^5 + C.$$

Recall that the Chain Rule states that

$$\frac{d}{dx}\left[f(g(x))\right] = f'(g(x)) \cdot g'(x).$$

Restating this relationship in terms of an indefinite integral,

$$\int f'(g(x))g'(x) \, dx = f(g(x)) + C. \tag{5.3.1}$$

Hence, Equation (5.3.1) tells us that if we can take a given function and view its algebraic structure as $f'(g(x))g'(x)$ for some appropriate choices of f and g, then we can antidifferentiate the function by reversing the Chain Rule. It is especially notable that both $g(x)$ and

$g'(x)$ appear in the form of $f'(g(x))g'(x)$; we will sometimes say that we seek to *identify a function-derivative pair* when trying to apply the rule in Equation (5.3.1).

In the situation where we can identify a function-derivative pair, we will introduce a new variable u to represent the function $g(x)$. Observing that with $u = g(x)$, it follows in Leibniz notation that $\frac{du}{dx} = g'(x)$, so that in terms of differentials[1], $du = g'(x)\,dx$. Now converting the indefinite integral of interest to a new one in terms of u, we have

$$\int f'(g(x))g'(x)\,dx = \int f'(u)\,du.$$

Provided that f' is an elementary function whose antiderivative is known, we can now easily evaluate the indefinite integral in u, and then go on to determine the desired overall antiderivative of $f'(g(x))g'(x)$. We call this process *u-substitution*. To see *u*-substitution at work, we consider the following example.

Example 5.3.1. Evaluate the indefinite integral

$$\int x^3 \cdot \sin(7x^4 + 3)\,dx$$

and check the result by differentiating.

Solution. We can make two key algebraic observations regarding the integrand, $x^3 \cdot \sin(7x^4 + 3)$. First, $\sin(7x^4 + 3)$ is a composite function; as such, we know we'll need a more sophisticated approach to antidifferentiating. Second, x^3 is almost the derivative of $(7x^4 + 3)$; the only issue is a missing constant. Thus, x^3 and $(7x^4 + 3)$ are nearly a function-derivative pair. Furthermore, we know the antiderivative of $f(u) = \sin(u)$. The combination of these observations suggests that we can evaluate the given indefinite integral by reversing the chain rule through u-substitution.

Letting u represent the inner function of the composite function $\sin(7x^4 + 3)$, we have $u = 7x^4 + 3$, and thus $\frac{du}{dx} = 28x^3$. In differential notation, it follows that $du = 28x^3\,dx$, and thus $x^3\,dx = \frac{1}{28}\,du$. We make this last observation because the original indefinite integral may now be written

$$\int \sin(7x^4 + 3) \cdot x^3\,dx,$$

and so by substituting the expressions in u for x (specifically u for $7x^4 + 3$ and $\frac{1}{28}\,du$ for $x^3\,dx$), it follows that

$$\int \sin(7x^4 + 3) \cdot x^3\,dx = \int \sin(u) \cdot \frac{1}{28}\,du.$$

Now we may evaluate the original integral by first evaluating the easier integral in u, followed by replacing u by the expression $7x^4 + 3$. Doing so, we find

$$\int \sin(7x^4 + 3) \cdot x^3\,dx = \int \sin(u) \cdot \frac{1}{28}\,du$$

---

[1] If we recall from the definition of the derivative that $\frac{du}{dx} \approx \frac{\Delta u}{\Delta x}$ and use the fact that $\frac{du}{dx} = g'(x)$, then we see that $g'(x) \approx \frac{\Delta u}{\Delta x}$. Solving for Δu, $\Delta u \approx g'(x)\Delta x$. It is this last relationship that, when expressed in "differential" notation enables us to write $du = g'(x)\,dx$ in the change of variable formula.

$$= \frac{1}{28} \int \sin(u) \, du$$

$$= \frac{1}{28}(-\cos(u)) + C$$

$$= -\frac{1}{28} \cos(7x^4 + 3) + C.$$

To check our work, we observe by the Chain Rule that

$$\frac{d}{dx}\left[-\frac{1}{28} \cos(7x^4 + 3) \right] = -\frac{1}{28} \cdot (-1)\sin(7x^4 + 3) \cdot 28x^3 = \sin(7x^4 + 3) \cdot x^3,$$

which is indeed the original integrand.

An essential observation about our work in Example 5.3.1 is that the u-substitution only worked because the function multiplying $\sin(7x^4 + 3)$ was x^3. If instead that function was x^2 or x^4, the substitution process may not (and likely would not) have worked. This is one of the primary challenges of antidifferentiation: slight changes in the integrand make tremendous differences. For instance, we can use u-substitution with $u = x^2$ and $du = 2x \, dx$ to find that

$$\int xe^{x^2} \, dx = \int e^u \cdot \frac{1}{2} \, du$$

$$= \frac{1}{2} \int e^u \, du$$

$$= \frac{1}{2} e^u + C$$

$$= \frac{1}{2} e^{x^2} + C.$$

If, however, we consider the similar indefinite integral

$$\int e^{x^2} \, dx,$$

the missing x to multiply e^{x^2} makes the u-substitution $u = x^2$ no longer possible. Hence, part of the lesson of u-substitution is just how specialized the process is: it only applies to situations where, up to a missing constant, the integrand that is present is the result of applying the Chain Rule to a different, related function.

Activity 5.3.3. Evaluate each of the following indefinite integrals by using these steps:

- Find two functions within the integrand that form (up to a possible missing constant) a function-derivative pair;

- Make a substitution and convert the integral to one involving u and du;

- Evaluate the new integral in u;

- Convert the resulting function of u back to a function of x by using your earlier substitution;

- Check your work by differentiating the function of x. You should come up with the integrand originally given.

a. $\int \frac{x^2}{5x^3+1} \, dx$

c. $\int \frac{\cos(\sqrt{x})}{\sqrt{x}} \, dx$

b. $\int e^x \sin(e^x) \, dx$

5.3.3 Evaluating Definite Integrals via u-substitution

We have just introduced u-substitution as a means to evaluate indefinite integrals of functions that can be written, up to a constant multiple, in the form $f(g(x))g'(x)$. This same technique can be used to evaluate definite integrals involving such functions, though we need to be careful with the corresponding limits of integration. Consider, for instance, the definite integral

$$\int_2^5 xe^{x^2} \, dx.$$

Whenever we write a definite integral, it is implicit that the limits of integration correspond to the variable of integration. To be more explicit, observe that

$$\int_2^5 xe^{x^2} \, dx = \int_{x=2}^{x=5} xe^{x^2} \, dx.$$

When we execute a u-substitution, we change the *variable* of integration; it is essential to note that this also changes the *limits* of integration. For instance, with the substitution $u = x^2$ and $du = 2x \, dx$, it also follows that when $x = 2$, $u = 2^2 = 4$, and when $x = 5$, $u = 5^2 = 25$. Thus, under the change of variables of u-substitution, we now have

$$\int_{x=2}^{x=5} xe^{x^2} \, dx = \int_{u=4}^{u=25} e^u \cdot \frac{1}{2} \, du$$

$$= \frac{1}{2} e^u \Big|_{u=4}^{u=25}$$

$$= \frac{1}{2} e^{25} - \frac{1}{2} e^4.$$

Alternatively, we could consider the related indefinite integral $\int xe^{x^2} \, dx$, find the antiderivative $\frac{1}{2} e^{x^2}$ through u-substitution, and then evaluate the original definite integral. From that perspective, we'd have

$$\int_2^5 xe^{x^2} \, dx = \frac{1}{2} e^{x^2} \Big|_2^5$$

$$= \frac{1}{2} e^{25} - \frac{1}{2} e^4,$$

which is, of course, the same result.

Activity 5.3.4. Evaluate each of the following definite integrals exactly through an appropriate u-substitution.

a. $\int_1^2 \frac{x}{1+4x^2}\, dx$

c. $\int_{2/\pi}^{4/\pi} \frac{\cos\left(\frac{1}{x}\right)}{x^2}\, dx$

b. $\int_0^1 e^{-x}(2e^{-x}+3)^9\, dx$

Summary

- To begin to find algebraic formulas for antiderivatives of more complicated algebraic functions, we need to think carefully about how we can reverse known differentiation rules. To that end, it is essential that we understand and recall known derivatives of basic functions, as well as the standard derivative rules.

- The indefinite integral provides notation for antiderivatives. When we write "$\int f(x)\, dx$," we mean "the general antiderivative of f." In particular, if we have functions f and F such that $F' = f$, the following two statements say the exact thing:

$$\frac{d}{dx}[F(x)] = f(x) \text{ and } \int f(x)\, dx = F(x) + C.$$

That is, f is the derivative of F, and F is an antiderivative of f.

- The technique of u-substitution helps us to evaluate indefinite integrals of the form $\int f(g(x))g'(x)\, dx$ through the substitutions $u = g(x)$ and $du = g'(x)\, dx$, so that

$$\int f(g(x))g'(x)\, dx = \int f(u)\, du.$$

A key part of choosing the expression in x to be represented by u is the identification of a function-derivative pair. To do so, we often look for an "inner" function $g(x)$ that is part of a composite function, while investigating whether $g'(x)$ (or a constant multiple of $g'(x)$) is present as a multiplying factor of the integrand.

Exercises

1. Find the following integral. Note that you can check your answer by differentiation.

$$\int t^3(t^4 - 3)^3\, dt = \rule{6cm}{0.4pt}$$

2. Find the the general antiderivative $F(x)$ of the function $f(x)$ given below. Note that you can check your answer by differentiation.

$f(x) = 2x^3 \sin(x^4)$

antiderivative $F(x) = \rule{5cm}{0.4pt}$

3. Find the following integral. Note that you can check your answer by differentiation.

$$\int \frac{\ln^7(z)}{z} \, dz = \boxed{}$$

4. Find the following integral. Note that you can check your answer by differentiation.

$$\int \frac{e^{5x}}{5 + e^{5x}} \, dx = \boxed{}$$

5. Find the following integral. Note that you can check your answer by differentiation.

$$\int \frac{7e^{2\sqrt{y}}}{\sqrt{y}} \, dy = \boxed{}$$

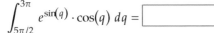

6. Use the Fundamental Theorem of Calculus to find

$$\int_{5\pi/2}^{3\pi} e^{\sin(q)} \cdot \cos(q) \, dq = \boxed{}$$

7. This problem centers on finding antiderivatives for the basic trigonometric functions other than $\sin(x)$ and $\cos(x)$.

a. Consider the indefinite integral $\int \tan(x) \, dx$. By rewriting the integrand as $\tan(x) = \frac{\sin(x)}{\cos(x)}$ and identifying an appropriate function-derivative pair, make a u-substitution and hence evaluate $\int \tan(x) \, dx$.

b. In a similar way, evaluate $\int \cot(x) \, dx$.

c. Consider the indefinite integral

$$\int \frac{\sec^2(x) + \sec(x)\tan(x)}{\sec(x) + \tan(x)} \, dx.$$

Evaluate this integral using the substitution $u = \sec(x) + \tan(x)$.

d. Simplify the integrand in (c) by factoring the numerator. What is a far simpler way to write the integrand?

e. Combine your work in (c) and (d) to determine $\int \sec(x) \, dx$.

f. Using (c)-(e) as a guide, evaluate $\int \csc(x) \, dx$.

8. Consider the indefinite integral $\int x\sqrt{x - 1} \, dx$.

a. At first glance, this integrand may not seem suited to substitution due to the presence of x in separate locations in the integrand. Nonetheless, using the composite function $\sqrt{x - 1}$ as a guide, let $u = x - 1$. Determine expressions for both x and dx in terms of u.

b. Convert the given integral in x to a new integral in u.

c. Evaluate the integral in (b) by noting that $\sqrt{u} = u^{1/2}$ and observing that it is now possible to rewrite the integrand in u by expanding through multiplication.

d. Evaluate each of the integrals $\int x^2\sqrt{x - 1} \, dx$ and $\int x\sqrt{x^2 - 1} \, dx$. Write a paragraph to discuss the similarities among the three indefinite integrals in this problem and the role of substitution and algebraic rearrangement in each.

9. Consider the indefinite integral $\int \sin^3(x)\,dx$.

   a. Explain why the substitution $u = \sin(x)$ will not work to help evaluate the given integral.

   b. Recall the Fundamental Trigonometric Identity, which states that $\sin^2(x) + \cos^2(x) = 1$. By observing that $\sin^3(x) = \sin(x) \cdot \sin^2(x)$, use the Fundamental Trigonometric Identity to rewrite the integrand as the product of $\sin(x)$ with another function.

   c. Explain why the substitution $u = \cos(x)$ now provides a possible way to evaluate the integral in (b).

   d. Use your work in (a)-(c) to evaluate the indefinite integral $\int \sin^3(x)\,dx$.

   e. Use a similar approach to evaluate $\int \cos^3(x)\,dx$.

10. For the town of Mathland, MI, residential power consumption has shown certain trends over recent years. Based on data reflecting average usage, engineers at the power company have modeled the town's rate of energy consumption by the function

$$r(t) = 4 + \sin(0.263t + 4.7) + \cos(0.526t + 9.4).$$

Here, t measures time in hours after midnight on a typical weekday, and r is the rate of consumption in megawatts[2] at time t. Units are critical throughout this problem.

   a. Sketch a carefully labeled graph of $r(t)$ on the interval $[0,24]$ and explain its meaning. Why is this a reasonable model of power consumption?

   b. Without calculating its value, explain the meaning of $\int_0^{24} r(t)\,dt$. Include appropriate units on your answer.

   c. Determine the exact amount of power Mathland consumes in a typical day.

   d. What is Mathland's average rate of energy consumption in a given 24-hour period? What are the units on this quantity?

---

[2]The unit *megawatt* is itself a rate, which measures energy consumption per unit time. A *megawatt-hour* is the total amount of energy that is equivalent to a constant stream of 1 megawatt of power being sustained for 1 hour.

5.4 Integration by Parts

Motivating Questions

- How do we evaluate indefinite integrals that involve products of basic functions such as $\int x \sin(x)\,dx$ and $\int xe^x\,dx$?

- What is the method of integration by parts and how can we consistently apply it to integrate products of basic functions?

- How does the algebraic structure of functions guide us in identifying u and dv in using integration by parts?

In Section 5.3, we learned the technique of u-substitution for evaluating indefinite integrals that involve certain composite functions. For example, the indefinite integral $\int x^3 \sin(x^4)\,dx$ is perfectly suited to u-substitution, since not only is there a composite function present, but also the inner function's derivative (up to a constant) is multiplying the composite function. Through u-substitution, we learned a general situation where recognizing the algebraic structure of a function can enable us to find its antiderivative.

It is natural to ask similar questions to those we considered in Section 5.3 about functions with a different elementary algebraic structure: those that are the product of basic functions. For instance, suppose we are interested in evaluating the indefinite integral

$$\int x \sin(x)\,dx.$$

Here, there is not a composite function present, but rather a product of the basic functions $f(x) = x$ and $g(x) = \sin(x)$. From our work in Section 2.3 with the Product Rule, we know that it is relatively complicated to compute the derivative of the product of two functions, so we should expect that antidifferentiating a product should be similarly involved. In addition, intuitively we expect that evaluating $\int x \sin(x)\,dx$ will involve somehow reversing the Product Rule.

To that end, in Preview Activity 5.4.1 we refresh our understanding of the Product Rule and then investigate some indefinite integrals that involve products of basic functions.

> **Preview Activity 5.4.1.** In Section 2.3, we developed the Product Rule and studied how it is employed to differentiate a product of two functions. In particular, recall that if f and g are differentiable functions of x, then
>
> $$\frac{d}{dx}\left[f(x) \cdot g(x)\right] = f(x) \cdot g'(x) + g(x) \cdot f'(x).$$

a. For each of the following functions, use the Product Rule to find the function's derivative. Be sure to label each derivative by name (e.g., the derivative of $g(x)$ should be labeled $g'(x)$).

i. $g(x) = x \sin(x)$

ii. $h(x) = xe^x$

iii. $p(x) = x \ln(x)$

iv. $q(x) = x^2 \cos(x)$

v. $r(x) = e^x \sin(x)$

b. Use your work in (a) to help you evaluate the following indefinite integrals. Use differentiation to check your work.

i. $\int xe^x + e^x \, dx$

ii. $\int e^x(\sin(x) + \cos(x)) \, dx$

iii. $\int 2x \cos(x) - x^2 \sin(x) \, dx$

iv. $\int x \cos(x) + \sin(x) \, dx$

v. $\int 1 + \ln(x) \, dx$

c. Observe that the examples in (b) work nicely because of the derivatives you were asked to calculate in (a). Each integrand in (b) is precisely the result of differentiating one of the products of basic functions found in (a). To see what happens when an integrand is still a product but not necessarily the result of differentiating an elementary product, we consider how to evaluate

$$\int x \cos(x) \, dx.$$

i. First, observe that

$$\frac{d}{dx}[x \sin(x)] = x \cos(x) + \sin(x).$$

Integrating both sides indefinitely and using the fact that the integral of a sum is the sum of the integrals, we find that

$$\int \left(\frac{d}{dx}[x \sin(x)] \right) dx = \int x \cos(x) \, dx + \int \sin(x) \, dx.$$

In this last equation, evaluate the indefinite integral on the left side as well as the rightmost indefinite integral on the right.

ii. In the most recent equation from (i.), solve the equation for the expression $\int x \cos(x) \, dx$.

iii. For which product of basic functions have you now found the antiderivative?

5.4.1 Reversing the Product Rule: Integration by Parts

Problem (c) in Preview Activity 5.4.1 provides a clue for how we develop the general technique known as Integration by Parts, which comes from reversing the Product Rule. Recall that the Product Rule states that

$$\frac{d}{dx}\left[f(x)g(x)\right] = f(x)g'(x) + g(x)f'(x).$$

Integrating both sides of this equation indefinitely with respect to x, it follows that

$$\int \frac{d}{dx}\left[f(x)g(x)\right]dx = \int f(x)g'(x)\,dx + \int g(x)f'(x)\,dx. \tag{5.4.1}$$

On the left in Equation (5.4.1), we recognize that we have the indefinite integral of the derivative of a function which, up to an additional constant, is the original function itself. Temporarily omitting the constant that may arise, we equivalently have

$$f(x)g(x) = \int f(x)g'(x)\,dx + \int g(x)f'(x)\,dx. \tag{5.4.2}$$

The most important thing to observe about Equation (5.4.2) is that it provides us with a choice of two integrals to evaluate. That is, in a situation where we can identify two functions f and g, if we can integrate $f(x)g'(x)$, then we know the indefinite integral of $g(x)f'(x)$, and vice versa. To that end, we choose the first indefinite integral on the left in Equation (5.4.2) and solve for it to generate the rule

$$\int f(x)g'(x)\,dx = f(x)g(x) - \int g(x)f'(x)\,dx. \tag{5.4.3}$$

Often we express Equation (5.4.3) in terms of the variables u and v, where $u = f(x)$ and $v = g(x)$. Note that in differential notation, $du = f'(x)\,dx$ and $dv = g'(x)\,dx$, and thus we can state the rule for Integration by Parts in its most common form as follows.

$$\int u\,dv = uv - \int v\,du.$$

To apply Integration by Parts, we look for a product of basic functions that we can identify as u and dv. If we can antidifferentiate dv to find v, and evaluating $\int v\,du$ is not more difficult than evaluating $\int u\,dv$, then this substitution usually proves to be fruitful. To demonstrate, we consider the following example.

Example 5.4.1. Evaluate the indefinite integral

$$\int x\cos(x)\,dx$$

using Integration by Parts.

Solution. Whenever we are trying to integrate a product of basic functions through Integration by Parts, we are presented with a choice for u and dv. In the current problem, we can either let $u = x$ and $dv = \cos(x)\,dx$, or let $u = \cos(x)$ and $dv = x\,dx$. While there is not a universal rule for how to choose u and dv, a good guideline is this: do so in a way that $\int v\,du$ is at least as simple as the original problem $\int u\,dv$.

In this setting, this leads us to choose[1] $u = x$ and $dv = \cos(x)\,dx$, from which it follows that $du = 1\,dx$ and $v = \sin(x)$. With this substitution, the rule for Integration by Parts tells us that

$$\int x\cos(x)\,dx = x\sin(x) - \int \sin(x) \cdot 1\,dx.$$

At this point, all that remains to do is evaluate the (simpler) integral $\int \sin(x) \cdot 1\,dx$. Doing so, we find

$$\int x\cos(x)\,dx = x\sin(x) - (-\cos(x)) + C = x\sin(x) + \cos(x) + C.$$

There are at least two additional important observations to make from Example 5.4.1. First, the general technique of Integration by Parts involves trading the problem of integrating the product of two functions for the problem of integrating the product of two related functions. In particular, we convert the problem of evaluating $\int u\,dv$ for that of evaluating $\int v\,du$. This perspective clearly shapes our choice of u and v. In Example 5.4.1, the original integral to evaluate was $\int x\cos(x)\,dx$, and through the substitution provided by Integration by Parts, we were instead able to evaluate $\int \sin(x) \cdot 1\,dx$. Note that the original function x was replaced by its derivative, while $\cos(x)$ was replaced by its antiderivative. Second, observe that when we get to the final stage of evaluating the last remaining antiderivative, it is at this step that we include the integration constant, $+C$.

> **Activity 5.4.2.** Evaluate each of the following indefinite integrals. Check each antiderivative that you find by differentiating.
>
> a. $\int te^{-t}\,dt$ c. $\int z\sec^2(z)\,dz$
>
> b. $\int 4x\sin(3x)\,dx$ d. $\int x\ln(x)\,dx$

5.4.2 Some Subtleties with Integration by Parts

There are situations where Integration by Parts is not an obvious choice, but the technique is appropriate nonetheless. One guide to understanding why is the observation that integration by parts allows us to replace one function in a product with its derivative while

---

[1] Observe that if we considered the alternate choice, and let $u = \cos(x)$ and $dv = x\,dx$, then $du = -\sin(x)\,dx$ and $v = \frac{1}{2}x^2$, from which we would write $\int x\cos(x)\,dx = \frac{1}{2}x^2\cos(x) - \int \frac{1}{2}x^2(-\sin(x))\,dx$. Thus we have replaced the problem of integrating $x\cos(x)$ with that of integrating $\frac{1}{2}x^2\sin(x)$; the latter is clearly more complicated, which shows that this alternate choice is not as helpful as the first choice.

replacing the other with its antiderivative. For instance, consider the problem of evaluating

$$\int \arctan(x)\,dx.$$

Initially, this problem seems ill-suited to Integration by Parts, since there does not appear to be a product of functions present. But if we note that $\arctan(x) = \arctan(x)\cdot 1$, and realize that we know the derivative of $\arctan(x)$ as well as the antiderivative of 1, we see the possibility for the substitution $u = \arctan(x)$ and $dv = 1\,dx$. We explore this substitution further in Activity 5.4.3.

In a related problem, if we consider $\int t^3 \sin(t^2)\,dt$, two key observations can be made about the algebraic structure of the integrand: there is a composite function present in $\sin(t^2)$, and there is not an obvious function-derivative pair, as we have t^3 present (rather than simply t) multiplying $\sin(t^2)$. This problem exemplifies the situation where we sometimes use both u-substitution and Integration by Parts in a single problem. If we write $t^3 = t \cdot t^2$ and consider the indefinite integral

$$\int t \cdot t^2 \cdot \sin(t^2)\,dt,$$

we can use a mix of the two techniques we have recently learned. First, let $z = t^2$ so that $dz = 2t\,dt$, and thus $t\,dt = \frac{1}{2}\,dz$. (We are using the variable z to perform a "z-substitution" since u will be used subsequently in executing Integration by Parts.) Under this z-substitution, we now have

$$\int t \cdot t^2 \cdot \sin(t^2)\,dt = \int z \cdot \sin(z) \cdot \frac{1}{2}\,dz.$$

The remaining integral is a standard one that can be evaluated by parts. This, too, is explored further in Activity 5.4.3.

The problems briefly introduced here exemplify that we sometimes must think creatively in choosing the variables for substitution in Integration by Parts, as well as that it is entirely possible that we will need to use the technique of substitution for an additional change of variables within the process of integrating by parts.

Activity 5.4.3. Evaluate each of the following indefinite integrals, using the provided hints.

a. Evaluate $\int \arctan(x)\,dx$ by using Integration by Parts with the substitution $u = \arctan(x)$ and $dv = 1\,dx$.

b. Evaluate $\int \ln(z)\,dz$. Consider a similar substitution to the one in (a).

c. Use the substitution $z = t^2$ to transform the integral $\int t^3 \sin(t^2)\,dt$ to a new integral in the variable z, and evaluate that new integral by parts.

d. Evaluate $\int s^5 e^{s^3}\,ds$ using an approach similar to that described in (c).

e. Evaluate $\int e^{2t} \cos(e^t)\,dt$. You will find it helpful to note that $e^{2t} = e^t \cdot e^t$.

5.4.3 Using Integration by Parts Multiple Times

We have seen that the technique of Integration by Parts is well suited to integrating the product of basic functions, and that it allows us to essentially trade a given integrand for a new one where one function in the product is replaced by its derivative, while the other is replaced by its antiderivative. The main goal in this trade of $\int u\,dv$ for $\int v\,du$ is to have the new integral not be more challenging to evaluate than the original one. At times, it turns out that it can be necessary to apply Integration by Parts more than once in order to ultimately evaluate a given indefinite integral.

For example, if we consider $\int t^2 e^t\,dt$ and let $u = t^2$ and $dv = e^t\,dt$, then it follows that $du = 2t\,dt$ and $v = e^t$, thus

$$\int t^2 e^t\,dt = t^2 e^t - \int 2t e^t\,dt.$$

The integral on the righthand side is simpler to evaluate than the one on the left, but it still requires Integration by Parts. Now letting $u = 2t$ and $dv = e^t\,dt$, we have $du = 2\,dt$ and $v = e^t$, so that

$$\int t^2 e^t\,dt = t^2 e^t - \left(2t e^t - \int 2e^t\,dt\right).$$

Note the key role of the parentheses, as it is essential to distribute the minus sign to the entire value of the integral $\int 2t e^t\,dt$. The final integral on the right in the most recent equation is a basic one; evaluating that integral and distributing the minus sign, we find

$$\int t^2 e^t\,dt = t^2 e^t - 2t e^t + 2e^t + C.$$

Of course, situations are possible where even more than two applications of Integration by Parts may be necessary. For instance, in the preceding example, it is apparent that if the integrand was $t^3 e^t$ instead, we would have to use Integration by Parts three times.

Next, we consider the slightly different scenario presented by the definite integral $\int e^t \cos(t)\,dt$. Here, we can choose to let u be either e^t or $\cos(t)$; we pick $u = \cos(t)$, and thus $dv = e^t\,dt$. With $du = -\sin(t)\,dt$ and $v = e^t$, Integration by Parts tells us that

$$\int e^t \cos(t)\,dt = e^t \cos(t) - \int e^t(-\sin(t))\,dt,$$

or equivalently that

$$\int e^t \cos(t)\,dt = e^t \cos(t) + \int e^t \sin(t)\,dt \qquad (5.4.4)$$

Observe that the integral on the right in Equation (5.4.4), $\int e^t \sin(t)\,dt$, while not being more complicated than the original integral we want to evaluate, it is essentially identical to $\int e^t \cos(t)\,dt$. While the overall situation isn't necessarily better than what we started with, the problem hasn't gotten worse. Thus, we proceed by integrating by parts again. This time we let $u = \sin(t)$ and $dv = e^t\, dt$, so that $du = \cos(t)\, dt$ and $v = e^t$, which implies

$$\int e^t \cos(t)\, dt = e^t \cos(t) + \left(e^t \sin(t) - \int e^t \cos(t)\, dt \right) \tag{5.4.5}$$

We seem to be back where we started, as two applications of Integration by Parts has led us back to the original problem, $\int e^t \cos(t)\, dt$. But if we look closely at Equation (5.4.5), we see that we can use algebra to solve for the value of the desired integral. In particular, adding $\int e^t \cos(t)\, dt$ to both sides of the equation, we have

$$2 \int e^t \cos(t)\, dt = e^t \cos(t) + e^t \sin(t),$$

and therefore

$$\int e^t \cos(t)\, dt = \frac{1}{2} \left(e^t \cos(t) + e^t \sin(t) \right) + C.$$

Note that since we never actually encountered an integral we could evaluate directly, we didn't have the opportunity to add the integration constant C until the final step, at which point we include it as part of the most general antiderivative that we sought from the outset in evaluating an indefinite integral.

> **Activity 5.4.4.** Evaluate each of the following indefinite integrals.
>
> a. $\int x^2 \sin(x)\, dx$
>
> b. $\int t^3 \ln(t)\, dt$
>
> c. $\int e^z \sin(z)\, dz$
>
> d. $\int s^2 e^{3s}\, ds$
>
> e. $\int t \arctan(t)\, dt$ (*Hint:* At a certain point in this problem, it is very helpful to note that $\frac{t^2}{1+t^2} = 1 - \frac{1}{1+t^2}$.)

5.4.4 Evaluating Definite Integrals Using Integration by Parts

Just as we saw with u-substitution in Section 5.3, we can use the technique of Integration by Parts to evaluate a definite integral. Say, for example, we wish to find the exact value of

$$\int_0^{\pi/2} t \sin(t)\, dt.$$

One option is to evaluate the related indefinite integral to find that $\int t \sin(t)\, dt = -t \cos(t) + \sin(t) + C$, and then use the resulting antiderivative along with the Fundamental Theorem of Calculus to find that

$$\int_0^{\pi/2} t\sin(t)\,dt = \left.(-t\cos(t) + \sin(t))\right|_0^{\pi/2}$$

$$= \left(-\frac{\pi}{2}\cos(\frac{\pi}{2}) + \sin(\frac{\pi}{2})\right) - (-0\cos(0) + \sin(0))$$

$$= 1.$$

Alternatively, we can apply Integration by Parts and work with definite integrals throughout. In this perspective, it is essential to remember to evaluate the product uv over the given limits of integration. To that end, using the substitution $u = t$ and $dv = \sin(t)\,dt$, so that $du = dt$ and $v = -\cos(t)$, we write

$$\int_0^{\pi/2} t\sin(t)\,dt = \left.-t\cos(t)\right|_0^{\pi/2} - \int_0^{\pi/2}(-\cos(t))\,dt$$

$$= \left.-t\cos(t)\right|_0^{\pi/2} + \left.\sin(t)\right|_0^{\pi/2}$$

$$= \left(-\frac{\pi}{2}\cos(\frac{\pi}{2}) + \sin(\frac{\pi}{2})\right) - (-0\cos(0) + \sin(0))$$

$$= 1.$$

As with any substitution technique, it is important to remember the overall goal of the problem, to use notation carefully and completely, and to think about our end result to ensure that it makes sense in the context of the question being answered.

5.4.5 When u-substitution and Integration by Parts Fail to Help

As we close this section, it is important to note that both integration techniques we have discussed apply in relatively limited circumstances. In particular, it is not hard to find examples of functions for which neither technique produces an antiderivative; indeed, there are many, many functions that appear elementary but that do not have an elementary algebraic antiderivative. For instance, if we consider the indefinite integrals

$$\int e^{x^2}\,dx \ \text{ and } \ \int x\tan(x)\,dx,$$

neither u-substitution nor Integration by Parts proves fruitful. While there are other integration techniques, some of which we will consider briefly, none of them enables us to find an algebraic antiderivative for e^{x^2} or $x\tan(x)$. There are at least two key observations to make: one, we do know from the Second Fundamental Theorem of Calculus that we can construct an integral antiderivative for each function; and two, antidifferentiation is much, much harder in general than differentiation. In particular, we observe that $F(x) = \int_0^x e^{t^2}\,dt$ is an antiderivative of $f(x) = e^{x^2}$, and $G(x) = \int_0^x t\tan(t)\,dt$ is an antiderivative of $g(x) = x\tan(x)$. But finding an elementary algebraic formula that doesn't involve integrals for either F or G turns out not only to be impossible through u-substitution or Integration by Parts, but indeed impossible altogether.

Summary

- Through the method of Integration by Parts, we can evaluate indefinite integrals that involve products of basic functions such as $\int x \sin(x)\,dx$ and $\int x \ln(x)\,dx$ through a substitution that enables us to effectively trade one of the functions in the product for its derivative, and the other for its antiderivative, in an effort to find a different product of functions that is easier to integrate.

- If we are given an integral whose algebraic structure we can identify as a product of basic functions in the form $\int f(x)g'(x)\,dx$, we can use the substitution $u = f(x)$ and $dv = g'(x)\,dx$ and apply the rule

$$\int u\,dv = uv - \int v\,du$$

in an effort to evaluate the original integral $\int f(x)g'(x)\,dx$ by instead evaluating $\int v\,du = \int f'(x)g(x)\,dx$.

- When deciding to integrate by parts, we normally have a product of functions present in the integrand and we have to select both u and dv. That selection is guided by the overall principal that we desire the new integral $\int v\,du$ to not be any more difficult or complicated than the original integral $\int u\,dv$. In addition, it is often helpful to recognize if one of the functions present is much easier to differentiate than antidifferentiate (such as $\ln(x)$), in which case that function often is best assigned the variable u. For sure, when choosing dv, the corresponding function must be one that we can antidifferentiate.

Exercises

1. For each of the following integrals, indicate whether integration by substitution or integration by parts is more appropriate, or if neither method is appropriate. Do not evaluate the integrals.

1. $\int x \sin x\,dx$
2. $\int \frac{x^2}{1+x^3}\,dx$
3. $\int x^2 e^{x^3}\,dx$
4. $\int x^2 \cos(x^3)\,dx$
5. $\int \frac{1}{\sqrt{3x+1}}\,dx$

(Note that because this is multiple choice, you will not be able to see which parts of the problem you got correct.)

2. Use integration by parts to evaluate the integral.

$$\int 3x \cos(2x)\,dx = \boxed{} +C$$

3. Find the integral

$$\int (z+1)\,e^{4z}\,dz = \boxed{}$$

4. Evaluate the definite integral.

$$\int_0^4 te^{-t}\,dt = \boxed{}$$

5. Let $f(t) = te^{-2t}$ and $F(x) = \int_0^x f(t)\,dt$.

a. Determine $F'(x)$.

b. Use the First FTC to find a formula for F that does not involve an integral.

c. Is F an increasing or decreasing function for $x > 0$? Why?

6. Consider the indefinite integral given by $\int e^{2x}\cos(e^x)\,dx$.

a. Noting that $e^{2x} = e^x \cdot e^x$, use the substitution $z = e^x$ to determine a new, equivalent integral in the variable z.

b. Evaluate the integral you found in (a) using an appropriate technique.

c. How is the problem of evaluating $\int e^{2x}\cos(e^{2x})\,dx$ different from evaluating the integral in (a)? Do so.

d. Evaluate each of the following integrals as well, keeping in mind the approach(es) used earlier in this problem:

- $\int e^{2x}\sin(e^x)\,dx$
- $\int e^{3x}\sin(e^{3x})\,dx$
- $\int xe^{x^2}\cos(e^{x^2})\sin(e^{x^2})\,dx$

7. For each of the following indefinite integrals, determine whether you would use u-substitution, integration by parts, neither*, or both to evaluate the integral. In each case, write one sentence to explain your reasoning, and include a statement of any substitutions used. (That is, if you decide in a problem to let $u = e^{3x}$, you should state that, as well as that $du = 3e^{3x}\,dx$.) Finally, use your chosen approach to evaluate each integral. (* one of the following problems does not have an elementary antiderivative and you are not expected to actually evaluate this integral; this will correspond with a choice of "neither" among those given.)

a. $\int x^2\cos(x^3)\,dx$

b. $\int x^5\cos(x^3)\,dx$ (Hint: $x^5 = x^2 \cdot x^3$)

c. $\int x\ln(x^2)\,dx$

d. $\int \sin(x^4)\,dx$

e. $\int x^3\sin(x^4)\,dx$

f. $\int x^7\sin(x^4)\,dx$

5.5 Other Options for Finding Algebraic Antiderivatives

Motivating Questions

- How does the method of partial fractions enable any rational function to be antidifferentiated?

- What role have integral tables historically played in the study of calculus and how can a table be used to evaluate integrals such as $\int \sqrt{a^2 + u^2}\, du$?

- What role can a computer algebra system play in the process of finding antiderivatives?

In the preceding sections, we have learned two very specific antidifferentiation techniques: u-substitution and integration by parts. The former is used to reverse the chain rule, while the latter to reverse the product rule. But we have seen that each only works in very specialized circumstances. For example, while $\int xe^{x^2}\, dx$ may be evaluated by u-substitution and $\int xe^x\, dx$ by integration by parts, neither method provides a route to evaluate $\int e^{x^2}\, dx$. That fact is not a particular shortcoming of these two antidifferentiation techniques, as it turns out there does not exist an elementary algebraic antiderivative for e^{x^2}. Said differently, no matter what antidifferentiation methods we could develop and learn to execute, none of them will be able to provide us with a simple formula that does not involve integrals for a function $F(x)$ that satisfies $F'(x) = e^{x^2}$.

In this section of the text, our main goals are to better understand some classes of functions that can always be antidifferentiated, as well as to learn some options for so doing. At the same time, we want to recognize that there are many functions for which an algebraic formula for an antiderivative does not exist, and also appreciate the role that computing technology can play in helping us find antiderivatives of other complicated functions. Throughout, it is helpful to remember what we have learned so far: how to reverse the chain rule through u-substitution, how to reverse the product rule through integration by parts, and that overall, there are subtle and challenging issues to address when trying to find antiderivatives.

> **Preview Activity 5.5.1.** For each of the indefinite integrals below, the main question is to decide whether the integral can be evaluated using u-substitution, integration by parts, a combination of the two, or neither. For integrals for which your answer is affirmative, state the substitution(s) you would use. It is not necessary to actually evaluate any of the integrals completely, unless the integral can be evaluated immediately using a familiar basic antiderivative.
>
> a. $\int x^2 \sin(x^3)\, dx, \ \int x^2 \sin(x)\, dx, \ \int \sin(x^3)\, dx, \ \int x^5 \sin(x^3)\, dx$
>
> b. $\int \frac{1}{1+x^2}\, dx, \ \int \frac{x}{1+x^2}\, dx, \ \int \frac{2x+3}{1+x^2}\, dx, \ \int \frac{e^x}{1+(e^x)^2}\, dx,$

   c. $\int x \ln(x)\,dx$, $\int \frac{\ln(x)}{x}\,dx$, $\int \ln(1 + x^2)\,dx$, $\int x \ln(1 + x^2)\,dx$,

   d. $\int x\sqrt{1 - x^2}\,dx$, $\int \frac{1}{\sqrt{1-x^2}}\,dx$, $\int \frac{x}{\sqrt{1-x^2}}\,dx$, $\int \frac{1}{x\sqrt{1-x^2}}\,dx$,

5.5.1 The Method of Partial Fractions

The method of partial fractions is used to integrate rational functions, and essentially involves reversing the process of finding a common denominator. For example, suppoes we have the function $R(x) = \frac{5x}{x^2-x-2}$ and want to evaluate

$$\int \frac{5x}{x^2 - x - 2}\,dx.$$

Thinking algebraically, if we factor the denominator, we can see how R might come from the sum of two fractions of the form $\frac{A}{x-2} + \frac{B}{x+1}$. In particular, suppose that

$$\frac{5x}{(x - 2)(x + 1)} = \frac{A}{x - 2} + \frac{B}{x + 1}.$$

Multiplying both sides of this last equation by $(x - 2)(x + 1)$, we find that

$$5x = A(x + 1) + B(x - 2).$$

Since we want this equation to hold for every value of x, we can use insightful choices of specific x-values to help us find A and B. Taking $x = -1$, we have

$$5(-1) = A(0) + B(-3),$$

and thus $B = \frac{5}{3}$. Choosing $x = 2$, it follows

$$5(2) = A(3) + B(0),$$

so $A = \frac{10}{3}$. Therefore, we now know that

$$\int \frac{5x}{x^2 - x - 2}\,dx = \int \frac{10/3}{x - 2} + \frac{5/3}{x + 1}\,dx.$$

This equivalent integral expression is straightforward to evaluate, and hence we find that

$$\int \frac{5x}{x^2 - x - 2}\,dx = \frac{10}{3}\ln|x - 2| + \frac{5}{3}\ln|x + 1| + C.$$

It turns out that for any rational function $R(x) = \frac{P(x)}{Q(x)}$ where the degree of the polynomial P is less than[1] the degree of the polynomial Q, the method of partial fractions can be used to rewrite the function as a sum of simpler rational functions of one of the following forms:

$$\frac{A}{x - c}, \quad \frac{A}{(x - c)^n}, \quad \frac{Ax + B}{x^2 + k}, \quad \text{or} \quad \frac{Ax + B}{(x^2 + k)^n}$$

---

[1] If the degree of P is greater than or equal to the degree of Q, long division may be used to write R as the sum of a polynomial plus a rational function where the numerator's degree is less than the denominator's.

where A, B, and c are real numbers, and k is a positive real number. Because each of these basic forms is one we can antidifferentiate, partial fractions enables us to antidifferentiate any rational function. A computer algebra system such as *Maple*, *Mathematica*, or *WolframAlpha* can be used to find the partial fraction decomposition of any rational function. In *WolframAlpha*, entering

```
partial fraction 5x/(x^2-x-2)
```

results in the output

$$\frac{5x}{x^2 - x - 2} = \frac{10}{3(x - 2)} + \frac{5}{3(x + 1)}.$$

We will primarily use technology to generate partial fraction decompositions of rational functions, and then work from there to evaluate the integrals of interest using established methods.

> **Activity 5.5.2.** For each of the following problems, evaluate the integral by using the partial fraction decomposition provided.
>
> a. $\int \frac{1}{x^2-2x-3} \, dx$, given that $\frac{1}{x^2-2x-3} = \frac{1/4}{x-3} - \frac{1/4}{x+1}$
>
> b. $\int \frac{x^2+1}{x^3-x^2} \, dx$, given that $\frac{x^2+1}{x^3-x^2} = -\frac{1}{x} - \frac{1}{x^2} + \frac{2}{x-1}$
>
> c. $\int \frac{x-2}{x^4+x^2} \, dx$, given that $\frac{x-2}{x^4+x^2} = \frac{1}{x} - \frac{2}{x^2} + \frac{-x+2}{1+x^2}$

5.5.2 Using an Integral Table

Calculus has a long history, with key ideas going back as far as Greek mathematicians in 400-300 BC. Its main foundations were first investigated and understood independently by Isaac Newton and Gottfried Wilhelm Leibniz in the late 1600s, making the modern ideas of calculus well over 300 years old. It is instructive to realize that until the late 1980s, the personal computer essentially did not exist, so calculus (and other mathematics) had to be done by hand for roughly 300 years. During the last 30 years, however, computers have revolutionized many aspects of the world we live in, including mathematics. In this section we take a short historical tour to precede the following discussion of the role computer algebra systems can play in evaluating indefinite integrals. In particular, we consider a class of integrals involving certain radical expressions that, until the advent of computer algebra systems, were often evaluated using an integral table.

As seen in the short table of integrals found in Appendix A, there are also many forms of integrals that involve $\sqrt{a^2 \pm w^2}$ and $\sqrt{w^2 - a^2}$. These integral rules can be developed using a technique known as *trigonometric substitution* that we choose to omit; instead, we will simply accept the results presented in the table. To see how these rules are needed and used, consider the differences among

$$\int \frac{1}{\sqrt{1 - x^2}} \, dx, \quad \int \frac{x}{\sqrt{1 - x^2}} \, dx, \quad \text{and} \quad \int \sqrt{1 - x^2} \, dx.$$

The first integral is a familiar basic one, and results in arcsin(x) + C. The second integral can be evaluated using a standard u-substitution with $u = 1 - x^2$. The third, however, is not familiar and does not lend itself to u-substitution.

In Appendix A, we find the rule

$$(h) \int \sqrt{a^2 - u^2} \, du = \frac{u}{2} \sqrt{a^2 - u^2} + \frac{a^2}{2} \arcsin \frac{u}{a} + C.$$

Using the substitutions $a = 1$ and $u = x$ (so that $du = dx$), it follows that

$$\int \sqrt{1 - x^2} \, dx = \frac{x}{2} \sqrt{1 - x^2} - \frac{1}{2} \arcsin x + C.$$

One important point to note is that whenever we are applying a rule in the table, we are doing a u-substitution. This is especially key when the situation is more complicated than allowing $u = x$ as in the last example. For instance, say we wish to evaluate the integral

$$\int \sqrt{9 + 64x^2} \, dx.$$

Here, we want to use Rule (c) from the table, and do so with $a = 3$ and $u = 8x$; we also choose the "+" option in the rule. With this substitution, it follows that $du = 8dx$, so $dx = \frac{1}{8} du$. Applying this substitution,

$$\int \sqrt{9 + 64x^2} \, dx = \int \sqrt{9 + u^2} \cdot \frac{1}{8} \, du = \frac{1}{8} \int \sqrt{9 + u^2} \, du.$$

By Rule (c), we now find that

$$\int \sqrt{9 + 64x^2} \, dx = \frac{1}{8} \left(\frac{u}{2} \sqrt{u^2 + 9} + \frac{9}{2} \ln |u + \sqrt{u^2 + 9}| + C \right)$$
$$= \frac{1}{8} \left(\frac{8x}{2} \sqrt{64x^2 + 9} + \frac{9}{2} \ln |8x + \sqrt{64x^2 + 9}| + C \right).$$

In problems such as this one, it is essential that we not forget to account for the factor of $\frac{1}{8}$ that must be present in the evaluation.

Activity 5.5.3. For each of the following integrals, evaluate the integral using u-substitution and/or an entry from the table found in Appendix A.

a. $\int \sqrt{x^2 + 4} \, dx$

b. $\int \frac{x}{\sqrt{x^2+4}} \, dx$

c. $\int \frac{2}{\sqrt{16+25x^2}} \, dx$

d. $\int \frac{1}{x^2 \sqrt{49-36x^2}} \, dx$

5.5.3 Using Computer Algebra Systems

A computer algebra system (CAS) is a computer program that is capable of executing symbolic mathematics. For a simple example, if we ask a CAS to solve the equation $ax^2 + bx + c = 0$ for the variable x, where a, b, and c are arbitrary constants, the program will return $x = \frac{-b \pm \sqrt{b^2 - 4ac}}{2a}$. While research to develop the first CAS dates to the 1960s, these programs became more common and publicly available in the early 1990s. Two prominent early examples are the programs *Maple* and *Mathematica*, which were among the first computer algebra systems to offer a graphical user interface. Today, *Maple* and *Mathematica* are exceptionally powerful professional software packages that are capable of executing an amazing array of sophisticated mathematical computations. They are also very expensive, as each is a proprietary program. The CAS *SAGE* is an open-source, free alternative to *Maple* and *Mathematica*.

For the purposes of this text, when we need to use a CAS, we are going to turn instead to a similar, but somewhat different computational tool, the web-based "computational knowledge engine" called *WolframAlpha*. There are two features of *WolframAlpha* that make it stand out from the CAS options mentioned above: (1) unlike *Maple* and *Mathematica*, *WolframAlpha* is free (provided we are willing to suffer through some pop-up advertising); and (2) unlike any of the three, the syntax in *WolframAlpha* is flexible. Think of *WolframAlpha* as being a little bit like doing a Google search: the program will interpret what is input, and then provide a summary of options.

If we want to have *WolframAlpha* evaluate an integral for us, we can provide it syntax such as

```
integrate x^2 dx
```

to which the program responds with

$$\int x^2 \, dx = \frac{x^3}{3} + \text{constant}.$$

While there is much to be enthusiastic about regarding CAS programs such as *WolframAlpha*, there are several things we should be cautious about: (1) a CAS only responds to exactly what is input; (2) a CAS can answer using powerful functions from highly advanced mathematics; and (3) there are problems that even a CAS cannot do without additional human insight.

Although (1) likely goes without saying, we have to be careful with our input: if we enter syntax that defines a function other than the problem of interest, the CAS will work with precisely the function we define. For example, if we are interested in evaluating the integral

$$\int \frac{1}{16 - 5x^2} \, dx,$$

and we mistakenly enter

```
integrate 1/16 - 5x^2 dx
```

a CAS will (correctly) reply with

$$\frac{1}{16}x - \frac{5}{3}x^3.$$

It is essential that we are sufficiently well-versed in antidifferentiation to recognize that this function cannot be the one that we seek: integrating a rational function such as $\frac{1}{16-5x^2}$, we expect the logarithm function to be present in the result.

Regarding (2), even for a relatively simple integral such as $\int \frac{1}{16-5x^2} \, dx$, some CASs will invoke advanced functions rather than simple ones. For instance, if we use *Maple* to execute the command

```
int(1/(16-5*x^2), x);
```

the program responds with

$$\int \frac{1}{16-5x^2} \, dx = \frac{\sqrt{5}}{20} \operatorname{arctanh}(\frac{\sqrt{5}}{4}x).$$

While this is correct (save for the missing arbitrary constant, which *Maple* never reports), the inverse hyperbolic tangent function is not a common nor familiar one; a simpler way to express this function can be found by using the partial fractions method, and happens to be the result reported by *WolframAlpha*:

$$\int \frac{1}{16-5x^2} \, dx = \frac{1}{8\sqrt{5}} \left(\log(4\sqrt{5} + 5\sqrt{x}) - \log(4\sqrt{5} - 5\sqrt{x}) \right) + \text{constant}.$$

Using sophisticated functions from more advanced mathematics is sometimes the way a CAS says to the user "I don't know how to do this problem." For example, if we want to evaluate

$$\int e^{-x^2} \, dx,$$

and we ask *WolframAlpha* to do so, the input

```
integrate exp(-x^2)dx
```

results in the output

$$\int e^{-x^2} \, dx = \frac{\sqrt{\pi}}{2} \operatorname{erf}(x) + \text{constant}.$$

The function "erf(x)" is the *error function*, which is actually defined by an integral:

$$\operatorname{erf}(x) = \frac{2}{\sqrt{\pi}} \int_0^x e^{-t^2} \, dt.$$

So, in producing output involving an integral, the CAS has basically reported back to us the very question we asked.

Finally, as remarked at (3) above, there are times that a CAS will actually fail without some additional human insight. If we consider the integral

$$\int (1+x)e^x \sqrt{1 + x^2 e^{2x}} \, dx$$

and ask *WolframAlpha* to evaluate

```
int (1+x)* exp(x)* sqrt(1+x^2 * exp(2x))dx,
```

the program thinks for a moment and then reports

(*no result found in terms of standard mathematical functions*)

But in fact this integral is not that difficult to evaluate. If we let $u = xe^x$, then $du = (1 + x)e^x\, dx$, which means that the preceding integral has form

$$\int (1+x)e^x \sqrt{1 + x^2 e^{2x}}\, dx = \int \sqrt{1 + u^2}\, du,$$

which is a straightforward one for any CAS to evaluate.

So, the above observations regarding computer algebra systems lead us to proceed with some caution: while any CAS is capable of evaluating a wide range of integrals (both definite and indefinite), there are times when the result can mislead us. We must think carefully about the meaning of the output, whether it is consistent with what we expect, and whether or not it makes sense to proceed.

Summary

- The method of partial fractions enables any rational function to be antidifferentiated, because any polynomial function can be factored into a product of linear and irreducible quadratic terms. This allows any rational function to be written as the sum of a polynomial plus rational terms of the form $\frac{A}{(x-c)^n}$ (where n is a natural number) and $\frac{Bx+C}{x^2+k}$ (where k is a positive real number).

- Until the development of computing algebra systems, integral tables enabled students of calculus to more easily evaluate integrals such as $\int \sqrt{a^2 + u^2}\, du$, where a is a positive real number. A short table of integrals may be found in Appendix A.

- Computer algebra systems can play an important role in finding antiderivatives, though we must be cautious to use correct input, to watch for unusual or unfamiliar advanced functions that the CAS may cite in its result, and to consider the possibility that a CAS may need further assistance or insight from us in order to answer a particular question.

Exercises

1. Calculate the integral below by partial fractions and by using the indicated substitution. Be sure that you can show how the results you obtain are the same.

$$\int \frac{2x}{x^2 - 25}\, dx$$

First, rewrite this with partial fractions:

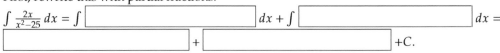

318

(Note that you should not include the +C in your entered answer, as it has been provided at the end of the expression.)

Next, use the substitution $w = x^2 - 25$ to find the integral:

$\int \frac{2x}{x^2-25} dx = \int$ [_____] $dw =$ [_____] $+C =$

[_____] $+C.$

(For the second answer blank, give your antiderivative in terms of the variable w. Again, note that you should not include the +C in your answer.)

2. Calculate the integral:

$$\int \frac{1}{(x + 6)(x + 8)} dx =$$ [_____]

3. Calculate the integral

$$\int \frac{7x + 3}{x^2 - 3x + 2} dx =$$ [_____]

4. Consider the following indefinite integral.

$$\int \frac{6x^3 + 8x^2 + 2x + 6}{x^4 + 1x^2} dx$$

The integrand has partial fractions decomposition:

$$\frac{a}{x^2} + \frac{b}{x} + \frac{cx + d}{x^2 + 1}$$

where

$a =$ [_____]

$b =$ [_____]

$c =$ [_____]

$d =$ [_____]

Now integrate term by term to evaluate the integral.

Answer: [_____] $+C$

5. The form of the partial fraction decomposition of a rational function is given below.

$$\frac{25x - 10x^2 - 45}{(x - 5)(x^2 + 9)} = \frac{A}{x - 5} + \frac{Bx + C}{x^2 + 9}$$

$A =$ [_____] $B =$ [_____] $C =$ [_____]

Now evaluate the indefinite integral.

$$\int \frac{25x - 10x^2 - 45}{(x - 5)(x^2 + 9)} dx =$$ [_____]

6. For each of the following integrals involving rational functions, (1) use a CAS to find the partial fraction decomposition of the integrand; (2) evaluate the integral of the resulting function without the assistance of technology; (3) use a CAS to evaluate the original integral to test and compare your result in (2).

a. $\int \frac{x^3+x+1}{x^4-1} \, dx$

b. $\int \frac{x^5+x^2+3}{x^3-6x^2+11x-6} \, dx$

c. $\int \frac{x^2-x-1}{(x-3)^3} \, dx$

7. For each of the following integrals involving radical functions, (1) use an appropriate *u*-substitution along with Appendix A to evaluate the integral without the assistance of technology, and (2) use a CAS to evaluate the original integral to test and compare your result in (1).

a. $\int \frac{1}{x\sqrt{9x^2+25}} \, dx$

b. $\int x\sqrt{1+x^4} \, dx$

c. $\int e^x \sqrt{4+e^{2x}} \, dx$

d. $\int \frac{\tan(x)}{\sqrt{9-\cos^2(x)}} \, dx$

8. Consider the indefinite integral given by

$$\int \frac{\sqrt{x+\sqrt{1+x^2}}}{x} \, dx.$$

a. Explain why *u*-substitution does not offer a way to simplify this integral by discussing at least two different options you might try for *u*.

b. Explain why integration by parts does not seem to be a reasonable way to proceed, either, by considering one option for *u* and *dv*.

c. Is there any line in the integral table in Appendix Athat is helpful for this integral?

d. Evaluate the given integral using *WolframAlpha*. What do you observe?

5.6 Numerical Integration

Motivating Questions

- How do we accurately evaluate a definite integral such as $\int_0^1 e^{-x^2}\, dx$ when we cannot use the First Fundamental Theorem of Calculus because the integrand lacks an elementary algebraic antiderivative? Are there ways to generate accurate estimates without using extremely large values of n in Riemann sums?

- What is the Trapezoid Rule, and how is it related to left, right, and middle Riemann sums?

- How are the errors in the Trapezoid Rule and Midpoint Rule related, and how can they be used to develop an even more accurate rule?

When we were first exploring the problem of finding the net-signed area bounded by a curve, we developed the concept of a Riemann sum as a helpful estimation tool and a key step in the definition of the definite integral. In particular, as we found in Section 4.2, recall that the left, right, and middle Riemann sums of a function f on an interval $[a, b]$ are denoted L_n, R_n, and M_n, with formulas

$$L_n = f(x_0)\Delta x + f(x_1)\Delta x + \cdots + f(x_{n-1})\Delta x = \sum_{i=0}^{n-1} f(x_i)\Delta x, \qquad (5.6.1)$$

$$R_n = f(x_1)\Delta x + f(x_2)\Delta x + \cdots + f(x_n)\Delta x = \sum_{i=1}^{n} f(x_i)\Delta x, \qquad (5.6.2)$$

$$M_n = f(\overline{x}_1)\Delta x + f(\overline{x}_2)\Delta x + \cdots + f(\overline{x}_n)\Delta x = \sum_{i=1}^{n} f(\overline{x}_i)\Delta x, \qquad (5.6.3)$$

where $x_0 = a$, $x_i = a + i\Delta x$, $x_n = b$, and $\Delta x = \frac{b-a}{n}$. For the middle sum, note that $\overline{x}_i = (x_{i-1} + x_i)/2$.

Further, recall that a Riemann sum is essentially a sum of (possibly signed) areas of rectangles, and that the value of n determines the number of rectangles, while our choice of left endpoints, right endpoints, or midpoints determines how we use the given function to find the heights of the respective rectangles we choose to use. Visually, we can see the similarities and differences among these three options in Figure 5.6.1, where we consider the function $f(x) = \frac{1}{20}(x-4)^3 + 7$ on the interval $[1, 8]$, and use 5 rectangles for each of the Riemann sums.

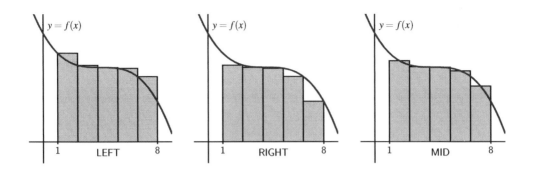

Figure 5.6.1: Left, right, and middle Riemann sums for $y = f(x)$ on $[1, 8]$ with 5 subintervals.

While it is a good exercise to compute a few Riemann sums by hand, just to ensure that we understand how they work and how varying the function, the number of subintervals, and the choice of endpoints or midpoints affects the result, it is of course the case that using computing technology is the best way to determine L_n, R_n, and M_n going forward. Any computer algebra system will offer this capability; as we saw in Preview Activity 4.3.1, a straightforward option that happens to also be freely available online is the applet[1] at http://gvsu.edu/s/a9.

Note that we can adjust the formula for $f(x)$, the window of x- and y-values of interest, the number of subintervals, and the method. See Preview Activity 4.3.1 for any needed reminders on how the applet works.

In what follows in this section we explore several different alternatives, including left, right, and middle Riemann sums, for estimating definite integrals. One of our main goals in the upcoming section is to develop formulas that enable us to estimate definite integrals accurately without having to use exceptionally large numbers of rectangles.

> **Preview Activity 5.6.1.** As we begin to investigate ways to approximate definite integrals, it will be insightful to compare results to integrals whose exact values we know. To that end, the following sequence of questions centers on $\int_0^3 x^2 \, dx$.
>
> a. Use the applet at http://gvsu.edu/s/a9 with the function $f(x) = x^2$ on the window of x values from 0 to 3 to compute L_3, the left Riemann sum with three subintervals.
>
> b. Likewise, use the applet to compute R_3 and M_3, the right and middle Riemann sums with three subintervals, respectively.
>
> c. Use the Fundamental Theorem of Calculus to compute the exact value of $I = \int_0^3 x^2 \, dx$.

---

[1]Marc Renault, Shippensburg University

d. We define the *error* in an approximation of a definite integral to be the difference between the integral's exact value and the approximation's value. What is the error that results from using L_3? From R_3? From M_3?

e. In what follows in this section, we will learn a new approach to estimating the value of a definite integral known as the Trapezoid Rule. The basic idea is to use trapezoids, rather than rectangles, to estimate the area under a curve. What is the formula for the area of a trapezoid with bases of length b_1 and b_2 and height h?

f. Working by hand, estimate the area under $f(x) = x^2$ on $[0, 3]$ using three subintervals and three corresponding trapezoids. What is the error in this approximation? How does it compare to the errors you calculated in (d)?

5.6.1 The Trapezoid Rule

Throughout our work to date with developing and estimating definite integrals, we have used the simplest possible quadrilaterals (that is, rectangles) to subdivide regions with complicated shapes. It is natural, however, to wonder if other familiar shapes might serve us even better. In particular, our goal is to be able to accurately estimate $\int_a^b f(x)\,dx$ without having to use extremely large values of n in Riemann sums.

To this end, we consider an alternative to L_n, R_n, and M_n, know as the *Trapezoid Rule*. The fundamental idea is simple: rather than using a rectangle to estimate the (signed) area bounded by $y = f(x)$ on a small interval, we use a trapezoid. For example, in Figure 5.6.2, we estimate the area under the pictured curve using three subintervals and the trapezoids that result from connecting the corresponding points on the curve with straight lines.

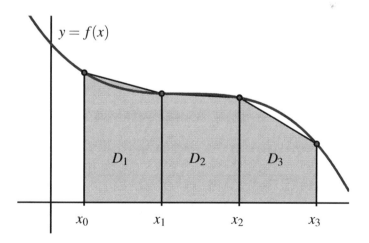

Figure 5.6.2: Estimating $\int_a^b f(x)\,dx$ using three subintervals and trapezoids, rather than rectangles, where $a = x_0$ and $b = x_3$.

The biggest difference between the Trapezoid Rule and a left, right, or middle Riemann sum is that on each subinterval, the Trapezoid Rule uses two function values, rather than one, to estimate the (signed) area bounded by the curve. For instance, to compute D_1, the area of the trapezoid generated by the curve $y = f(x)$ in Figure 5.6.2 on $[x_0, x_1]$, we observe that the left base of this trapezoid has length $f(x_0)$, while the right base has length $f(x_1)$. In addition, the height of this trapezoid is $x_1 - x_0 = \Delta x = \frac{b-a}{3}$. Since the area of a trapezoid is the average of the bases times the height, we have

$$D_1 = \frac{1}{2}(f(x_0) + f(x_1)) \cdot \Delta x.$$

Using similar computations for D_2 and D_3, we find that T_3, the trapezoidal approximation to $\int_a^b f(x)\, dx$ is given by

$$T_3 = D_1 + D_2 + D_3$$
$$= \frac{1}{2}(f(x_0) + f(x_1)) \cdot \Delta x + \frac{1}{2}(f(x_1) + f(x_2)) \cdot \Delta x + \frac{1}{2}(f(x_2) + f(x_3)) \cdot \Delta x.$$

Because both left and right endpoints are being used, we recognize within the trapezoidal approximation the use of both left and right Riemann sums. In particular, rearranging the expression for T_3 by removing factors of $\frac{1}{2}$ and Δx, grouping the left endpoint evaluations of f, and grouping the right endpoint evaluations of f, we see that

$$T_3 = \frac{1}{2}\left[(f(x_0) + f(x_1) + f(x_2))\right]\Delta x + (f(x_1) + \frac{1}{2}\left[f(x_2) + f(x_3))\right]\Delta x. \qquad (5.6.4)$$

At this point, we observe that two familiar sums have arisen. Since the left Riemann sum L_3 is $L_3 = f(x_0)\Delta x + f(x_1)\Delta x + f(x_2)\Delta x$, and the right Riemann sum is $R_3 = f(x_1)\Delta x + f(x_2)\Delta x + f(x_3)\Delta x$, substituting L_3 and R_3 for the corresponding expressions in Equation (5.6.4), it follows that $T_3 = \frac{1}{2}[L_3 + R_3]$. We have thus seen the main ideas behind a very important result: using trapezoids to estimate the (signed) area bounded by a curve is the same as averaging the estimates generated by using left and right endpoints.

The Trapezoid Rule

The trapezoidal approximation, T_n, of the definite integral $\int_a^b f(x)\, dx$ using n subintervals is given by the rule

$$T_n = \left[\frac{1}{2}(f(x_0) + f(x_1)) + \frac{1}{2}(f(x_1) + f(x_2)) + \cdots + \frac{1}{2}(f(x_{n-1}) + f(x_n))\right]\Delta x.$$
$$= \sum_{i=0}^{n-1} \frac{1}{2}(f(x_i) + f(x_{i+1}))\Delta x.$$

Moreover, $T_n = \frac{1}{2}[L_n + R_n]$.

Activity 5.6.2. In this activity, we explore the relationships among the errors generated by left, right, midpoint, and trapezoid approximations to the definite integral $\int_1^2 \frac{1}{x^2}\, dx$

a. Use the First FTC to evaluate $\int_1^2 \frac{1}{x^2}\, dx$ exactly.

b. Use appropriate computing technology to compute the following approximations for $\int_1^2 \frac{1}{x^2}\, dx$: T_4, M_4, T_8, and M_8.

c. Let the *error* of an approximation be the difference between the exact value of the definite integral and the resulting approximation. For instance, if we let $E_{T,4}$ represent the error that results from using the trapezoid rule with 4 subintervals to estimate the integral, we have

$$E_{T,4} = \int_1^2 \frac{1}{x^2}\, dx - T_4.$$

Similarly, we compute the error of the midpoint rule approximation with 8 subintervals by the formula

$$E_{M,8} = \int_1^2 \frac{1}{x^2}\, dx - M_8.$$

Based on your work in (a) and (b) above, compute $E_{T,4}$, $E_{T,8}$, $E_{M,4}$, $E_{M,8}$.

d. Which rule consistently over-estimates the exact value of the definite integral? Which rule consistently under-estimates the definite integral?

e. What behavior(s) of the function $f(x) = \frac{1}{x^2}$ lead to your observations in (d)?

5.6.2 Comparing the Midpoint and Trapezoid Rules

We know from the definition of the definite integral of a continuous function f, that if we let n be large enough, we can make the value of any of the approximations L_n, R_n, and M_n as close as we'd like (in theory) to the exact value of $\int_a^b f(x)\, dx$. Thus, it may be natural to wonder why we ever use any rule other than L_n or R_n (with a sufficiently large n value) to estimate a definite integral. One of the primary reasons is that as $n \to \infty$, $\Delta x = \frac{b-a}{n} \to 0$, and thus in a Riemann sum calculation with a large n value, we end up multiplying by a number that is very close to zero. Doing so often generates roundoff error, as representing numbers close to zero accurately is a persistent challenge for computers.

Hence, we are exploring ways by which we can estimate definite integrals to high levels of precision, but without having to use extremely large values of n. Paying close attention to patterns in errors, such as those observed in Activity 5.6.2, is one way to begin to see some alternate approaches.

To begin, we make a comparison of the errors in the Midpoint and Trapezoid rules from two different perspectives. First, consider a function of consistent concavity on a given interval, and picture approximating the area bounded on that interval by both the Midpoint and

Trapezoid rules using a single subinterval.

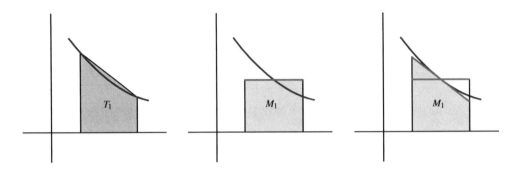

Figure 5.6.3: Estimating $\int_a^b f(x)\,dx$ using a single subinterval: at left, the trapezoid rule; in the middle, the midpoint rule; at right, a modified way to think about the midpoint rule.

As seen in Figure 5.6.3, it is evident that whenever the function is concave up on an interval, the Trapezoid Rule with one subinterval, T_1, will overestimate the exact value of the definite integral on that interval. Moreover, from a careful analysis of the line that bounds the top of the rectangle for the Midpoint Rule (shown in magenta), we see that if we rotate this line segment until it is tangent to the curve at the point on the curve used in the Midpoint Rule (as shown at right in Figure 5.6.3), the resulting trapezoid has the same area as M_1, and this value is less than the exact value of the definite integral. Hence, when the function is concave up on the interval, M_1 underestimates the integral's true value.

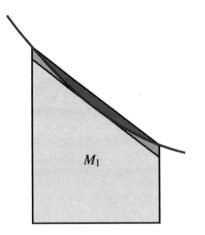

Figure 5.6.4: Comparing the error in estimating $\int_a^b f(x)\,dx$ using a single subinterval: in red, the error from the Trapezoid rule; in light red, the error from the Midpoint rule.

These observations extend easily to the situation where the function's concavity remains consistent but we use higher values of n in the Midpoint and Trapezoid Rules. Hence, whenever f is concave up on $[a, b]$, T_n will overestimate the value of $\int_a^b f(x)\, dx$, while M_n will underestimate $\int_a^b f(x)\, dx$. The reverse observations are true in the situation where f is concave down.

Next, we compare the size of the errors between M_n and T_n. Again, we focus on M_1 and T_1 on an interval where the concavity of f is consistent. In Figure 5.6.4, where the error of the Trapezoid Rule is shaded in red, while the error of the Midpoint Rule is shaded lighter red, it is visually apparent that the error in the Trapezoid Rule is more significant. To see how much more significant, let's consider two examples and some particular computations.

If we let $f(x) = 1 - x^2$ and consider $\int_0^1 f(x)\, dx$, we know by the First FTC that the exact value of the integral is

$$\int_0^1 (1 - x^2)\, dx = x - \frac{x^3}{3}\Big|_0^1 = \frac{2}{3}.$$

Using appropriate technology to compute M_4, M_8, T_4, and T_8, as well as the corresponding errors $E_{M,4}$, $E_{M,8}$, $E_{T,4}$, and $E_{T,8}$, as we did in Activity 5.6.2, we find the results summarized in Table 5.6.5. Note that in the table, we also include the approximations and their errors for the example $\int_1^2 \frac{1}{x^2}\, dx$ from Activity 5.6.2.

Rule	$\int_0^1 (1 - x^2)\, dx = 0.\overline{6}$	error	$\int_1^2 \frac{1}{x^2}\, dx = 0.5$	error
T_4	0.65625	−0.0104166667	0.5089937642	0.0089937642
M_4	0.671875	0.0052083333	0.4955479365	−0.0044520635
T_8	0.6640625	−0.0026041667	0.5022708502	0.0022708502
M_8	0.66796875	0.0013020833	0.4988674899	−0.0011325101

Table 5.6.5: Calculations of T_4, M_4, T_8, and M_8, along with corresponding errors, for the definite integrals $\int_0^1 (1 - x^2)\, dx$ and $\int_1^2 \frac{1}{x^2}\, dx$.

Recall that for a given function f and interval $[a, b]$, $E_{T,4} = \int_a^b f(x)\, dx - T_4$ calculates the difference between the exact value of the definite integral and the approximation generated by the Trapezoid Rule with $n = 4$. If we look at not only $E_{T,4}$, but also the other errors generated by using T_n and M_n with $n = 4$ and $n = 8$ in the two examples noted in Table 5.6.5, we see an evident pattern. Not only is the sign of the error (which measures whether the rule generates an over- or under-estimate) tied to the rule used and the function's concavity, but the magnitude of the errors generated by T_n and M_n seems closely connected. In particular, the errors generated by the Midpoint Rule seem to be about half the size of those generated by the Trapezoid Rule.

That is, we can observe in both examples that $E_{M,4} \approx -\frac{1}{2}E_{T,4}$ and $E_{M,8} \approx -\frac{1}{2}E_{T,8}$, which demonstrates a property of the Midpoint and Trapezoid Rules that turns out to hold in general: for a function of consistent concavity, the error in the Midpoint Rule has the opposite sign and approximately half the magnitude of the error of the Trapezoid Rule. Said symbolically,

$$E_{M,n} \approx -\frac{1}{2}E_{T,n}.$$

This important relationship suggests a way to combine the Midpoint and Trapezoid Rules to create an even more accurate approximation to a definite integral.

5.6.3 Simpson's Rule

When we first developed the Trapezoid Rule, we observed that it can equivalently be viewed as resulting from the average of the Left and Right Riemann sums:

$$T_n = \frac{1}{2}(L_n + R_n).$$

Whenever a function is always increasing or always decreasing on the interval $[a, b]$, one of L_n and R_n will over-estimate the true value of $\int_a^b f(x)\,dx$, while the other will under-estimate the integral. Said differently, the errors found in L_n and R_n will have opposite signs; thus, averaging L_n and R_n eliminates a considerable amount of the error present in the respective approximations. In a similar way, it makes sense to think about averaging M_n and T_n in order to generate a still more accurate approximation.

At the same time, we've just observed that M_n is typically about twice as accurate as T_n. Thus, we instead choose to use the weighted average

$$S_{2n} = \frac{2M_n + T_n}{3}. \tag{5.6.5}$$

The rule for S_{2n} giving by Equation (5.6.5) is usually known as *Simpson's Rule*.[2] Note that we use "S_{2n}" rather that "S_n" since the n points the Midpoint Rule uses are different from the n points the Trapezoid Rule uses, and thus Simpson's Rule is using $2n$ points at which to evaluate the function. We build upon the results in Table 5.6.5 to see the approximations generated by Simpson's Rule. In particular, in Table 5.6.6, we include all of the results in Table 5.6.5, but include additional results for $S_8 = \frac{2M_4+T_4}{3}$ and $S_{16} = \frac{2M_8+T_8}{3}$.

---

[2]Thomas Simpson was an 18th century mathematician; his idea was to extend the Trapezoid rule, but rather than using straight lines to build trapezoids, to use quadratic functions to build regions whose area was bounded by parabolas (whose areas he could find exactly). Simpson's Rule is often developed from the more sophisticated perspective of using interpolation by quadratic functions.

Rule	$\int_0^1 (1 - x^2)\,dx = 0.\overline{6}$	error	$\int_1^2 \frac{1}{x^2}\,dx = 0.5$	error
T_4	0.65625	−0.0104166667	0.5089937642	0.0089937642
M_4	0.671875	0.0052083333	0.4955479365	−0.0044520635
S_8	0.6666666667	0	0.5000298792	0.0000298792
T_8	0.6640625	−0.0026041667	0.5022708502	0.0022708502
M_8	0.66796875	0.0013020833	0.4988674899	−0.0011325101
S_{16}	0.6666666667	0	0.5000019434	0.0000019434

Table 5.6.6: Table 5.6.5 updated to include S_8, S_{16}, and the corresponding errors.

The results seen in Table 5.6.6 are striking. If we consider the S_{16} approximation of $\int_1^2 \frac{1}{x^2}\,dx$, the error is only $E_{S,16} = 0.0000019434$. By contrast, $L_8 = 0.5491458502$, so the error of that estimate is $E_{L,8} = -0.0491458502$. Moreover, we observe that generating the approximations for Simpson's Rule is almost no additional work: once we have L_n, R_n, and M_n for a given value of n, it is a simple exercise to generate T_n, and from there to calculate S_{2n}. Finally, note that the error in the Simpson's Rule approximations of $\int_0^1 (1 - x^2)\,dx$ is zero![3]

These rules are not only useful for approximating definite integrals such as $\int_0^1 e^{-x^2}\,dx$, for which we cannot find an elementary antiderivative of e^{-x^2}, but also for approximating definite integrals in the setting where we are given a function through a table of data.

Activity 5.6.3. A car traveling along a straight road is braking and its velocity is measured at several different points in time, as given in the following table. Assume that v is continuous, always decreasing, and always decreasing at a decreasing rate, as is suggested by the data.

seconds, t	0	0.3	0.6	0.9	1.2	1.5	1.8
Velocity in ft/sec, $v(t)$	100	99	96	90	80	50	0

Table 5.6.7: Data for the braking car.

a. Plot the given data on the set of axes provided in Figure 5.6.8 with time on the horizontal axis and the velocity on the vertical axis.

b. What definite integral will give you the exact distance the car traveled on $[0, 1.8]$?

c. Estimate the total distance traveled on $[0, 1.8]$ by computing L_3, R_3, and T_3. Which of these under-estimates the true distance traveled?

---

[3]Similar to how the Midpoint and Trapezoid approximations are exact for linear functions, Simpson's Rule approximations are exact for quadratic and cubic functions. See additional discussion on this issue later in the section and in the exercises.

d. Estimate the total distance traveled on $[0, 1.8]$ by computing M_3. Is this an over- or under-estimate? Why?

e. Using your results from (c) and (d), improve your estimate further by using Simpson's Rule.

f. What is your best estimate of the average velocity of the car on $[0, 1.8]$? Why? What are the units on this quantity?

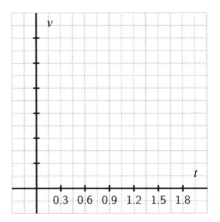

Figure 5.6.8: Axes for plotting the data in Activity 5.6.3.

5.6.4 Overall observations regarding L_n, R_n, T_n, M_n, and S_{2n}.

As we conclude our discussion of numerical approximation of definite integrals, it is important to summarize general trends in how the various rules over- or under-estimate the true value of a definite integral, and by how much. To revisit some past observations and see some new ones, we consider the following activity.

Activity 5.6.4. Consider the functions $f(x) = 2 - x^2$, $g(x) = 2 - x^3$, and $h(x) = 2 - x^4$, all on the interval $[0, 1]$. For each of the questions that require a numerical answer in what follows, write your answer exactly in fraction form.

a. On the three sets of axes provided in Figure 5.6.9, sketch a graph of each function on the interval $[0, 1]$, and compute L_1 and R_1 for each. What do you observe?

b. Compute M_1 for each function to approximate $\int_0^1 f(x)\,dx$, $\int_0^1 g(x)\,dx$, and $\int_0^1 h(x)\,dx$, respectively.

c. Compute T_1 for each of the three functions, and hence compute S_2 for each of the three functions.

d. Evaluate each of the integrals $\int_0^1 f(x)\,dx$, $\int_0^1 g(x)\,dx$, and $\int_0^1 h(x)\,dx$ exactly using the First FTC.

e. For each of the three functions f, g, and h, compare the results of L_1, R_1, M_1, T_1, and S_2 to the true value of the corresponding definite integral. What patterns do you observe?

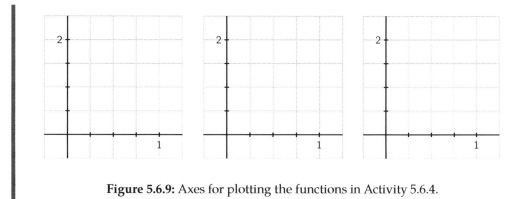

Figure 5.6.9: Axes for plotting the functions in Activity 5.6.4.

The results seen in the examples in Activity 5.6.4 generalize nicely. For instance, for any function f that is decreasing on $[a, b]$, L_n will over-estimate the exact value of $\int_a^b f(x)\,dx$, and for any function f that is concave down on $[a, b]$, M_n will over-estimate the exact value of the integral. An excellent exercise is to write a collection of scenarios of possible function behavior, and then categorize whether each of L_n, R_n, T_n, and M_n is an over- or under-estimate.

Finally, we make two important notes about Simpson's Rule. When T. Simpson first developed this rule, his idea was to replace the function f on a given interval with a quadratic function that shared three values with the function f. In so doing, he guaranteed that this new approximation rule would be exact for the definite integral of any quadratic polynomial. In one of the pleasant surprises of numerical analysis, it turns out that even though it was designed to be exact for quadratic polynomials, Simpson's Rule is exact for any cubic polynomial: that is, if we are interested in an integral such as $\int_2^5 (5x^3 - 2x^2 + 7x - 4)\,dx$, S_{2n} will always be exact, regardless of the value of n. This is just one more piece of evidence that shows how effective Simpson's Rule is as an approximation tool for estimating definite integrals.[4]

Summary

- For a definite integral such as $\int_0^1 e^{-x^2}\,dx$ when we cannot use the First Fundamental Theorem of Calculus because the integrand lacks an elementary algebraic antiderivative, we can estimate the integral's value by using a sequence of Riemann sum approximations. Typically, we start by computing L_n, R_n, and M_n for one or more chosen values of n.

- The Trapezoid Rule, which estimates $\int_a^b f(x)\,dx$ by using trapezoids, rather than rectangles, can also be viewed as the average of Left and Right Riemann sums. That is, $T_n = \frac{1}{2}(L_n + R_n)$.

---

[4]One reason that Simpson's Rule is so effective is that S_{2n} benefits from using $2n + 1$ points of data. Because it combines M_n, which uses n midpoints, and T_n, which uses the $n + 1$ endpoints of the chosen subintervals, S_{2n} takes advantage of the maximum amount of information we have when we know function values at the endpoints and midpoints of n subintervals.

- The Midpoint Rule is typically twice as accurate as the Trapezoid Rule, and the signs of the respective errors of these rules are opposites. Hence, by taking the weighted average $S_n = \frac{2M_n + T_n}{3}$, we can build a much more accurate approximation to $\int_a^b f(x)\,dx$ by using approximations we have already computed. The rule for S_n is known as Simpson's Rule, which can also be developed by approximating a given continuous function with pieces of quadratic polynomials.

Exercises

1. Note: for this problem, because later answers depend on earlier ones, you must enter answers for all answer blanks for the problem to be correctly graded. If you would like to get feedback before you completed all computations, enter a "1" for each answer you did not yet compute and then submit the problem. (But note that this will, obviously, result in a problem submission.)

(a) What is the exact value of $\int_0^3 e^x\,dx$?

$\int_0^3 e^x\,dx =$ [   ]

(b)
Find LEFT(2), RIGHT(2), TRAP(2), MID(2), and SIMP(2); compute the error for each.

	LEFT(2)	RIGHT(2)	TRAP(2)	MID(2)	SIMP(2)
value					
error					

(c)
Repeat part (b) with $n = 4$ (instead of $n = 2$).

	LEFT(4)	RIGHT(4)	TRAP(4)	MID(4)	SIMP(4)
value					
error					

(d)
For each rule in part (b), as n goes from $n = 2$ to $n = 4$, does the error go down approximately as you would expect? Explain by calculating the ratios of the errors:

Error LEFT(2)/Error LEFT(4) = [   ]

Error RIGHT(2)/Error RIGHT(4) = [   ]

Error TRAP(2)/Error TRAP(4) = [   ]

Error MID(2)/Error MID(4) = [   ]

Error SIMP(2)/Error SIMP(4) = [   ]

(Be sure that you can explain in words why these do (or don't) make sense.)

2. Using the figure showing $f(x)$ below, order the following approximations to the integral $\int_0^3 f(x)\,dx$ and its exact value from smallest to largest.

Enter each of "LEFT(n)", "RIGHT(n)", "TRAP(n)", "MID(n)" and "Exact" in one of the following answer blanks to indicate the correct ordering:

	<		<		<		<	

3. Using a fixed number of subdivisions, we approximate the integrals of f and g on the interval shown in the figure below.

(The function $f(x)$ is shown in blue, and $g(x)$ in black.)
For which function, f or g is LEFT more accurate?
For which function, f or g is RIGHT more accurate?
For which function, f or g is MID more accurate?
For which function, f or g is TRAP more accurate?

4. Consider the four functions shown below. On the first two, an approximation for $\int_a^b f(x)\,dx$ is shown.

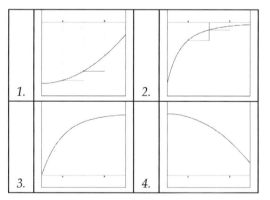

1. For graph number 1, Which integration method is shown?
Is this method an over- or underestimate?
2. For graph number 2, Which integration method is shown?
Is this method an over- or underestimate?
3. On a copy of graph number 3, sketch an estimate with $n = 2$ subdivisions using the midpoint rule.
Is this method an over- or underestimate?

4. On a copy of graph number 4, sketch an estimate with $n = 2$ subdivisions using the trapezoid rule.
Is this method an over- or underestimate?

5. Consider the definite integral $\int_0^1 x \tan(x)\,dx$.

a. Explain why this integral cannot be evaluated exactly by using either u-substitution or by integrating by parts.

b. Using 4 subintervals, compute L_4, R_4, M_4, T_4, and S_4.

c. Which of the approximations in (b) is an over-estimate to the true value of $\int_0^1 x \tan(x)\,dx$? Which is an under-estimate? How do you know?

6. For an unknown function $f(x)$, the following information is known.

- f is continuous on $[3, 6]$;
- f is either always increasing or always decreasing on $[3, 6]$;
- f has the same concavity throughout the interval $[3, 6]$;
- As approximations to $\int_3^6 f(x)\,dx$, $L_4 = 7.23$, $R_4 = 6.75$, and $M_4 = 7.05$.

a. Is f increasing or decreasing on $[3, 6]$? What data tells you?

b. Is f concave up or concave down on $[3, 6]$? Why?

c. Determine the best possible estimate you can for $\int_3^6 f(x)\,dx$, based on the given information.

7. The rate at which water flows through Table Rock Dam on the White River in Branson, MO, is measured in thousands of cubic feet per second (TCFS). As engineers open the floodgates, flow rates are recorded according to the following chart.

seconds, t	0	10	20	30	40	50	60
flow in TCFS, $r(t)$	2000	2100	2400	3000	3900	5100	6500

Table 5.6.10: Water flow data.

a. What definite integral measures the total volume of water to flow through the dam in the 60 second time period provided by the table above?

b. Use the given data to calculate M_n for the largest possible value of n to approximate the integral you stated in (a). Do you think M_n over- or under-estimates the exact value of the integral? Why?

c. Approximate the integral stated in (a) by calculating S_n for the largest possible value of n, based on the given data.

d. Compute $\frac{1}{60} S_n$ and $\frac{2000+2100+2400+3000+3900+5100+6500}{7}$. What quantity do both of these values estimate? Which is a more accurate approximation?

CHAPTER 6

Using Definite Integrals

6.1 Using Definite Integrals to Find Area and Length

Motivating Questions

- How can we use definite integrals to measure the area between two curves?

- How do we decide whether to integrate with respect to x or with respect to y when we try to find the area of a region?

- How can a definite integral be used to measure the length of a curve?

Early on in our work with the definite integral, we learned that if we have a nonnegative velocity function, v, for an object moving along an axis, the area under the velocity function between a and b tells us the distance the object traveled on that time interval. Moreover, based on the definition of the definite integral, that area is given precisely by $\int_a^b v(t)\, dt$. Indeed, for any nonnegative function f on an interval $[a, b]$, we know that $\int_a^b f(x)\, dx$ measures the area bounded by the curve and the x-axis between $x = a$ and $x = b$.

Through our upcoming work in the present section and chapter, we will explore how definite integrals can be used to represent a variety of different physically important properties. In Preview Activity 6.1.1, we begin this investigation by seeing how a single definite integral may be used to represent the area between two curves.

> **Preview Activity 6.1.1.** Consider the functions given by $f(x) = 5 - (x - 1)^2$ and $g(x) = 4 - x$.
>
> a. Use algebra to find the points where the graphs of f and g intersect.
>
> b. Sketch an accurate graph of f and g on the axes provided, labeling the curves by name and the intersection points with ordered pairs.
>
> c. Find and evaluate exactly an integral expression that represents the area between $y = f(x)$ and the x-axis on the interval between the intersection points of f and g.

d. Find and evaluate exactly an integral expression that represents the area between $y = g(x)$ and the x-axis on the interval between the intersection points of f and g.

e. What is the exact area between f and g between their intersection points? Why?

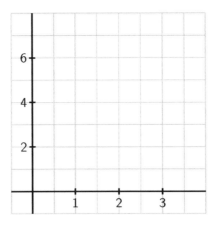

Figure 6.1.1: Axes for plotting f and g in Preview Activity 6.1.1

6.1.1 The Area Between Two Curves

Through Preview Activity 6.1.1, we encounter a natural way to think about the area between two curves: the area between the curves is the area beneath the upper curve minus the area below the lower curve. For the functions $f(x) = (x - 1)^2 + 1$ and $g(x) = x + 2$, shown in Figure 6.1.2,

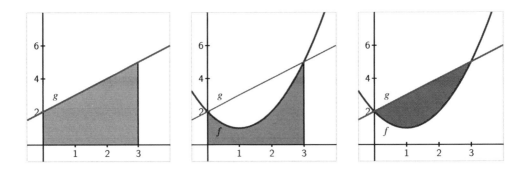

Figure 6.1.2: The areas bounded by the functions $f(x) = (x - 1)^2 + 1$ and $g(x) = x + 2$ on the interval $[0, 3]$.

we see that the upper curve is $g(x) = x + 2$, and that the graphs intersect at $(0, 2)$ and $(3, 5)$.

Note that we can find these intersection points by solving the system of equations given by $y = (x - 1)^2 + 1$ and $y = x + 2$ through substitution: substituting $x + 2$ for y in the first equation yields $x + 2 = (x - 1)^2 + 1$, so $x + 2 = x^2 - 2x + 1 + 1$, and thus

$$x^2 - 3x = x(x - 3) = 0,$$

from which it follows that $x = 0$ or $x = 3$. Using $y = x + 2$, we find the corresponding y-values of the intersection points.

On the interval $[0, 3]$, the area beneath g is

$$\int_0^3 (x + 2)\, dx = \frac{21}{2},$$

while the area under f on the same interval is

$$\int_0^3 [(x - 1)^2 + 1]\, dx = 6.$$

Thus, the area between the curves is

$$A = \int_0^3 (x + 2)\, dx - \int_0^3 [(x - 1)^2 + 1]\, dx = \frac{21}{2} - 6 = \frac{9}{2}. \tag{6.1.1}$$

A slightly different perspective is also helpful here: if we take the region between two curves and slice it up into thin vertical rectangles (in the same spirit as we originally sliced the region between a single curve and the x-axis in Section 4.2), then we see that the height of a typical rectangle is given by the difference between the two functions. For example, for the rectangle shown at left in Figure 6.1.3,

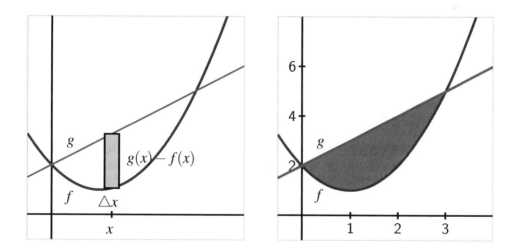

Figure 6.1.3: The area bounded by the functions $f(x) = (x - 1)^2 + 1$ and $g(x) = x + 2$ on the interval $[0, 3]$.

we see that the rectangle's height is $g(x) - f(x)$, while its width can be viewed as Δx, and thus the area of the rectangle is

$$A_{\text{rect}} = (g(x) - f(x))\Delta x.$$

The area between the two curves on $[0, 3]$ is thus approximated by the Riemann sum

$$A \approx \sum_{i=1}^{n} (g(x_i) - f(x_i))\Delta x,$$

and then as we let $n \to \infty$, it follows that the area is given by the single definite integral

$$A = \int_0^3 (g(x) - f(x))\,dx. \tag{6.1.2}$$

In many applications of the definite integral, we will find it helpful to think of a "representative slice" and how the definite integral may be used to add these slices to find the exact value of a desired quantity. Here, the integral essentially sums the areas of thin rectangles.

Finally, whether we think of the area between two curves as the difference between the area bounded by the individual curves (as in (6.1.1)) or as the limit of a Riemann sum that adds the areas of thin rectangles between the curves (as in (6.1.2)), these two results are the same, since the difference of two integrals is the integral of the difference:

$$\int_0^3 g(x)\,dx - \int_0^3 f(x)\,dx = \int_0^3 (g(x) - f(x))\,dx.$$

Moreover, our work so far in this section exemplifies the following general principle.

If two curves $y = g(x)$ and $y = f(x)$ intersect at $(a, g(a))$ and $(b, g(b))$, and for all x such that $a \leq x \leq b$, $g(x) \geq f(x)$, then the area between the curves is $A = \int_a^b (g(x) - f(x))\,dx$.

Activity 6.1.2. In each of the following problems, our goal is to determine the area of the region described. For each region, (i) determine the intersection points of the curves, (ii) sketch the region whose area is being found, (iii) draw and label a representative slice, and (iv) state the area of the representative slice. Then, state a definite integral whose value is the exact area of the region, and evaluate the integral to find the numeric value of the region's area.

a. The finite region bounded by $y = \sqrt{x}$ and $y = \frac{1}{4}x$.

b. The finite region bounded by $y = 12 - 2x^2$ and $y = x^2 - 8$.

c. The area bounded by the y-axis, $f(x) = \cos(x)$, and $g(x) = \sin(x)$, where we consider the region formed by the first positive value of x for which f and g intersect.

d. The finite regions between the curves $y = x^3 - x$ and $y = x^2$.

6.1.2 Finding Area with Horizontal Slices

At times, the shape of a geometric region may dictate that we need to use horizontal rectangular slices, rather than vertical ones. For instance, consider the region bounded by the parabola $x = y^2 - 1$ and the line $y = x - 1$, pictured in Figure 6.1.4. First, we observe that by solving the second equation for x and writing $x = y + 1$, we can eliminate a variable through substitution and find that $y + 1 = y^2 - 1$, and hence the curves intersect where $y^2 - y - 2 = 0$. Thus, we find $y = -1$ or $y = 2$, so the intersection points of the two curves are $(0, -1)$ and $(3, 2)$.

We see that if we attempt to use vertical rectangles to slice up the area, at certain values of x (specifically from $x = -1$ to $x = 0$, as seen in the center graph of Figure 6.1.4), the curves that govern the top and bottom of the rectangle are one and the same. This suggests, as shown in the rightmost graph in the figure, that we try using horizontal rectangles as a way to think about the area of the region.

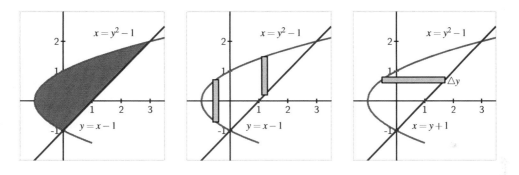

Figure 6.1.4: The area bounded by the functions $x = y^2 - 1$ and $y = x - 1$ (at left), with the region sliced vertically (center) and horizontally (at right).

For such a horizontal rectangle, note that its width depends on y, the height at which the rectangle is constructed. In particular, at a height y between $y = -1$ and $y = 2$, the right end of a representative rectangle is determined by the line, $x = y + 1$, while the left end of the rectangle is determined by the parabola, $x = y^2 - 1$, and the thickness of the rectangle is Δy.

Therefore, the area of the rectangle is

$$A_{\text{rect}} = [(y + 1) - (y^2 - 1)]\Delta y,$$

from which it follows that the area between the two curves on the y-interval $[-1, 2]$ is approximated by the Riemann sum

$$A \approx \sum_{i=1}^{n} [(y_i + 1) - (y_i^2 - 1)]\Delta y.$$

Taking the limit of the Riemann sum, it follows that the area of the region is

$$A = \int_{y=-1}^{y=2} [(y+1) - (y^2 - 1)] \, dy. \tag{6.1.3}$$

We emphasize that we are integrating with respect to y; this is dictated by the fact that we chose to use horizontal rectangles whose widths depend on y and whose thickness is denoted Δy. It is a straightforward exercise to evaluate the integral in Equation (6.1.3) and find that $A = \frac{9}{2}$.

Just as with the use of vertical rectangles of thickness Δx, we have a general principle for finding the area between two curves, which we state as follows.

> If two curves $x = g(y)$ and $x = f(y)$ intersect at $(g(c), c)$ and $(g(d), d)$, and for all y such that $c \leq y \leq d$, $g(y) \geq f(y)$, then the area between the curves is
>
> $$A = \int_{y=c}^{y=d} (g(y) - f(y)) \, dy.$$

Activity 6.1.3. In each of the following problems, our goal is to determine the area of the region described. For each region, (i) determine the intersection points of the curves, (ii) sketch the region whose area is being found, (iii) draw and label a representative slice, and (iv) state the area of the representative slice. Then, state a definite integral whose value is the exact area of the region, and evaluate the integral to find the numeric value of the region's area. *Note well:* At the step where you draw a representative slice, you need to make a choice about whether to slice vertically or horizontally.

    a. The finite region bounded by $x = y^2$ and $x = 6 - 2y^2$.

    b. The finite region bounded by $x = 1 - y^2$ and $x = 2 - 2y^2$.

    c. The area bounded by the x-axis, $y = x^2$, and $y = 2 - x$.

    d. The finite regions between the curves $x = y^2 - 2y$ and $y = x$.

6.1.3 Finding the length of a curve

In addition to being able to use definite integrals to find the areas of certain geometric regions, we can also use the definite integral to find the length of a portion of a curve. We use the same fundamental principle: we take a curve whose length we cannot easily find, and slice it up into small pieces whose lengths we can easily approximate. In particular, we take a given curve and subdivide it into small approximating line segments, as shown at left in Figure 6.1.5.

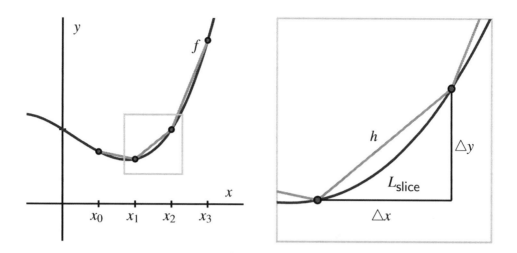

Figure 6.1.5: At left, a continuous function $y = f(x)$ whose length we seek on the interval $a = x_0$ to $b = x_3$. At right, a close up view of a portion of the curve.

To see how we find such a definite integral that measures arc length on the curve $y = f(x)$ from $x = a$ to $x = b$, we think about the portion of length, L_{slice}, that lies along the curve on a small interval of length Δx, and estimate the value of Lslice using a well-chosen triangle. In particular, if we consider the right triangle with legs parallel to the coordinate axes and hypotenuse connecting two points on the curve, as seen at right in Figure 6.1.5, we see that the length, h, of the hypotenuse approximates the length, L_{slice}, of the curve between the two selected points. Thus,

$$L_{slice} \approx h = \sqrt{(\Delta x)^2 + (\Delta y)^2}.$$

By algebraically rearranging the expression for the length of the hypotenuse, we see how a definite integral can be used to compute the length of a curve. In particular, observe that by removing a factor of $(\Delta x)^2$, we find that

$$L_{slice} \approx \sqrt{(\Delta x)^2 + (\Delta y)^2}$$

$$= \sqrt{(\Delta x)^2 \left(1 + \frac{(\Delta y)^2}{(\Delta x)^2}\right)}$$

$$= \sqrt{1 + \frac{(\Delta y)^2}{(\Delta x)^2}} \cdot \Delta x.$$

Furthermore, as $n \to \infty$ and $\Delta x \to 0$, it follows that $\frac{\Delta y}{\Delta x} \to \frac{dy}{dx} = f'(x)$. Thus, we can say that

$$L_{slice} \approx \sqrt{1 + f'(x)^2} \Delta x.$$

Taking a Riemann sum of all of these slices and letting $n \to \infty$, we arrive at the following fact.

Given a differentiable function f on an interval $[a, b]$, the total arc length, L, along the curve $y = f(x)$ from $x = a$ to $x = b$ is given by

$$L = \int_a^b \sqrt{1 + f'(x)^2} \, dx.$$

Activity 6.1.4. Each of the following questions somehow involves the arc length along a curve.

a. Use the definition and appropriate computational technology to determine the arc length along $y = x^2$ from $x = -1$ to $x = 1$.

b. Find the arc length of $y = \sqrt{4 - x^2}$ on the interval $-2 \le x \le 2$. Find this value in two different ways: (a) by using a definite integral, and (b) by using a familiar property of the curve.

c. Determine the arc length of $y = xe^{3x}$ on the interval $[0, 1]$.

d. Will the integrals that arise calculating arc length typically be ones that we can evaluate exactly using the First FTC, or ones that we need to approximate? Why?

e. A moving particle is traveling along the curve given by $y = f(x) = 0.1x^2 + 1$, and does so at a constant rate of 7 cm/sec, where both x and y are measured in cm (that is, the curve $y = f(x)$ is the path along which the object actually travels; the curve is not a "position function"). Find the position of the particle when $t = 4$ sec, assuming that when $t = 0$, the particle's location is $(0, f(0))$.

Summary

- To find the area between two curves, we think about slicing the region into thin rectangles. If, for instance, the area of a typical rectangle on the interval $x = a$ to $x = b$ is given by $A_{\text{rect}} = (g(x) - f(x))\Delta x$, then the exact area of the region is given by the definite integral

$$A = \int_a^b (g(x) - f(x)) \, dx.$$

- The shape of the region usually dictates whether we should use vertical rectangles of thickness Δx or horizontal rectangles of thickness Δy. We desire to have the height of the rectangle governed by the difference between two curves: if those curves are best thought of as functions of y, we use horizontal rectangles, whereas if those curves are best viewed as functions of x, we use vertical rectangles.

- The arc length, L, along the curve $y = f(x)$ from $x = a$ to $x = b$ is given by

$$L = \int_a^b \sqrt{1 + f'(x)^2}\, dx.$$

Exercises

1. Find the area of the region between $y = x^{1/2}$ and $y = x^{1/4}$ for $0 \le x \le 1$.

area =

2. Find the area between $y = 7\sin x$ and $y = 10\cos x$ over the interval $[0, \pi]$. Sketch the curves if necessary.

$A =$

3. Sketch the region enclosed by $x + y^2 = 42$ and $x + y = 0$.
Decide whether to integrate with respect to x or y, and then find the area of the region.

The area is .

4. Find the arc length of the graph of the function $f(x) = 9\sqrt{x^3}$ from $x = 5$ to $x = 8$.

arc length =

5. Find the exact area of each described region.

  a. The finite region between the curves $x = y(y - 2)$ and $x = -(y - 1)(y - 3)$.

  b. The region between the sine and cosine functions on the interval $[\frac{\pi}{4}, \frac{3\pi}{4}]$.

  c. The finite region between $x = y^2 - y - 2$ and $y = 2x - 1$.

  d. The finite region between $y = mx$ and $y = x^2 - 1$, where m is a positive constant.

6. Let $f(x) = 1 - x^2$ and $g(x) = ax^2 - a$, where a is an unknown positive real number. For what value(s) of a is the area between the curves f and g equal to 2?

7. Let $f(x) = 2 - x^2$. Recall that the average value of any continuous function f on an interval $[a, b]$ is given by $\frac{1}{b-a} \int_a^b f(x)\, dx$.

  a. Find the average value of $f(x) = 2 - x^2$ on the interval $[0, \sqrt{2}]$. Call this value r.

  b. Sketch a graph of $y = f(x)$ and $y = r$. Find their intersection point(s).

  c. Show that on the interval $[0, \sqrt{2}]$, the amount of area that lies below $y = f(x)$ and above $y = r$ is equal to the amount of area that lies below $y = r$ and above $y = f(x)$.

  d. Will the result of (c) be true for any continuous function and its average value on any interval? Why?

6.2 Using Definite Integrals to Find Volume

Motivating Questions

- How can we use a definite integral to find the volume of a three-dimensional solid of revolution that results from revolving a two-dimensional region about a particular axis?

- In what circumstances do we integrate with respect to y instead of integrating with respect to x?

- What adjustments do we need to make if we revolve about a line other than the x- or y-axis?

Just as we can use definite integrals to add the areas of rectangular slices to find the exact area that lies between two curves, we can also employ integrals to determine the volume of certain regions that have cross-sections of a particular consistent shape.

As a very elementary example, consider a cylinder of radius 2 and height 3, as pictured in Figure 6.2.1. While we know that we can compute the area of any circular cylinder by the formula $V = \pi r^2 h$, if we think about slicing the cylinder into thin pieces, we see that each is a cylinder of radius $r = 2$ and height (thickness) Δx. Hence, the volume of a representative slice is

$$V_{\text{slice}} = \pi \cdot 2^2 \cdot \Delta x.$$

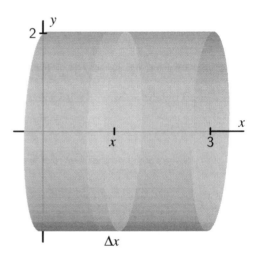

Figure 6.2.1: A right circular cylinder.

Letting $\Delta x \to 0$ and using a definite integral to add the volumes of the slices, we find that

$$V = \int_0^3 \pi \cdot 2^2 \, dx.$$

Moreover, since $\int_0^3 4\pi \, dx = 12\pi$, we have found that the volume of the cylinder is 12π. The principal problem of interest in our upcoming work will be to find the volume of certain solids whose cross-sections are all thin cylinders (or washers) and to do so by using a definite integral. To that end, we first consider another familiar shape in Preview Activity 6.2.1: a circular cone.

Preview Activity 6.2.1. Consider a circular cone of radius 3 and height 5, which we view horizontally as pictured in Figure 6.2.2. Our goal in this activity is to use a definite integral to determine the volume of the cone.

a. Find a formula for the linear function $y = f(x)$ that is pictured in Figure 6.2.2.

b. For the representative slice of thickness Δx that is located horizontally at a location x (somewhere between $x = 0$ and $x = 5$), what is the radius of the representative slice? Note that the radius depends on the value of x.

c. What is the volume of the representative slice you found in (b)?

d. What definite integral will sum the volumes of the thin slices across the full horizontal span of the cone? What is the exact value of this definite integral?

e. Compare the result of your work in (d) to the volume of the cone that comes from using the formula $V_{\text{cone}} = \frac{1}{3}\pi r^2 h$.

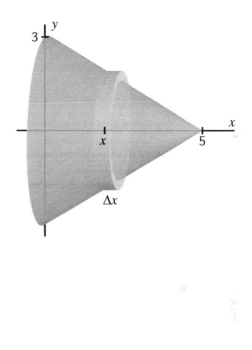

Figure 6.2.2: The circular cone described in Preview Activity 6.2.1

6.2.1 The Volume of a Solid of Revolution

A solid of revolution is a three dimensional solid that can be generated by revolving one or more curves around a fixed axis. For example, we can think of a circular cylinder as a solid of revolution: in Figure 6.2.1, this could be accomplished by revolving the line segment from $(0, 2)$ to $(3, 2)$ about the x-axis. Likewise, the circular cone in Figure 6.2.2 is the solid of revolution generated by revolving the portion of the line $y = 3 - \frac{3}{5}x$ from $x = 0$ to $x = 5$ about the x-axis. It is particularly important to notice in any solid of revolution that if we slice the solid perpendicular to the axis of revolution, the resulting cross-section is circular.

We consider two examples to highlight some of the natural issues that arise in determining the volume of a solid of revolution.

Example 6.2.3. Find the volume of the solid of revolution generated when the region R bounded by $y = 4 - x^2$ and the x-axis is revolved about the x-axis.

Solution. First, we observe that $y = 4 - x^2$ intersects the x-axis at the points $(-2, 0)$ and $(2, 0)$. When we take the region R that lies between the curve and the x-axis on this interval and revolve it about the x-axis, we get the three-dimensional solid pictured in Figure 6.2.4.

Taking a representative slice of the solid located at a value x that lies between $x = -2$ and $x = 2$, we see that the thickness of such a slice is Δx (which is also the height of the cylinder-shaped slice), and that the radius of the slice is determined by the curve $y = 4 - x^2$. Hence, we find that

$$V_{\text{slice}} = \pi(4 - x^2)^2 \Delta x,$$

since the volume of a cylinder of radius r and height h is $V = \pi r^2 h$.

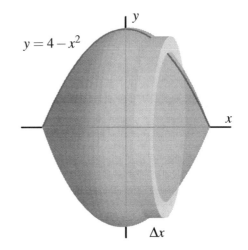

Figure 6.2.4: The solid of revolution in Example 6.2.3.

Using a definite integral to sum the volumes of the representative slices, it follows that

$$V = \int_{-2}^{2} \pi(4 - x^2)^2 \, dx.$$

It is straightforward to evaluate the integral and find that the volume is $V = \frac{512}{15}\pi$.

For a solid such as the one in Example 6.2.3, where each cross-section is a cylindrical disk, we first find the volume of a typical cross-section (noting particularly how this volume depends on x), and then we integrate over the range of x-values through which we slice the solid in order to find the exact total volume. Often, we will be content with simply finding the integral that represents the sought volume; if we desire a numeric value for the integral, we typically use a calculator or computer algebra system to find that value.

The general principle we are using to find the volume of a solid of revolution generated by a single curve is often called the *disk method*.

The Disk Method

If $y = r(x)$ is a nonnegative continuous function on $[a, b]$, then the volume of the solid of revolution generated by revolving the curve about the x-axis over this interval is given by

$$V = \int_a^b \pi r(x)^2 \, dx.$$

A different type of solid can emerge when two curves are involved, as we see in the following example.

Example 6.2.5. Find the volume of the solid of revolution generated when the finite region R that lies between $y = 4 - x^2$ and $y = x + 2$ is revolved about the x-axis.

Solution. First, we must determine where the curves $y = 4 - x^2$ and $y = x + 2$ intersect. Substituting the expression for y from the second equation into the first equation, we find that $x + 2 = 4 - x^2$. Rearranging, it follows that

$$x^2 + x - 2 = 0,$$

and the solutions to this equation are $x = -2$ and $x = 1$. The curves therefore cross at $(-2, 0)$ and $(1, 1)$.

When we take the region R that lies between the curves and revolve it about the x-axis, we get the three-dimensional solid pictured at left in Figure 6.2.6.

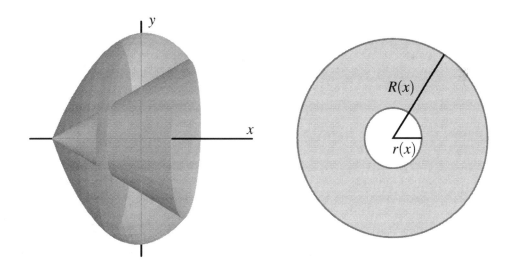

Figure 6.2.6: At left, the solid of revolution in Example 6.2.5. At right, a typical slice with inner radius $r(x)$ and outer radius $R(x)$.

Immediately we see a major difference between the solid in this example and the one in

Example 6.2.3: here, the three-dimensional solid of revolution isn't "solid" in the sense that it has open space in its center. If we slice the solid perpendicular to the axis of revolution, we observe that in this setting the resulting representative slice is not a solid disk, but rather a *washer*, as pictured at right in Figure 6.2.6. Moreover, at a given location x between $x = -2$ and $x = 1$, the small radius $r(x)$ of the inner circle is determined by the curve $y = x + 2$, so $r(x) = x + 2$. Similarly, the big radius $R(x)$ comes from the function $y = 4 - x^2$, and thus $R(x) = 4 - x^2$.

Thus, to find the volume of a representative slice, we compute the volume of the outer disk and subtract the volume of the inner disk. Since

$$\pi R(x)^2 \Delta x - \pi r(x)^2 \Delta x = \pi [R(x)^2 - r(x)^2] \Delta x,$$

it follows that the volume of a typical slice is

$$V_{\text{slice}} = \pi [(4 - x^2)^2 - (x + 2)^2] \Delta x.$$

Hence, using a definite integral to sum the volumes of the respective slices across the integral, we find that

$$V = \int_{-2}^{1} \pi [(4 - x^2)^2 - (x + 2)^2] \, dx.$$

Evaluating the integral, the volume of the solid of revolution is $V = \frac{108}{5} \pi$.

The general principle we are using to find the volume of a solid of revolution generated by a single curve is often called the *washer method*.

The Washer Method

If $y = R(x)$ and $y = r(x)$ are nonnegative continuous functions on $[a, b]$ that satisfy $R(x) \geq r(x)$ for all x in $[a, b]$, then the volume of the solid of revolution generated by revolving the region between them about the x-axis over this interval is given by

$$V = \int_{a}^{b} \pi [R(x)^2 - r(x)^2] \, dx.$$

Activity 6.2.2. In each of the following questions, draw a careful, labeled sketch of the region described, as well as the resulting solid that results from revolving the region about the stated axis. In addition, draw a representative slice and state the volume of that slice, along with a definite integral whose value is the volume of the entire solid. It is not necessary to evaluate the integrals you find.

   a. The region S bounded by the x-axis, the curve $y = \sqrt{x}$, and the line $x = 4$; revolve S about the x-axis.

   b. The region S bounded by the y-axis, the curve $y = \sqrt{x}$, and the line $y = 2$; revolve S about the x-axis.

c. The finite region S bounded by the curves $y = \sqrt{x}$ and $y = x^3$; revolve S about the x-axis.

d. The finite region S bounded by the curves $y = 2x^2 + 1$ and $y = x^2 + 4$; revolve S about the x-axis.

e. The region S bounded by the y-axis, the curve $y = \sqrt{x}$, and the line $y = 2$; revolve S about the y-axis. How is this problem different from the one posed in part (b)?

6.2.2 Revolving about the y-axis

As seen in Activity 6.2.2, problem (e), the problem changes considerably when we revolve a given region about the y-axis. Foremost, this is due to the fact that representative slices now have thickness Δy, which means that it becomes necessary to integrate with respect to y. Let's consider a particular example to demonstrate some of the key issues.

Example 6.2.7. Find the volume of the solid of revolution generated when the finite region R that lies between $y = \sqrt{x}$ and $y = x^4$ is revolved about the y-axis.

Solution. We observe that these two curves intersect when $x = 1$, hence at the point $(1, 1)$. When we take the region R that lies between the curves and revolve it about the y-axis, we get the three-dimensional solid pictured at left in Figure 6.2.8.

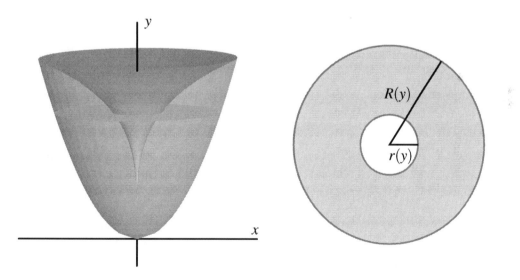

Figure 6.2.8: At left, the solid of revolution in Example 6.2.7. At right, a typical slice with inner radius $r(y)$ and outer radius $R(y)$.

Now, it is particularly important to note that the thickness of a representative slice is Δy, and that the slices are only cylindrical washers in nature when taken perpendicular to the

y-axis. Hence, we envision slicing the solid horizontally, starting at $y = 0$ and proceeding up to $y = 1$. Because the inner radius is governed by the curve $y = \sqrt{x}$, but from the perspective that x is a function of y, we solve for x and get $x = y^2 = r(y)$. In the same way, we need to view the curve $y = x^4$ (which governs the outer radius) in the form where x is a function of y, and hence $x = \sqrt[4]{y}$. Therefore, we see that the volume of a typical slice is

$$V_{\text{slice}} = \pi[R(y)^2 - r(y)^2] = \pi[\sqrt[4]{y}^2 - (y^2)^2]\Delta y.$$

Using a definite integral to sum the volume of all the representative slices from $y = 0$ to $y = 1$, the total volume is

$$V = \int_{y=0}^{y=1} \pi \left[\sqrt[4]{y}^2 - (y^2)^2 \right] dy.$$

It is straightforward to evaluate the integral and find that $V = \frac{7}{15}\pi$.

Activity 6.2.3. In each of the following questions, draw a careful, labeled sketch of the region described, as well as the resulting solid that results from revolving the region about the stated axis. In addition, draw a representative slice and state the volume of that slice, along with a definite integral whose value is the volume of the entire solid. It is not necessary to evaluate the integrals you find.

a. The region S bounded by the y-axis, the curve $y = \sqrt{x}$, and the line $y = 2$; revolve S about the y-axis.

b. The region S bounded by the x-axis, the curve $y = \sqrt{x}$, and the line $x = 4$; revolve S about the y-axis.

c. The finite region S in the first quadrant bounded by the curves $y = 2x$ and $y = x^3$; revolve S about the x-axis.

d. The finite region S in the first quadrant bounded by the curves $y = 2x$ and $y = x^3$; revolve S about the y-axis.

e. The finite region S bounded by the curves $x = (y - 1)^2$ and $y = x - 1$; revolve S about the y-axis

---

6.2.3 Revolving about horizontal and vertical lines other than the coordinate axes

Just as we can revolve about one of the coordinate axes ($y = 0$ or $x = 0$), it is also possible to revolve around any horizontal or vertical line. Doing so essentially adjusts the radii of cylinders or washers involved by a constant value. A careful, well-labeled plot of the solid of revolution will usually reveal how the different axis of revolution affects the definite integral we set up. Again, an example is instructive.

Example 6.2.9. Find the volume of the solid of revolution generated when the finite region S that lies between $y = x^2$ and $y = x$ is revolved about the line $y = -1$.

Solution.

Graphing the region between the two curves in the first quadrant between their points of intersection $((0,0)$ and $(1,1))$ and then revolving the region about the line $y = -1$, we see the solid shown in Figure 6.2.10. Each slice of the solid perpendicular to the axis of revolution is a washer, and the radii of each washer are governed by the curves $y = x^2$ and $y = x$. But we also see that there is one added change: the axis of revolution adds a fixed length to each radius. In particular, the inner radius of a typical slice, $r(x)$, is given by $r(x) = x^2 + 1$, while the outer radius is $R(x) = x + 1$.

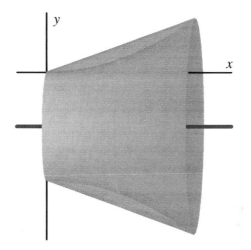

Figure 6.2.10: The solid of revolution described in Example 6.2.9.

Therefore, the volume of a typical slice is

$$V_{\text{slice}} = \pi[R(x)^2 - r(x)^2]\Delta x = \pi\left[(x+1)^2 - (x^2+1)^2\right]\Delta x.$$

Finally, we integrate to find the total volume, and

$$V = \int_0^1 \pi\left[(x+1)^2 - (x^2+1)^2\right] dx = \frac{7}{15}\pi.$$

Activity 6.2.4. In each of the following questions, draw a careful, labeled sketch of the region described, as well as the resulting solid that results from revolving the region about the stated axis. In addition, draw a representative slice and state the volume of that slice, along with a definite integral whose value is the volume of the entire solid. It is not necessary to evaluate the integrals you find. For each prompt, use the finite region S in the first quadrant bounded by the curves $y = 2x$ and $y = x^3$.

a. Revolve S about the line $y = -2$.

b. Revolve S about the line $y = 4$.

c. Revolve S about the line $x = -1$.

d. Revolve S about the line $x = 5$.

Summary

- We can use a definite integral to find the volume of a three-dimensional solid of revolution that results from revolving a two-dimensional region about a particular axis by taking slices perpendicular to the axis of revolution which will then be circular disks or washers.

- If we revolve about a vertical line and slice perpendicular to that line, then our slices are horizontal and of thickness Δy. This leads us to integrate with respect to y, as opposed to with respect to x when we slice a solid vertically.

- If we revolve about a line other than the x- or y-axis, we need to carefully account for the shift that occurs in the radius of a typical slice. Normally, this shift involves taking a sum or difference of the function along with the constant connected to the equation for the horizontal or vertical line; a well-labeled diagram is usually the best way to decide the new expression for the radius.

Exercises

1. The region bounded by $y = e^{2x}$, $y = 0$, $x = -2$, $x = 0$ is rotated around the x-axis. Find the volume.

volume =

2. Find the volume of the solid obtained by rotating the region in the first quadrant bounded by $y = x^6$, $y = 1$, and the y-axis around the y-axis.

Volume =

3. Find the volume of the solid obtained by rotating the region in the first quadrant bounded by $y = x^6$, $y = 1$, and the y-axis around the x-axis.

Volume =

4. Find the volume of the solid obtained by rotating the region in the first quadrant bounded by $y = x^6$, $y = 1$, and the y-axis about the line $y = -5$.

Volume =

5. Find the volume of the solid obtained by rotating the region bounded by the curves

$$y = x^2, \quad y = 1$$

about the line $y = 4$.

Answer:

6. Find the volume of the solid obtained by rotating the region bounded by the given curves about the line $x = -5$

$$y = x^2, \ x = y^2$$

Answer:

7. Consider the curve $f(x) = 3\cos(\frac{x^3}{4})$ and the portion of its graph that lies in the first quadrant between the y-axis and the first positive value of x for which $f(x) = 0$. Let R denote the region bounded by this portion of f, the x-axis, and the y-axis.

   a. Set up a definite integral whose value is the exact arc length of f that lies along the upper boundary of R. Use technology appropriately to evaluate the integral you find.

   b. Set up a definite integral whose value is the exact area of R. Use technology appropriately to evaluate the integral you find.

   c. Suppose that the region R is revolved around the x-axis. Set up a definite integral whose value is the exact volume of the solid of revolution that is generated. Use technology appropriately to evaluate the integral you find.

   d. Suppose instead that R is revolved around the y-axis. If possible, set up an integral expression whose value is the exact volume of the solid of revolution and evaluate the integral using appropriate technology. If not possible, explain why.

8. Consider the curves given by $y = \sin(x)$ and $y = \cos(x)$. For each of the following problems, you should include a sketch of the region/solid being considered, as well as a labeled representative slice.

   a. Sketch the region R bounded by the y-axis and the curves $y = \sin(x)$ and $y = \cos(x)$ up to the first positive value of x at which they intersect. What is the exact intersection point of the curves?

   b. Set up a definite integral whose value is the exact area of R.

   c. Set up a definite integral whose value is the exact volume of the solid of revolution generated by revolving R about the x-axis.

   d. Set up a definite integral whose value is the exact volume of the solid of revolution generated by revolving R about the y-axis.

   e. Set up a definite integral whose value is the exact volume of the solid of revolution generated by revolving R about the line $y = 2$.

   f. Set up a definite integral whose value is the exact volume of the solid of revolution generated by revolving R about the $x = -1$.

9. Consider the finite region R that is bounded by the curves $y = 1 + \frac{1}{2}(x-2)^2$, $y = \frac{1}{2}x^2$, and $x = 0$.

   a. Determine a definite integral whose value is the area of the region enclosed by the two curves.

   b. Find an expression involving one or more definite integrals whose value is the volume of the solid of revolution generated by revolving the region R about the line $y = -1$.

   c. Determine an expression involving one or more definite integrals whose value is the volume of the solid of revolution generated by revolving the region R about the y-axis.

   d. Find an expression involving one or more definite integrals whose value is the perimeter of the region R.

6.3 Density, Mass, and Center of Mass

Motivating Questions

- How are mass, density, and volume related?

- How is the mass of an object with varying density computed?

- What is is the center of mass of an object, and how are definite integrals used to compute it?

We have seen in several different circumstances how studying the units on the integrand and variable of integration enables us to better understand the meaning of a definite integral. For instance, if $v(t)$ is the velocity of an object moving along an axis, measured in feet per second, while t measures time in seconds, then both the definite integral and its Riemann sum approximation,

$$\int_a^b v(t)\,dt \approx \sum_{i=1}^n v(t_i)\Delta t,$$

have their overall units given by the product of the units of $v(t)$ and t:

$$(\text{feet}/\text{sec}) \cdot (\text{sec}) = \text{feet}.$$

Thus, $\int_a^b v(t)\,dt$ measures the total change in position (in feet) of the moving object.

This type of unit analysis will be particularly helpful to us in what follows. To begin, in the following preview activity we consider two different definite integrals where the integrand is a function that measures how a particular quantity is distributed over a region and think about how the units on the integrand and the variable of integration indicate the meaning of the integral.

> **Preview Activity 6.3.1.** In each of the following scenarios, we consider the distribution of a quantity along an axis.
>
> a. Suppose that the function $c(x) = 200 + 100e^{-0.1x}$ models the density of traffic on a straight road, measured in cars per mile, where x is number of miles east of a major interchange, and consider the definite integral $\int_0^2 (200 + 100e^{-0.1x})\,dx$.
>
> i. What are the units on the product $c(x) \cdot \Delta x$?
>
> ii. What are the units on the definite integral and its Riemann sum approximation given by
>
> $$\int_0^2 c(x)\,dx \approx \sum_{i=1}^n c(x_i)\Delta x?$$

iii. Evaluate the definite integral $\int_0^2 c(x)\,dx = \int_0^2 \left(200 + 100e^{-0.1x}\right)dx$ and write one sentence to explain the meaning of the value you find.

b. On a 6 foot long shelf filled with books, the function B models the distribution of the weight of the books, in pounds per inch, where x is the number of inches from the left end of the bookshelf. Let $B(x)$ be given by the rule $B(x) = 0.5 + \frac{1}{(x+1)^2}$.

i. What are the units on the product $B(x) \cdot \Delta x$?

ii. What are the units on the definite integral and its Riemann sum approximation given by

$$\int_{12}^{36} B(x)\,dx \approx \sum_{i=1}^{n} B(x_i)\Delta x?$$

iii. Evaluate the definite integral $\int_0^{72} B(x)\,dx = \int_0^{72}\left(0.5 + \frac{1}{(x+1)^2}\right)dx$ and write one sentence to explain the meaning of the value you find.

6.3.1 Density

The *mass* of a quantity, typically measured in metric units such as grams or kilograms, is a measure of the amount of a quantity. In a corresponding way, the *density* of an object measures the distribution of mass per unit volume. For instance, if a brick has mass 3 kg and volume 0.002 m^3, then the density of the brick is

$$\frac{3\text{kg}}{0.002\text{m}^3} = 1500\frac{\text{kg}}{\text{m}^3}.$$

As another example, the mass density of water is 1000 kg/m^3. Each of these relationships demonstrate the following general principle.

For an object of constant density d, with mass m and volume V,

$$d = \frac{m}{V}, \text{ or } m = d \cdot V.$$

But what happens when the density is not constant?

If we consider the formula $m = d \cdot V$, it is reminiscent of two other equations that we have used frequently in recent work: for a body moving in a fixed direction, distance = rate · time, and, for a rectangle, its area is given by $A = l \cdot w$. These formulas hold when the principal quantities involved, such as the rate the body moves and the height of the rectangle, are *constant*. When these quantities are not constant, we have turned to the definite integral for assistance. The main idea in each situation is that by working with small slices of the quantity that is varying, we can use a definite integral to add up the values of small pieces on which the quantity of interest (such as the velocity of a moving object) are approximately constant.

For example, in the setting where we have a nonnegative velocity function that is not constant, over a short time interval Δt we know that the distance traveled is approximately $v(t)\Delta t$, since $v(t)$ is almost constant on a small interval, and for a constant rate, distance = rate · time. Similarly, if we are thinking about the area under a nonnegative function f whose value is changing, on a short interval Δx the area under the curve is approximately the area of the rectangle whose height is $f(x)$ and whose width is Δx: $f(x)\Delta x$. Both of these principles are represented visually in Figure 6.3.1.

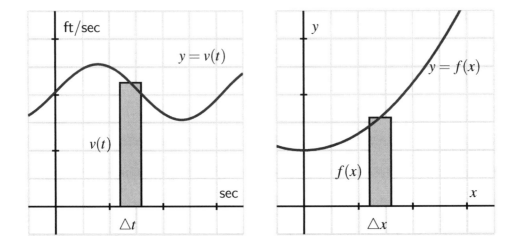

Figure 6.3.1: At left, estimating a small amount of distance traveled, $v(t)\Delta t$, and at right, a small amount of area under the curve, $f(x)\Delta x$.

In a similar way, if we consider the setting where the density of some quantity is not constant, the definite integral enables us to still compute the overall mass of the quantity. Throughout, we will focus on problems where the density varies in only one dimension, say along a single axis, and think about how mass is distributed relative to location along the axis.

Let's consider a thin bar of length b that is situated so its left end is at the origin, where $x = 0$, and assume that the bar has constant cross-sectional area of 1 cm². We let the function $\rho(x)$ represent the mass density function of the bar, measured in grams per cubic centimeter. That is, given a location x, $\rho(x)$ tells us approximately how much mass will be found in a one-centimeter wide slice of the bar at x.

Figure 6.3.2: A thin bar of constant cross-sectional area 1 cm² with density function $\rho(x)$ g/cm³.

If we now consider a thin slice of the bar of width Δx, as pictured in Figure 6.3.2, the volume of such a slice is the cross-sectional area times Δx. Since the cross-sections each have constant area 1 cm^2, it follows that the volume of the slice is $1\Delta x$ cm^3. Moreover, since mass is the product of density and volume (when density is constant), we see that the mass of this given slice is approximately

$$\text{mass}_{\text{slice}} \approx \rho(x) \frac{\text{g}}{\text{cm}^3} \cdot 1\Delta x \text{ cm}^3 = \rho(x) \cdot \Delta x \text{ g.}$$

Hence, for the corresponding Riemann sum (and thus for the integral that it approximates),

$$\sum_{i=1}^{n} \rho(x_i)\Delta x \approx \int_0^b \rho(x)\,dx,$$

we see that these quantities measure the mass of the bar between 0 and b. (The Riemann sum is an approximation, while the integral will be the exact mass.)

At this point, we note that we will be focused primarily on situations where mass is distributed relative to horizontal location, x, for objects whose cross-sectional area is constant. In that setting, it makes sense to think of the density function $\rho(x)$ with units "mass per unit length," such as g/cm. Thus, when we compute $\rho(x) \cdot \Delta x$ on a small slice Δx, the resulting units are g/cm $\cdot$ cm = g, which thus measures the mass of the slice. The general principle follows.

For an object of constant cross-sectional area whose mass is distributed along a single axis according to the function $\rho(x)$ (whose units are units of mass per unit of length), the total mass, M of the object between $x = a$ and $x = b$ is given by

$$M = \int_a^b \rho(x)\,dx.$$

Activity 6.3.2. Consider the following situations in which mass is distributed in a non-constant manner.

a. Suppose that a thin rod with constant cross-sectional area of 1 cm^2 has its mass distributed according to the density function $\rho(x) = 2e^{-0.2x}$, where x is the distance in cm from the left end of the rod, and the units on $\rho(x)$ are g/cm. If the rod is 10 cm long, determine the exact mass of the rod.

b. Consider the cone that has a base of radius 4 m and a height of 5 m. Picture the cone lying horizontally with the center of its base at the origin and think of the cone as a solid of revolution.

    i. Write and evaluate a definite integral whose value is the volume of the cone.

    ii. Next, suppose that the cone has uniform density of 800 kg/m^3. What is the mass of the solid cone?

    iii. Now suppose that the cone's density is not uniform, but rather that the cone is most dense at its base. In particular, assume that the density of the cone is uniform across cross sections parallel to its base, but that in each such cross section that is a distance x units from the origin, the density of the cross section is given by the function $\rho(x) = 400 + \frac{200}{1+x^2}$, measured in kg/m^3. Determine and evaluate a definite integral whose value is the mass of this cone of non-uniform density. Do so by first thinking about the mass of a given slice of the cone x units away from the base; remember that in such a slice, the density will be *essentially constant*.

c. Let a thin rod of constant cross-sectional area 1 cm^2 and length 12 cm have its mass be distributed according to the density function $\rho(x) = \frac{1}{25}(x - 15)^2$, measured in g/cm. Find the exact location z at which to cut the bar so that the two pieces will each have identical mass.

6.3.2 Weighted Averages

The concept of an average is a natural one, and one that we have used repeatedly as part of our understanding of the meaning of the definite integral. If we have n values $a_1, a_2, \ldots, a_n$, we know that their average is given by

$$\frac{a_1 + a_2 + \cdots + a_n}{n},$$

and for a quantity being measured by a function f on an interval $[a, b]$, the average value of the quantity on $[a, b]$ is

$$\frac{1}{b-a} \int_a^b f(x)\,dx.$$

As we continue to think about problems involving the distribution of mass, it is natural to consider the idea of a *weighted* average, where certain quantities involved are counted more in the average.

A common use of weighted averages is in the computation of a student's GPA, where grades are weighted according to credit hours. Let's consider the scenario in Table 6.3.3.

class	grade	grade points	credits
chemistry	B+	3.3	5
calculus	A-	3.7	4
history	B-	2.7	3
psychology	B-	2.7	3

Table 6.3.3: A college student's semester grades.

If all of the classes were of the same weight (i.e., the same number of credits), the student's GPA would simply be calculated by taking the average

$$\frac{3.3 + 3.7 + 2.7 + 2.7}{4} = 3.1.$$

But since the chemistry and calculus courses have higher weights (of 5 and 4 credits respectively), we actually compute the GPA according to the weighted average

$$\frac{3.3 \cdot 5 + 3.7 \cdot 4 + 2.7 \cdot 3 + 2.7 \cdot 3}{5 + 4 + 3 + 3} = 3.1\overline{6}.$$

The weighted average reflects the fact that chemistry and calculus, as courses with higher credits, have a greater impact on the students' grade point average. Note particularly that in the weighted average, each grade gets multiplied by its weight, and we divide by the sum of the weights.

In the following activity, we explore further how weighted averages can be used to find the balancing point of a physical system.

Activity 6.3.3. For quantities of equal weight, such as two children on a teeter-totter, the balancing point is found by taking the average of their locations. When the weights of the quantities differ, we use a weighted average of their respective locations to find the balancing point.

a. Suppose that a shelf is 6 feet long, with its left end situated at $x = 0$. If one book of weight 1 lb is placed at $x_1 = 0$, and another book of weight 1 lb is placed at $x_2 = 6$, what is the location of $\overline{x}$, the point at which the shelf would (theoretically) balance on a fulcrum?

b. Now, say that we place four books on the shelf, each weighing 1 lb: at $x_1 = 0$, at $x_2 = 2$, at $x_3 = 4$, and at $x_4 = 6$. Find $\overline{x}$, the balancing point of the shelf.

c. How does $\overline{x}$ change if we change the location of the third book? Say the locations of the 1-lb books are $x_1 = 0$, $x_2 = 2$, $x_3 = 3$, and $x_4 = 6$.

d. Next, suppose that we place four books on the shelf, but of varying weights: at $x_1 = 0$ a 2-lb book, at $x_2 = 2$ a 3-lb book, at $x_3 = 4$ a 1-lb book, and at $x_4 = 6$ a 1-lb book. Use a weighted average of the locations to find $\overline{x}$, the balancing point of the shelf. How does the balancing point in this scenario compare to that found in (b)?

e. What happens if we change the location of one of the books? Say that we keep everything the same in (d), except that $x_3 = 5$. How does $\overline{x}$ change?

f. What happens if we change the weight of one of the books? Say that we keep everything the same in (d), except that the book at $x_3 = 4$ now weighs 2 lbs. How does $\overline{x}$ change?

g. Experiment with a couple of different scenarios of your choosing where you move one of the books to the left, or you decrease the weight of one of the books.

h. Write a couple of sentences to explain how adjusting the location of one of the books or the weight of one of the books affects the location of the balancing point of the shelf. Think carefully here about how your changes should be considered relative to the location of the balancing point $\overline{x}$ of the current scenario.

6.3.3 Center of Mass

In Activity 6.3.3, we saw that the balancing point of a system of point-masses[1] (such as books on a shelf) is found by taking a weighted average of their respective locations. In the activity, we were computing the *center of mass* of a system of masses distributed along an axis, which is the balancing point of the axis on which the masses rest.

Center of Mass (point-masses)

For a collection of n masses $m_1, \ldots, m_n$ that are distributed along a single axis at the locations $x_1, \ldots, x_n$, the *center of mass* is given by

$$\overline{x} = \frac{x_1 m_1 + x_2 m_2 + \cdots + x_n m_n}{m_1 + m_2 + \cdots + m_n}.$$

What if we instead consider a thin bar over which density is distributed continuously? If the density is constant, it is obvious that the balancing point of the bar is its midpoint. But if density is not constant, we must compute a weighted average. Let's say that the function $\rho(x)$ tells us the density distribution along the bar, measured in g/cm. If we slice the bar into small sections, this enables us to think of the bar as holding a collection of adjacent point-masses. For a slice of thickness Δx at location x_i, note that the mass of the slice, m_i, satisfies $m_i \approx \rho(x_i)\Delta x$.

Taking n slices of the bar, we can approximate its center of mass by

$$\overline{x} \approx \frac{x_1 \cdot \rho(x_1)\Delta x + x_2 \cdot \rho(x_2)\Delta x + \cdots + x_n \cdot \rho(x_n)\Delta x}{\rho(x_1)\Delta x + \rho(x_2)\Delta x + \cdots + \rho(x_n)\Delta x}.$$

Rewriting the sums in sigma notation, it follows that

$$\overline{x} \approx \frac{\sum_{i=1}^{n} x_i \cdot \rho(x_i)\Delta x}{\sum_{i=1}^{n} \rho(x_i)\Delta x}. \tag{6.3.1}$$

Moreover, it is apparent that the greater the number of slices, the more accurate our estimate of the balancing point will be, and that the sums in Equation (6.3.1) can be viewed as Riemann sums. Hence, in the limit as $n \to \infty$, we find that the center of mass is given by the quotient of two integrals.

Center of Mass (continuous mass distribution)

For a thin rod of density $\rho(x)$ distributed along an axis from $x = a$ to $x = b$, the center of mass of the rod is given by

$$\overline{x} = \frac{\int_a^b x\rho(x)\,dx}{\int_a^b \rho(x)\,dx}.$$
---

[1] In the activity, we actually used *weight* rather than *mass*. Since weight is proportional to mass, the computations for the balancing point result in the same location regardless of whether we use weight or mass. The gravitational constant is present in both the numerator and denominator of the weighted average.

Note particularly that the denominator of $\overline{x}$ is the mass of the bar, and that this quotient of integrals is simply the continuous version of the weighted average of locations, x, along the bar.

> **Activity 6.3.4.** Consider a thin bar of length 20 cm whose density is distributed according to the function $\rho(x) = 4 + 0.1x$, where $x = 0$ represents the left end of the bar. Assume that ρ is measured in g/cm and x is measured in cm.
>
> a. Find the total mass, M, of the bar.
>
> b. Without doing any calculations, do you expect the center of mass of the bar to be equal to 10, less than 10, or greater than 10? Why?
>
> c. Compute $\overline{x}$, the exact center of mass of the bar.
>
> d. What is the average density of the bar?
>
> e. Now consider a different density function, given by $p(x) = 4e^{0.020732x}$, also for a bar of length 20 cm whose left end is at $x = 0$. Plot both $\rho(x)$ and $p(x)$ on the same axes. Without doing any calculations, which bar do you expect to have the greater center of mass? Why?
>
> f. Compute the exact center of mass of the bar described in (e) whose density function is $p(x) = 4e^{0.020732x}$. Check the result against the prediction you made in (e).

Summary

- For an object of constant density D, with volume V and mass m, we know that $m = D \cdot V$.

- If an object with constant cross-sectional area (such as a thin bar) has its density distributed along an axis according to the function $\rho(x)$, then we can find the mass of the object between $x = a$ and $x = b$ by

$$m = \int_a^b \rho(x)\,dx.$$

- For a system of point-masses distributed along an axis, say $m_1, \ldots, m_n$ at locations $x_1, \ldots, x_n$, the center of mass, $\overline{x}$, is given by the weighted average

$$\overline{x} = \frac{\sum_{i=1}^n x_i m_i}{\sum_{i=1}^n m_i}.$$

If instead we have mass continuously distributed along an axis, such as by a density function $\rho(x)$ for a thin bar of constant cross-sectional area, the center of mass of the portion of the bar between $x = a$ and $x = b$ is given by

$$\overline{x} = \frac{\int_a^b x\rho(x)\,dx}{\int_a^b \rho(x)\,dx}.$$

In each situation, $\bar{x}$ represents the balancing point of the system of masses or of the portion of the bar.

Exercises

1. A rod has length 3 meters. At a distance x meters from its left end, the density of the rod is given by $\delta(x) = 4 + 5x$ g/m.

(a) Complete the Riemann sum for the total mass of the rod (use Dx in place of Δx):

mass = Σ [   ]

(b) Convert the Riemann sum to an integral and find the exact mass.

mass = [   ]

(include units)

2. A rod with density $\delta(x) = 7 + \sin(x)$ lies on the x-axis between $x = 0$ and $x = \pi$. Find the mass and center of mass of the rod.

mass = [   ]

center of mass = [   ]

3. Suppose that the density of cars (in cars per mile) down a 20-mile stretch of the Pennsylvania Turnpike is approximated by $\delta(x) = 350 \left(2 + \sin\left(4\sqrt{x + 0.175}\right)\right)$, at a distance x miles from the Breezewood toll plaza. Sketch a graph of this function for $0 \le x \le 20$.

(a) Complete the Riemann sum that approximates the total number of cars on this 20-mile stretch (use Dx instead of Δx):

Number = Σ [   ]

(b) Find the total number of cars on the 20-mile stretch.

Number = [   ]

4. A point mass of 1 grams located 6 centimeters to the left of the origin and a point mass of 3 grams located 7 centimeters to the right of the origin are connected by a thin, light rod. Find the center of mass of the system.

Center of Mass = [   ]

[Choose: to the left of the origin | to the right of the origin | (at the origin)]

(include units in your center of mass)

5. Let a thin rod of length a have density distribution function $\rho(x) = 10e^{-0.1x}$, where x is measured in cm and ρ in grams per centimeter.

  a. If the mass of the rod is 30 g, what is the value of a?

  b. For the 30g rod, will the center of mass lie at its midpoint, to the left of the midpoint, or to the right of the midpoint? Why?

  c. For the 30g rod, find the center of mass, and compare your prediction in (b).

  d. At what value of x should the 30g rod be cut in order to form two pieces of equal mass?

6. Consider two thin bars of constant cross-sectional area, each of length 10 cm, with respective mass density functions $\rho(x) = \frac{1}{1+x^2}$ and $p(x) = e^{-0.1x}$.

    a. Find the mass of each bar.

    b. Find the center of mass of each bar.

    c. Now consider a new 10 cm bar whose mass density function is $f(x) = \rho(x) + p(x)$.

        a. Explain how you can easily find the mass of this new bar with little to no additional work.

        b. Similarly, compute $\int_0^{10} x f(x)\,dx$ as simply as possible, in light of earlier computations.

        c. True or false: the center of mass of this new bar is the average of the centers of mass of the two earlier bars. Write at least one sentence to say why your conclusion makes sense.

7. Consider the curve given by $y = f(x) = 2xe^{-1.25x} + (30 - x)e^{-0.25(30-x)}$.

    a. Plot this curve in the window $x = 0 \ldots 30$, $y = 0 \ldots 3$ (with constrained scaling so the units on the x and y axis are equal), and use it to generate a solid of revolution about the x-axis. Explain why this curve could generate a reasonable model of a baseball bat.

    b. Let x and y be measured in inches. Find the total volume of the baseball bat generated by revolving the given curve about the x-axis. Include units on your answer

    c. Suppose that the baseball bat has constant weight density, and that the weight density is 0.6 ounces per cubic inch. Find the total weight of the bat whose volume you found in (b).

    d. Because the baseball bat does not have constant cross-sectional area, we see that the amount of weight concentrated at a location x along the bat is determined by the volume of a slice at location x. Explain why we can think about the function $\rho(x) = 0.6\pi f(x)^2$ (where f is the function given at the start of the problem) as being the weight density function for how the weight of the baseball bat is distributed from $x = 0$ to $x = 30$.

    e. Compute the center of mass of the baseball bat.

6.4 Physics Applications: Work, Force, and Pressure

Motivating Questions

- How do we measure the work accomplished by a varying force that moves an object a certain distance?

- What is the total force exerted by water against a dam?

- How are both of the above concepts and their corresponding use of definite integrals similar to problems we have encountered in the past involving formulas such as "distance equals rate times time" and "mass equals density times volume"?

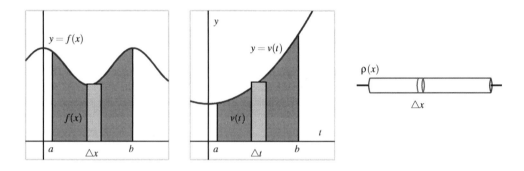

Figure 6.4.1: Three settings where we compute the accumulation of a varying quantity: the area under $y = f(x)$, the distance traveled by an object with velocity $y = v(t)$, and the mass of a bar with density function $y = \rho(x)$.

In our work to date with the definite integral, we have seen several different circumstances where the integral enables us to measure the accumulation of a quantity that varies, provided the quantity is approximately constant over small intervals. For instance, based on the fact that the area of a rectangle is $A = l \cdot w$, if we wish to find the area bounded by a nonnegative curve $y = f(x)$ and the x-axis on an interval $[a, b]$, a representative slice of width Δx has area $A_{\text{slice}} = f(x)\Delta x$, and thus as we let the width of the representative slice tend to zero, we find that the exact area of the region is

$$A = \int_a^b f(x)\,dx.$$

In a similar way, if we know that the velocity of a moving object is given by the function $y = v(t)$, and we wish to know the distance the object travels on an interval $[a, b]$ where $v(t)$

is nonnegative, we can use a definite integral to generalize the fact that $d = r \cdot t$ when the rate, r, is constant. More specifically, on a short time interval Δt, $v(t)$ is roughly constant, and hence for a small slice of time, $d_{\text{slice}} = v(t)\Delta t$, and so as the width of the time interval Δt tends to zero, the exact distance traveled is given by the definite integral

$$d = \int_a^b v(t)\,dt.$$

Finally, when we recently learned about the mass of an object of non-constant density, we saw that since $M = D \cdot V$ (mass equals density times volume, provided that density is constant), if we can consider a small slice of an object on which the density is approximately constant, a definite integral may be used to determine the exact mass of the object. For instance, if we have a thin rod whose cross sections have constant density, but whose density is distributed along the x axis according to the function $y = \rho(x)$, it follows that for a small slice of the rod that is Δx thick, $M_{\text{slice}} = \rho(x)\Delta x$. In the limit as $\Delta x \to 0$, we then find that the total mass is given by

$$M = \int_a^b \rho(x)\,dx.$$

Note that all three of these situations are similar in that we have a basic rule ($A = l \cdot w$, $d = r \cdot t$, $M = D \cdot V$) where one of the two quantities being multiplied is no longer constant; in each, we consider a small interval for the other variable in the formula, calculate the approximate value of the desired quantity (area, distance, or mass) over the small interval, and then use a definite integral to sum the results as the length of the small intervals is allowed to approach zero. It should be apparent that this approach will work effectively for other situations where we have a quantity of interest that varies.

We next turn to the notion of *work*: from physics, a basic principal is that work is the product of force and distance. For example, if a person exerts a force of 20 pounds to lift a 20-pound weight 4 feet off the ground, the total work accomplished is

$$W = F \cdot d = 20 \cdot 4 = 80 \text{ foot-pounds.}$$

If force and distance are measured in English units (pounds and feet), then the units on work are *foot-pounds*. If instead we work in metric units, where forces are measured in Newtons and distances in meters, the units on work are *Newton-meters*.

Of course, the formula $W = F \cdot d$ only applies when the force is constant while it is exerted over the distance d. In Preview Activity 6.4.1, we explore one way that we can use a definite integral to compute the total work accomplished when the force exerted varies.

Preview Activity 6.4.1. A bucket is being lifted from the bottom of a 50-foot deep well; its weight (including the water), B, in pounds at a height h feet above the water is given by the function $B(h)$. When the bucket leaves the water, the bucket and water together weigh $B(0) = 20$ pounds, and when the bucket reaches the top of the well, $B(50) = 12$ pounds. Assume that the bucket loses water at a constant rate (as a function of height,

h) throughout its journey from the bottom to the top of the well.

    a. Find a formula for $B(h)$.

    b. Compute the value of the product $B(5)\Delta h$, where $\Delta h = 2$ feet. Include units on your answer. Explain why this product represents the approximate work it took to move the bucket of water from $h = 5$ to $h = 7$.

    c. Is the value in (b) an over- or under-estimate of the actual amount of work it took to move the bucket from $h = 5$ to $h = 7$? Why?

    d. Compute the value of the product $B(22)\Delta h$, where $\Delta h = 0.25$ feet. Include units on your answer. What is the meaning of the value you found?

    e. More generally, what does the quantity $W_{\text{slice}} = B(h)\Delta h$ measure for a given value of h and a small positive value of Δh?

    f. Evaluate the definite integral $\int_0^{50} B(h)\,dh$. What is the meaning of the value you find? Why?

6.4.1 Work

Because work is calculated by the rule $W = F \cdot d$, whenever the force F is constant, it follows that we can use a definite integral to compute the work accomplished by a varying force. For example, suppose that in a setting similar to the problem posed in Preview Activity 6.4.1, we have a bucket being lifted in a 50-foot well whose weight at height h is given by $B(h) = 12 + 8e^{-0.1h}$.

In contrast to the problem in the preview activity, this bucket is not leaking at a constant rate; but because the weight of the bucket and water is not constant, we have to use a definite integral to determine the total work that results from lifting the bucket. Observe that at a height h above the water, the approximate work to move the bucket a small distance Δh is

$$W_{\text{slice}} = B(h)\Delta h = (12 + 8e^{-0.1h})\Delta h.$$

Hence, if we let Δh tend to 0 and take the sum of all of the slices of work accomplished on these small intervals, it follows that the total work is given by

$$W = \int_0^{50} B(h)\,dh = \int_0^{50} (12 + 8e^{-0.1h})\,dh.$$

While is a straightforward exercise to evaluate this integral exactly using the First Fundamental Theorem of Calculus, in applied settings such as this one we will typically use computing technology to find accurate approximations of integrals that are of interest to us. Here, it turns out that $W = \int_0^{50} (12 + 8e^{-0.1h})\,dh \approx 679.461$ foot-pounds.

Our work in Preview Activity 6.4.1 and in the most recent example above employs the following important general principle.

For an object being moved in the positive direction along an axis, x, by a force $F(x)$, the total work to move the object from a to b is given by

$$W = \int_a^b F(x)\,dx.$$

Activity 6.4.2. Consider the following situations in which a varying force accomplishes work.

a. Suppose that a heavy rope hangs over the side of a cliff. The rope is 200 feet long and weighs 0.3 pounds per foot; initially the rope is fully extended. How much work is required to haul in the entire length of the rope? (Hint: set up a function $F(h)$ whose value is the weight of the rope remaining over the cliff after h feet have been hauled in.)

b. A leaky bucket is being hauled up from a 100 foot deep well. When lifted from the water, the bucket and water together weigh 40 pounds. As the bucket is being hauled upward at a constant rate, the bucket leaks water at a constant rate so that it is losing weight at a rate of 0.1 pounds per foot. What function $B(h)$ tells the weight of the bucket after the bucket has been lifted h feet? What is the total amount of work accomplished in lifting the bucket to the top of the well?

c. Now suppose that the bucket in (b) does not leak at a constant rate, but rather that its weight at a height h feet above the water is given by $B(h) = 25 + 15e^{-0.05h}$. What is the total work required to lift the bucket 100 feet? What is the average force exerted on the bucket on the interval $h = 0$ to $h = 100$?

d. From physics, *Hooke's Law* for springs states that the amount of force required to hold a spring that is compressed (or extended) to a particular length is proportionate to the distance the spring is compressed (or extended) from its natural length. That is, the force to compress (or extend) a spring x units from its natural length is $F(x) = kx$ for some constant k (which is called the *spring constant*.) For springs, we choose to measure the force in pounds and the distance the spring is compressed in feet. Suppose that a force of 5 pounds extends a particular spring 4 inches (1/3 foot) beyond its natural length.

   i. Use the given fact that $F(1/3) = 5$ to find the spring constant k.

   ii. Find the work done to extend the spring from its natural length to 1 foot beyond its natural length.

   iii. Find the work required to extend the spring from 1 foot beyond its natural length to 1.5 feet beyond its natural length.

6.4.2 Work: Pumping Liquid from a Tank

In certain geographic locations where the water table is high, residential homes with basements have a peculiar feature: in the basement, one finds a large hole in the floor, and in the hole, there is water. For example, in Figure 6.4.2 we see a *sump crock*[1]. Essentially, a sump crock provides an outlet for water that may build up beneath the basement floor; of course, as that water rises, it is imperative that the water not flood the basement.

Hence, in the crock we see the presence of a floating pump that sits on the surface of the water: this pump is activated by elevation, so when the water level reaches a particular height, the pump turns on and pumps a certain portion of the water out of the crock, hence relieving the water buildup beneath the foundation. One of the questions we'd like to answer is: how much work does a sump pump accomplish?

Figure 6.4.2: A sump crock.

To that end, let's suppose that we have a sump crock that has the shape of a frustum of a cone, as pictured in Figure 6.4.3. Assume that the crock has a diameter of 3 feet at its surface, a diameter of 1.5 feet at its base, and a depth of 4 feet. In addition, suppose that the sump pump is set up so that it pumps the water vertically up a pipe to a drain that is located at ground level just outside a basement window. To accomplish this, the pump must send the water to a location 9 feet above the surface of the sump crock.

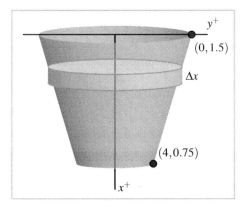

Figure 6.4.3: A sump crock with approximately cylindrical cross-sections that is 4 feet deep, 1.5 feet in diameter at its base, and 3 feet in diameter at its top.

It turns out to be advantageous to think of the depth below the surface of the crock as being the independent variable, so, in problems such as this one we typically let the positive x-axis

[1]Image credit to www.warreninspect.com/basement-moisture.

point down, and the positive y-axis to the right, as pictured in the figure. As we think about the work that the pump does, we first realize that the pump sits on the surface of the water, so it makes sense to think about the pump moving the water one "slice" at a time, where it takes a thin slice from the surface, pumps it out of the tank, and then proceeds to pump the next slice below.

For the sump crock described in this example, each slice of water is cylindrical in shape. We see that the radius of each approximately cylindrical slice varies according to the linear function $y = f(x)$ that passes through the points $(0, 1.5)$ and $(4, 0.75)$, where x is the depth of the particular slice in the tank; it is a straightforward exercise to find that $f(x) = 1.5 - 0.1875x$. Now we are prepared to think about the overall problem in several steps: (a) determining the volume of a typical slice; (b) finding the weight[2] of a typical slice (and thus the force that must be exerted on it); (c) deciding the distance that a typical slice moves; and (d) computing the work to move a representative slice. Once we know the work it takes to move one slice, we use a definite integral over an appropriate interval to find the total work.

Consider a representative cylindrical slice that sits on the surface of the water at a depth of x feet below the top of the crock. It follows that the approximate volume of that slice is given by

$$V_{\text{slice}} = \pi f(x)^2 \Delta x = \pi(1.5 - 0.1875x)^2 \Delta x.$$

Since water weighs 62.4 lb/ft^3, it follows that the approximate weight of a representative slice, which is also the approximate force the pump must exert to move the slice, is

$$F_{\text{slice}} = 62.4 \cdot V_{\text{slice}} = 62.4\pi(1.5 - 0.1875x)^2 \Delta x.$$

Because the slice is located at a depth of x feet below the top of the crock, the slice being moved by the pump must move x feet to get to the level of the basement floor, and then, as stated in the problem description, be moved another 9 feet to reach the drain at ground level outside a basement window. Hence, the total distance a representative slice travels is

$$d_{\text{slice}} = x + 9.$$

Finally, we note that the work to move a representative slice is given by

$$W_{\text{slice}} = F_{\text{slice}} \cdot d_{\text{slice}} = 62.4\pi(1.5 - 0.1875x)^2 \Delta x \cdot (x + 9),$$

since the force to move a particular slice is constant.

We sum the work required to move slices throughout the tank (from $x = 0$ to $x = 4$), let $\Delta x \to 0$, and hence

$$W = \int_0^4 62.4\pi(1.5 - 0.1875x)^2(x + 9)\, dx,$$

which, when evaluated using appropriate technology, shows that the total work is $W = 10970.5\pi$ foot-pounds.

The preceding example demonstrates the standard approach to finding the work required to empty a tank filled with liquid. The main task in each such problem is to determine

---

[2]We assume that the weight density of water is 62.4 pounds per cubic foot.

the volume of a representative slice, followed by the force exerted on the slice, as well as the distance such a slice moves. In the case where the units are metric, there is one key difference: in the metric setting, rather than weight, we normally first find the mass of a slice. For instance, if distance is measured in meters, the mass density of water is 1000 kg/m^3. In that setting, we can find the mass of a typical slice (in kg). To determine the force required to move it, we use $F = ma$, where m is the object's mass and a is the gravitational constant 9.81 N/kg^3. That is, in metric units, the weight density of water is 9810 N/m^3.

> **Activity 6.4.3.** In each of the following problems, determine the total work required to accomplish the described task. In parts (b) and (c), a key step is to find a formula for a function that describes the curve that forms the side boundary of the tank.

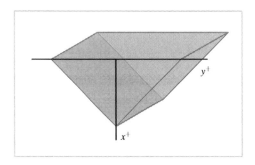

Figure 6.4.4: A trough with triangular ends, as described in Activity 6.4.3, part (c).

a. Consider a vertical cylindrical tank of radius 2 meters and depth 6 meters. Suppose the tank is filled with 4 meters of water of mass density 1000 kg/m^3, and the top 1 meter of water is pumped over the top of the tank.

b. Consider a hemispherical tank with a radius of 10 feet. Suppose that the tank is full to a depth of 7 feet with water of weight density 62.4 pounds/ft^3, and the top 5 feet of water are pumped out of the tank to a tanker truck whose height is 5 feet above the top of the tank.

c. Consider a trough with triangular ends, as pictured in Figure 6.4.4, where the tank is 10 feet long, the top is 5 feet wide, and the tank is 4 feet deep. Say that the trough is full to within 1 foot of the top with water of weight density 62.4 pounds/ft^3, and a pump is used to empty the tank until the water remaining in the tank is 1 foot deep.

6.4.3 Force due to Hydrostatic Pressure

When a dam is built, it is imperative to for engineers to understand how much force water will exert against the face of the dam. The first thing we realize is the force exerted by the fluid is related to the natural concept of pressure. The pressure a force exerts on a region is measured in units of force per unit of area: for example, the air pressure in a tire is often measured in pounds per square inch (PSI). Hence, we see that the general relationship is given by

$$P = \frac{F}{A}, \text{ or } F = P \cdot A,$$

where P represents pressure, F represents force, and A the area of the region being considered. Of course, in the equation $F = PA$, we assume that the pressure is constant over the entire region A.

Most people know from experience that the deeper one dives underwater while swimming, the greater the pressure that is exerted by the water. This is due to the fact that the deeper one dives, the more water there is right on top of the swimmer: it is the force that "column" of water exerts that determines the pressure the swimmer experiences. To get water pressure measured in its standard units (pounds per square foot), we say that the total water pressure is found by computing the total weight of the column of water that lies above a region of area 1 square foot at a fixed depth. Such a rectangular column with a 1×1 base and a depth of d feet has volume $V = 1 \cdot 1 \cdot d$ ft^3, and thus the corresponding weight of the water overhead is $62.4d$. Since this is also the amount of force being exerted on a 1 square foot region at a depth d feet underwater, we see that $P = 62.4d$ (lbs/ft^2) is the pressure exerted by water at depth d.

The understanding that $P = 62.4d$ will tell us the pressure exerted by water at a depth of d, along with the fact that $F = PA$, will now enable us to compute the total force that water exerts on a dam, as we see in the following example.

Example 6.4.5. Consider a trapezoid-shaped dam that is 60 feet wide at its base and 90 feet wide at its top, and assume the dam is 25 feet tall with water that rises to within 5 feet of the top of its face. Water weighs 62.4 pounds per cubic foot. How much force does the water exert against the dam?

Solution. First, we sketch a picture of the dam, as shown in Figure 6.4.6. Note that, as in problems involving the work to pump out a tank, we let the positive x-axis point down.

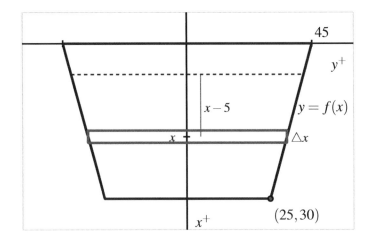

Figure 6.4.6: A trapezoidal dam that is 25 feet tall, 60 feet wide at its base, 90 feet wide at its top, with the water line 5 feet down from the top of its face.

It is essential to use the fact that pressure is constant at a fixed depth. Hence, we consider a slice of water at constant depth on the face, such as the one shown in the figure. First, the approximate area of this slice is the area of the pictured rectangle. Since the width of that rectangle depends on the variable x (which represents the how far the slice lies from the top of the dam), we find a formula for the function $y = f(x)$ that determines one side of the face of the dam. Since f is linear, it is straightforward to find that $y = f(x) = 45 - \frac{3}{5}x$. Hence, the approximate area of a representative slice is

$$A_{\text{slice}} = 2f(x)\Delta x = 2(45 - \frac{3}{5}x)\Delta x.$$

At any point on this slice, the depth is approximately constant, and thus the pressure can be considered constant. In particular, we note that since x measures the distance to the top of the dam, and because the water rises to within 5 feet of the top of the dam, the depth of any point on the representative slice is approximately $(x - 5)$. Now, since pressure is given by $P = 62.4d$, we have that at any point on the representative slice

$$P_{\text{slice}} = 62.4(x - 5).$$

Knowing both the pressure and area, we can find the force the water exerts on the slice. Using $F = PA$, it follows that

$$F_{\text{slice}} = P_{\text{slice}} \cdot A_{\text{slice}} = 62.4(x - 5) \cdot 2(45 - \frac{3}{5}x)\Delta x.$$

Finally, we use a definite integral to sum the forces over the appropriate range of x-values. Since the water rises to within 5 feet of the top of the dam, we start at $x = 5$ and slice all the way to the bottom of the dam, where $x = 30$. Hence,

$$F = \int_{x=5}^{x=30} 62.4(x-5) \cdot 2\left(45 - \frac{3}{5}x\right) dx.$$

Using technology to evaluate the integral, we find $F \approx 1.248 \times 10^6$ pounds.

Activity 6.4.4. In each of the following problems, determine the total force exerted by water against the surface that is described.

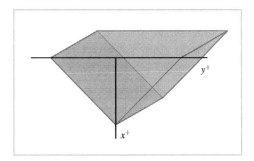
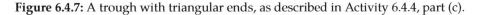

Figure 6.4.7: A trough with triangular ends, as described in Activity 6.4.4, part (c).

a. Consider a rectangular dam that is 100 feet wide and 50 feet tall, and suppose that water presses against the dam all the way to the top.

b. Consider a semicircular dam with a radius of 30 feet. Suppose that the water rises to within 10 feet of the top of the dam.

c. Consider a trough with triangular ends, as pictured in Figure 6.4.7, where the tank is 10 feet long, the top is 5 feet wide, and the tank is 4 feet deep. Say that the trough is full to within 1 foot of the top with water of weight density 62.4 pounds/ft^3. How much force does the water exert against one of the triangular ends?

While there are many different formulas that we use in solving problems involving work, force, and pressure, it is important to understand that the fundamental ideas behind these problems are similar to several others that we've encountered in applications of the definite integral. In particular, the basic idea is to take a difficult problem and somehow slice it into more manageable pieces that we understand, and then use a definite integral to add up these simpler pieces.

Summary

- To measure the work accomplished by a varying force that moves an object, we subdivide the problem into pieces on which we can use the formula $W = F \cdot d$, and then use a definite integral to sum the work accomplished on each piece.

- To find the total force exerted by water against a dam, we use the formula $F = P \cdot A$ to measure the force exerted on a slice that lies at a fixed depth, and then use a definite integral to sum the forces across the appropriate range of depths.

- Because work is computed as the product of force and distance (provided force is constant), and the force water exerts on a dam can be computed as the product of pressure and area (provided pressure is constant), problems involving these concepts are similar to earlier problems we did using definite integrals to find distance (via "distance equals rate times time") and mass ("mass equals density times volume").

Exercises

1. A tank in the shape of an inverted right circular cone has height 4 meters and radius 2 meters. It is filled with 2 meters of hot chocolate. Find the work required to empty the tank by pumping the hot chocolate over the top of the tank. The density of hot chocolate is $\delta = 1080$ kg/m^3. Your answer must include the correct units.

Work =

2. A fuel oil tank is an upright cylinder, buried so that its circular top is 10 feet beneath ground level. The tank has a radius of 7 feet and is 21 feet high, although the current oil level is only 17 feet deep. Calculate the work required to pump all of the oil to the surface. Oil weighs 50 lb/ft^3.

Work =
(include units)

3. A rectangular swimming pool 40 ft long, 15 ft wide, and 16 ft deep is filled with water to a depth of 15 ft. Use an integral to find the work required to pump all the water out over the top. (Take as the density of water $\delta = 62.4$lb/ft^3.)

Work =
(include units)

4. Water in a cylinder of height 10 ft and radius 3 ft is to be pumped out. The density of water is 62.4 lb/ft^3. Find the work required if
(*a*) The tank is full of water and the water is to be pumped over the top of the tank.

Work =
(include units)
(*b*) The tank is full of water and the water must be pumped to a height 4 ft above the top of the tank.

Work =

(include units)

(c) The depth of water in the tank is 8 ft and the water must be pumped over the top of the tank.

Work =

(include units)

5. A lobster tank in a restaurant is 1.25 m long by 1 m wide by 80 cm deep. Taking the density of water to be $1000\text{kg}/\text{m}^3$, find the water forces

on the bottom of the tank: Force =

on each of the larger sides of the tank: Force =

on each of the smaller sides of the tank: Force =

(include units for each, and use $g = 9.8 \text{ m}/\text{s}^2$)

6. Consider the curve $f(x) = 3\cos(\frac{x^3}{4})$ and the portion of its graph that lies in the first quadrant between the y-axis and the first positive value of x for which $f(x) = 0$. Let R denote the region bounded by this portion of f, the x-axis, and the y-axis. Assume that x and y are each measured in feet.

  a. Picture the coordinate axes rotated 90 degrees clockwise so that the positive x-axis points straight down, and the positive y-axis points to the right. Suppose that R is rotated about the x axis to form a solid of revolution, and we consider this solid as a storage tank. Suppose that the resulting tank is filled to a depth of 1.5 feet with water weighing 62.4 pounds per cubic foot. Find the amount of work required to lower the water in the tank until it is 0.5 feet deep, by pumping the water to the top of the tank.

  b. Again picture the coordinate axes rotated 90 degrees clockwise so that the positive x-axis points straight down, and the positive y-axis points to the right. Suppose that R, together with its reflection across the x-axis, forms one end of a storage tank that is 10 feet long. Suppose that the resulting tank is filled completely with water weighing 62.4 pounds per cubic foot. Find a formula for a function that tells the amount of work required to lower the water by h feet.

  c. Suppose that the tank described in (b) is completely filled with water. Find the total force due to hydrostatic pressure exerted by the water on one end of the tank.

7. A cylindrical tank, buried on its side, has radius 3 feet and length 10 feet. It is filled completely with water whose weight density is 62.4 lbs/ft³, and the top of the tank is two feet underground.

  a. Set up, but do not evaluate, an integral expression that represents the amount of work required to empty the top half of the water in the tank to a truck whose tank lies 4.5 feet above ground.

  b. With the tank now only half-full, set up, but do not evaluate an integral expression that represents the total force due to hydrostatic pressure against one end of the tank.

6.5 Improper Integrals

Motivating Questions

- What are improper integrals and why are they important?

- What does it mean to say that an improper integral converges or diverges?

- What are some typical improper integrals that we can classify as convergent or divergent?

Another important application of the definite integral regards how the likelihood of certain events can be measured. For example, consider a company that manufactures incandescent light bulbs, and suppose that based on a large volume of test results, they have determined that the fraction of light bulbs that fail between times $t = a$ and $t = b$ of use (where t is measured in months) is given by

$$\int_a^b 0.3e^{-0.3t}\, dt.$$

For example, the fraction of light bulbs that fail during their third month of use is given by

$$\int_2^3 0.3e^{-0.3t}\, dt = -e^{-0.3t}\Big|_2^3$$
$$= -e^{-0.9} + e^{-0.6}$$
$$\approx 0.1422.$$

Thus about 14.22% of all lightbulbs fail between $t = 2$ and $t = 3$. Clearly we could adjust the limits of integration to measure the fraction of light bulbs that fail during any time period of interest.

Preview Activity 6.5.1. A company with a large customer base has a call center that receives thousands of calls a day. After studying the data that represents how long callers wait for assistance, they find that the function $p(t) = 0.25e^{-0.25t}$ models the time customers wait in the following way: the fraction of customers who wait between $t = a$ and $t = b$ minutes is given by

$$\int_a^b p(t)\, dt.$$

Use this information to answer the following questions.

    a. Determine the fraction of callers who wait between 5 and 10 minutes.

    b. Determine the fraction of callers who wait between 10 and 20 minutes.

    c. Next, let's study the fraction who wait up to a certain number of minutes:

  i. What is the fraction of callers who wait between 0 and 5 minutes?

  ii. What is the fraction of callers who wait between 0 and 10 minutes?

  iii. Between 0 and 15 minutes? Between 0 and 20?

d. Let $F(b)$ represent the fraction of callers who wait between 0 and b minutes. Find a formula for $F(b)$ that involves a definite integral, and then use the First FTC to find a formula for $F(b)$ that does not involve a definite integral.

e. What is the value of the limit $\lim_{b \to \infty} F(b)$? What is its meaning in the context of the problem?

6.5.1 Improper Integrals Involving Unbounded Intervals

In light of our example with light bulbs that fail, as well as with the problem involving customer wait time in Preview Activity 6.5.1, we see that it is natural to consider questions where we desire to integrate over an interval whose upper limit grows without bound. For example, if we are interested in the fraction of light bulbs that fail within the first b months of use, we know that the expression

$$\int_0^b 0.3e^{-0.3t}\, dt$$

measures this value. To think about the fraction of light bulbs that fail *eventually*, we understand that we wish to find

$$\lim_{b \to \infty} \int_0^b 0.3e^{-0.3t}\, dt,$$

for which we will also use the notation

$$\int_0^\infty 0.3e^{-0.3t}\, dt. \tag{6.5.1}$$

Note particularly that we are studying the area of an unbounded region, as pictured in Figure 6.5.1.

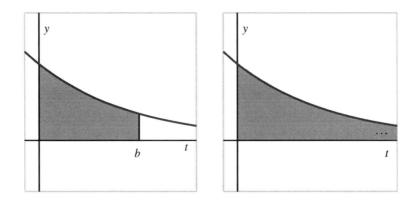

Figure 6.5.1: At left, the area bounded by $p(t) = 0.3e^{-0.3t}$ on the finite interval $[0, b]$; at right, the result of letting $b \to \infty$. By "$\cdots$" in the righthand figure, we mean that the region extends to the right without bound.

Anytime we are interested in an integral for which the interval of integration is unbounded (that is, one for which at least one of the limits of integration involves ∞), we say that the integral is *improper*. For instance, the integrals

$$\int_1^\infty \frac{1}{x^2}\,dx, \quad \int_{-\infty}^0 \frac{1}{1+x^2}\,dx, \quad \text{and} \quad \int_{-\infty}^\infty e^{-x^2}\,dx$$

are all improper due to having limits of integration that involve ∞. We investigate the value of any such integral by replacing the improper integral with a limit of proper integrals; for an improper integral such as $\int_0^\infty f(x)\,dx$, we write

$$\int_0^\infty f(x)\,dx = \lim_{b \to \infty} \int_0^b f(x)\,dx.$$

We can then attempt to evaluate $\int_0^b f(x)\,dx$ using the First FTC, after which we can evaluate the limit. An immediate and important question arises: is it even possible for the area of such an unbounded region to be finite? The following activity explores this issue and others in more detail.

Activity 6.5.2. In this activity we explore the improper integrals $\int_1^\infty \frac{1}{x}\,dx$ and $\int_1^\infty \frac{1}{x^{3/2}}\,dx$.

a. First we investigate $\int_1^\infty \frac{1}{x}\,dx$.

   i. Use the First FTC to determine the exact values of $\int_1^{10} \frac{1}{x}\,dx$, $\int_1^{1000} \frac{1}{x}\,dx$, and $\int_1^{100000} \frac{1}{x}\,dx$. Then, use your calculator to compute a decimal approximation of each result.

ii. Use the First FTC to evaluate the definite integral $\int_1^b \frac{1}{x}\,dx$ (which results in an expression that depends on b).

iii. Now, use your work from (ii.) to evaluate the limit given by

$$\lim_{b \to \infty} \int_1^b \frac{1}{x}\,dx.$$

b. Next, we investigate $\int_1^\infty \frac{1}{x^{3/2}}\,dx$.

i. Use the First FTC to determine the exact values of $\int_1^{10} \frac{1}{x^{3/2}}\,dx$, $\int_1^{1000} \frac{1}{x^{3/2}}\,dx$, and $\int_1^{100000} \frac{1}{x^{3/2}}\,dx$. Then, use your calculator to compute a decimal approximation of each result.

ii. Use the First FTC to evaluate the definite integral $\int_1^b \frac{1}{x^{3/2}}\,dx$ (which results in an expression that depends on b).

iii. Now, use your work from (ii.) to evaluate the limit given by

$$\lim_{b \to \infty} \int_1^b \frac{1}{x^{3/2}}\,dx.$$

c. Plot the functions $y = \frac{1}{x}$ and $y = \frac{1}{x^{3/2}}$ on the same coordinate axes for the values $x = 0 \ldots 10$. How would you compare their behavior as x increases without bound? What is similar? What is different?

d. How would you characterize the value of $\int_1^\infty \frac{1}{x}\,dx$? of $\int_1^\infty \frac{1}{x^{3/2}}\,dx$? What does this tell us about the respective areas bounded by these two curves for $x \geq 1$?

6.5.2 Convergence and Divergence

Our work so far has suggested that when we consider a nonnegative function f on an interval $[1, \infty]$, such as $f(x) = \frac{1}{x}$ or $f(x) = \frac{1}{x^{3/2}}$, there are at least two possibilities for the value of $\lim_{b \to \infty} \int_1^b f(x)\,dx$: the limit is finite or infinite. With these possibilities in mind, we introduce the following terminology.

If $f(x)$ is nonnegative for $x \geq a$, then we say that the improper integral $\int_a^\infty f(x)\,dx$ *converges* provided that

$$\lim_{b \to \infty} \int_a^b f(x)\,dx$$

exists and is finite. Otherwise, we say that $\int_a^\infty f(x)\,dx$ *diverges*.

We normally restrict our interest to improper integrals for which the integrand is nonnegative. Further, we note that our primary interest is in functions f for which $\lim_{x \to \infty} f(x) = 0$, for if the function f does not approach 0 as $x \to \infty$, then it is impossible for $\int_a^\infty f(x)\,dx$ to converge.

Activity 6.5.3. Determine whether each of the following improper integrals converges or diverges. For each integral that converges, find its exact value.

a. $\int_1^\infty \frac{1}{x^2}\,dx$

b. $\int_0^\infty e^{-x/4}\,dx$

c. $\int_2^\infty \frac{9}{(x+5)^{2/3}}\,dx$

d. $\int_4^\infty \frac{3}{(x+2)^{5/4}}\,dx$

e. $\int_0^\infty xe^{-x/4}\,dx$

f. $\int_1^\infty \frac{1}{x^p}\,dx$, where p is a positive real number

6.5.3 Improper Integrals Involving Unbounded Integrands

It is also possible for an integral to be improper due to the integrand being unbounded on the interval of integration. For example, if we consider

$$\int_0^1 \frac{1}{\sqrt{x}}\,dx,$$

we see that because $f(x) = \frac{1}{\sqrt{x}}$ has a vertical asymptote at $x = 0$, f is not continuous on $[0, 1]$, and the integral is attempting to represent the area of the unbounded region shown at right in Figure 6.5.2.

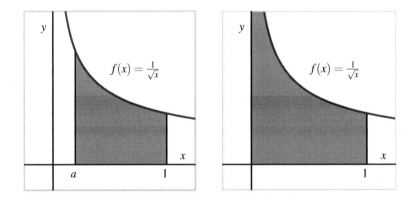

Figure 6.5.2: At left, the area bounded by $f(x) = \frac{1}{\sqrt{x}}$ on the finite interval $[a, 1]$; at right, the result of letting $a \to 0^+$, where we see that the shaded region will extend vertically without bound.

Just as we did with improper integrals involving infinite limits, we address the problem of the integrand being unbounded by replacing such an improper integral with a limit of proper integrals. For example, to evaluate $\int_0^1 \frac{1}{\sqrt{x}}\,dx$, we replace 0 with a and let a approach 0 from the right. Thus,

$$\int_0^1 \frac{1}{\sqrt{x}}\,dx = \lim_{a \to 0^+} \int_a^1 \frac{1}{\sqrt{x}}\,dx,$$

and then we evaluate the proper integral $\int_a^1 \frac{1}{\sqrt{x}}\,dx$, followed by taking the limit. In the same way as with improper integrals involving unbounded regions, we will say that the improper integral converges provided that this limit exists, and diverges otherwise. In the present example, we observe that

$$\int_0^1 \frac{1}{\sqrt{x}}\,dx = \lim_{a \to 0^+} \int_a^1 \frac{1}{\sqrt{x}}\,dx$$
$$= \lim_{a \to 0^+} 2\sqrt{x}\Big|_a^1$$
$$= \lim_{a \to 0^+} 2\sqrt{1} - 2\sqrt{a}$$
$$= 2,$$

and therefore the improper integral $\int_0^1 \frac{1}{\sqrt{x}}\,dx$ converges (to the value 2).

We have to be particularly careful with unbounded integrands, for they may arise in ways that may not initially be obvious. Consider, for instance, the integral

$$\int_1^3 \frac{1}{(x-2)^2}\,dx.$$

At first glance we might think that we can simply apply the Fundamental Theorem of Calculus by antidifferentiating $\frac{1}{(x-2)^2}$ to get $-\frac{1}{x-2}$ and then evaluate from 1 to 3. Were we to do so, we would be erroneously applying the FTC because $f(x) = \frac{1}{(x-2)^2}$ fails to be continuous throughout the interval, as seen in Figure 6.5.3.

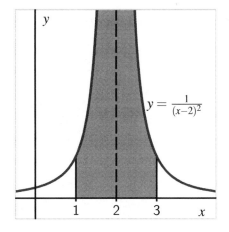

Figure 6.5.3: The function $f(x) = \frac{1}{(x-2)^2}$ on an interval including $x = 2$.

Such an incorrect application of the FTC leads to an impossible result (-2), which would itself suggest that something we did must be wrong. Indeed, we must address the vertical asymptote in $f(x) = \frac{1}{(x-2)^2}$ at $x = 2$ by writing

$$\int_1^3 \frac{1}{(x-2)^2}\, dx = \lim_{a \to 2^-} \int_1^a \frac{1}{(x-2)^2}\, dx + \lim_{b \to 2^+} \int_b^3 \frac{1}{(x-2)^2}\, dx$$

and then evaluate two separate limits of proper integrals. For instance, doing so for the integral with a approaching 2 from the left, we find

$$\int_1^2 \frac{1}{(x-2)^2}\, dx = \lim_{a \to 2^-} \int_1^a \frac{1}{(x-2)^2}\, dx$$
$$= \lim_{a \to 2^-} -\frac{1}{(x-2)}\Big|_1^a$$
$$= \lim_{a \to 2^-} -\frac{1}{(a-2)} + \frac{1}{1-2}$$
$$= \infty,$$

since $\frac{1}{a-2} \to -\infty$ as a approaches 2 from the left. Thus, the improper integral $\int_1^2 \frac{1}{(x-2)^2}\, dx$ diverges; similar work shows that $\int_2^3 \frac{1}{(x-2)^2}\, dx$ also diverges. From either of these two results, we can conclude that that the original integral, $\int_1^3 \frac{1}{(x-2)^2}\, dx$ diverges, too.

> **Activity 6.5.4.** For each of the following definite integrals, decide whether the integral is improper or not. If the integral is proper, evaluate it using the First FTC. If the integral is improper, determine whether or not the integral converges or diverges; if the integral converges, find its exact value.
>
> a. $\int_0^1 \frac{1}{x^{1/3}}\, dx$
>
> b. $\int_0^2 e^{-x}\, dx$
>
> c. $\int_1^4 \frac{1}{\sqrt{4-x}}\, dx$
>
> d. $\int_{-2}^2 \frac{1}{x^2}\, dx$
>
> e. $\int_0^{\pi/2} \tan(x)\, dx$
>
> f. $\int_0^1 \frac{1}{\sqrt{1-x^2}}\, dx$

Summary

- An integral $\int_a^b f(x)\, dx$ can be improper if at least one of a or b is $\pm\infty$, making the interval unbounded, or if f has a vertical asymptote at $x = c$ for some value of c that satisfies $a \le c \le b$. One reason that improper integrals are important is that certain probabilities can be represented by integrals that involve infinite limits.

- When we encounter an improper integral, we work to understand it by replacing the improper integral with a limit of proper integrals. For instance, we write

$$\int_a^\infty f(x)\,dx = \lim_{b\to\infty} \int_a^b f(x)\,dx,$$

and then work to determine whether the limit exists and is finite. For any improper integral, if the resulting limit of proper integrals exists and is finite, we say the improper integral converges. Otherwise, the improper integral diverges.

- An important class of improper integrals is given by

$$\int_1^\infty \frac{1}{x^p}\,dx$$

where p is a positive real number. We can show that this improper integral converges whenever $p > 1$, and diverges whenever $0 < p \le 1$. A related class of improper integrals is $\int_0^1 \frac{1}{x^p}\,dx$, which converges for $0 < p < 1$, and diverges for $p \ge 1$.

Exercises

1. Consider the integral

$$\int_0^3 \frac{-8}{x\sqrt{x}}\,dx$$

If the integral is divergent, type an upper-case "D". Otherwise, evaluate the integral.

2. Calculate the integral below, if it converges. If it does not converge, enter *diverges* for your answer.

$\int_2^\infty 1x^2 e^{-x^3}\,dx =$ [   ]

3. Calculate the integral, if it converges. If it diverges, enter *diverges* for your answer.

$\int_{-\infty}^1 \frac{e^{5x}}{1+e^{5x}}\,dx =$ [   ]

4. Calculate the integral, if it converges. If it diverges, enter *diverges* for your answer.

$\int_{-1}^1 \frac{1}{v}\,dv =$ [   ]

5. Find the area under the curve $y = \frac{1}{\cos^2(t)}$ between $t = 0$ and $t = \pi/2$. Enter *diverges* if the area is not bounded.

area = [   ]

6. Determine, with justification, whether each of the following improper integrals converges or diverges.

a. $\int_e^\infty \frac{\ln(x)}{x}\,dx$

b. $\int_e^\infty \frac{1}{x\ln(x)}\,dx$

c. $\int_e^\infty \frac{1}{x(\ln(x))^2}\,dx$

d. $\int_e^\infty \frac{1}{x(\ln(x))^p}\,dx$, where p is a positive real number

e. $\int_0^1 \frac{\ln(x)}{x}\,dx$

f. $\int_0^1 \ln(x)\,dx$

7. Sometimes we may encounter an improper integral for which we cannot easily evaluate the limit of the corresponding proper integrals. For instance, consider $\int_1^\infty \frac{1}{1+x^3}\,dx$. While it is hard (or perhaps impossible) to find an antiderivative for $\frac{1}{1+x^3}$, we can still determine whether or not the improper integral converges or diverges by comparison to a simpler one. Observe that for all $x > 0$, $1 + x^3 > x^3$, and therefore

$$\frac{1}{1+x^3} < \frac{1}{x^3}.$$

It therefore follows that

$$\int_1^b \frac{1}{1+x^3}\,dx < \int_1^b \frac{1}{x^3}\,dx$$

for every $b > 1$. If we let $b \to \infty$ so as to consider the two improper integrals $\int_1^\infty \frac{1}{1+x^3}\,dx$ and $\int_1^\infty \frac{1}{x^3}\,dx$, we know that the larger of the two improper integrals converges. And thus, since the smaller one lies below a convergent integral, it follows that the smaller one must converge, too. In particular, $\int_1^\infty \frac{1}{1+x^3}\,dx$ must converge, even though we never explicitly evaluated the corresponding limit of proper integrals. We use this idea and similar ones in the exercises that follow.

a. Explain why $x^2 + x + 1 > x^2$ for all $x \geq 1$, and hence show that $\int_1^\infty \frac{1}{x^2+x+1}\,dx$ converges by comparison to $\int_1^\infty \frac{1}{x^2}\,dx$.

b. Observe that for each $x > 1$, $\ln(x) < x$. Explain why

$$\int_2^b \frac{1}{x}\,dx < \int_2^b \frac{1}{\ln(x)}\,dx$$

for each $b > 2$. Why must it be true that $\int_2^b \frac{1}{\ln(x)}\,dx$ diverges?

c. Explain why $\sqrt{\frac{x^4+1}{x^4}} > 1$ for all $x > 1$. Then, determine whether or not the improper integral

$$\int_1^\infty \frac{1}{x} \cdot \sqrt{\frac{x^4+1}{x^4}}\,dx$$

converges or diverges.

CHAPTER 7

Differential Equations

7.1 An Introduction to Differential Equations

Motivating Questions

- What is a differential equation and what kinds of information can it tell us?

- How do differential equations arise in the world around us?

- What do we mean by a solution to a differential equation?

In previous chapters, we have seen that a function's derivative tells us the rate at which the function is changing. More recently, the Fundamental Theorem of Calculus helped us to determine the total change of a function over an interval when we know the function's rate of change. For instance, an object's velocity tells us the rate of change of that object's position. By integrating the velocity over a time interval, we may determine by how much the position changes over that time interval. In particular, if we know where the object is at the beginning of that interval, then we have enough information to accurately predict where it will be at the end of the interval.

In this chapter, we will introduce the concept of *differential equations* and explore this idea in more depth. Simply said, a differential equation is an equation that provides a description of a function's derivative, which means that it tells us the function's rate of change. Using this information, we would like to learn as much as possible about the function itself. For instance, we would ideally like to have an algebraic description of the function. As we'll see, this may be too much to ask in some situations, but we will still be able to make accurate approximations.

> **Preview Activity 7.1.1.** The position of a moving object is given by the function $s(t)$, where s is measured in feet and t in seconds. We determine that the velocity is $v(t) = 4t + 1$ feet per second.
>
> a. How much does the position change over the time interval $[0, 4]$?

b. Does this give you enough information to determine $s(4)$, the position at time $t = 4$? If so, what is $s(4)$? If not, what additional information would you need to know to determine $s(4)$?

c. Suppose you are told that the object's initial position $s(0) = 7$. Determine $s(2)$, the object's position 2 seconds later.

d. If you are told instead that the object's initial position is $s(0) = 3$, what is $s(2)$?

e. If we only know the velocity $v(t) = 4t + 1$, is it possible that the object's position at all times is $s(t) = 2t^2 + t - 4$? Explain how you know.

f. Are there other possibilities for $s(t)$? If so, what are they?

g. If, in addition to knowing the velocity function is $v(t) = 4t + 1$, we know the initial position $s(0)$, how many possibilities are there for $s(t)$?

7.1.1 What is a differential equation?

A differential equation is an equation that describes the derivative, or derivatives, of a function that is unknown to us. For instance, the equation

$$\frac{dy}{dx} = x \sin x$$

is a differential equation since it describes the derivative of a function $y(x)$ that is unknown to us.

As many important examples of differential equations involve quantities that change in time, the independent variable in our discussion will frequently be time t. For instance, in the preview activity, we considered the differential equation

$$\frac{ds}{dt} = 4t + 1.$$

Knowing the velocity and the starting position of the object, we were able to find the position at any later time.

Because differential equations describe the derivative of a function, they give us information about how that function changes. Our goal will be to take this information and use it to predict the value of the function in the future; in this way, differential equations provide us with something like a crystal ball.

Differential equations arise frequently in our every day world. For instance, you may hear a bank advertising:

Your money will grow at a 3% annual interest rate with us.

This innocuous statement is really a differential equation. Let's translate: $A(t)$ will be amount of money you have in your account at time t. On one hand, the rate at which your money grows is the derivative dA/dt. On the other hand, we are told that this rate is $0.03A$. This leads to the differential equation

$$\frac{dA}{dt} = 0.03A.$$

This differential equation has a slightly different feel than the previous equation $\frac{ds}{dt} = 4t + 1$. In the earlier example, the rate of change depends only on the independent variable t, and we may find $s(t)$ by integrating the velocity $4t + 1$. In the banking example, however, the rate of change depends on the dependent variable A, so we'll need some new techniques in order to find $A(t)$.

> **Activity 7.1.2.** Express the following statements as differential equations. In each case, you will need to introduce notation to describe the important quantities in the statement so be sure to clearly state what your notation means.
>
> a. The population of a town grows continuously at an annual rate of 1.25%.
>
> b. A radioactive sample loses 5.6% of its mass every day.
>
> c. You have a bank account that continuously earns 4% interest every year. At the same time, you withdraw money continually from the account at the rate of $1000 per year.
>
> d. A cup of hot chocolate is sitting in a 70° room. The temperature of the hot chocolate cools continuously by 10% of the difference between the hot chocolate's temperature and the room temperature every minute.
>
> e. A can of cold soda is sitting in a 70° room. The temperature of the soda warms continuously at the rate of 10% of the difference between the soda's temperature and the room's temperature every minute.

7.1.2 Differential equations in the world around us

As we have noted, differential equations give a natural way to describe phenomena we see in the real world. For instance, physical principles are frequently expressed as a description of how a quantity changes. A good example is Newton's Second Law, an important physcial principle that says:

The product of an object's mass and acceleration equals the force applied to it.

For instance, when gravity acts on an object near the earth's surface, it exerts a force equal to mg, the mass of the object times the gravitational constant g. We therefore have

$$ma = mg, \text{ or}$$

$$\frac{dv}{dt} = g,$$

where v is the velocity of the object, and $g = 9.8$ meters per second squared. Notice that this physical principle does not tell us what the object's velocity is, but rather how the object's velocity changes.

Activity 7.1.3. Shown below are two graphs depicting the velocity of falling objects. On the left is the velocity of a skydiver, while on the right is the velocity of a meteorite entering the Earth's atmosphere.

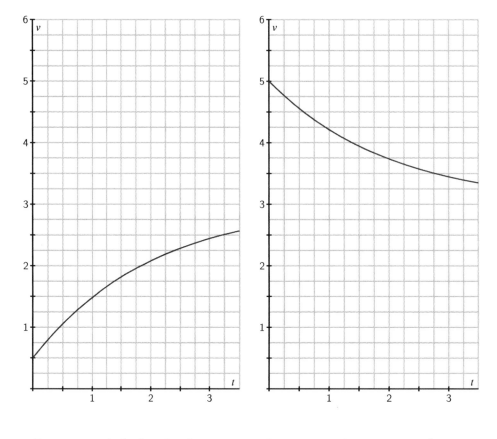

Figure 7.1.1: A skydiver's velocity.   **Figure 7.1.2:** A meteorite's velocity.

a. Begin with the skydiver's velocity and use the given graph to measure the rate of change dv/dt when the velocity is $v = 0.5, 1.0, 1.5, 2.0$, and 2.5. Plot your values on the graph below. You will want to think carefully about this: you are plotting the derivative dv/dt as a function of *velocity*.

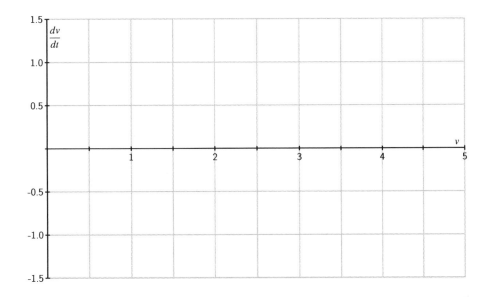

b. Now do the same thing with the meteorite's velocity: use the given graph to measure the rate of change dv/dt when the velocity is $v = 3.5, 4.0, 4.5$, and 5.0. Plot your values on the graph above.

c. You should find that all your points lie on a line. Write the equation of this line being careful to use proper notation for the quantities on the horizontal and vertical axes.

d. The relationship you just found is a differential equation. Write a complete sentence that explains its meaning.

e. By looking at the differential equation, determine the values of the velocity for which the velocity increases.

f. By looking at the differential equation, determine the values of the velocity for which the velocity decreases.

g. By looking at the differential equation, determine the values of the velocity for which the velocity remains constant.

The point of this activity is to demonstrate how differential equations model processes in the real world. In this example, two factors are influencing the velocities: gravity and wind resistance. The differential equation describes how these factors influence the rate of change of the objects' velocities.

7.1.3 Solving a differential equation

We have said that a differential equation is an equation that describes the derivative, or derivatives, of a function that is unknown to us. By a *solution* to a differential equation, we mean simply a function that satisies this description.

For instance, the first differential equation we looked at is

$$\frac{ds}{dt} = 4t + 1,$$

which describes an unknown function $s(t)$. We may check that $s(t) = 2t^2 + t$ is a solution because it satisfies this description. Notice that $s(t) = 2t^2 + t + 4$ is also a solution.

If we have a candidate for a solution, it is straightforward to check whether it is a solution or not. Before we demonstrate, however, let's consider the same issue in a simpler context. Suppose we are given the equation $2x^2 - 2x = 2x + 6$ and asked whether $x = 3$ is a solution. To answer this question, we could rewrite the variable x in the equation with the symbol □:

$$2\square^2 - 2\square = 2\square + 6.$$

To determine whether $x = 3$ is a solution, we can investigate the value of each side of the equation separately when the value 3 is placed in □ and see if indeed the two resulting values are equal. Doing so, we observe that

$$2\square^2 - 2\square = 2 \cdot 3^2 - 2 \cdot 3 = 12,$$

and

$$2\square + 6 = 2 \cdot 3 + 6 = 12.$$

Therefore, $x = 3$ is indeed a solution.

We will do the same thing with differential equations. Consider the differential equation

$$\frac{dv}{dt} = 1.5 - 0.5v, \text{ or}$$

$$\frac{d\square}{dt} = 1.5 - 0.5\square.$$

Let's ask whether $v(t) = 3 - 2e^{-0.5t}$ is a solution[1]. Using this formula for v, observe first that

$$\frac{dv}{dt} = \frac{d\square}{dt} = \frac{d}{dt}[3 - 2e^{-0.5t}] = -2e^{-0.5t} \cdot (-0.5) = e^{-0.5t}$$

and

$$1.5 - 0.5v = 1.5 - 0.5\square = 1.5 - 0.5(3 - 2e^{-0.5t}) = 1.5 - 1.5 + e^{-0.5t} = e^{-0.5t}.$$

Since $\frac{dv}{dt}$ and $1.5 - 0.5v$ agree for all values of t when $v = 3 - 2e^{-0.5t}$, we have indeed found a solution to the differential equation.

---

[1]At this time, don't worry about why we chose this function; we will learn techniques for finding solutions to differential equations soon enough.

Activity 7.1.4. Consider the differential equation

$$\frac{dv}{dt} = 1.5 - 0.5v.$$

Which of the following functions are solutions of this differential equation?

a. $v(t) = 1.5t - 0.25t^2$.

b. $v(t) = 3 + 2e^{-0.5t}$.

c. $v(t) = 3$.

d. $v(t) = 3 + Ce^{-0.5t}$ where C is any constant.

This activity shows us something interesting. Notice that the differential equation has infinitely many solutions, which are parametrized by the constant C in $v(t) = 3 + Ce^{-0.5t}$. In Figure 7.1.3, we see the graphs of these solutions for a few values of C, as labeled.

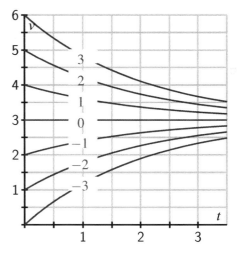

Figure 7.1.3: The family of solutions to the differential equation $\frac{dv}{dt} = 1.5 - 0.5v$.

Notice that the value of C is connected to the initial value of the velocity $v(0)$, since $v(0) = 3 + C$. In other words, while the differential equation describes how the velocity changes as a function of the velocity itself, this is not enough information to determine the velocity uniquely: we also need to know the initial velocity. For this reason, differential equations will typically have infinitely many solutions, one corresponding to each initial value. We have seen this phenomenon before, such as when given the velocity of a moving object $v(t)$, we were not able to uniquely determine the object's position unless we also know its initial position.

If we are given a differential equation and an initial value for the unknown function, we say that we have an *initial value problem*. For instance,

$$\frac{dv}{dt} = 1.5 - 0.5v, \; v(0) = 0.5$$

is an initial value problem. In this situation, we know the value of v at one time and we know how v is changing. Consequently, there should be exactly one function v that satisfies the initial value problem.

This demonstrates the following important general property of initial value problems.

> Initial value problems that are "well behaved" have exactly one solution, which exists in some interval around the initial point.

We won't worry about what "well behaved" means—it is a technical condition that will be satisfied by all the differential equations we consider.

To close this section, we note that differential equations may be classified based on certain characteristics they may possess. Indeed, you may see many different types of differential equations in a later course in differential equations. For now, we would like to introduce a few terms that are used to describe differential equations.

A *first-order* differential equation is one in which only the first derivative of the function occurs. For this reason,

$$\frac{dv}{dt} = 1.5 - 0.5v$$

is a first-order equation while

$$\frac{d^2y}{dt^2} = -10y$$

is a second-order equation.

A differential equation is *autonomous* if the independent variable does not appear in the description of the derivative. For instance,

$$\frac{dv}{dt} = 1.5 - 0.5v$$

is autonomous because the description of the derivative dv/dt does not depend on time. The equation

$$\frac{dy}{dt} = 1.5t - 0.5y,$$

however, is not autonomous.

Summary

- A differential equation is simply an equation that describes the derivative(s) of an unknown function.

- Physical principles, as well as some everyday situations, often describe how a quantity changes, which lead to differential equations.

- A solution to a differential equation is a function whose derivatives satisfy the equation's description. Differential equations typically have infinitely many solutions, parametrized by the initial values.

Exercises

1. Match the solutions to the differential equations. If there is more than one solution to an equation, select the answer that includes all solutions.

1. $\frac{d^2y}{dx^2} = 9y$

2. $\frac{dy}{dx} = 3y$

3. $\frac{d^2y}{dx^2} = -9y$

4. $\frac{dy}{dx} = -3y$

A. $y = e^{-3x}$ or $y = e^{3x}$

B. $y = e^{3x}$

C. $y = 3\sin(x)$

D. $y = \sin(3x)$ or $y = 3\sin(x)$

E. $y = e^{-3x}$

F. $y = \sin(3x)$

2. Find a positive value of k for which $y = \cos(kt)$ satisfies

$$\frac{d^2y}{dt^2} + 4y = 0.$$

$k = \boxed{}$

3. Let A and k be positive constants.

Which of the given functions is a solution to $\frac{dy}{dt} = k(A - y)$?

4. Suppose that $T(t)$ represents the temperature of a cup of coffee set out in a room, where T is expressed in degrees Fahrenheit and t in minutes. A physical principle known as Newton's Law of Cooling tells us that

$$\frac{dT}{dt} = -\frac{1}{15}T + 5.$$

a. Supposes that $T(0) = 105$. What does the differential equation give us for the value of $\frac{dT}{dt}\big|_{T=105}$? Explain in a complete sentence the meaning of these two facts.

b. Is T increasing or decreasing at $t = 0$?

c. What is the approximate temperature at $t = 1$?

d. On the graph below, make a plot of dT/dt as a function of T.

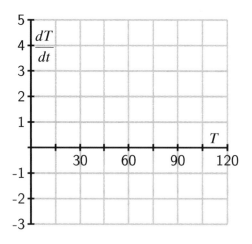

e. For which values of T does T increase? For which values of T does T decrease?

f. What do you think is the temperature of the room? Explain your thinking.

g. Verify that $T(t) = 75 + 30e^{-t/15}$ is the solution to the differential equation with initial value $T(0) = 105$. What happens to this solution after a long time?

5. Suppose that the population of a particular species is described by the function $P(t)$, where P is expressed in millions. Suppose further that the population's rate of change is governed by the differential equation

$$\frac{dP}{dt} = f(P)$$

where $f(P)$ is the function graphed below.

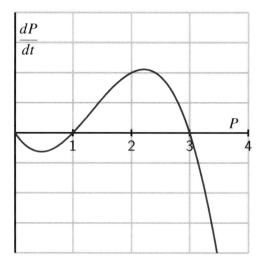

a. For which values of the population P does the population increase?

b. For which values of the population P does the population decrease?

c. If $P(0) = 3$, how will the population change in time?

d. If the initial population satisfies $0 < P(0) < 1$, what will happen to the population after a very long time?

e. If the initial population satisfies $1 < P(0) < 3$, what will happen to the population after a very long time?

f. If the initial population satisfies $3 < P(0)$, what will happen to the population after a very long time?

g. This model for a population's growth is sometimes called "growth with a threshold." Explain why this is an appropriate name.

6. In this problem, we test further what it means for a function to be a solution to a given differential equation.

a. Consider the differential equation

$$\frac{dy}{dt} = y - t.$$

Determine whether the following functions are solutions to the given differential equation.

- $y(t) = t + 1 + 2e^t$
- $y(t) = t + 1$
- $y(t) = t + 2$

b. When you weigh bananas in a scale at the grocery store, the height h of the bananas is described by the differential equation

$$\frac{d^2h}{dt^2} = -kh$$

where k is the *spring constant*, a constant that depends on the properties of the spring in the scale. After you put the bananas in the scale, you (cleverly) observe that the height of the bananas is given by $h(t) = 4\sin(3t)$. What is the value of the spring constant?

7.2 Qualitative behavior of solutions to DEs

Motivating Questions

- What is a slope field?

- How can we use a slope field to obtain qualitative information about the solutions of a differential equation?

- What are stable and unstable equilibrium solutions of an autonomous differential equation?

In earlier work, we have used the tangent line to the graph of a function f at a point a to approximate the values of f near a. The usefulness of this approximation is that we need to know very little about the function; armed with only the value $f(a)$ and the derivative $f'(a)$, we may find the equation of the tangent line and the approximation

$$f(x) \approx f(a) + f'(a)(x - a).$$

Remember that a first-order differential equation gives us information about the derivative of an unknown function. Since the derivative at a point tells us the slope of the tangent line at this point, a differential equation gives us crucial information about the tangent lines to the graph of a solution. We will use this information about the tangent lines to create a *slope field* for the differential equation, which enables us to sketch solutions to initial value problems. Our aim will be to understand the solutions qualitatively. That is, we would like to understand the basic nature of solutions, such as their long-range behavior, without precisely determining the value of a solution at a particular point.

Preview Activity 7.2.1. Let's consider the initial value problem

$$\frac{dy}{dt} = t - 2, \quad y(0) = 1.$$

a. Use the differential equation to find the slope of the tangent line to the solution $y(t)$ at $t = 0$. Then use the initial value to find the equation of the tangent line at $t = 0$. Sketch this tangent line over the interval $-0.25 \leq t \leq 0.25$ on the axes provided in Figure 7.2.1.

b. Also shown in Figure 7.2.1 are the tangent lines to the solution $y(t)$ at the points $t = 1, 2$, and 3 (we will see how to find these later). Use the graph to measure the slope of each tangent line and verify that each agrees with the value specified by the differential equation.

c. Using these tangent lines as a guide, sketch a graph of the solution $y(t)$ over the interval $0 \leq t \leq 3$ so that the lines are tangent to the graph of $y(t)$.

d. Use the Fundamental Theorem of Calculus to find $y(t)$, the solution to this initial value problem.

e. Graph the solution you found in (d) on the axes provided, and compare it to the sketch you made using the tangent lines.

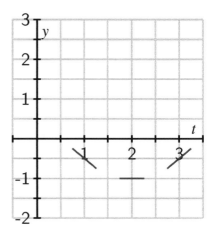

Figure 7.2.1: Grid for plotting partial tangent lines.

7.2.1 Slope fields

Preview Activity 7.2.1 shows that we may sketch the solution to an initial value problem if we know an appropriate collection of tangent lines. Because we may use a given differential equation to determine the slope of the tangent line at any point of interest, by plotting a useful collection of these, we can get an accurate sense of how certain solution curves must behave.

Let's continue looking at the differential equation $\frac{dy}{dt} = t - 2$. If $t = 0$, this equation says that $dy/dt = 0 - 2 = -2$. Note that this value holds regardless of the value of y. We will therefore sketch tangent lines for several values of y and $t = 0$ with a slope of -2.

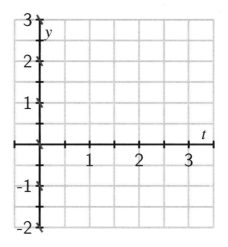

Figure 7.2.2: Tangent lines at points with $t = 0$.

Let's continue in the same way: if $t = 1$, the differential equation tells us that $dy/dt = 1 - 2 = -1$, and this holds regardless of the value of y. We now sketch tangent lines for several values of y and $t = 1$ with a slope of -1.

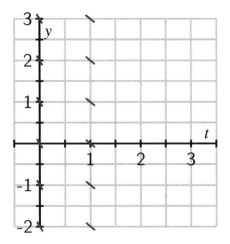

Figure 7.2.3: Adding tangent lines at points with $t = 1$.

Similarly, we see that when $t = 2$, $dy/dt = 0$ and when $t = 3$, $dy/dt = 1$. We may therefore add to our growing collection of tangent line plots to achieve the next figure.

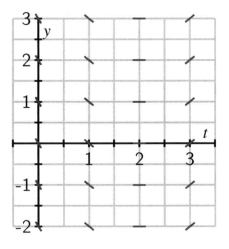

Figure 7.2.4: Adding tangent lines at points with $t = 2$ and $t = 3$.

In this figure, we begin to see the solutions to the differential equation emerge. However, for the sake of clarity, we add more tangent lines to provide the even more complete picture shown below.

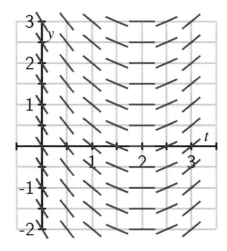

Figure 7.2.5: A completed slope field.

This most recent figure, which is called a *slope field* for the differential equation, allows us to sketch solutions just as we did in the preview activity. Here, we will begin with the initial value $y(0) = 1$ and start sketching the solution by following the tangent line, as shown in the Figure 7.2.6.

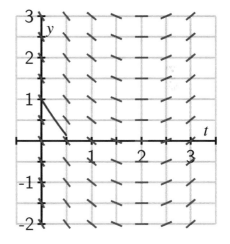

Figure 7.2.6: Beginning to sketch the solution function.

We then continue using this principle: whenever the solution passes through a point at which a tangent line is drawn, that line is tangent to the solution. Doing so leads us to the following sequence of images.

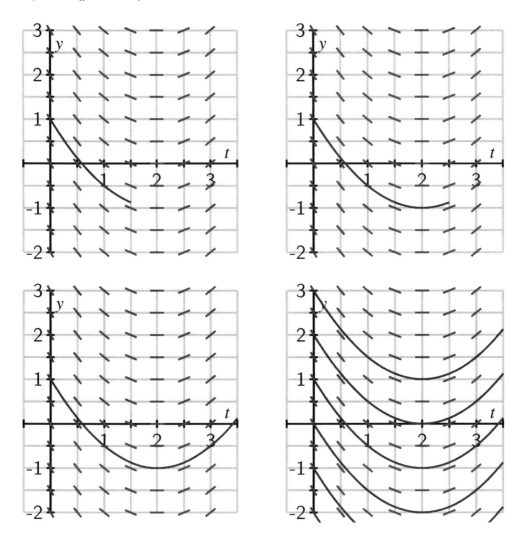

Figure 7.2.7: The third image in a sequence that show how to sketch the IVP solution that satisfies $y(0) = 1$.

Figure 7.2.8: Different solutions to $\frac{dy}{dt} = t - 2$ that correspond to different initial values.

In fact, we may draw solutions for any possible initial value, and doing this for several different initial values for $y(0)$ results in the graphs shown at bottom right in Figure 7.2.8.

Just as we have done for the most recent example with $\frac{dy}{dt} = t - 2$, we can construct a slope field for any differential equation of interest. The slope field provides us with visual information about how we expect solutions to the differential equation to behave.

Activity 7.2.2. Consider the autonomous differential equation

$$\frac{dy}{dt} = -\frac{1}{2}(y - 4).$$

a. Make a plot of $\frac{dy}{dt}$ versus y on the axes provided. Looking at the graph, for what values of y does y increase and for what values of y does y decrease?

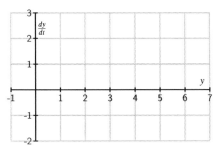

b. Next, sketch the slope field for this differential equation on the axes provided.

c. Use your work in (b) to sketch (on the axes in Figure 7.2.9.) solutions that satisfy $y(0) = 0$, $y(0) = 2$, $y(0) = 4$ and $y(0) = 6$.

d. Verify that $y(t) = 4+2e^{-t/2}$ is a solution to the given differential equation with the initial value $y(0) = 6$. Compare its graph to the one you sketched in (c).

e. What is special about the solution where $y(0) = 4$?

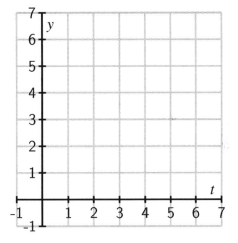

Figure 7.2.9: Axes for plotting the slope field for $\frac{dy}{dt} = -\frac{1}{2}(y - 4)$.

7.2.2 Equilibrium solutions and stability

As our work in Activity 7.2.2 demonstrates, first-order autonomous equations may have solutions that are constant. These are quite easy to detect by inspecting the differential equation $dy/dt = f(y)$: constant solutions necessarily have a zero derivative so $dy/dt = 0 = f(y)$.

For example, in Activity 7.2.2, we considered the equation $\frac{dy}{dt} = f(y) = -\frac{1}{2}(y-4)$. Constant solutions are found by setting $f(y) = -\frac{1}{2}(y-4) = 0$, which we immediately see implies that $y = 4$.

Values of y for which $f(y) = 0$ in an autonomous differential equation $\frac{dy}{dt} = f(y)$ are usually called or *equilibrium solutions* of the differential equation.

Activity 7.2.3. Consider the autonomous differential equation

$$\frac{dy}{dt} = -\frac{1}{2}y(y-4).$$

a. Make a plot of $\frac{dy}{dt}$ versus y on the axes provided. Looking at the graph, for what values of y does y increase and for what values of y does y decrease?

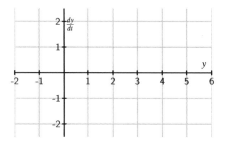

b. Identify any equilibrium solutions of the given differential equation.

c. Now sketch the slope field for the given differential equation on the axes provided in Figure 7.2.10.

d. Sketch the solutions to the given differential equation that correspond to initial values $y(0) = -1, 0, 1, \ldots, 5$.

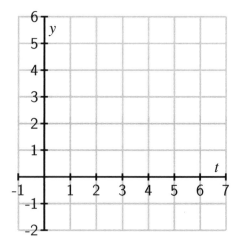

Figure 7.2.10: Axes for plotting the slope field for $\frac{dy}{dt} = -\frac{1}{2}y(y-4)$.

e. An equilibrium solution $\overline{y}$ is called *stable* if nearby solutions converge to $\overline{y}$. This means that if the initial condition varies slightly from $\overline{y}$, then $\lim_{t\to\infty} y(t) = \overline{y}$. Conversely, an equilibrium solution $\overline{y}$ is called *unstable* if nearby solutions are pushed away from $\overline{y}$. Using your work above, classify the equilibrium solutions you found in (b) as either stable or unstable.

f. Suppose that $y(t)$ describes the population of a species of living organisms and that the initial value $y(0)$ is positive. What can you say about the eventual fate of this population?

g. Now consider a general autonomous differential equation of the form $dy/dt = f(y)$. Remember that an equilibrium solution $\overline{y}$ satisfies $f(\overline{y}) = 0$. If we graph $dy/dt = f(y)$ as a function of y, for which of the differential equations represented in Figure 7.2.11 and Figure 7.2.12 is $\overline{y}$ a stable equilibrium and for which is $\overline{y}$ unstable? Why?

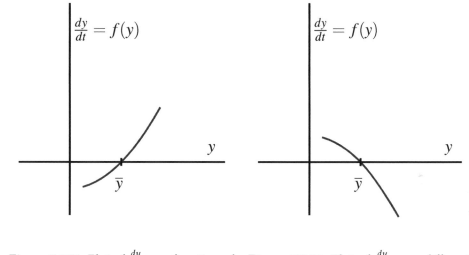

Figure 7.2.11: Plot of $\frac{dy}{dt}$ as a function of y.

Figure 7.2.12: Plot of $\frac{dy}{dt}$ as a different function of y.

Summary

- A slope field is a plot created by graphing the tangent lines of many different solutions to a differential equation.

- Once we have a slope field, we may sketch the graph of solutions by drawing a curve that is always tangent to the lines in the slope field.

- Autonomous differential equations sometimes have constant solutions that we call equilibrium solutions. These may be classified as stable or unstable, depending on the behavior of nearby solutions.

Exercises

1. Consider the direction field below for a differential equation. Use the graph to find the equilibrium solutions.

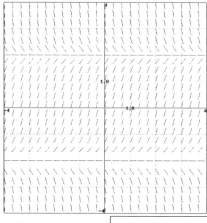

Answer (separate by commas): $y =$

Note: You can click on the graph to enlarge the image.

2. Consider the two slope fields shown, in figures 1 and 2 below.

figure 1	figure 2

On a print-out of these slope fields, sketch for each three solution curves to the differential equations that generated them. Then complete the following statements:

For the slope field in figure 1, a solution passing through the point (3,-1) has a [Choose: positive | negative | zero | undefined] slope.

For the slope field in figure 1, a solution passing through the point (-3,2) has a [Choose: positive | negative | zero | undefined] slope.
For the slope field in figure 2, a solution passing through the point (2,-1) has a [Choose: positive | negative | zero | undefined] slope.
For the slope field in figure 2, a solution passing through the point (0,3) has a [Choose: positive | negative | zero | undefined] slope.

3. Match the following equations with their direction field. Clicking on each picture will give you an enlarged view. While you can probably solve this problem by guessing, it is useful to try to predict characteristics of the direction field and then match them to the picture.
Here are some handy characteristics to start with – you will develop more as you practice.

A. Set y equal to zero and look at how the derivative behaves along the x-axis.

B. Do the same for the y-axis by setting x equal to 0

C. Consider the curve in the plane defined by setting $y' = 0$ – this should correspond to the points in the picture where the slope is zero.

D. Setting y' equal to a constant other than zero gives the curve of points where the slope is that constant. These are called isoclines, and can be used to construct the direction field picture by hand.

1. $y' = 2xy + 2xe^{-x^2}$

2. $y' = 2\sin(3x) + 1 + y$

3. $y' = 2y - 2$

4. $y' = -\dfrac{(2x + y)}{(2y)}$

Error: PGbasicmacros: imageRow: Unknown displayMode: PTX.
Error: PGbasicmacros: imageRow: Unknown displayMode: PTX.

4. Given the differential equation $x'(t) = x^4 - 5x^3 - 2x^2 + 24x + 0$.
List the constant (or equilibrium) solutions to this differential equation in increasing order and indicate whether or not these equations are stable, semi-stable, or unstable. (It helps to sketch the graph. xFunctions will plot functions as well as phase planes.)

	[Choose:	stable	unstable	semi-stable]
	[Choose:	stable	unstable	semi-stable]
	[Choose:	stable	unstable	semi-stable]
	[Choose:	stable	unstable	semi-stable]

5. Consider the differential equation

$$\frac{dy}{dt} = t - y.$$

a. Sketch a slope field on the plot below:

b. Sketch the solutions whose initial values are $y(0) = -4, -3, \ldots, 4$.

c. What do your sketches suggest is the solution whose initial value is $y(0) = -1$? Verify that this is indeed the solution to this initial value problem.

d. By considering the differential equation and the graphs you have sketched, what is the relationship between t and y at a point where a solution has a local minimum?

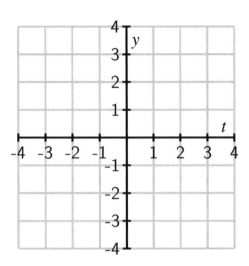

6. Consider the situation from problem 2 of Section 7.1: Suppose that the population of a particular species is described by the function $P(t)$, where P is expressed in millions. Suppose further that the population's rate of change is governed by the differential equation

$$\frac{dP}{dt} = f(P)$$

where $f(P)$ is the function graphed below.

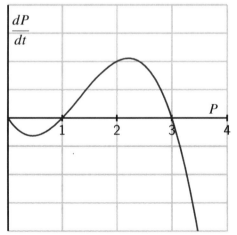

a. Sketch a slope field for this differential equation. You do not have enough information to determine the actual slopes, but you should have enough information to determine where slopes are positive, negative, zero, large, or small, and hence determine the qualitative behavior of solutions.

b. Sketch some solutions to this differential equation when the initial population $P(0) > 0$.

c. Identify any equilibrium solutions to the differential equation and classify them as stable or unstable.

d. If $P(0) > 1$, what is the eventual fate of the species?

e. if $P(0) < 1$, what is the eventual fate of the species?

f. Remember that we referred to this model for population growth as "growth with a threshold." Explain why this characterization makes sense by considering solutions whose inital value is close to 1.

7. The population of a species of fish in a lake is $P(t)$ where P is measured in thousands of fish and t is measured in months. The growth of the population is described by the differential equation

$$\frac{dP}{dt} = f(P) = P(6 - P).$$

a. Sketch a graph of $f(P) = P(6 - P)$ and use it to determine the equilibrium solutions and whether they are stable or unstable. Write a complete sentence that describes the long-term behavior of the fish population.

b. Suppose now that the owners of the lake allow fishers to remove 1000 fish from the lake every month (remember that $P(t)$ is measured in *thousands* of fish). Modify the differential equation to take this into account. Sketch the new graph of dP/dt versus P. Determine the new equilibrium solutions and decide whether they are stable or unstable.

c. Given the situation in part (b), give a description of the long-term behavior of the fish population.

d. Suppose that fishermen remove h thousand fish per month. How is the differential equation modified?

e. What is the largest number of fish that can be removed per month without eliminating the fish population? If fish are removed at this maximum rate, what is the eventual population of fish?

8. Let $y(t)$ be the number of thousands of mice that live on a farm; assume time t is measured in years.[1]

a. The population of the mice grows at a yearly rate that is twenty times the number of mice. Express this as a differential equation.

b. At some point, the farmer brings C cats to the farm. The number of mice that the cats can eat in a year is

$$M(y) = C\frac{y}{2 + y}$$

thousand mice per year. Explain how this modifies the differential equation that you found in part a).

---

[1]This problem is based on an ecological analysis presented in a research paper by C.S. Hollings: The Components of Predation as Revealed by a Study of Small Mammal Predation of the European Pine Sawfly, *Canadian Entomology 91*: 283-320.

c. Sketch a graph of the function $M(y)$ for a single cat $C = 1$ and explain its features by looking, for instance, at the behavior of $M(y)$ when y is small and when y is large.

d. Suppose that $C = 1$. Find the equilibrium solutions and determine whether they are stable or unstable. Use this to explain the long-term behavior of the mice population depending on the initial population of the mice.

e. Suppose that $C = 60$. Find the equilibrium solutions and determine whether they are stable or unstable. Use this to explain the long-term behavior of the mice population depending on the initial population of the mice.

f. What is the smallest number of cats you would need to keep the mice population from growing arbitrarily large?

7.3 Euler's method

Motivating Questions

- What is Euler's method and how can we use it to approximate the solution to an initial value problem?

- How accurate is Euler's method?

In Section 7.2, we saw how a slope field can be used to sketch solutions to a differential equation. In particular, the slope field is a plot of a large collection of tangent lines to a large number of solutions of the differential equation, and we sketch a single solution by simply following these tangent lines. With a little more thought, we may use this same idea to numerically approximate the solutions of a differential equation.

> **Preview Activity 7.3.1.** Consider the initial value problem
>
> $$\frac{dy}{dt} = \frac{1}{2}(y + 1), \ y(0) = 0.$$
>
> a. Use the differential equation to find the slope of the tangent line to the solution $y(t)$ at $t = 0$. Then use the given initial value to find the equation of the tangent line at $t = 0$.
>
> b. Sketch the tangent line on the axes provided in Figure 7.3.1 on the interval $0 \le t \le 2$ and use it to approximate $y(2)$, the value of the solution at $t = 2$.
>
>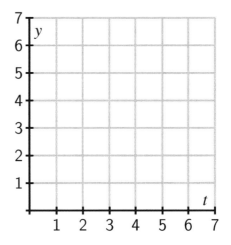
>
> **Figure 7.3.1:** Grid for plotting the tangent line.
>
> c. Assuming that your approximation for $y(2)$ is the actual value of $y(2)$, use the differential equation to find the slope of the tangent line to $y(t)$ at $t = 2$. Then,

write the equation of the tangent line at $t = 2$.

d. Add a sketch of this tangent line on the interval $2 \leq t \leq 4$ to your plot Figure 7.3.1; use this new tangent line to approximate $y(4)$, the value of the solution at $t = 4$.

e. Repeat the same step to find an approximation for $y(6)$.

7.3.1 Euler's Method

Preview Activity 7.3.1 demonstrates the essence of an algorithm known as Euler's[1] Method that generates a numerical approximation to the solution of an initial value problem. In this algorithm, we will approximate the solution by taking horizontal steps of a fixed size that we denote by Δt.

Before explaining the algorithm in detail, let's remember how we compute the slope of a line: the slope is the ratio of the vertical change to the horizontal change, as shown in Figure 7.3.2.

In other words, $m = \frac{\Delta y}{\Delta t}$. From another perspective, the vertical change is the product of the slope and the horizontal change, or

$$\Delta y = m \Delta t.$$

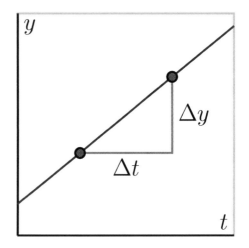

Figure 7.3.2: The role of slope in Euler's Method.

Now, suppose that we would like to solve the initial value problem

$$\frac{dy}{dt} = t - y, \ y(0) = 1.$$

While there is an algorithm by which we can find an algebraic formula for the solution to this initial value problem, and we can check that this solution is $y(t) = t - 1 + 2e^{-t}$, we are instead interested in generating an approximate solution by creating a sequence of points (t_i, y_i), where $y_i \approx y(t_i)$. For this first example, we choose $\Delta t = 0.2$.

Since we know that $y(0) = 1$, we will take the initial point to be $(t_0, y_0) = (0, 1)$ and move horizontally by $\Delta t = 0.2$ to the point (t_1, y_1). Therefore, $t_1 = t_0 + \Delta t = 0.2$. The differential

---

[1]"Euler" is pronounced "Oy-ler." Among other things, Euler is the mathematician credited with the famous number e; if you incorrectly pronounce his name "You-ler," you fail to appreciate his genius and legacy.

equation tells us that the slope of the tangent line at this point is

$$m = \frac{dy}{dt}\bigg|_{(0,1)} = 0 - 1 = -1.$$

Therefore, if we move along the tangent line by taking a horizontal step of size $\Delta t = 0.2$, we must also move vertically by

$$\Delta y = m\Delta t = -1 \cdot 0.2 = -0.2.$$

We then have the approximation $y(0.2) \approx y_1 = y_0 + \Delta y = 1 - 0.2 = 0.8$. At this point, we have executed one step of Euler's method, as seen graphically in Figure 7.3.3.

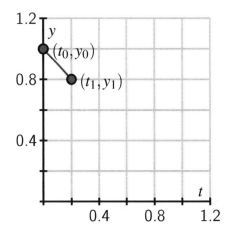

Figure 7.3.3: One step of Euler's method.

Now we repeat this process: at $(t_1, y_1) = (0.2, 0.8)$, the differential equation tells us that the slope is

$$m = \frac{dy}{dt}\bigg|_{(0.2,0.8)} = 0.2 - 0.8 = -0.6.$$

If we move forward horizontally by Δt to $t_2 = t_1 + \Delta = 0.4$, we must move vertically by

$$\Delta y = -0.6 \cdot 0.2 = -0.12.$$

We consequently arrive at $y_2 = y_1 + \Delta y = 0.8 - 0.12 = 0.68$, which gives $y(0.2) \approx 0.68$. Now we have completed the second step of Euler's method, as shown in Figure 7.3.4.

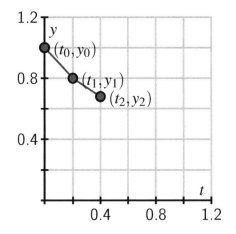

Figure 7.3.4: Two steps of Euler's method.

If we continue in this way, we may generate the points (t_i, y_i) shown in Figure 7.3.5. In situations where we are able to find a formula for the actual solution $y(t)$, we can graph $y(t)$ to compare it to the points generated by Euler's method, as shown in Figure 7.3.6.

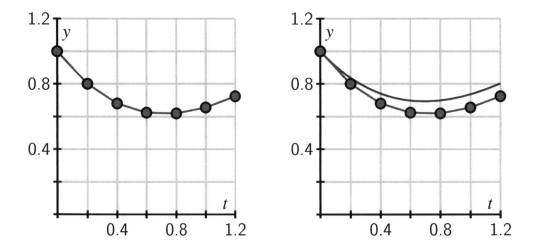

Figure 7.3.5: The points and piecewise linear approximate solution generated by Euler's method.

Figure 7.3.6: The approximate solution compared to the exact solution (shown in blue).

Because we need to generate a large number of points (t_i, y_i), it is convenient to organize the implementation of Euler's method in a table as shown. We begin with the given initial data.

t_i	y_i	dy/dt	Δy
0.0000	1.0000		

From here, we compute the slope of the tangent line $m = dy/dt$ using the formula for dy/dt from the differential equation, and then we find Δy, the change in y, using the rule $\Delta y = m\Delta t$.

t_i	y_i	dy/dt	Δy
0.0000	1.0000	−1.0000	−0.2000

Next, we increase t_i by Δt and y_i by Δy to get

t_i	y_i	dy/dt	Δy
0.0000	1.0000	−1.0000	−0.2000
0.2000	0.8000		

and then we simply continue the process for however many steps we decide, eventually generating a table like Table 7.3.7.

t_i	y_i	dy/dt	Δy
0.0000	1.0000	−1.0000	−0.2000
0.2000	0.8000	−0.6000	−0.1200
0.4000	0.6800	−0.2800	−0.0560
0.6000	0.6240	−0.0240	−0.0048
0.8000	0.6192	0.1808	0.0362
1.0000	0.6554	0.3446	0.0689
1.2000	0.7243	0.4757	0.0951

Table 7.3.7: Euler's method for 6 steps with $\Delta t = 0.2$.

Activity 7.3.2. Consider the initial value problem

$$\frac{dy}{dt} = 2t - 1, \ y(0) = 0$$

a. Use Euler's method with $\Delta t = 0.2$ to approximate the solution at $t_i = 0.2, 0.4, 0.6, 0.8$, and 1.0. Record your work in the following table, and sketch the points (t_i, y_i) on the axes provided.

t_i	y_i	dy/dt	Δy
0.0000	0.0000		
0.2000			
0.4000			
0.6000			
0.8000			
1.0000			

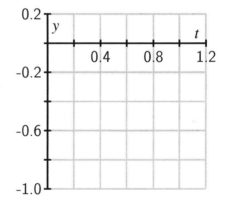

Table 7.3.8: Table for recording results of Euler's method.

Figure 7.3.9: Grid for plotting points generated by Euler's method.

b. Find the exact solution to the original initial value problem and use this function to find the error in your approximation at each one of the points t_i.

c. Explain why the value y_5 generated by Euler's method for this initial value problem produces the same value as a left Riemann sum for the definite integral $\int_0^1 (2t - 1) \, dt$.

d. How would your computations differ if the initial value was $y(0) = 1$? What does this mean about different solutions to this differential equation?

Activity 7.3.3. Consider the differential equation $\frac{dy}{dt} = 6y - y^2$.

a. Sketch the slope field for this differential equation on the axes provided in Figure 7.3.10.

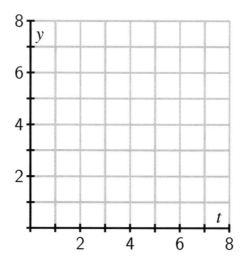

Figure 7.3.10: Grid for plotting the slope field of the given differential equation.

b. Identify any equilibrium solutions and determine whether they are stable or unstable.

c. What is the long-term behavior of the solution that satisfies the initial value $y(0) = 1$?

d. Using the initial value $y(0) = 1$, use Euler's method with $\Delta t = 0.2$ to approximate the solution at $t_i = 0.2, 0.4, 0.6, 0.8$, and 1.0. Record your results in Table 7.3.11 and sketch the corresponding points (t_i, y_i) on the axes provided in Figure 7.3.12. Note the different horizontal scale on the axes in Figure 7.3.12 compared to Figure 7.3.10.

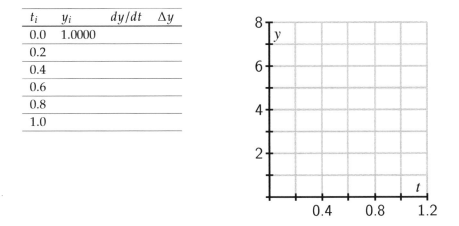

t_i	y_i	dy/dt	Δy
0.0	1.0000		
0.2			
0.4			
0.6			
0.8			
1.0			

Table 7.3.11: Table for recording results of Euler's method with $\Delta t = 0.2$.

Figure 7.3.12: Axes for plotting the results of Euler's method.

e. What happens if we apply Euler's method to approximate the solution with $y(0) = 6$?

7.3.2 The error in Euler's method

Since we are approximating the solutions to an initial value problem using tangent lines, we should expect that the error in the approximation will be less when the step size is smaller. To explore this observation quantitatively, let's consider the initial value problem

$$\frac{dy}{dt} = y, \ y(0) = 1$$

whose solution we can easily find.

Consider the question posed by this initial value problem: "what function do we know that is the same as its own derivative and has value 1 when $t = 0$?" It is not hard to see that the solution is $y(t) = e^t$. We now apply Euler's method to approximate $y(1) = e$ using several values of Δt. These approximations will be denoted by $E_{\Delta t}$, and these estimates provide us a way to see how accurate Euler's Method is.

To begin, we apply Euler's method with a step size of $\Delta t = 0.2$. In that case, we find that $y(1) \approx E_{0.2} = 2.4883$. The error is therefore

$$y(1) - E_{0.2} = e - 2.4883 \approx 0.2300.$$

Repeatedly halving Δt gives the following results, expressed in both tabular and graphical form.

Δt	$E_{\Delta t}$	Error
0.200	2.4883	0.2300
0.100	2.5937	0.1245
0.050	2.6533	0.0650
0.025	2.6851	0.0332

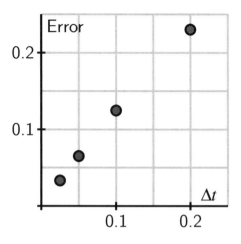

Table 7.3.13: Errors that correspond to different Δt values.

Figure 7.3.14: A plot of the error as a function of Δt.

Notice, both numerically and graphically, that the error is roughly halved when Δt is halved. This example illustrates the following general principle.

If Euler's method is to approximate the solution to an initial value problem at a point $\bar{t}$, then the error is proportional to Δt. That is,

$$y(\bar{t}) - E_{\Delta t} \approx K\Delta t$$

for some constant of proportionality K.

Summary

- Euler's method is an algorithm for approximating the solution to an initial value problem by following the tangent lines while we take horizontal steps across the t-axis.

- If we wish to approximate $y(\bar{t})$ for some fixed $\bar{t}$ by taking horizontal steps of size Δt, then the error in our approximation is proportional to Δt.

Exercises

1. Consider the differential equation $y' = -x - y$.
Use Euler's method with $\Delta x = 0.1$ to estimate y when $x = 1.4$ for the solution curve satisfying

$y(1) = 1$: Euler's approximation gives $y(1.4) \approx$ [   ]
Use Euler's method with $\Delta x = 0.1$ to estimate y when $x = 2.4$ for the solution curve satisfy-

ing $y(1) = 0$: Euler's approximation gives $y(2.4) \approx$ [   ]

2. Consider the solution of the differential equation $y' = -3y$ passing through $y(0) = 1.5$.

A. Sketch the slope field for this differential equation, and sketch the solution passing through the point $(0, 1.5)$.

B. Use Euler's method with step size $\Delta x = 0.2$ to estimate the solution at $x = 0.2, 0.4, \ldots, 1$, using these to fill in the following table. *(Be sure not to round your answers at each step!)*

$x =$	0	0.2	0.4	0.6	0.8	1.0
$y \approx$	1.5					

C. Plot your estimated solution on your slope field. Compare the solution and the slope field. Is the estimated solution an over or under estimate for the actual solution?

D. Check that $y = 1.5e^{-3x}$ is a solution to $y' = -3y$ with $y(0) = 1.5$.

3. Use Euler's method to solve

$$\frac{dB}{dt} = 0.04B$$

with initial value $B = 1100$ when $t = 0$.

A. $\Delta t = 1$ and 1 step: $B(1) \approx$

B. $\Delta t = 0.5$ and 2 steps: $B(1) \approx$

C. $\Delta t = 0.25$ and 4 steps: $B(1) \approx$

D. Suppose B is the balance in a bank account earning interest. Be sure that you can explain why the result of your calculation in part (a) is equivalent to compounding the interest once a year instead of continuously. Then interpret the result of your calculations in parts (b) and (c) in terms of compound interest.

4. Newton's Law of Cooling says that the rate at which an object, such as a cup of coffee, cools is proportional to the difference in the object's temperature and room temperature. If $T(t)$ is the object's temperature and T_r is room temperature, this law is expressed at

$$\frac{dT}{dt} = -k(T - T_r),$$

where k is a constant of proportionality. In this problem, temperature is measured in degrees Fahrenheit and time in minutes.

a. Two calculus students, Alice and Bob, enter a $70°$ classroom at the same time. Each has a cup of coffee that is $100°$. The differential equation for Alice has a constant of proportionality $k = 0.5$, while the constant of proportionality for Bob is $k = 0.1$. What is the initial rate of change for Alice's coffee? What is the initial rate of change for Bob's coffee?

b. What feature of Alice's and Bob's cups of coffee could explain this difference?

c. As the heating unit turns on and off in the room, the temperature in the room is

$$T_r = 70 + 10 \sin t.$$

Implement Euler's method with a step size of $\Delta t = 0.1$ to approximate the temperature of Alice's coffee over the time interval $0 \leq t \leq 50$. This will most easily be performed using a spreadsheet such as *Excel*. Graph the temperature of her coffee and room temperature over this interval.

d. In the same way, implement Euler's method to approximate the temperature of Bob's coffee over the same time interval. Graph the temperature of his coffee and room temperature over the interval.

e. Explain the similarities and differences that you see in the behavior of Alice's and Bob's cups of coffee.

5. We have seen that the error in approximating the solution to an initial value problem is proportional to Δt. That is, if $E_{\Delta t}$ is the Euler's method approximation to the solution to an initial value problem at $\bar{t}$, then

$$y(\bar{t}) - E_{\Delta t} \approx K\Delta t$$

for some constant of proportionality K.

In this problem, we will see how to use this fact to improve our estimates, using an idea called *accelerated convergence*.

a. We will create a new approximation by assuming the error is *exactly* proportional to Δt, according to the formula

$$y(\bar{t}) - E_{\Delta t} = K\Delta t.$$

Using our earlier results from the initial value problem $dy/dt = y$ and $y(0) = 1$ with $\Delta t = 0.2$ and $\Delta t = 0.1$, we have

$$y(1) - 2.4883 = 0.2K$$
$$y(1) - 2.5937 = 0.1K.$$

This is a system of two linear equations in the unknowns $y(1)$ and K. Solve this system to find a new approximation for $y(1)$. (You may remember that the exact value is $y(1) = e = 2.71828\ldots$.)

b. Use the other data, $E_{0.05} = 2.6533$ and $E_{0.025} = 2.6851$ to do similar work as in (a) to obtain another approximation. Which gives the better approximation? Why do you think this is?

c. Let's now study the initial value problem

$$\frac{dy}{dt} = t - y, \ y(0) = 0.$$

Approximate $y(0.3)$ by applying Euler's method to find approximations $E_{0.1}$ and $E_{0.05}$. Now use the idea of accelerated convergence to obtain a better approximation. (For the sake of comparison, you want to note that the actual value is $y(0.3) = 0.0408$.)

6. In this problem, we'll modify Euler's method to obtain better approximations to solutions of initial value problems. This method is called the *Improved Euler's method.*

In Euler's method, we walk across an interval of width Δt using the slope obtained from the differential equation at the left endpoint of the interval. Of course, the slope of the solution will most likely change over this interval. We can improve our approximation by trying to incorporate the change in the slope over the interval.

Let's again consider the initial value problem $dy/dt = y$ and $y(0) = 1$, which we will approximate using steps of width $\Delta t = 0.2$. Our first interval is therefore $0 \le t \le 0.2$. At $t = 0$, the differential equation tells us that the slope is 1, and the approximation we obtain from Euler's method is that $y(0.2) \approx y_1 = 1 + 1(0.2) = 1.2$.

This gives us some idea for how the slope has changed over the interval $0 \le t \le 0.2$. We know the slope at $t = 0$ is 1, while the slope at $t = 0.2$ is 1.2, trusting in the Euler's method approximation. We will therefore refine our estimate of the initial slope to be the average of these two slopes; that is, we will estimate the slope to be $(1 + 1.2)/2 = 1.1$. This gives the new approximation $y(1) = y_1 = 1 + 1.1(0.2) = 1.22$.

The first few steps look like what is found in Table 7.3.15.

t_i	y_i	Slope at (t_{i+1}, y_{i+1})	Average slope
0.0	1.0000	1.2000	1.1000
0.2	1.2200	1.4640	1.3420
0.4	1.4884	1.7861	1.6372
$\vdots$	$\vdots$	$\vdots$	$\vdots$

Table 7.3.15: The first several steps of the improved Euler's method

a. Continue with this method to obtain an approximation for $y(1) = e$.

b. Repeat this method with $\Delta t = 0.1$ to obtain a better approximation for $y(1)$.

c. We saw that the error in Euler's method is proportional to Δt. Using your results from parts (a) and (b), what power of Δt appears to be proportional to the error in the Improved Euler's Method?

7.4 Separable differential equations

Motivating Questions

- What is a separable differential equation?
- How can we find solutions to a separable differential equation?
- Are some of the differential equations that arise in applications separable?

In Sections 7.2 and Section 7.3, we have seen several ways to approximate the solution to an initial value problem. Given the frequency with which differential equations arise in the world around us, we would like to have some techniques for finding explicit algebraic solutions of certain initial value problems. In this section, we focus on a particular class of differential equations (called *separable*) and develop a method for finding algebraic formulas for solutions to these equations.

A *separable differential equation* is a differential equation whose algebraic structure permits the variables present to be separated in a particular way. For instance, consider the equation

$$\frac{dy}{dt} = ty.$$

We would like to separate the variables t and y so that all occurrences of t appear on the right-hand side, and all occurrences of y appears on the left and multiply dy/dt. We may do this in the preceding differential equation by dividing both sides by y:

$$\frac{1}{y}\frac{dy}{dt} = t.$$

Note particularly that when we attempt to separate the variables in a differential equation, we require that the left-hand side be a product in which the derivative dy/dt is one term.

Not every differential equation is separable. For example, if we consider the equation

$$\frac{dy}{dt} = t - y,$$

it may seem natural to separate it by writing

$$y + \frac{dy}{dt} = t.$$

As we will see, this will not be helpful since the left-hand side is not a product of a function of y with $\frac{dy}{dt}$.

Preview Activity 7.4.1. In this preview activity, we explore whether certain differential equations are separable or not, and then revisit some key ideas from earlier work in integral calculus.

a. Which of the following differential equations are separable? If the equation is separable, write the equation in the revised form $g(y)\frac{dy}{dt} = h(t)$.

    i. $\dfrac{dy}{dt} = -3y$.

    ii. $\dfrac{dy}{dt} = ty - y$.

    iii. $\dfrac{dy}{dt} = t + 1$.

    iv. $\dfrac{dy}{dt} = t^2 - y^2$.

b. Explain why any autonomous differential equation is guaranteed to be separable.

c. Why do we include the term "+C" in the expression

$$\int x \, dx = \frac{x^2}{2} + C?$$

d. Suppose we know that a certain function f satisfies the equation

$$\int f'(x) \, dx = \int x \, dx.$$

What can you conclude about f?

7.4.1 Solving separable differential equations

Before we discuss a general approach to solving a separable differential equation, it is instructive to consider an example.

Example 7.4.1. Find all functions y that are solutions to the differential equation

$$\frac{dy}{dt} = \frac{t}{y^2}.$$

Solution. We begin by separating the variables and writing

$$y^2 \frac{dy}{dt} = t.$$

Integrating both sides of the equation with respect to the independent variable t shows that

$$\int y^2 \frac{dy}{dt}\, dt = \int t\, dt.$$

Next, we notice that the left-hand side allows us to change the variable of antidifferentiation[1] from t to y. In particular, $dy = \frac{dy}{dt}\, dt$, so we now have

$$\int y^2\, dy = \int t\, dt.$$

This most recent equation says that two families of antiderivatives are equal to one another. Therefore, when we find representative antiderivatives of both sides, we know they must differ by arbitrary constant C. Antidifferentiating and including the integration constant C on the right, we find that

$$\frac{y^3}{3} = \frac{t^2}{2} + C.$$

Again, note that it is not necessary to include an arbitrary constant on both sides of the equation; we know that $y^3/3$ and $t^2/2$ are in the same family of antiderivatives and must therefore differ by a single constant.

Finally, we may now solve the last equation above for y as a function of t, which gives

$$y(t) = \sqrt[3]{\frac{3}{2} t^2 + 3C}.$$

Of course, the term $3C$ on the right-hand side represents 3 times an unknown constant. It is, therefore, still an unknown constant, which we will rewrite as C. We thus conclude that the funtion

$$y(t) = \sqrt[3]{\frac{3}{2} t^2 + C}$$

is a solution to the original differential equation for any value of C.

Notice that because this solution depends on the arbitrary constant C, we have found an infinite family of solutions. This makes sense because we expect to find a unique solution that corresponds to any given initial value.

For example, if we want to solve the initial value problem

$$\frac{dy}{dt} = \frac{t}{y^2}, \quad y(0) = 2,$$

we know that the solution has the form $y(t) = \sqrt[3]{\frac{3}{2} t^2 + C}$ for some constant C. We therefore must find the appropriate value for C that gives the initial value $y(0) = 2$. Hence,

$$2 = y(0) = \sqrt[3]{\frac{3}{2} 0^2 + C} = \sqrt[3]{C},$$

---

[1]This is why we required that the left-hand side be written as a product in which dy/dt is one of the terms.

which shows that $C = 2^3 = 8$. The solution to the initial value problem is then

$$y(t) = \sqrt[3]{\frac{3}{2} t^2 + 8}.$$

The strategy of Example 7.4.1 may be applied to any differential equation of the form $\frac{dy}{dt} = g(y) \cdot h(t)$, and any differential equation of this form is said to be *separable*. We work to solve a separable differential equation by writing

$$\frac{1}{g(y)} \frac{dy}{dt} = h(t),$$

and then integrating both sides with respect to t. After integrating, we strive to solve algebraically for y in order to write y as a function of t.

We consider one more example before doing further exploration in some activities.

Example 7.4.2. Solve the differential equation

$$\frac{dy}{dt} = 3y.$$

Solution.   Following the same strategy as in Example 7.4.1, we have

$$\frac{1}{y} \frac{dy}{dt} = 3.$$

Integrating both sides with respect to t,

$$\int \frac{1}{y} \frac{dy}{dt} \, dt = \int 3 \, dt,$$

and thus

$$\int \frac{1}{y} \, dy = \int 3 \, dt.$$

Antidifferentiating and including the integration constant, we find that

$$\ln |y| = 3t + C.$$

Finally, we need to solve for y. Here, one point deserves careful attention. By the definition of the natural logarithm function, it follows that

$$|y| = e^{3t+C} = e^{3t} e^{C}.$$

Since C is an unknown constant, e^{C} is as well, though we do know that it is positive (because e^x is positive for any x). When we remove the absolute value in order to solve for y, however, this constant may be either positive or negative. We will denote this updated constant (that accounts for a possible $+$ or $-$) by C to obtain

$$y(t) = Ce^{3t}.$$

There is one more slightly technical point to make. Notice that $y = 0$ is an equilibrium solution to this differential equation. In solving the equation above, we begin by dividing both sides by y, which is not allowed if $y = 0$. To be perfectly careful, therefore, we will typically consider the equilibrium solutions separately. In this case, notice that the final form of our solution captures the equilibrium solution by allowing $C = 0$.

Activity 7.4.2. Suppose that the population of a town is growing continuously at an annual rate of 3% per year.

a. Let $P(t)$ be the population of the town in year t. Write a differential equation that describes the annual growth rate.

b. Find the solutions of this differential equation.

c. If you know that the town's population in year 0 is 10,000, find the population $P(t)$.

d. How long does it take for the population to double? This time is called the *doubling time*.

e. Working more generally, find the doubling time if the annual growth rate is k times the population.

Activity 7.4.3. Suppose that a cup of coffee is initially at a temperature of 105° F and is placed in a 75° F room. Newton's law of cooling says that

$$\frac{dT}{dt} = -k(T - 75),$$

where k is a constant of proportionality.

a. Suppose you measure that the coffee is cooling at one degree per minute at the time the coffee is brought into the room. Use the differential equation to determine the value of the constant k.

b. Find all the solutions of this differential equation.

c. What happens to all the solutions as $t \to \infty$? Explain how this agrees with your intuition.

d. What is the temperature of the cup of coffee after 20 minutes?

e. How long does it take for the coffee to cool to 80°?

Activity 7.4.4. Solve each of the following differential equations or initial value problems.

a. $\frac{dy}{dt} - (2 - t)y = 2 - t$

b. $\frac{1}{t}\frac{dy}{dt} = e^{t^2-2y}$

c. $y' = 2y + 2, \ y(0) = 2$

d. $y' = 2y^2, \ y(-1) = 2$

e. $\frac{dy}{dt} = \frac{-2ty}{t^2+1}, \ y(0) = 4$

Summary

- A separable differential equation is one that may be rewritten with all occurrences of the dependent variable multiplying the derivative and all occurrences of the independent variable on the other side of the equation.

- We may find the solutions to certain separable differential equations by separating variables, integrating with respect to t, and ultimately solving the resulting algebraic equation for y.

- This technique allows us to solve many important differential equations that arise in the world around us. For instance, questions of growth and decay and Newton's Law of Cooling give rise to separable differential equations. Later, we will learn in Section 7.6 that the important logistic differential equation is also separable.

Exercises

1. Find the equation of the solution to $\dfrac{dy}{dx} = x^2y$ through the point $(x, y) = (1, 3)$.

| help (equations)

2. Find the solution to the differential equation

$$\frac{dy}{dt} = 0.5(y - 250)$$

if $y = 70$ when $t = 0$.

$y = $

3. Find the solution to the differential equation

$$\frac{dy}{dt} = y^2(5 + t),$$

$y = 5$ when $t = 1$.

$y = $

4. Solve the separable differential equation for u

$$\frac{du}{dt} = e^{2u+8t}.$$

Use the following initial condition: $u(0) = 13$.

$u = $ [   ].

5. Find an equation of the curve that satisfies

$$\frac{dy}{dx} = 45yx^4$$

and whose y-intercept is 5.

$y(x) = $ [   ].

6. The mass of a radioactive sample decays at a rate that is proportional to its mass.

a. Express this fact as a differential equation for the mass $M(t)$ using k for the constant of proportionality.

b. If the initial mass is M_0, find an expression for the mass $M(t)$.

c. The *half-life* of the sample is the amount of time required for half of the mass to decay. Knowing that the half-life of Carbon-14 is 5730 years, find the value of k for a sample of Carbon-14.

d. How long does it take for a sample of Carbon-14 to be reduced to one-quarter its original mass?

e. Carbon-14 naturally occurs in our environment; any living organism takes in Carbon-14 when it eats and breathes. Upon dying, however, the organism no longer takes in Carbon-14. Suppose that you find remnants of a pre-historic firepit. By analyzing the charred wood in the pit, you determine that the amount of Carbon-14 is only 30% of the amount in living trees. Estimate the age of the firepit.[2]

7. Consider the initial value problem

$$\frac{dy}{dt} = -\frac{t}{y}, \quad y(0) = 8$$

a. Find the solution of the initial value problem and sketch its graph.

b. For what values of t is the solution defined?

c. What is the value of y at the last time that the solution is defined?

d. By looking at the differential equation, explain why we should not expect to find solutions with the value of y you noted in (c).

8. Suppose that a cylindrical water tank with a hole in the bottom is filled with water. The water, of course, will leak out and the height of the water will decrease. Let $h(t)$ denote the height of the water. A physical principle called *Torricelli's Law* implies that the height decreases at a rate proportional to the square root of the height.

a. Express this fact using k as the constant of proportionality.

---

[2]This approach is the basic idea behind radiocarbon dating.

b. Suppose you have two tanks, one with $k = -1$ and another with $k = -10$. What physical differences would you expect to find?

c. Suppose you have a tank for which the height decreases at 20 inches per minute when the water is filled to a depth of 100 inches. Find the value of k.

d. Solve the initial value problem for the tank in part (c), and graph the solution you determine.

e. How long does it take for the water to run out of the tank?

f. Is the solution that you found valid for all time t? If so, explain how you know this. If not, explain why not.

9. The *Gompertz equation* is a model that is used to describe the growth of certain populations. Suppose that $P(t)$ is the population of some organism and that

$$\frac{dP}{dt} = -P \ln\left(\frac{P}{3}\right) = -P(\ln P - \ln 3).$$

a. Sketch a slope field for $P(t)$ over the range $0 \le P \le 6$.

b. Identify any equilibrium solutions and determine whether they are stable or unstable.

c. Find the population $P(t)$ assuming that $P(0) = 1$ and sketch its graph. What happens to $P(t)$ after a very long time?

d. Find the population $P(t)$ assuming that $P(0) = 6$ and sketch its graph. What happens to $P(t)$ after a very long time?

e. Verify that the long-term behavior of your solutions agrees with what you predicted by looking at the slope field.

7.5 Modeling with differential equations

Motivating Questions

- How can we use differential equations to describe phenomena in the world around us?

- How can we use differential equations to better understand these phenomena?

In our work to date, we have seen several ways that differential equations arise in the natural world, from the growth of a population to the temperature of a cup of coffee. In this section, we will look more closely at how differential equations give us a natural way to describe various phenoma. As we'll see, the key is to focus on understanding the different factors that cause a quantity to change.

> **Preview Activity 7.5.1.** Any time that the rate of change of a quantity is related to the amount of a quantity, a differential equation naturally arises. In the following two problems, we see two such scenarios; for each, we want to develop a differential equation whose solution is the quantity of interest.
>
> a. Suppose you have a bank account in which money grows at an annual rate of 3%.
>
>   - If you have $10,000 in the account, at what rate is your money growing?
>
>   - Suppose that you are also withdrawing money from the account at $1,000 per year. What is the rate of change in the amount of money in the account? What are the units on this rate of change?
>
> b. Suppose that a water tank holds 100 gallons and that a salty solution, which contains 20 grams of salt in every gallon, enters the tank at 2 gallons per minute.
>
>   - How much salt enters the tank each minute?
>
>   - Suppose that initially there are 300 grams of salt in the tank. How much salt is in each gallon at this point in time?
>
>   - Finally, suppose that evenly mixed solution is pumped out of the tank at the rate of 2 gallons per minute. How much salt leaves the tank each minute?
>
>   - What is the total rate of change in the amount of salt in the tank?

7.5.1 Developing a differential equation

Preview activity 7.5.1 demonstrates the kind of thinking we will be doing in this section. In each of the two examples we considered, there is a quantity, such as the amount of money in the bank account or the amount of salt in the tank, that is changing due to several factors.

The governing differential equation results from the total rate of change being the difference between the rate of increase and the rate of decrease.

Example 7.5.1. In the Great Lakes region, rivers flowing into the lakes carry a great deal of pollution in the form of small pieces of plastic averaging 1 millimeter in diameter. In order to understand how the amount of plastic in Lake Michigan is changing, construct a model for how this type pollution has built up in the lake.

Solution. First, some basic facts about Lake Michigan.

- The volume of the lake is $5 \cdot 10^{12}$ cubic meters.

- Water flows into the lake at a rate of $5 \cdot 10^{10}$ cubic meters per year. It flows out of the lake at the same rate.

- Each cubic meter flowing into the lake contains roughly $3 \cdot 10^{-8}$ cubic meters of plastic pollution.

Let's denote the amount of pollution in the lake by $P(t)$, where P is measured in cubic meters of plastic and t in years. Our goal is to describe the rate of change of this function; in other words, we want to develop a differential equation describing $P(t)$.

First, we will measure how $P(t)$ increases due to pollution flowing into the lake. We know that $5 \cdot 10^{10}$ cubic meters of water enters the lake every year and each cubic meter of water contains $3 \cdot 10^{-8}$ cubic meters of pollution. Therefore, pollution enters the lake at the rate of

$$\left(5 \cdot 10^{10} \frac{m^3 \text{ water}}{\text{year}}\right)\left(3 \cdot 10^{-8} \frac{m^3 \text{ plastic}}{m^3 \text{ water}}\right) = 1.5 \cdot 10^3 \text{cubic m of plastic per year.}$$

Second, we will measure how $P(t)$ decreases due to pollution flowing out of the lake. If the total amount of pollution is P cubic meters and the volume of Lake Michigan is $5 \cdot 10^{12}$ cubic meters, then the concentration of plastic pollution in Lake Michigan is

$$\frac{P}{5 \cdot 10^{12}} \text{cubic m of plastic per cubic m of water.}$$

Since $5 \cdot 10^{10}$ cubic meters of water flow out each year[1], then the plastic pollution leaves the lake at the rate of

$$\left(\frac{P}{5 \cdot 10^{12}} \frac{m^3 \text{ plastic}}{m^3 \text{ water}}\right)\left(5 \cdot 10^{10} \frac{m^3 \text{ water}}{\text{year}}\right) = \frac{P}{100} \text{cubic m of plastic per year.}$$

The total rate of change of P is thus the difference between the rate at which pollution enters the lake minus the rate at which pollution leaves the lake; that is,

$$\frac{dP}{dt} = 1.5 \cdot 10^3 - \frac{P}{100}$$

$$= \frac{1}{100}(1.5 \cdot 10^5 - P).$$

---

[1]and we assume that each cubic meter of water that flows out carries with it the plastic pollution it contains

We have now found a differential equation that describes the rate at which the amount of pollution is changing. To better understand the behavior of $P(t)$, we now apply some of the techniques we have recently developed.

Since this is an autonomous differential equation, we can sketch dP/dt as a function of P and then construct a slope field, as shown in Figure 7.5.2 and Figure 7.5.3.

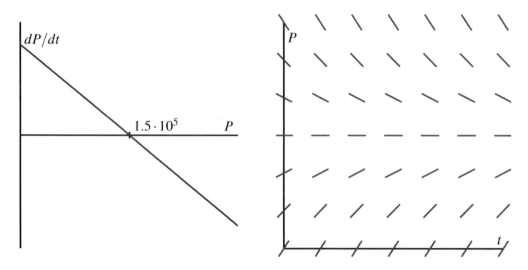

Figure 7.5.2: Plot of $\frac{dP}{dt}$ vs. P for $\frac{dP}{dt} = \frac{1}{100}(1.5 \cdot 10^5 - P)$.

Figure 7.5.3: Plot of the slope field for $\frac{dP}{dt} = \frac{1}{100}(1.5 \cdot 10^5 - P)$.

These plots both show that $P = 1.5 \cdot 10^5$ is a stable equilibrium. Therefore, we should expect that the amount of pollution in Lake Michigan will stabilize near $1.5 \cdot 10^5$ cubic meters of pollution.

Next, assuming that there is initially no pollution in the lake, we will solve the initial value problem

$$\frac{dP}{dt} = \frac{1}{100}(1.5 \cdot 10^5 - P), \ P(0) = 0.$$

Separating variables, we find that

$$\frac{1}{1.5 \cdot 10^5 - P}\frac{dP}{dt} = \frac{1}{100}.$$

Integrating with respect to t, we have

$$\int \frac{1}{1.5 \cdot 10^5 - P}\frac{dP}{dt}\, dt = \int \frac{1}{100}\, dt,$$

and thus changing variables on the left and antidifferentiating on both sides, we find that

$$\int \frac{dP}{1.5 \cdot 10^5 - P} = \int \frac{1}{100}\, dt$$

$$-\ln|1.5 \cdot 10^5 - P| = \frac{1}{100}t + C$$

Finally, multiplying both sides by -1 and using the definition of the logarithm, we find that

$$1.5 \cdot 10^5 - P = Ce^{-t/100}. \tag{7.5.1}$$

This is a good time to determine the constant C. Since $P = 0$ when $t = 0$, we have

$$1.5 \cdot 10^5 - 0 = Ce^0 = C.$$

In other words, $C = 1.5 \cdot 10^5$.

Using this value of C in Equation (7.5.1) and solving for P, we arrive at the solution

$$P(t) = 1.5 \cdot 10^5(1 - e^{-t/100}).$$

Superimposing the graph of P on the slope field we saw in Figure 7.5.3, we see, as shown in Figure 7.5.4

We see that, as expected, the amount of plastic pollution stabilizes around $1.5 \cdot 10^5$ cubic meters.

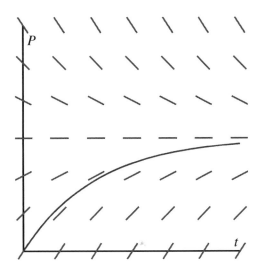

Figure 7.5.4: The solution $P(t)$ and the slope field for the differential equation $\frac{dP}{dt} = \frac{1}{100}(1.5 \cdot 10^5 - P)$.

There are many important lessons to learn from Example 7.5.1. Foremost is how we can develop a differential equation by thinking about the "total rate = rate in - rate out" model. In addition, we note how we can bring together all of our available understanding (plotting $\frac{dP}{dt}$ vs. P, creating a slope field, solving the differential equation) to see how the differential equation describes the behavior of a changing quantity.

Of course, we can also explore what happens when certain aspects of the problem change. For instance, let's suppose we are at a time when the plastic pollution entering Lake Michigan has stabilized at $1.5 \cdot 10^5$ cubic meters, and that new legislation is passed to prevent this type of pollution entering the lake. So, there is no longer any inflow of plastic pollution to the lake. How does the amount of plastic pollution in Lake Michigan now change? For example, how long does it take for the amount of plastic pollution in the lake to halve?

Restarting the problem at time $t = 0$, we now have the modified initial value problem

$$\frac{dP}{dt} = -\frac{1}{100}P, \; P(0) = 1.5 \cdot 10^5.$$

It is a straightforward and familiar exercise to find that the solution to this equation is $P(t) = 1.5 \cdot 10^5 e^{-t/100}$. The time that it takes for half of the pollution to flow out of the lake is given by T where $P(T) = 0.75 \cdot 10^5$. Thus, we must solve the equation

$$0.75 \cdot 10^5 = 1.5 \cdot 10^5 e^{-T/100},$$

or

$$\frac{1}{2} = e^{-T/100}.$$

It follows that

$$T = -100 \ln\left(\frac{1}{2}\right) \approx 69.3 \text{years.}$$

In the upcoming activities, we explore some other natural settings in which differential equations model changing quantities.

Activity 7.5.2. Suppose you have a bank account that grows by 5% every year. Let $A(t)$ be the amount of money in the account in year t.

a. What is the rate of change of A with respect to t?

b. Suppose that you are also withdrawing $10,000 per year. Write a differential equation that expresses the total rate of change of A.

c. Sketch a slope field for this differential equation, find any equilibrium solutions, and identify them as either stable or unstable. Write a sentence or two that describes the significance of the stability of the equilibrium solution.

d. Suppose that you initially deposit $100,000 into the account. How long does it take for you to deplete the account?

e. What is the smallest amount of money you would need to have in the account to guarantee that you never deplete the money in the account?

f. If your initial deposit is $300,000, how much could you withdraw every year without depleting the account?

Activity 7.5.3. A dose of morphine is absorbed from the bloodstream of a patient at a rate proportional to the amount in the bloodstream.

a. Write a differential equation for $M(t)$, the amount of morphine in the patient's bloodstream, using k as the constant proportionality.

b. Assuming that the initial dose of morphine is M_0, solve the initial value problem to find $M(t)$. Use the fact that the half-life for the absorption of morphine is two hours to find the constant k.

c. Suppose that a patient is given morphine intravenously at the rate of 3 milligrams per hour. Write a differential equation that combines the intravenous administration of morphine with the body's natural absorption.

d. Find any equilibrium solutions and determine their stability.

e. Assuming that there is initially no morphine in the patient's bloodstream, solve the initial value problem to determine $M(t)$. What happens to $M(t)$ after a very long time?

f. To what rate should a doctor reduce the intravenous rate so that there is eventually 7 milligrams of morphine in the patient's bloodstream?

Summary

- Differential equations arise in a situation when we understand how various factors cause a quantity to change.

- We may use the tools we have developed so far—slope fields, Euler's methods, and our method for solving separable equations—to understand a quantity described by a differential equation.

Exercises

1. A tank contains 2860 L of pure water. A solution that contains 0.04 kg of sugar per liter enters the tank at the rate 6 L/min. The solution is mixed and drains from the tank at the same rate.
(a) How much sugar is in the tank at the beginning?

$y(0) = \boxed{}$ (include units)
(b) With S representing the amount of sugar (in kg) at time t (in minutes) write a differential equation which models this situation.

$S' = f(t, S) = \boxed{}$.
Note: Make sure you use a capital S, (and don't use S(t), it confuses the computer). Don't enter units for this function.
(c) Find the amount of sugar (in kg) after t minutes.

$S(t) = \boxed{}$ (function of t)
(d) Find the amount of the sugar after 78 minutes.

$S(78) = \boxed{}$ (include units)

2. A tank contains 50 kg of salt and 1000 L of water. A solution of a concentration 0.025 kg of salt per liter enters a tank at the rate 10 L/min. The solution is mixed and drains from the tank at the same rate.
(a) What is the concentration of our solution in the tank initially?

concentration = [   ] (kg/L)

(b) Find the amount of salt in the tank after 3 hours.

amount = [   ] (kg)

(c) Find the concentration of salt in the solution in the tank as time approaches infinity.

concentration = [   ] (kg/L)

3. A bacteria culture starts with 220 bacteria and grows at a rate proportional to its size. After 5 hours there will be 1100 bacteria.

(a) Express the population after t hours as a function of t.

population: [   ] (function of t)

(b) What will be the population after 5 hours?

(c) How long will it take for the population to reach 1510 ?

4. An unknown radioactive element decays into non-radioactive substances. In 320 days the radioactivity of a sample decreases by 58 percent.

(a) What is the half-life of the element?

half-life: [   ] (days)

(b) How long will it take for a sample of 100 mg to decay to 88 mg?

time needed: [   ] (days)

5. A young person with no initial capital invests k dollars per year in a retirement account at an annual rate of return 0.05. Assume that investments are made continuously and that the return is compounded continuously.

Determine a formula for the sum $S(t)$ – (this will involve the parameter k):

$S(t) =$ [   ]

What value of k will provide 1820000 dollars in 47 years?

$k =$ [   ]

6. Congratulations, you just won the lottery! In one option presented to you, you will be paid one million dollars a year for the next 25 years. You can deposit this money in an account that will earn 5% each year.

   a. Set up a differential equation that describes the rate of change in the amount of money in the account. Two factors cause the amount to grow—first, you are depositing one millon dollars per year and second, you are earning 5% interest.

   b. If there is no amount of money in the account when you open it, how much money will you have in the account after 25 years?

   c. The second option presented to you is to take a lump sum of 10 million dollars, which you will deposit into a similar account. How much money will you have in that account after 25 years?

   d. Do you prefer the first or second option? Explain your thinking.

   e. At what time does the amount of money in the account under the first option overtake the amount of money in the account under the second option?

7. When a skydiver jumps from a plane, gravity causes her downward velocity to increase at the rate of $g \approx 9.8$ meters per second squared. At the same time, wind resistance causes her velocity to decrease at a rate proportional to the velocity.

   a. Using k to represent the constant of proportionality, write a differential equation that describes the rate of change of the skydiver's velocity.

   b. Find any equilibrium solutions and decide whether they are stable or unstable. Your result should depend on k.

   c. Suppose that the initial velocity is zero. Find the velocity $v(t)$.

   d. A typical terminal velocity for a skydiver falling face down is 54 meters per second. What is the value of k for this skydiver?

   e. How long does it take to reach 50% of the terminal velocity?

8. During the first few years of life, the rate at which a baby gains weight is proportional to the reciprocal of its weight.

   a. Express this fact as a differential equation.

   b. Suppose that a baby weighs 8 pounds at birth and 9 pounds one month later. How much will he weigh at one year?

   c. Do you think this is a realistic model for a long time?

9. Suppose that you have a water tank that holds 100 gallons of water. A briny solution, which contains 20 grams of salt per gallon, enters the tank at the rate of 3 gallons per minute. At the same time, the solution is well mixed, and water is pumped out of the tank at the rate of 3 gallons per minute.

   a. Since 3 gallons enters the tank every minute and 3 gallons leaves every minute, what can you conclude about the volume of water in the tank.

   b. How many grams of salt enters the tank every minute?

   c. Suppose that $S(t)$ denotes the number of grams of salt in the tank in minute t. How many grams are there in each gallon in minute t?

   d. Since water leaves the tank at 3 gallons per minute, how many grams of salt leave the tank each minute?

   e. Write a differential equation that expresses the total rate of change of S.

   f. Identify any equilibrium solutions and determine whether they are stable or unstable.

   g. Suppose that there is initially no salt in the tank. Find the amount of salt $S(t)$ in minute t.

   h. What happens to $S(t)$ after a very long time? Explain how you could have predicted this only knowing how much salt there is in each gallon of the briny solution that enters the tank.

7.6 Population Growth and the Logistic Equation

Motivating Questions

- How can we use differential equations to realistically model the growth of a population?

- How can we assess the accuracy of our models?

The growth of the earth's population is one of the pressing issues of our time. Will the population continue to grow? Or will it perhaps level off at some point, and if so, when? In this section, we will look at two ways in which we may use differential equations to help us address questions such as these.

Before we begin, let's consider again two important differential equations that we have seen in earlier work this chapter.

Preview Activity 7.6.1. Recall that one model for population growth states that a population grows at a rate proportional to its size.

a. We begin with the differential equation

$$\frac{dP}{dt} = \frac{1}{2}P.$$

Sketch a slope field below as well as a few typical solutions on the axes provided.

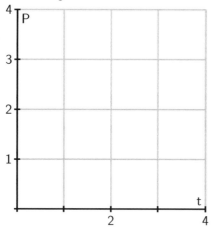

b. Find all equilibrium solutions of the equation $\frac{dP}{dt} = \frac{1}{2}P$ and classify them as stable or unstable.

c. If $P(0)$ is positive, describe the long-term behavior of the solution to $\frac{dP}{dt} = \frac{1}{2}P$.

d. Let's now consider a modified differential equation given by

$$\frac{dP}{dt} = \frac{1}{2}P(3 - P).$$

As before, sketch a slope field as well as a few typical solutions on the following

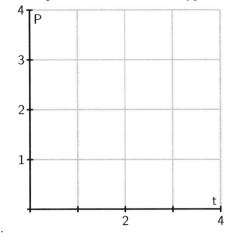

axes provided.

e. Find any equilibrium solutions and classify them as stable or unstable.

f. If $P(0)$ is positive, describe the long-term behavior of the solution.

7.6.1 The earth's population

We will now begin studying the earth's population. To get started, in Table 7.6.1 are some data for the earth's population in recent years that we will use in our investigations.

Year	1998	1999	2000	2001	2002	2005	2006	2007	2008	2009	2010
Pop (billions)	5.932	6.008	6.084	6.159	6.234	6.456	6.531	6.606	6.681	6.756	6.831

Table 7.6.1: Some recent population data for planet Earth.

Activity 7.6.2. Our first model will be based on the following assumption:

The rate of change of the population is proportional to the population.

On the face of it, this seems pretty reasonable. When there is a relatively small number of people, there will be fewer births and deaths so the rate of change will be small. When there is a larger number of people, there will be more births and deaths so we expect a larger rate of change.

If $P(t)$ is the population t years after the year 2000, we may express this assumption as

$$\frac{dP}{dt} = kP$$

where k is a constant of proportionality.

    a. Use the data in the table to estimate the derivative $P'(0)$ using a central difference. Assume that $t = 0$ corresponds to the year 2000.

    b. What is the population $P(0)$?

    c. Use these two facts to estimate the constant of proportionality k in the differential equation.

    d. Now that we know the value of k, we have the initial value problem

$$\frac{dP}{dt} = kP, \ P(0) = 6.084.$$

    Find the solution to this initial value problem.

    e. What does your solution predict for the population in the year 2010? Is this close to the actual population given in the table?

    f. When does your solution predict that the population will reach 12 billion?

    g. What does your solution predict for the population in the year 2500?

    h. Do you think this is a reasonable model for the earth's population? Why or why not? Explain your thinking using a couple of complete sentences.

Our work in Activity 7.6.2 shows that that the exponential model is fairly accurate for years relatively close to 2000. However, if we go too far into the future, the model predicts increasingly large rates of change, which causes the population to grow arbitrarily large. This does not make much sense since it is unrealistic to expect that the earth would be able to support such a large population.

The constant k in the differential equation has an important interpretation. Let's rewrite the differential equation $\frac{dP}{dt} = kP$ by solving for k, so that we have

$$k = \frac{dP/dt}{P}.$$

Viewed in this light, k is the ratio of the rate of change to the population; in other words, it is the contribution to the rate of change from a single person. We call this the *per capita growth rate*.

In the exponential model we introduced in Activity 7.6.2, the per capita growth rate is constant. In particular, we are assuming that when the population is large, the per capita growth rate is the same as when the population is small. It is natural to think that the per capita growth rate should decrease when the population becomes large, since there will not be enough resources to support so many people. In other words, we expect that a more realistic model would hold if we assume that the per capita growth rate depends on the population P.

In the previous activity, we computed the per capita growth rate in a single year by computing k, the quotient of $\frac{dP}{dt}$ and P (which we did for $t = 0$). If we return to the data in Table 7.6.1 and compute the per capita growth rate over a range of years, we generate the data shown in Figure 7.6.2, which shows how the per capita growth rate is a function of the population, P.

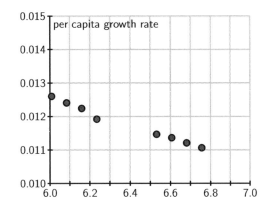

Figure 7.6.2: A plot of per capita growth rate vs. population P.

From the data, we see that the per capita growth rate appears to decrease as the population increases. In fact, the points seem to lie very close to a line, which is shown at two different scales in Figure 7.6.3.

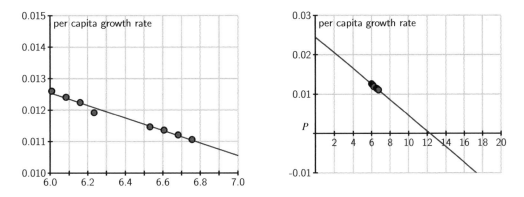

Figure 7.6.3: The line that approximates per capita growth as a function of population, P.

Looking at this line carefully, we can find its equation to be

$$\frac{dP/dt}{P} = 0.025 - 0.002P.$$

If we multiply both sides by P, we arrive at the differential equation

$$\frac{dP}{dt} = P(0.025 - 0.002P).$$

Graphing the dependence of dP/dt on the population P, we see that this differential equation demonstrates a quadratic relationship between $\frac{dP}{dt}$ and P, as shown in Figure 7.6.4.

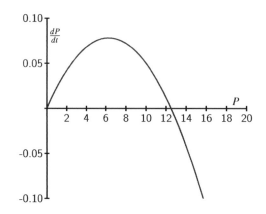

Figure 7.6.4: A plot of $\frac{dP}{dt}$ vs. P for the differential equation $\frac{dP}{dt} = P(0.025 - 0.002P)$.

The equation $\frac{dP}{dt} = P(0.025 - 0.002P)$ is an example of the *logistic equation*, and is the second model for population growth that we will consider. We have reason to believe that it will be more realistic since the per capita growth rate is a decreasing function of the population.

Indeed, the graph in Figure 7.6.4 shows that there are two equilibrium solutions, $P = 0$, which is unstable, and $P = 12.5$, which is a stable equilibrium. The graph shows that any solution with $P(0) > 0$ will eventually stabilize around 12.5. In other words, our model predicts the world's population will eventually stabilize around 12.5 billion.

A prediction for the long-term behavior of the population is a valuable conclusion to draw from our differential equation. We would, however, like to answer some quantitative questions. For instance, how long will it take to reach a population of 10 billion? To determine this, we need to find an explicit solution of the equation.

7.6.2 Solving the logistic differential equation

Since we would like to apply the logistic model in more general situations, we state the logistic equation in its more general form,

$$\frac{dP}{dt} = kP(N - P). \tag{7.6.1}$$

The equilibrium solutions here are when $P = 0$ and $1 - \frac{P}{N} = 0$, which shows that $P = N$. The equilibrium at $P = N$ is called the *carrying capacity* of the population for it represents the stable population that can be sustained by the environment.

We now solve the logistic equation (7.6.1). The equation is separable, so we separate the variables

$$\frac{1}{P(N-P)}\frac{dP}{dt} = k,$$

and integrate to find that

$$\int \frac{1}{P(N-P)}\,dP = \int k\,dt.$$

To find the antiderivative on the left, we use the partial fraction decomposition

$$\frac{1}{P(N-P)} = \frac{1}{N}\left[\frac{1}{P} + \frac{1}{N-P}\right].$$

Now we are ready to integrate, with

$$\int \frac{1}{N}\left[\frac{1}{P} + \frac{1}{N-P}\right]\,dP = \int k\,dt.$$

On the left, observe that N is constant, so we can remove the factor of $\frac{1}{N}$ and antidifferentiate to find that

$$\frac{1}{N}(\ln|P| - \ln|N-P|) = kt + C.$$

Multiplying both sides of this last equation by N and using an important rule of logarithms, we next find that

$$\ln\left|\frac{P}{N-P}\right| = kNt + C.$$

From the definition of the logarithm, replacing e^C with C, and letting C absorb the absolute value signs, we now know that

$$\frac{P}{N-P} = Ce^{kNt}.$$

At this point, all that remains is to determine C and solve algebraically for P.

If the initial population is $P(0) = P_0$, then it follows that $C = \frac{P_0}{N-P_0}$, so

$$\frac{P}{N-P} = \frac{P_0}{N-P_0}e^{kNt}.$$

We will solve this most recent equation for P by multiplying both sides by $(N-P)(N-P_0)$ to obtain

$$P(N-P_0) = P_0(N-P)e^{kNt}$$
$$= P_0Ne^{kNt} - P_0Pe^{kNt}.$$

Swapping the left and right sides, expanding, and factoring, it follows that

$$P_0 N e^{kNt} = P(N - P_0) + P_0 P e^{kNt}$$
$$= P(N - P_0 + P_0 e^{kNt}).$$

Dividing to solve for P, we see that

$$P = \frac{P_0 N e^{kNt}}{N - P_0 + P_0 e^{kNt}}.$$

Finally, we choose to multiply the numerator and denominator by $\frac{1}{P_0} e^{-kNt}$ to obtain

$$P(t) = \frac{N}{\left(\frac{N-P_0}{P_0}\right) e^{-kNt} + 1}.$$

While that was a lot of algebra, notice the result: we have found an explicit solution to the initial value problem

$$\frac{dP}{dt} = kP(N - P), \ P(0) = P_0,$$

and that solution is

$$P(t) = \frac{N}{\left(\frac{N-P_0}{P_0}\right) e^{-kNt} + 1}. \tag{7.6.2}$$

For the logistic equation describing the earth's population that we worked with earlier in this section, we have

$$k = 0.002, N = 12.5, \text{and} P_0 = 6.084.$$

This gives the solution

$$P(t) = \frac{12.5}{1.0546 e^{-0.025t} + 1},$$

whose graph is shown in Figure 7.6.5.

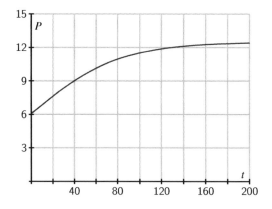

Figure 7.6.5: The solution to the logistic equation modeling the earth's population.

Notice that the graph shows the population leveling off at 12.5 billion, as we expected, and that the population will be around 10 billion in the year 2050. These results, which we have found using a relatively simple mathematical model, agree fairly well with predictions made using a much more sophisticated model developed by the United Nations.

The logistic equation is useful in other situations, too, as it is good for modeling any situation in which limited growth is possible. For instance, it could model the spread of a flu virus through a population contained on a cruise ship, the rate at which a rumor spreads within a small town, or the behavior of an animal population on an island. Again, it is important to realize that through our work in this section, we have completely solved the logistic equation, regardless of the values of the constants N, k, and P_0. Anytime we encounter a logistic equation, we can apply the formula we found in Equation (7.6.2).

Activity 7.6.3. Consider the logistic equation

$$\frac{dP}{dt} = kP(N - P)$$

with the graph of $\frac{dP}{dt}$ vs. P shown in Figure 7.6.6.

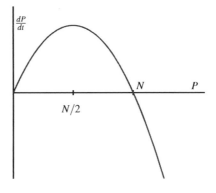

Figure 7.6.6: Plot of $\frac{dP}{dt}$ vs. P.

a. At what value of P is the rate of change greatest?

b. Consider the model for the earth's population that we created. At what value of P is the rate of change greatest? How does that compare to the population in recent years?

c. According to the model we developed, what will the population be in the year 2100?

d. According to the model we developed, when will the population reach 9 billion?

e. Now consider the general solution to the general logistic initial value problem that we found, given by

$$P(t) = \frac{N}{\left(\frac{N-P_0}{P_0}\right)e^{-kNt} + 1}.$$

Verify algebraically that $P(0) = P_0$ and that $\lim_{t \to \infty} P(t) = N$.

Summary

- If we assume that the rate of growth of a population is proportional to the population, we are led to a model in which the population grows without bound and at a rate that grows without bound.

- By assuming that the per capita growth rate decreases as the population grows, we are led to the logistic model of population growth, which predicts that the population will eventually stabilize at the carrying capacity.

Exercises

1. The slope field for a population P modeled by $dP/dt = 3P - 3P^2$ is shown in the figure below.

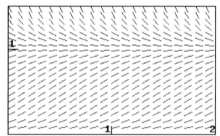

(a) On a print-out of the slope field, sketch three non-zero solution curves showing different types of behavior for the population P. Give an initial condition that will produce each:

$P(0) =$ [_____],

$P(0) =$ [_____], and

$P(0) =$ [_____].

(b) Is there a stable value of the population? If so, give the value; if not, enter *none*:

Stable value = [_____]

(c) Considering the shape of solutions for the population, give any intervals for which the following are true. If no such interval exists, enter *none*, and if there are multiple intervals, give them as a list. (*Thus, if solutions are increasing when P is between 1 and 3, enter (1,3) for that answer; if they are decreasing when P is between 1 and 2 or between 3 and 4, enter (1,2),(3,4). Note that your answers may reflect the fact that P is a population.*)

P is increasing when P is in [_____]

P is decreasing when P is in [   ]
Think about what these conditions mean for the population, and be sure that you are able to explain that.
In the long-run, what is the most likely outcome for the population?

$P \rightarrow$ [   ]
(*Enter* infinity *if the population grows without bound.*)
Are there any inflection points in the solutions for the population? If so, give them as a comma-separated list (e.g., 1,3); if not, enter *none*.

Inflection points are at $P =$ [   ]
Be sure you can explain what the meaning of the inflection points is for the population.
(*d*) Sketch a graph of dP/dt against P. Use your graph to answer the following questions.
When is dP/dt positive?

When P is in [   ]
When is dP/dt negative?

When P is in [   ]
(*Give your answers as intervals or a list of intervals.*)
When is dP/dt zero?

When $P =$ [   ]
(*If there is more than one answer, give a list of answers, e.g., 1,2.*)
When is dP/dt at a maximum?

When $P =$ [   ]
Be sure that you can see how the shape of your graph of dP/dt explains the shape of solution curves to the differential equation.

2. The table below gives the percentage, P, of households with a VCR, as a function of year.

Year	1978	1979	1980	1981	1982	1983	1984
P	0.3	0.5	1.1	1.8	3.1	5.5	10.6
Year	1985	1986	1987	1988	1989	1990	1991
P	20.8	36.0	48.7	58.0	64.6	71.9	71.9

(*a*) A logistic model is a good one to use for these data. Explain why this might be the case: logically, how large would the growth in VCR ownership be when they are first introduced? How large can the ownership ever be?
We can also investigate this by estimating the growth rate of P for the given data. Do this at the beginning, middle, and near the end of the data:

$P'(1980) \approx$ [   ]

$P'(1985) \approx$ [   ]

$P'(1990) \approx$ [   ]
Be sure you can explain why this suggests that a logistic model is appropriate.
(*b*) Use the data to estimate the year when the point of inflection of P occurs.

The inflection point occurs approximately at [   ].

(Give the year in which it occurs.)

What percent of households had VCRs then? $P =$ [   ]

What limiting value L does this point of inflection predict (note that if the logistic model is reasonable, this prediction should agree with the data for 1990 and 1991)?

$L =$ [   ]

(c) The best logistic equation (solution to the logistic differential equation) for these data turns out to be the following.

$$P = \frac{75}{1 + 316.75e^{-0.699t}}.$$

What limiting value does this predict?

$L =$ [   ]

3. The total number of people infected with a virus often grows like a logistic curve. Suppose that 20 people originally have the virus, and that in the early stages of the virus (with time, t, measured in weeks), the number of people infected is increasing exponentially with $k = 2$. It is estimated that, in the long run, approximately 6000 people become infected.

(a) Use this information to find a logistic function to model this situation.

$P =$ [   ]

(b) Sketch a graph of your answer to part (a). Use your graph to estimate the length of time until the rate at which people are becoming infected starts to decrease. What is the vertical coordinate at this point?

vertical coordinate = [   ]

4. Any population, P, for which we can ignore immigration, satisfies

$$\frac{dP}{dt} = \text{Birth rate} - \text{Death rate}.$$

For organisms which need a partner for reproduction but rely on a chance encounter for meeting a mate, the birth rate is proportional to the square of the population. Thus, the population of such a type of organism satisfies a differential equation of the form

$$\frac{dP}{dt} = aP^2 - bP \quad \text{with } a, b > 0.$$

This problem investigates the solutions to such an equation.

(a) Sketch a graph of dP/dt against P. Note when dP/dt is positive and negative.

$dP/dt < 0$ when P is in [   ]

$dP/dt > 0$ when P is in [   ]

(Your answers may involve a and b. Give your answers as an interval or list of intervals: thus, if dP/dt is less than zero for P between 1 and 3 and P greater than 4, enter (1,3),(4,infinity).)

(b) Use this graph to sketch the shape of solution curves with various initial values: use your answers in part (a), and where dP/dt is increasing and decreasing to decide what the shape of the curves has to be. Based on your solution curves, why is $P = b/a$ called the threshold population?

If $P(0) > b/a$, what happens to P in the long run?

$P \rightarrow$ [   ]

If $P(0) = b/a$, what happens to P in the long run?

$P \rightarrow$ [   ]

If $P(0) < b/a$, what happens to P in the long run?

$P \rightarrow$ [   ]

5. The logistic equation may be used to model how a rumor spreads through a group of people. Suppose that $p(t)$ is the fraction of people that have heard the rumor on day t. The equation

$$\frac{dp}{dt} = 0.2p(1 - p)$$

describes how p changes. Suppose initially that one-tenth of the people have heard the rumor; that is, $p(0) = 0.1$.

  a. What happens to $p(t)$ after a very long time?

  b. Determine a formula for the function $p(t)$.

  c. At what time is p changing most rapidly?

  d. How long does it take before 80% of the people have heard the rumor?

6. Suppose that $b(t)$ measures the number of bacteria living in a colony in a Petri dish, where b is measured in thousands and t is measured in days. One day, you measure that there are 6,000 bacteria and the per capita growth rate is 3. A few days later, you measure that there are 9,000 bacteria and the per capita growth rate is 2.

  a. Assume that the per capita growth rate $\frac{db/dt}{b}$ is a linear function of b. Use the measurements to find this function and write a logistic equation to describe $\frac{db}{dt}$.

  b. What is the carrying capacity for the bacteria?

  c. At what population is the number of bacteria increasing most rapidly?

  d. If there are initially 1,000 bacteria, how long will it take to reach 80% of the carrying capacity?

7. Suppose that the population of a species of fish is controlled by the logistic equation

$$\frac{dP}{dt} = 0.1P(10 - P),$$

where P is measured in thousands of fish and t is measured in years.

  a. What is the carrying capacity of this population?

  b. Suppose that a long time has passed and that the fish population is stable at the carrying capacity. At this time, humans begin harvesting 20% of the fish every year. Modify the differential equation by adding a term to incorporate the harvesting of fish.

  c. What is the new carrying capacity?

  d. What will the fish population be one year after the harvesting begins?

  e. How long will it take for the population to be within 10% of the carrying capacity?

CHAPTER 8

Sequences and Series

8.1 Sequences

Motivating Questions

- What is a sequence?
- What does it mean for a sequence to converge?
- What does it mean for a sequence to diverge?

We encounter sequences every day. Your monthly utility payments, the annual interest you earn on investments, the amount you spend on groceries each week; all are examples of sequences. Other sequences with which you may be familiar include the Fibonacci sequence $1, 1, 2, 3, 5, 8, \ldots$ in which each entry is the sum of the two preceding entries, and the triangular numbers $1, 3, 6, 10, 15, 21, 28, 36, 45, 55, \ldots$ which are numbers that correspond to the number of vertices seen in the triangles in Figure 8.1.1.

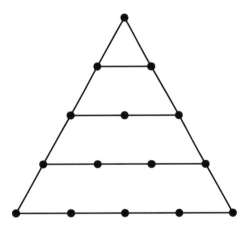

Figure 8.1.1: Triangular numbers

Sequences of integers are of such interest to mathematicians and others that they have a journal[1] devoted to them and an on-line encyclopedia[2] that catalogs a huge number of integer

---

[1]*The Journal of Integer Sequences* at http://www.cs.uwaterloo.ca/journals/JIS/
[2]The On-Line Encyclopedia of Integer Sequences at http://oeis.org/

sequences and their connections. Sequences are also used in digital recordings and images.

To this point, most of our studies in calculus have dealt with continuous information (e.g., continuous functions). The major difference we will see now is that sequences model *discrete* instead of continuous information. We will study ways to represent and work with discrete information in this chapter as we investigate *sequences* and *series*, and ultimately see key connections between the discrete and continuous.

Preview Activity 8.1.1. Suppose you receive $5000 through an inheritance. You decide to invest this money into a fund that pays 8% annually, compounded monthly. That means that each month your investment earns $\frac{0.08}{12} \cdot P$ additional dollars, where P is your principal balance at the start of the month. So in the first month your investment earns

$$5000 \left(\frac{0.08}{12} \right)$$

or $33.33. If you reinvest this money, you will then have $5033.33 in your account at the end of the first month. From this point on, assume that you reinvest all of the interest you earn.

a. How much interest will you earn in the second month? How much money will you have in your account at the end of the second month?

b. Complete Table 8.1.2 to determine the interest earned and total amount of money in this investment each month for one year.

c. As we will see later, the amount of money P_n in the account after month n is given by

$$P_n = 5000 \left(1 + \frac{0.08}{12} \right)^n .$$

Use this formula to check your calculations in Table 8.1.2. Then find the amount of money in the account after 5 years.

Month	Interest earned	Total amount of money in the account
0	$0.00	$5000.00
1	$33.33	$5033.33
2		
3		
4		
5		
6		
7		
8		
9		
10		
11		
12		

Table 8.1.2: Interest

c. How many years will it be before the account has doubled in value to $10000?

8.1.1 Sequences

As our discussion in the introduction and Preview Activity 8.1.1 illustrate, many discrete phenomena can be represented as lists of numbers (like the amount of money in an account over a period of months). We call these any such list a *sequence*. In other words, a sequence is nothing more than list of terms in some order. To be able to refer to a sequence in a general sense, we often list the entries of the sequence with subscripts,

$$s_1, s_2, \ldots, s_n \ldots,$$

where the subscript denotes the position of the entry in the sequence. More formally,

Definition 8.1.3. A sequence is a list of terms $s_1, s_2, s_3, \ldots$ in a specified order.

As an alternative to Definition 8.1.3, we can also consider a sequence to be a function f whose domain is the set of positive integers. In this context, the sequence $s_1, s_2, s_3, \ldots$ would correspond to the function f satisfying $f(n) = s_n$ for each positive integer n. This alternative view will be be useful in many situations.

We will often write the sequence

$$s_1, s_2, s_3, \ldots$$

using the shorthand notation $\{s_n\}$. The value s_n (alternatively $s(n)$) is called the nth *term* in the sequence. If the terms are all 0 after some fixed value of n, we say the sequence is finite. Otherwise the sequence is infinite. We will work with both finite and infinite sequences, but focus more on the infinite sequences. With infinite sequences, we are often interested in their end behavior and the idea of *convergent* sequences.

Activity 8.1.2.

a. Let s_n be the nth term in the sequence $1, 2, 3, \ldots$. Find a formula for s_n and use appropriate technological tools to draw a graph of entries in this sequence by plotting points of the form (n, s_n) for some values of n. Most graphing calculators can plot sequences; directions follow for the TI-84.

- In the MODE menu, highlight SEQ in the FUNC line and press ENTER.

- In the Y= menu, you will now see lines to enter sequences. Enter a value for nMin (where the sequence starts), a function for u(n) (the nth term in the sequence), and the value of u_{n Min}.

- Set your window coordinates (this involves choosing limits for n as well as the window coordinates XMin, XMax, YMin, and YMax.

- The GRAPH key will draw a plot of your sequence.

Using your knowledge of limits of continuous functions as $x \to \infty$, decide if this sequence $\{s_n\}$ has a limit as $n \to \infty$. Explain your reasoning.

b. Let s_n be the nth term in the sequence $1, \frac{1}{2}, \frac{1}{3}, \ldots$. Find a formula for s_n. Draw a graph of some points in this sequence. Using your knowledge of limits of continuous functions as $x \to \infty$, decide if this sequence $\{s_n\}$ has a limit as $n \to \infty$. Explain your reasoning.

c. Let s_n be the nth term in the sequence $2, \frac{3}{2}, \frac{4}{3}, \frac{5}{4}, \ldots$. Find a formula for s_n. Using your knowledge of limits of continuous functions as $x \to \infty$, decide if this sequence $\{s_n\}$ has a limit as $n \to \infty$. Explain your reasoning.

Next we formalize the ideas from Activity 8.1.2.

Activity 8.1.3.

a. Recall our earlier work with limits involving infinity in Section 2.8. State clearly what it means for a continuous function f to have a limit L as $x \to \infty$.

b. Given that an infinite sequence of real numbers is a function from the integers to the real numbers, apply the idea from part (a) to explain what you think it means for a sequence $\{s_n\}$ to have a limit as $n \to \infty$.

c. Based on your response to the part (b), decide if the sequence $\{\frac{1+n}{2+n}\}$ has a limit as $n \to \infty$. If so, what is the limit? If not, why not?

In Activities 8.1.2 and Activity 8.1.3 we investigated the notion of a sequence $\{s_n\}$ having a limit as n goes to infinity. If a sequence $\{s_n\}$ has a limit as n goes to infinity, we say that the sequence *converges* or is a *convergent sequence*. If the limit of a convergent sequence is the number L, we use the same notation as we did for continuous functions and write

$$\lim_{n \to \infty} s_n = L.$$

If a sequence $\{s_n\}$ does not converge then we say that the sequence $\{s_n\}$ *diverges*. Convergence of sequences is a major idea in this section and we describe it more formally as follows.

A sequence $\{s_n\}$ of real numbers converges to a number L if we can make all values of s_k for $k \geq n$ as close to L as we want by choosing n to be sufficiently large.

Remember, the idea of sequence having a limit as $n \to \infty$ is the same as the idea of a continuous function having a limit as $x \to \infty$. The only new wrinkle here is that our sequences are discrete instead of continuous.

We conclude this section with a few more examples in the following activity.

Activity 8.1.4. Use graphical and/or algebraic methods to determine whether each of the following sequences converges or diverges.

a. $\left\{\frac{1+2n}{3n-2}\right\}$

b. $\left\{\frac{5+3^n}{10+2^n}\right\}$

c. $\left\{\frac{10^n}{n!}\right\}$ (where ! is the *factorial* symbol and $n! = n(n-1)(n-2)\cdots(2)(1)$ for any positive integer n (as convention we define $0!$ to be 1)).

Summary

- A sequence is a list of objects in a specified order. We will typically work with sequences of real numbers and can also think of a sequence as a function from the positive integers to the set of real numbers.

- A sequence $\{s_n\}$ of real numbers converges to a number L if we can make every value of s_k for $k \geq n$ as close as we want to L by choosing n sufficiently large.

- A sequence diverges if it does not converge.

Exercises

1. Match the formulas with the descriptions of the behavior of the sequence as $n \to \infty$.

1. $s_n = n(n+1) - 1$
2. $s_n = 1/(n+1)$
3. $s_n = 3 - 1/n$
4. $s_n = n\sin(n)/(n+1)$
5. $s_n = (n+1)/n$

A. does not converge, but doesn't go to $\pm\infty$

B. converges to three from below

C. diverges to ∞

D. converges to one from above

E. converges to zero through positive numbers

2. Find a formula for s_n, $n \geq 1$ for the sequence -3, 5, -7, 9, -11...

$s_n = $

3. For each of the sequences below, enter either *diverges* if the sequence diverges, or the limit of the sequence if the sequence converges as $n \to \infty$. (*Note that to avoid this becoming a "multiple guess" problem you will not see partial correct answers.*)

A. $\frac{4n+8}{n}$:

B. 4^n :

453

C. $\frac{4n+8}{n^2}$:

D. $\frac{\sin n}{4n}$:

4. In electrical engineering, a continuous function like $f(t) = \sin t$, where t is in seconds, is referred to as an analog signal. To digitize the signal, we sample $f(t)$ every Δt seconds to form the sequence $s_n = f(n\Delta t)$. For example, sampling f every $1/10$ second produces the sequence $\sin(1/10), \sin(2/10), \sin(3/10),...$
Suppose that the analog signal is given by

$$f(t) = (t - 0.5)^2.$$

Give the first 6 terms of a sampling of the signal every $\Delta t = 0.25$ seconds:
(Enter your answer as a comma-separated list.)

5. Finding limits of convergent sequences can be a challenge. However, there is a useful tool we can adapt from our study of limits of continuous functions at infinity to use to find limits of sequences. We illustrate in this exercise with the example of the sequence

$$\frac{\ln(n)}{n}.$$

a. Calculate the first 10 terms of this sequence. Based on these calculations, do you think the sequence converges or diverges? Why?

b. For this sequence, there is a corresponding continuous function f defined by

$$f(x) = \frac{\ln(x)}{x}.$$

Draw the graph of $f(x)$ on the interval $[0, 10]$ and then plot the entries of the sequence on the graph. What conclusion do you think we can draw about the sequence $\left\{\frac{\ln(n)}{n}\right\}$ if $\lim_{x\to\infty} f(x) = L$? Explain.

c. Note that $f(x)$ has the indeterminate form $\frac{\infty}{\infty}$ as x goes to infinity. What idea from differential calculus can we use to calculate $\lim_{x\to\infty} f(x)$? Use this method to find $\lim_{x\to\infty} f(x)$. What, then, is $\lim_{n\to\infty} \frac{\ln(n)}{n}$?

6. We return to the example begun in Preview Activity 8.1.1 to see how to derive the formula for the amount of money in an account at a given time. We do this in a general setting. Suppose you invest P dollars (called the *principal*) in an account paying $r\%$ interest compounded monthly. In the first month you will receive $\frac{r}{12}$ (here r is in decimal form; e.g., if we have 8% interest, we write $\frac{0.08}{12}$) of the principal P in interest, so you earn

$$P\left(\frac{r}{12}\right)$$

dollars in interest. Assume that you reinvest all interest. Then at the end of the first month your account will contain the original principal P plus the interest, or a total of

$$P_1 = P + P\left(\frac{r}{12}\right) = P\left(1 + \frac{r}{12}\right)$$

dollars.

a. Given that your principal is now P_1 dollars, how much interest will you earn in the second month? If P_2 is the total amount of money in your account at the end of the second month, explain why

$$P_2 = P_1 \left(1 + \frac{r}{12}\right) = P \left(1 + \frac{r}{12}\right)^2.$$

b. Find a formula for P_3, the total amount of money in the account at the end of the third month in terms of the original investment P.

c. There is a pattern to these calculations. Let P_n the total amount of money in the account at the end of the third month in terms of the original investment P. Find a formula for P_n.

7. Sequences have many applications in mathematics and the sciences. In a recent paper[3] the authors write

> The incretin hormone glucagon-like peptide-1 (GLP-1) is capable of ameliorating glucose-dependent insulin secretion in subjects with diabetes. However, its very short half-life (1.5-5 min) in plasma represents a major limitation for its use in the clinical setting.

The half-life of GLP-1 is the time it takes for half of the hormone to decay in its medium. For this exercise, assume the half-life of GLP-1 is 5 minutes. So if A is the amount of GLP-1 in plasma at some time t, then only $\frac{A}{2}$ of the hormone will be present after $t + 5$ minutes. Suppose $A_0 = 100$ grams of the hormone are initially present in plasma.

a. Let A_1 be the amount of GLP-1 present after 5 minutes. Find the value of A_1.

b. Let A_2 be the amount of GLP-1 present after 10 minutes. Find the value of A_2.

c. Let A_3 be the amount of GLP-1 present after 15 minutes. Find the value of A_3.

d. Let A_4 be the amount of GLP-1 present after 20 minutes. Find the value of A_4.

e. Let A_n be the amount of GLP-1 present after $5n$ minutes. Find a formula for A_n.

f. Does the sequence $\{A_n\}$ converge or diverge? If the sequence converges, find its limit and explain why this value makes sense in the context of this problem.

g. Determine the number of minutes it takes until the amount of GLP-1 in plasma is 1 gram.

8. Continuous data is the basis for analog information, like music stored on old cassette tapes or vinyl records. A digital signal like on a CD or MP3 file is obtained by sampling an analog signal at some regular time interval and storing that information. For example, the sampling rate of a compact disk is 44,100 samples per second. So a digital recording is only an approximation of the actual analog information. Digital information can be manipulated in many useful ways that allow for, among other things, noisy signals to be cleaned up and large collections of information to be compressed and stored in much smaller space. While we won't investigate these techniques in this chapter, this exercise is intended to give an idea of the importance of discrete (digital) techniques.

---

[3]Hui H, Farilla L, Merkel P, Perfetti R. The short half-life of glucagon-like peptide-1 in plasma does not reflect its long-lasting beneficial effects, *Eur J Endocrinol* 2002 Jun;146(6):863-9.

Let f be the continuous function defined by $f(x) = \sin(4x)$ on the interval $[0, 10]$. A graph of f is shown in Figure 8.1.4.

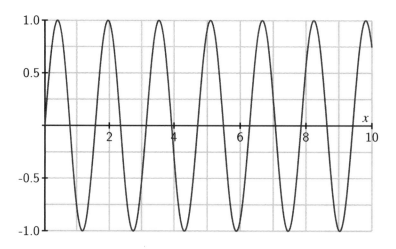

Figure 8.1.4: The graph of $f(x) = \sin(4x)$ on the interval $[0, 10]$

We approximate f by *sampling*, that is by partitioning the interval $[0, 10]$ into uniform subintervals and recording the values of f at the endpoints.

a. Ineffective sampling can lead to several problems in reproducing the original signal. As an example, partition the interval $[0, 10]$ into 8 equal length subintervals and create a list of points (the *sample*) using the endpoints of each subinterval. Plot your sample on graph of f in Figure Figure 8.1.4. What can you say about the period of your sample as compared to the period of the original function?

b. The sampling rate is the number of samples of a signal taken per second. As the part (a) illustrates, sampling at too small a rate can cause serious problems with reproducing the original signal (this problem of inefficient sampling leading to an inaccurate approximation is called *aliasing*). There is an elegant theorem called the Nyquist-Shannon Sampling Theorem that says that human perception is limited, which allows that replacement of a continuous signal with a digital one without any perceived loss of information. This theorem also provides the *lowest* rate at which a signal can be sampled (called the Nyquist rate) without such a loss of information. The theorem states that we should sample at double the maximum desired frequency so that every cycle of the original signal will be sampled at at least two points. Recall that the frequency of a sinusoidal function is the reciprocal of the period. Identify the frequency of the function f and determine the number of partitions of the interval $[0, 10]$ that give us the Nyquist rate.

c. Humans cannot typically pick up signals above 20 kHz. Explain why, then, that information on a compact disk is sampled at 44,100 Hz.

8.2 Geometric Series

Motivating Questions

- What is a geometric series?

- What is a partial sum of a geometric series? What is a simplified form of the nth partial sum of a geometric series?

- Under what conditions does a geometric series converge? What is the sum of a convergent geometric series?

Many important sequences are generated through the process of addition. In Preview Activity 8.2.1, we see a particular example of a special type of sequence that is connected to a sum.

> **Preview Activity 8.2.1.** Warfarin is an anticoagulant that prevents blood clotting; often it is prescribed to stroke victims in order to help ensure blood flow. The level of warfarin has to reach a certain concentration in the blood in order to be effective.
>
> Suppose warfarin is taken by a particular patient in a 5 mg dose each day. The drug is absorbed by the body and some is excreted from the system between doses. Assume that at the end of a 24 hour period, 8% of the drug remains in the body. Let $Q(n)$ be the amount (in mg) of warfarin in the body before the $(n + 1)$st dose of the drug is administered.
>
> a. Explain why $Q(1) = 5 \times 0.08$ mg.
>
> b. Explain why $Q(2) = (5 + Q(1)) \times 0.08$ mg. Then show that
> $$Q(2) = (5 \times 0.08)(1 + 0.08) \text{ mg.}$$
>
> c. Explain why $Q(3) = (5 + Q(2)) \times 0.08$ mg. Then show that
> $$Q(3) = (5 \times 0.08)\left(1 + 0.08 + 0.08^2\right) \text{ mg.}$$
>
> d. Explain why $Q(4) = (5 + Q(3)) \times 0.08$ mg. Then show that
> $$Q(4) = (5 \times 0.08)\left(1 + 0.08 + 0.08^2 + 0.08^3\right) \text{ mg.}$$
>
> e. There is a pattern that you should see emerging. Use this pattern to find a formula for $Q(n)$, where n is an arbitrary positive integer.

f. Complete Table 8.2.1 with values of $Q(n)$ for the provided n-values (reporting $Q(n)$ to 10 decimal places). What appears to be happening to the sequence $Q(n)$ as n increases?

$$Q(1) = \quad 0.40$$
$$Q(2) =$$
$$Q(3) =$$
$$Q(4) =$$
$$Q(5) =$$
$$Q(6) =$$
$$Q(7) =$$
$$Q(8) =$$
$$Q(9) =$$
$$Q(10) =$$

Table 8.2.1: Values of $Q(n)$ for selected values of n

8.2.1 Geometric Sums

In Preview Activity 8.2.1 we encountered the sum

$$(5 \times 0.08)\left(1 + 0.08 + 0.08^2 + 0.08^3 + \cdots + 0.08^{n-1}\right).$$

In order to evaluate the long-term level of Warfarin in the patient's system, we will want to fully understand the sum in this expression. This sum has the form

$$a + ar + ar^2 + \cdots + ar^{n-1} \tag{8.2.1}$$

where $a = 5 \times 0.08$ and $r = 0.08$. Such a sum is called a *geometric sum* with ratio r. We will analyze this sum in more detail in the next activity.

Activity 8.2.2. Let a and r be real numbers (with $r \neq 1$) and let

$$S_n = a + ar + ar^2 + \cdots + ar^{n-1}.$$

In this activity we will find a shortcut formula for S_n that does not involve a sum of n terms.

a. Multiply S_n by r. What does the resulting sum look like?

b. Subtract rS_n from S_n and explain why

$$S_n - rS_n = a - ar^n. \tag{8.2.2}$$

c. Solve equation (8.2.2) for S_n to find a simple formula for S_n that does not involve adding n terms.

We can summarize the result of Activity 8.2.2 in the following way.

A geometric sum S_n is a sum of the form

$$S_n = a + ar + ar^2 + \cdots + ar^{n-1}, \tag{8.2.3}$$

where a and r are real numbers such that $r \neq 1$. The geometric sum S_n can be written more simply as

$$S_n = a + ar + ar^2 + \cdots + ar^{n-1} = \frac{a(1 - r^n)}{1 - r}. \tag{8.2.4}$$

We now apply equation (8.2.4) to the example involving warfarin from Preview Activity 8.2.1. Recall that

$$Q(n) = (5 \times 0.08)\left(1 + 0.08 + 0.08^2 + 0.08^3 + \cdots + 0.08^{n-1}\right) \text{mg},$$

so $Q(n)$ is a geometric sum with $a = 5 \times 0.08 = 0.4$ and $r = 0.08$. Thus,

$$Q(n) = 0.4\left(\frac{1 - 0.08^n}{1 - 0.08}\right) = \frac{1}{2.3}\left(1 - 0.08^n\right).$$

Notice that as n goes to infinity, the value of 0.08^n goes to 0. So,

$$\lim_{n \to \infty} Q(n) = \lim_{n \to \infty} \frac{1}{2.3}\left(1 - 0.08^n\right) = \frac{1}{2.3} \approx 0.435.$$

Therefore, the long-term level of Warfarin in the blood under these conditions is $\frac{1}{2.3}$, which is approximately 0.435 mg.

To determine the long-term effect of Warfarin, we considered a geometric sum of n terms, and then considered what happened as n was allowed to grow without bound. In this sense, we were actually interested in an infinite geometric sum (the result of letting n go to infinity in the finite sum). We call such an infinite geometric sum a *geometric series*.

Definition 8.2.2. A geometric series is an infinite sum of the form

$$a + ar + ar^2 + \cdots = \sum_{n=0}^{\infty} ar^n. \tag{8.2.5}$$

The value of r in the geometric series (8.2.5) is called the *common ratio* of the series because the ratio of the $(n + 1)$st term ar^n to the nth term ar^{n-1} is always r.

Geometric series are very common in mathematics and arise naturally in many different situations. As a familiar example, suppose we want to write the number with repeating decimal expansion

$$N = 0.12\overline{12}$$

as a rational number. Observe that

$$N = 0.12 + 0.0012 + 0.000012 + \cdots$$

$$= \left(\frac{12}{100}\right) + \left(\frac{12}{100}\right)\left(\frac{1}{100}\right) + \left(\frac{12}{100}\right)\left(\frac{1}{100}\right)^2 + \cdots,$$

which is an infinite geometric series with $a = \frac{12}{100}$ and $r = \frac{1}{100}$. In the same way that we were able to find a shortcut formula for the value of a (finite) geometric sum, we would like to develop a formula for the value of a (infinite) geometric series. We explore this idea in the following activity.

Activity 8.2.3. Let $r \neq 1$ and a be real numbers and let

$$S = a + ar + ar^2 + \cdots ar^{n-1} + \cdots$$

be an infinite geometric series. For each positive integer n, let

$$S_n = a + ar + ar^2 + \cdots + ar^{n-1}.$$

Recall that

$$S_n = a\frac{1 - r^n}{1 - r}.$$

a. What should we allow n to approach in order to have S_n approach S?

b. What is the value of $\lim_{n \to \infty} r^n$ for $|r| > 1$? for $|r| < 1$? Explain.

c. If $|r| < 1$, use the formula for S_n and your observations in (a) and (b) to explain why S is finite and find a resulting formula for S.

From our work in Activity 8.2.3, we can now find the value of the geometric series $N = \left(\frac{12}{100}\right) + \left(\frac{12}{100}\right)\left(\frac{1}{100}\right) + \left(\frac{12}{100}\right)\left(\frac{1}{100}\right)^2 + \cdots$. In particular, using $a = \frac{12}{100}$ and $r = \frac{1}{100}$, we see that

$$N = \frac{12}{100}\left(\frac{1}{1 - \frac{1}{100}}\right) = \frac{12}{100}\left(\frac{100}{99}\right) = \frac{4}{33}.$$

It is important to notice that a geometric sum is simply the sum of a finite number of terms of a geometric series. In other words, the geometric sum S_n for the geometric series

$$\sum_{k=0}^{\infty} ar^k$$

is

$$S_n = a + ar + ar^2 + \cdots + ar^{n-1} = \sum_{k=0}^{n-1} ar^k.$$

We also call this sum S_n the nth *partial sum* of the geometric series. We summarize our recent work with geometric series as follows.

- A geometric series is an infinite sum of the form

$$a + ar + ar^2 + \cdots = \sum_{n=0}^{\infty} ar^n, \tag{8.2.6}$$

  where a and r are real numbers such that $r \neq 0$.

- The nth partial sum S_n of the geometric series is

$$S_n = a + ar + ar^2 + \cdots + ar^{n-1}.$$

- If $|r| < 1$, then using the fact that $S_n = a\frac{1-r^n}{1-r}$, it follows that the sum S of the geometric series (8.2.6) is

$$S = \lim_{n \to \infty} S_n = \lim_{n \to \infty} a\frac{1-r^n}{1-r} = \frac{a}{1-r}$$

Activity 8.2.4. The formulas we have derived for the geometric series and its partial sum so far have assumed we begin indexing our sums at $n = 0$. If instead we have a sum that does not begin at $n = 0$, we can factor out common terms and use our established formulas. This process is illustrated in the examples in this activity.

a. Consider the sum

$$\sum_{k=1}^{\infty} (2)\left(\frac{1}{3}\right)^k = (2)\left(\frac{1}{3}\right) + (2)\left(\frac{1}{3}\right)^2 + (2)\left(\frac{1}{3}\right)^3 + \cdots.$$

Remove the common factor of $(2)\left(\frac{1}{3}\right)$ from each term and hence find the sum of the series.

b. Next let a and r be real numbers with $-1 < r < 1$. Consider the sum

$$\sum_{k=3}^{\infty} ar^k = ar^3 + ar^4 + ar^5 + \cdots.$$

Remove the common factor of ar^3 from each term and find the sum of the series.

c. Finally, we consider the most general case. Let a and r be real numbers with $-1 < r < 1$, let n be a positive integer, and consider the sum

$$\sum_{k=n}^{\infty} ar^k = ar^n + ar^{n+1} + ar^{n+2} + \cdots.$$

| Remove the common factor of ar^n from each term to find the sum of the series.

Summary

- A geometric series is an infinite sum of the form

$$\sum_{k=0}^{\infty} ar^k$$

where a and r are real numbers and $r \neq 0$.

- For the geometric series $\sum_{k=0}^{\infty} ar^k$, its nth partial sum is

$$S_n = \sum_{k=0}^{n-1} ar^k.$$

An alternate formula for the nth partial sum is

$$S_n = a\frac{1 - r^n}{1 - r}.$$

Whenever $|r| < 1$, the infinite geometric series $\sum_{k=0}^{\infty} ar^k$ has the finite sum $\frac{a}{1-r}$.

Exercises

1. Find the 7th term of the geometric sequence
$-3, -16.5, -90.75, \dots$

Answer: _____

2. Find the sum of the series
$$1 + \frac{1}{3} + \frac{1}{9} + \dots + \frac{1}{3^{n-1}} + \dots.$$

Answer: _____

3. Determine the sum of the following series.

$$\sum_{n=1}^{\infty} \left(\frac{3^n + 5^n}{9^n}\right)$$

_____

4. Find the sum of each of the geometric series given below. For the value of the sum, enter an expression that gives the exact value, rather than entering an approximation.

A. $-16 + 4 - \frac{4}{4} + \frac{4}{16} - \frac{4}{64} + \frac{4}{256} - \dots =$ _____

B. $\sum_{n=3}^{14} \left(\frac{1}{2}\right)^n =$ _____

5. There is an old question that is often used to introduce the power of geometric growth. Here is one version. Suppose you are hired for a one month (30 days, working every day) job and are given two options to be paid.

Option 1. You can be paid $500 per day or

Option 2. You can be paid 1 cent the first day, 2 cents the second day, 4 cents the third day, 8 cents the fourth day, and so on, doubling the amount you are paid each day.

    a. How much will you be paid for the job in total under Option 1?

    b. Complete Table 8.2.3 to determine the pay you will receive under Option 2 for the first 10 days.

Day	Pay on this day	Total amount paid to date
1	$0.01	$0.01
2	$0.02	$0.03
3		
4		
5		
6		
7		
8		
9		
10		

Table 8.2.3: Option 2 payments

    c. Find a formula for the amount paid on day n, as well as for the total amount paid by day n. Use this formula to determine which option (1 or 2) you should take.

6. Suppose you drop a golf ball onto a hard surface from a height h. The collision with the ground causes the ball to lose energy and so it will not bounce back to its original height. The ball will then fall again to the ground, bounce back up, and continue. Assume that at each bounce the ball rises back to a height $\frac{3}{4}$ of the height from which it dropped. Let h_n be the height of the ball on the nth bounce, with $h_0 = h$. In this exercise we will determine the distance traveled by the ball and the time it takes to travel that distance.

    a. Determine a formula for h_1 in terms of h.

    b. Determine a formula for h_2 in terms of h.

    c. Determine a formula for h_3 in terms of h.

    d. Determine a formula for h_n in terms of h.

    e. Write an infinite series that represents the total distance traveled by the ball. Then determine the sum of this series.

    f. Next, let's determine the total amount of time the ball is in the air.

    i) When the ball is dropped from a height H, if we assume the only force acting on it is the acceleration due to gravity, then the height of the ball at time t is given by

$$H - \frac{1}{2}gt^2.$$

Use this formula to determine the time it takes for the ball to hit the ground after being dropped from height H.

    ii) Use your work in the preceding item, along with that in (a)-(e) above to determine the total amount of time the ball is in the air.

7. Suppose you play a game with a friend that involves rolling a standard six-sided die. Before a player can participate in the game, he or she must roll a six with the die. Assume that you roll first and that you and your friend take alternate rolls. In this exercise we will determine the probability that you roll the first six.

  a. Explain why the probability of rolling a six on any single roll (including your first turn) is $\frac{1}{6}$.

  b. If you don't roll a six on your first turn, then in order for you to roll the first six on your second turn, both you and your friend had to fail to roll a six on your first turns, and then you had to succeed in rolling a six on your second turn. Explain why the probability of this event is

$$\left(\frac{5}{6}\right)\left(\frac{5}{6}\right)\left(\frac{1}{6}\right) = \left(\frac{5}{6}\right)^2\left(\frac{1}{6}\right).$$

  c. Now suppose you fail to roll the first six on your second turn. Explain why the probability is

$$\left(\frac{5}{6}\right)\left(\frac{5}{6}\right)\left(\frac{5}{6}\right)\left(\frac{5}{6}\right)\left(\frac{1}{6}\right) = \left(\frac{5}{6}\right)^4\left(\frac{1}{6}\right)$$

that you to roll the first six on your third turn.

  d. The probability of you rolling the first six is the probability that you roll the first six on your first turn plus the probability that you roll the first six on your second turn plus the probability that your roll the first six on your third turn, and so on. Explain why this probability is

$$\frac{1}{6} + \left(\frac{5}{6}\right)^2\left(\frac{1}{6}\right) + \left(\frac{5}{6}\right)^4\left(\frac{1}{6}\right) + \cdots.$$

Find the sum of this series and determine the probability that you roll the first six.

8. The goal of a federal government stimulus package is to positively affect the economy. Economists and politicians quote numbers like "k million jobs and a net stimulus to the economy of n billion of dollars." Where do they get these numbers? Let's consider one aspect of a stimulus package: tax cuts. Economists understand that tax cuts or rebates can result in long-term spending that is many times the amount of the rebate. For example, assume that for a typical person, 75% of her entire income is spent (that is, put back into the economy). Further, assume the government provides a tax cut or rebate that totals P dollars for each person.

a. The tax cut of P dollars is income for its recipient. How much of this tax cut will be spent?

b. In this simple model, we will say that the spent portion of the tax cut/rebate from part (a) then becomes income for another person who, in turn, spends 75% of this income. After this "second round" of spent income, how many total dollars have been added to the economy as a result of the original tax cut/rebate?

c. This second round of spending becomes income for another group who spend 75% of this income, and so on. In economics this is called the *multiplier effect*. Explain why an original tax cut/rebate of P dollars will result in multiplied spending of

$$0.75P(1 + 0.75 + 0.75^2 + \cdots).$$

dollars.

d. Based on these assumptions, how much stimulus will a 200 billion dollar tax cut/rebate to consumers add to the economy, assuming consumer spending remains consistent forever.

9. Like stimulus packages, home mortgages and foreclosures also impact the economy. A problem for many borrowers is the adjustable rate mortgage, in which the interest rate can change (and usually increases) over the duration of the loan, causing the monthly payments to increase beyond the ability of the borrower to pay. Most financial analysts recommend fixed rate loans, ones for which the monthly payments remain constant throughout the term of the loan. In this exercise we will analyze fixed rate loans.

When most people buy a large ticket item like car or a house, they have to take out a loan to make the purchase. The loan is paid back in monthly installments until the entire amount of the loan, plus interest, is paid. With a loan, we borrow money, say P dollars (called the principal), and pay off the loan at an interest rate of r%. To pay back the loan we make regular monthly payments, some of which goes to pay off the principal and some of which is charged as interest. In most cases, the interest is computed based on the amount of principal that remains at the beginning of the month. We assume a fixed rate loan, that is one in which we make a constant monthly payment M on our loan, beginning in the original month of the loan.

Suppose you want to buy a house. You have a certain amount of money saved to make a down payment, and you will borrow the rest to pay for the house. Of course, for the privilege of loaning you the money, the bank will charge you interest on this loan, so the amount you pay back to the bank is more than the amount you borrow. In fact, the amount you ultimately pay depends on three things: the amount you borrow (called the *principal*), the interest rate, and the length of time you have to pay off the loan plus interest (called the *duration* of the loan). For this example, we assume that the interest rate is fixed at r%.

To pay off the loan, each month you make a payment of the same amount (called *installments*). Suppose we borrow P dollars (our principal) and pay off the loan at an interest rate of r% with regular monthly installment payments of M dollars. So in month 1 of the loan, before we make any payments, our principal is P dollars. Our goal in this exercise is to find a formula that relates these three parameters to the time duration of the loan.

We are charged interest every month at an annual rate of r%, so each month we pay $\frac{r}{12}$% interest on the principal that remains. Given that the original principal is P dollars, we will pay $\left(\frac{0.0r}{12}\right) P$ dollars in interest on our first payment. Since we paid M dollars in total for our first payment, the remainder of the payment $\left(M - \left(\frac{r}{12}\right) P\right)$ goes to pay down the principal. So the principal remaining after the first payment (let's call it P_1) is the original principal minus what we paid on the principal, or

$$P_1 = P - \left(M - \left(\frac{r}{12}\right) P\right) = \left(1 + \frac{r}{12}\right) P - M.$$

As long as P_1 is positive, we still have to keep making payments to pay off the loan.

a. Recall that the amount of interest we pay each time depends on the principal that remains. How much interest, in terms of P_1 and r, do we pay in the second installment?

b. How much of our second monthly installment goes to pay off the principal? What is the principal P_2, or the balance of the loan, that we still have to pay off after making the second installment of the loan? Write your response in the form $P_2 = (\ )P_1 - (\ )M$, where you fill in the parentheses.

c. Show that $P_2 = \left(1 + \frac{r}{12}\right)^2 P - \left[1 + \left(1 + \frac{r}{12}\right)\right] M.$

d. Let P_3 be the amount of principal that remains after the third installment. Show that

$$P_3 = \left(1 + \frac{r}{12}\right)^3 P - \left[1 + \left(1 + \frac{r}{12}\right) + \left(1 + \frac{r}{12}\right)^2\right] M.$$

e. If we continue in the manner described in the problems above, then the remaining principal of our loan after n installments is

$$P_n = \left(1 + \frac{r}{12}\right)^n P - \left[\sum_{k=0}^{n-1} \left(1 + \frac{r}{12}\right)^k\right] M. \tag{8.2.7}$$

This is a rather complicated formula and one that is difficult to use. However, we can simplify the sum if we recognize part of it as a partial sum of a geometric series. Find a formula for the sum

$$\sum_{k=0}^{n-1} \left(1 + \frac{r}{12}\right)^k. \tag{8.2.8}$$

and then a general formula for P_n that does not involve a sum.

f. It is usually more convenient to write our formula for P_n in terms of years rather than months. Show that $P(t)$, the principal remaining after t years, can be written as

$$P(t) = \left(P - \frac{12M}{r}\right)\left(1 + \frac{r}{12}\right)^{12t} + \frac{12M}{r}. \tag{8.2.9}$$

g. Now that we have analyzed the general loan situation, we apply formula (8.2.9) to an actual loan. Suppose we charge $1,000 on a credit card for holiday expenses. If our credit card charges 20% interest and we pay only the minimum payment of $25 each month, how long will it take us to pay off the $1,000 charge? How much in total will we have paid on this $1,000 charge? How much total interest will we pay on this loan?

h. Now we consider larger loans, e.g., automobile loans or mortgages, in which we borrow a specified amount of money over a specified period of time. In this situation, we need to determine the amount of the monthly payment we need to make to pay off the loan in the specified amount of time. In this situation, we need to find the monthly payment M that will take our outstanding principal to 0 in the specified amount of time. To do so, we want to know the value of M that makes $P(t) = 0$ in formula (8.2.9). If we set $P(t) = 0$ and solve for M, it follows that

$$M = \frac{rP\left(1 + \frac{r}{12}\right)^{12t}}{12\left(\left(1 + \frac{r}{12}\right)^{12t} - 1\right)}.$$

i) Suppose we want to borrow $15,000 to buy a car. We take out a 5 year loan at 6.25%. What will our monthly payments be? How much in total will we have paid for this $15,000 car? How much total interest will we pay on this loan?

ii) Suppose you charge your books for winter semester on your credit card. The total charge comes to $525. If your credit card has an interest rate of 18% and you pay $20 per month on the card, how long will it take before you pay off this debt? How much total interest will you pay?

iii) Say you need to borrow $100,000 to buy a house. You have several options on the loan:

- 30 years at 6.5%
- 25 years at 7.5%
- 15 years at 8.25%.

a. What are the monthly payments for each loan?

b. Which mortgage is ultimately the best deal (assuming you can afford the monthly payments)? In other words, for which loan do you pay the least amount of total interest?

8.3 Series of Real Numbers

Motivating Questions

- What is an infinite series?

- What is the nth partial sum of an infinite series?

- How do we add up an infinite number of numbers? In other words, what does it mean for an infinite series of real numbers to converge?

- What does it mean for an infinite series of real numbers to diverge?

In Section 8.2, we encountered several situations where we naturally considered an infinite sum of numbers called a geometric series. For example, by writing

$$N = 0.1212121212\cdots = \frac{12}{100} + \frac{12}{100} \cdot \frac{1}{100} + \frac{12}{100} \cdot \frac{1}{100^2} + \cdots$$

as a geometric series, we found a way to write the repeating decimal expansion of N as a single fraction: $N = \frac{4}{33}$. There are many other situations in mathematics where infinite sums of numbers arise, but often these are not geometric. In this section, we begin exploring these other types of infinite sums. Preview Activity 8.3.1 provides a context in which we see how one such sum is related to the famous number e.

> **Preview Activity 8.3.1.** Have you ever wondered how your calculator can produce a numeric approximation for complicated numbers like e, π or $\ln(2)$? After all, the only operations a calculator can really perform are addition, subtraction, multiplication, and division, the operations that make up polynomials. This activity provides the first steps in understanding how this process works. Throughout the activity, let $f(x) = e^x$.
>
> a. Find the tangent line to f at $x = 0$ and use this linearization to approximate e. That is, find a formula $L(x)$ for the tangent line, and compute $L(1)$, since $L(1) \approx f(1) = e$.
>
> b. The linearization of e^x does not provide a good approximation to e since 1 is not very close to 0. To obtain a better approximation, we alter our approach a bit. Instead of using a straight line to approximate e, we put an appropriate bend in our estimating function to make it better fit the graph of e^x for x close to 0. With the linearization, we had both $f(x)$ and $f'(x)$ share the same value as the linearization at $x = 0$. We will now use a quadratic approximation $P_2(x)$ to $f(x) = e^x$ centered at $x = 0$ which has the property that $P_2(0) = f(0)$, $P_2'(0) = f'(0)$, and $P_2''(0) = f''(0)$.
>
> > i) Let $P_2(x) = 1 + x + \frac{x^2}{2}$. Show that $P_2(0) = f(0)$, $P_2'(0) = f'(0)$, and $P_2''(0) = f''(0)$. Then, use $P_2(x)$ to approximate e by observing that $P_2(1) \approx f(1)$.

ii) We can continue approximating e with polynomials of larger degree whose higher derivatives agree with those of f at 0. This turns out to make the polynomials fit the graph of f better for more values of x around 0. For example, let $P_3(x) = 1 + x + \frac{x^2}{2} + \frac{x^3}{6}$. Show that $P_3(0) = f(0)$, $P_3'(0) = f'(0)$, $P_3''(0) = f''(0)$, and $P_3'''(0) = f'''(0)$. Use $P_3(x)$ to approximate e in a way similar to how you did so with $P_2(x)$ above.

8.3.1 ·Infinite Series

Preview Activity 8.3.1 shows that an approximation to e using a linear polynomial is 2, an approximation to e using a quadratic polynomial is 2.5, and an approximation using a cubic polynomial is 2.6667. As we will see later, if we continue this process we can obtain approximations from quartic (degree 4), quintic (degree 5), and higher degree polynomials giving us the following approximations to e:

linear	$1 + 1$	2
quadratic	$1 + 1 + \frac{1}{2}$	2.5
cubic	$1 + 1 + \frac{1}{2} + \frac{1}{6}$	$2.\overline{6}$
quartic	$1 + 1 + \frac{1}{2} + \frac{1}{6} + \frac{1}{24}$	$2.708\overline{3}$
quintic	$1 + 1 + \frac{1}{2} + \frac{1}{6} + \frac{1}{24} + \frac{1}{120}$	$2.716\overline{6}$

We see an interesting pattern here. The number e is being approximated by the sum

$$1 + 1 + \frac{1}{2} + \frac{1}{6} + \frac{1}{24} + \frac{1}{120} + \cdots + \frac{1}{n!} \qquad (8.3.1)$$

for increasing values of n.

And just as we did with Riemann sums, we can use the summation notation as a shorthand[1] for writing the sum in Equation (8.3.1) so that

$$e \approx 1 + 1 + \frac{1}{2} + \frac{1}{6} + \frac{1}{24} + \frac{1}{120} + \cdots + \frac{1}{n!} = \sum_{k=0}^{n} \frac{1}{k!}. \qquad (8.3.2)$$

We can calculate this sum using as large an n as we want, and the larger n is the more accurate the approximation (8.3.2) is. Ultimately, this argument shows that we can write the number e as the infinite sum

$$e = \sum_{k=0}^{\infty} \frac{1}{k!}. \qquad (8.3.3)$$

This sum is an example of a *series* (or an *infinite series*). Note that the series (8.3.3) is the sum of the terms of the (infinite) sequence $\{\frac{1}{n!}\}$. In general, we use the following notation and terminology.

---

[1]Note that 0! appears in Equation (8.3.2). By definition, $0! = 1$.

Definition 8.3.1. An **infinite series** of real numbers is the sum of the entries in an infinite sequence of real numbers. In other words, an infinite series is sum of the form

$$a_1 + a_2 + \cdots + a_n + \cdots = \sum_{k=1}^{\infty} a_k,$$

where $a_1, a_2, \ldots,$ are real numbers.

We will normally use summation notation to identify a series. If the series adds the entries of a sequence $\{a_n\}_{n \geq 1}$, then we will write the series as

$$\sum_{k \geq 1} a_k$$

or

$$\sum_{k=1}^{\infty} a_k.$$

Note well: each of these notations is simply shorthand for the infinite sum $a_1 + a_2 + \cdots + a_n + \cdots$.

Is it even possible to sum an infinite list of numbers? This question is one whose answer shouldn't come as a surprise. After all, we have used the definite integral to add up continuous (infinite) collections of numbers, so summing the entries of a sequence might be even easier. Moreover, we have already examined the special case of geometric series in the previous section. The next activity provides some more insight into how we make sense of the process of summing an infinite list of numbers.

Activity 8.3.2. Consider the series

$$\sum_{k=1}^{\infty} \frac{1}{k^2}.$$

While it is physically impossible to add an infinite collection of numbers, we can, of course, add any finite collection of them. In what follows, we investigate how understanding how to find the nth partial sum (that is, the sum of the first n terms) enables us to make sense of the infinite sum.

a. Sum the first two numbers in this series. That is, find a numeric value for

$$\sum_{k=1}^{2} \frac{1}{k^2}$$

b. Next, add the first three numbers in the series.

c. Continue adding terms in this series to complete Table 8.3.2. Carry each sum to at least 8 decimal places.

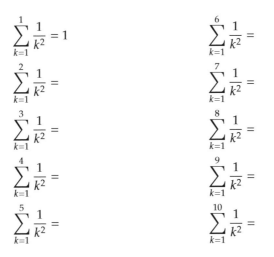

$$\sum_{k=1}^{1} \frac{1}{k^2} = 1 \qquad\qquad \sum_{k=1}^{6} \frac{1}{k^2} =$$

$$\sum_{k=1}^{2} \frac{1}{k^2} = \qquad\qquad \sum_{k=1}^{7} \frac{1}{k^2} =$$

$$\sum_{k=1}^{3} \frac{1}{k^2} = \qquad\qquad \sum_{k=1}^{8} \frac{1}{k^2} =$$

$$\sum_{k=1}^{4} \frac{1}{k^2} = \qquad\qquad \sum_{k=1}^{9} \frac{1}{k^2} =$$

$$\sum_{k=1}^{5} \frac{1}{k^2} = \qquad\qquad \sum_{k=1}^{10} \frac{1}{k^2} =$$

Table 8.3.2: Sums of some of the first terms of the series $\sum_{k=1}^{\infty} \frac{1}{k^2}$

d. The sums in the table in (c) form a sequence whose nth term is $S_n = \sum_{k=1}^{n} \frac{1}{k^2}$. Based on your calculations in the table, do you think the sequence $\{S_n\}$ converges or diverges? Explain. How do you think this sequence $\{S_n\}$ is related to the series $\sum_{k=1}^{\infty} \frac{1}{k^2}$?

The example in Activity 8.3.2 illustrates how we define the sum of an infinite series. We can add up the first n terms of the series to obtain a new sequence of numbers (called the *sequence of partial sums*). Provided that sequence converges, the corresponding infinite series is said to converge, and we say that we can find the sum of the series.

Definition 8.3.3. The n**th partial sum** of the series $\sum_{k=1}^{\infty} a_k$ is the finite sum $S_n = \sum_{k=1}^{n} a_k$.

In other words, the nth partial sum S_n of a series is the sum of the first n terms in the series, or

$$S_n = a_1 + a_2 + \cdots + a_n.$$

We then investigate the behavior of a given series by examining the sequence

$$S_1, S_2, \ldots, S_n, \ldots$$

of its partial sums. If the sequence of partial sums converges to some finite number, then we say that the corresponding series *converges*. Otherwise, we say the series *diverges*. From our work in Activity 8.3.2, the series

$$\sum_{k=1}^{\infty} \frac{1}{k^2}$$

appears to converge to some number near 1.54977. We formalize the concept of convergence and divergence of an infinite series in the following definition.

Definition 8.3.4. The infinite series

$$\sum_{k=1}^{\infty} a_k$$

converges (or is **convergent**) if the sequence $\{S_n\}$ of partial sums converges, where

$$S_n = \sum_{k=1}^{n} a_k.$$

If $\lim_{n \to \infty} S_n = S$, then we call S the sum of the series $\sum_{k=1}^{\infty} a_k$. That is,

$$\sum_{k=1}^{\infty} a_k = \lim_{n \to \infty} S_n = S.$$

If the sequence of partial sums does not converge, then the series

$$\sum_{k=1}^{\infty} a_k$$

diverges (or is **divergent**).

The early terms in a series do not contribute to whether or not the series converges or diverges. Rather, the convergence or divergence of a series

$$\sum_{k=1}^{\infty} a_k$$

is determined by what happens to the terms a_k for very large values of k. To see why, suppose that m is some constant larger than 1. Then

$$\sum_{k=1}^{\infty} a_k = (a_1 + a_2 + \cdots + a_m) + \sum_{k=m+1}^{\infty} a_k.$$

Since $a_1 + a_2 + \cdots + a_m$ is a finite number, the series $\sum_{k=1}^{\infty} a_k$ will converge if and only if the series $\sum_{k=m+1}^{\infty} a_k$ converges. Because the starting index of the series doesn't affect whether the series converges or diverges, we will often just write

$$\sum a_k$$

when we are interested in questions of convergence/divergence and not necessarily the exact sum of a series.

In Section 8.2 we encountered the special family of infinite geometric series whose convergence or divergence we completely determined. Recall that a geometric series is a special series of the form $\sum_{k=0}^{\infty} ar^k$ where a and r are real numbers (and $r \neq 1$). We found that the nth partial sum S_n of a geometric series is given by the convenient formula

$$S_n = \frac{1 - r^n}{1 - r},$$

and thus a geometric series converges if $|r| < 1$. Geometric series diverge for all other values of r. While we have completely determined the convergence or divergence of geometric series, it is generally a difficult question to determine if a given nongeometric series converges or diverges. There are several tests we can use that we will consider in the following sections.

8.3.2 The Divergence Test

The first question we ask about any infinite series is usually "Does the series converge or diverge?" There is a straightforward way to check that certain series diverge; we explore this test in the next activity.

> **Activity 8.3.3.** If the *series* $\sum a_k$ converges, then an important result necessarily follows regarding the *sequence* $\{a_n\}$. This activity explores this result.
>
> Assume that the series $\sum_{k=1}^{\infty} a_k$ converges and has sum equal to L.
>
> a. What is the nth partial sum S_n of the series $\sum_{k=1}^{\infty} a_k$?
>
> b. What is the $(n-1)$st partial sum S_{n-1} of the series $\sum_{k=1}^{\infty} a_k$?
>
> c. What is the difference between the nth partial sum and the $(n-1)$st partial sum of the series $\sum_{k=1}^{\infty} a_k$?
>
> d. Since we are assuming that $\sum_{k=1}^{\infty} a_k = L$, what does that tell us about $\lim_{n \to \infty} S_n$? Why? What does that tell us about $\lim_{n \to \infty} S_{n-1}$? Why?
>
> e. Combine the results of the previous two parts of this activity to determine $\lim_{n \to \infty} a_n = \lim_{n \to \infty}(S_n - S_{n-1})$.

The result of Activity 8.3.3 is the following important conditional statement:

> If the series $\sum_{k=1}^{\infty} a_k$ converges, then the sequence $\{a_k\}$ of kth terms converges to 0.

It is logically equivalent to say that if the sequence $\{a_k\}$ of n terms does not converge to 0, then the series $\sum_{k=1}^{\infty} a_k$ cannot converge. This statement is called the Divergence Test.

The Divergence Test

If $\lim_{k \to \infty} a_k \neq 0$, then the series $\sum a_k$ diverges.

> **Activity 8.3.4.** Determine if the Divergence Test applies to the following series. If the test does not apply, explain why. If the test does apply, what does it tell us about the series?
>
> a. $\sum \frac{k}{k+1}$ b. $\sum (-1)^k$ c. $\sum \frac{1}{k}$

Note well: be very careful with the Divergence Test. This test only tells us what happens to a series if the terms of the corresponding sequence do not converge to 0. If the sequence of the terms of the series does converge to 0, the Divergence Test does not apply: indeed, as we will soon see, a series whose terms go to zero may either converge or diverge.

8.3.3 The Integral Test

The Divergence Test settles the questions of divergence or convergence of series $\sum a_k$ in which $\lim_{k \to \infty} a_k \neq 0$. Determining the convergence or divergence of series $\sum a_k$ in which $\lim_{k \to \infty} a_k = 0$ turns out to be more complicated. Often, we have to investigate the sequence of partial sums or apply some other technique.

As an example, consider the *harmonic* series[2]

$$\sum_{k=1}^{\infty} \frac{1}{k}.$$

Table 8.3.5 shows some partial sums of this series.

$$\sum_{k=1}^{1} \frac{1}{k} = 1$$

$$\sum_{k=1}^{2} \frac{1}{k} = 1.5$$

$$\sum_{k=1}^{3} \frac{1}{k} = 1.833333333$$

$$\sum_{k=1}^{4} \frac{1}{k} = 2.083333333$$

$$\sum_{k=1}^{5} \frac{1}{k} = 2.283333333$$

$$\sum_{k=1}^{6} \frac{1}{k} = 2.450000000$$

$$\sum_{k=1}^{7} \frac{1}{k} = 2.592857143$$

$$\sum_{k=1}^{8} \frac{1}{k} = 2.717857143$$

$$\sum_{k=1}^{9} \frac{1}{k} = 2.828968254$$

$$\sum_{k=1}^{10} \frac{1}{k} = 2.928968254$$

Table 8.3.5: Sums of some of the first terms of the sequence $\sum_{k=1}^{\infty} \frac{1}{k}$.

This information doesn't seem to be enough to tell us if the series $\sum_{k=1}^{\infty} \frac{1}{k}$ converges or diverges. The partial sums could eventually level off to some fixed number or continue to grow without bound. Even if we look at larger partial sums, such as $\sum_{n=1}^{1000} \frac{1}{k} \approx 7.485470861$, the result doesn't particularly sway us one way or another. The Integral Test is one way to determine whether or not the harmonic series converges, and we explore this further in the next activity.

Activity 8.3.5. Consider the harmonic series $\sum_{k=1}^{\infty} \frac{1}{k}$. Recall that the harmonic series will converge provided that its sequence of partial sums converges. The nth partial sum S_n of the series $\sum_{k=1}^{\infty} \frac{1}{k}$ is

$$S_n = \sum_{k=1}^{n} \frac{1}{k}$$
$$= 1 + \frac{1}{2} + \frac{1}{3} + \cdots + \frac{1}{n}$$
$$= 1(1) + (1)\left(\frac{1}{2}\right) + (1)\left(\frac{1}{3}\right) + \cdots + (1)\left(\frac{1}{n}\right).$$

Through this last expression for S_n, we can visualize this partial sum as a sum of areas of rectangles with heights $\frac{1}{m}$ and bases of length 1, as shown in Figure 8.3.6, which uses the 9th partial sum.

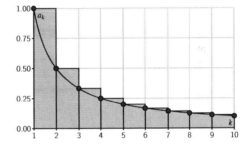

Figure 8.3.6: A picture of the 9th partial sum of the harmonic series as a sum of areas of rectangles.

The graph of the continuous function f defined by $f(x) = \frac{1}{x}$ is overlaid on this plot.

a. Explain how this picture represents a particular Riemann sum.

b. What is the definite integral that corresponds to the Riemann sum you considered in (a)?

c. Which is larger, the definite integral in (b), or the corresponding partial sum S_9 of the series? Why?

d. If instead of considering the 9th partial sum, we consider the nth partial sum, and we let n go to infinity, we can then compare the series $\sum_{k=1}^{\infty} \frac{1}{k}$ to the improper integral $\int_1^{\infty} \frac{1}{x} \, dx$. Which of these quantities is larger? Why?

e. Does the improper integral $\int_1^{\infty} \frac{1}{x} \, dx$ converge or diverge? What does that result, together with your work in (d), tell us about the series $\sum_{k=1}^{\infty} \frac{1}{k}$?

The ideas from Activity 8.3.5 hold more generally. Suppose that f is a continuous decreasing function and that $a_k = f(k)$ for each value of k. Consider the corresponding series $\sum_{k=1}^{\infty} a_k$. The partial sum

$$S_n = \sum_{k=1}^{n} a_k$$

can always be viewed as a left hand Riemann sum of $f(x)$ using rectangles with heights given by the values a_k and bases of length 1. A representative picture is shown at left in Figure 8.3.7. Since f is a decreasing function, we have that

$$S_n > \int_{1}^{n} f(x)\, dx.$$

Taking limits as n goes to infinity shows that

$$\sum_{k=1}^{\infty} a_k > \int_{1}^{\infty} f(x)\, dx.$$

Therefore, if the improper integral $\int_{1}^{\infty} f(x)\, dx$ diverges, so does the series $\sum_{k=1}^{\infty} a_k$.

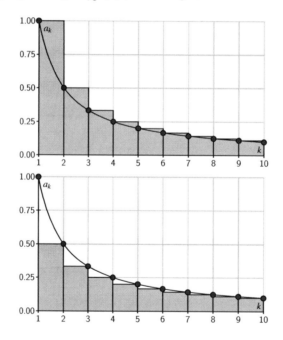

Figure 8.3.7: Comparing an improper integral to a series

What's more, if we look at the right hand Riemann sums of f on $[1, n]$ as shown at right in Figure 8.3.7, we see that

$$\int_{1}^{\infty} f(x)\, dx > \sum_{k=2}^{\infty} a_k.$$

So if $\int_1^\infty f(x)\,dx$ converges, then so does $\sum_{k=2}^\infty a_k$, which also means that the series $\sum_{k=1}^\infty a_k$ converges. Our preceding discussion has demonstrated the truth of the Integral Test.

The Integral Test

Let f be a real valued function and assume f is decreasing and positive for all x larger than some number c. Let $a_k = f(k)$ for each positive integer k.

   a. If the improper integral $\int_c^\infty f(x)\,dx$ converges, then the series $\sum_{k=1}^\infty a_k$ converges.

   b. If the improper integral $\int_c^\infty f(x)\,dx$ diverges, then the series $\sum_{k=1}^\infty a_k$ diverges.

Note that the Integral Test compares a given infinite series to a natural, corresponding improper integral and basically says that the infinite series and corresponding improper integral both have the same convergence status. In the next activity, we apply the Integral Test to determine the convergence or divergence of a class of important series.

Activity 8.3.6. The series $\sum \frac{1}{k^p}$ are special series called p-series. We have already seen that the p-series with $p = 1$ (the harmonic series) diverges. We investigate the behavior of other p-series in this activity.

   a. Evaluate the improper integral $\int_1^\infty \frac{1}{x^2}\,dx$. Does the series $\sum_{k=1}^\infty \frac{1}{k^2}$ converge or diverge? Explain.

   b. Evaluate the improper integral $\int_1^\infty \frac{1}{x^p}\,dx$ where $p > 1$. For which values of p can we conclude that the series $\sum_{k=1}^\infty \frac{1}{k^p}$ converges?

   c. Evaluate the improper integral $\int_1^\infty \frac{1}{x^p}\,dx$ where $p < 1$. What does this tell us about the corresponding p-series $\sum_{k=1}^\infty \frac{1}{k^p}$?

   d. Summarize your work in this activity by completing the following statement.

   The p-series $\sum_{k=1}^\infty \frac{1}{k^p}$ converges if and only if _____.

8.3.4 The Limit Comparison Test

The Integral Test allows us to determine the convergence of an entire family of series: the p-series. However, we have seen that it is, in general, difficult to integrate functions, so the Integral Test is not one that we can use all of the time. In fact, even for a relatively simple series like $\sum \frac{k^2+1}{k^4+2k+2}$, the Integral Test is not an option. In this section we will develop a test that we can use to apply to series of rational functions like this by comparing their behavior to the behavior of p-series.

Activity 8.3.7. Consider the series $\sum \frac{k+1}{k^3+2}$. Since the convergence or divergence of a series only depends on the behavior of the series for large values of k, we might examine the terms of this series more closely as k gets large.

    a. By computing the value of $\frac{k+1}{k^3+2}$ for $k = 100$ and $k = 1000$, explain why the terms $\frac{k+1}{k^3+2}$ are essentially $\frac{k}{k^3}$ when k is large.

    b. Let's formalize our observations in (a) a bit more. Let $a_k = \frac{k+1}{k^3+2}$ and $b_k = \frac{k}{k^3}$. Calculate

$$\lim_{k \to \infty} \frac{a_k}{b_k}.$$

    What does the value of the limit tell you about a_k and b_k for large values of k? Compare your response from part (a).

    c. Does the series $\sum \frac{k}{k^3}$ converge or diverge? Why? What do you think that tells us about the convergence or divergence of the series $\sum \frac{k+1}{k^3+2}$? Explain.

Activity 8.3.7 illustrates how we can compare one series with positive terms to another whose behavior (that is, whether the series converges or diverges) we know. More generally, suppose we have two series $\sum a_k$ and $\sum b_k$ with positive terms and we know the behavior of the series $\sum a_k$. Recall that the convergence or divergence of a series depends only on what happens to the terms of the series for large values of k, so if we know that a_k and b_k are essentially proportional to each other for large k, then the two series $\sum a_k$ and $\sum b_k$ should behave the same way. In other words, if there is a positive finite constant c such that

$$\lim_{k \to \infty} \frac{b_k}{a_k} = c,$$

then $b_k \approx ca_k$ for large values of k. So

$$\sum b_k \approx \sum ca_k = c \sum a_k.$$

Since multiplying by a nonzero constant does not affect the convergence or divergence of a series, it follows that the series $\sum a_k$ and $\sum b_k$ either both converge or both diverge. The formal statement of this fact is called the Limit Comparison Test.

> ### The Limit Comparison Test
>
> Let $\sum a_k$ and $\sum b_k$ be series with positive terms. If
>
> $$\lim_{k \to \infty} \frac{b_k}{a_k} = c$$
>
> for some positive (finite) constant c, then $\sum a_k$ and $\sum b_k$ either both converge or both diverge.

In essence, the Limit Comparison Test shows that if we have a series $\sum \frac{p(k)}{q(k)}$ of rational functions where $p(k)$ is a polynomial of degree m and $q(k)$ a polynomial of degree l, then the series $\sum \frac{p(k)}{q(k)}$ will behave like the series $\sum \frac{k^m}{k^l}$. So this test allows us to quickly and easily determine the convergence or divergence of series whose summands are rational functions.

Activity 8.3.8. Use the Limit Comparison Test to determine the convergence or divergence of the series

$$\sum \frac{3k^2 + 1}{5k^4 + 2k + 2}.$$

by comparing it to the series $\sum \frac{1}{k^2}$.

8.3.5 The Ratio Test

The Limit Comparison Test works well if we can find a series with known behavior to compare. But such series are not always easy to find. In this section we will examine a test that allows us to examine the behavior of a series by comparing it to a geometric series, without knowing in advance which geometric series we need.

Activity 8.3.9. Consider the series defined by

$$\sum_{k=1}^{\infty} \frac{2^k}{3^k - k}. \qquad (8.3.4)$$

This series is not a geometric series, but this activity will illustrate how we might compare this series to a geometric one. Recall that a series $\sum a_k$ is geometric if the ratio $\frac{a_{k+1}}{a_k}$ is always the same. For the series in (8.3.4), note that $a_k = \frac{2^k}{3^k - k}$.

a. To see if $\sum \frac{2^k}{3^k - k}$ is comparable to a geometric series, we analyze the ratios of successive terms in the series. Complete Table 8.3.8, listing your calculations to at least 8 decimal places.

k	$\dfrac{a_{k+1}}{a_k}$
5	
10	
20	
21	
22	
23	
24	
25	

Table 8.3.8: Ratios of successive terms in the series $\sum \frac{2^k}{3^k - k}$

b. Based on your calculations in Table 8.3.8, what can we say about the ratio $\frac{a_{k+1}}{a_k}$ if k

is large?

c. Do you agree or disagree with the statement: "the series $\sum \frac{2^k}{3^k - k}$ is approximately geometric when k is large"? If not, why not? If so, do you think the series $\sum \frac{2^k}{3^k - k}$ converges or diverges? Explain.

We can generalize the argument in Activity 8.3.9 in the following way. Consider the series $\sum a_k$. If

$$\frac{a_{k+1}}{a_k} \approx r$$

for large values of k, then $a_{k+1} \approx r a_k$ for large k and the series $\sum a_k$ is approximately the geometric series $\sum a r^k$ for large k. Since the geometric series with ratio r converges only for $-1 < r < 1$, we see that the series $\sum a_k$ will converge if

$$\lim_{k \to \infty} \frac{a_{k+1}}{a_k} = r$$

for a value of r such that $|r| < 1$. This result is known as the Ratio Test.

The Ratio Test

Let $\sum a_k$ be an infinite series. Suppose

$$\lim_{k \to \infty} \frac{|a_{k+1}|}{|a_k|} = r.$$

a. If $0 \leq r < 1$, then the series $\sum a_k$ converges.

b. If $1 < r$, then the series $\sum a_k$ diverges.

c. If $r = 1$, then the test is inconclusive.

Note well: The Ratio Test takes a given series and looks at the limit of the ratio of consecutive terms; in so doing, the test is essentially asking, "is this series approximately geometric?" If the series can be thought of as essentially geometric, the test use the limiting common ratio to determine if the given series converges.

We have now encountered several tests for determining convergence or divergence of series. The Divergence Test can be used to show that a series diverges, but never to prove that a series converges. We used the Integral Test to determine the convergence status of an entire class of series, the p-series. The Limit Comparison Test works well for series that involve rational functions and which can therefore by compared to p-series. Finally, the Ratio Test allows us to compare our series to a geometric series; it is particularly useful for series that involve nth powers and factorials. Two other tests, the Direct Comparison Test and the Root Test, are discussed in the exercises. Now it is time for some practice.

Activity 8.3.10. Determine whether each of the following series converges or diverges. Explicitly state which test you use.

a. $\sum \frac{k}{2^k}$

c. $\sum \frac{10^k}{k!}$

b. $\sum \frac{k^3+2}{k^2+1}$

d. $\sum \frac{k^3-2k^2+1}{k^6+4}$

Summary

- An infinite series is a sum of the elements in an infinite sequence. In other words, an infinite series is a sum of the form

$$a_1 + a_2 + \cdots + a_n + \cdots = \sum_{k=1}^{\infty} a_k$$

where a_k is a real number for each positive integer k.

- The nth partial sum S_n of the series $\sum_{k=1}^{\infty} a_k$ is the sum of the first n terms of the series. That is,

$$S_n = a_1 + a_2 + \cdots + a_n = \sum_{k=1}^{n} a_k.$$

- The sequence $\{S_n\}$ of partial sums of a series $\sum_{k=1}^{\infty} a_k$ tells us about the convergence or divergence of the series. In particular

  o The series $\sum_{k=1}^{\infty} a_k$ converges if the sequence $\{S_n\}$ of partial sums converges. In this case we say that the series is the limit of the sequence of partial sums and write

$$\sum_{k=1}^{\infty} a_k = \lim_{n \to \infty} S_n.$$

  o The series $\sum_{k=1}^{\infty} a_k$ diverges if the sequence $\{S_n\}$ of partial sums diverges.

Exercises

1. Given:
$$A_n = \frac{80}{8^n}$$
Determine:

(a) whether $\sum_{n=1}^{\infty} (A_n)$ is convergent.

(b) whether $\{A_n\}$ is convergent.
If convergent, enter the limit of convergence. If not, enter DIV.

2. Consider the series $\sum_{n=1}^{\infty} \frac{10}{n+2}$. Let s_n be the n-th partial sum; that is,

$$s_n = \sum_{i=1}^{n} \frac{10}{i+2}.$$

481

Find s_4 and s_8

$s_4 =$

$s_8 =$

3. Let

$$a_n = \frac{2n}{10n + 7}.$$

For the following answer blanks, decide whether the given sequence or series is convergent or divergent. If convergent, enter the limit (for a sequence) or the sum (for a series). If divergent, enter 'infinity' if it diverges to ∞, '-infinity' if it diverges to $-\infty$ or 'DNE' otherwise.

(a) The series $\displaystyle\sum_{n=1}^{\infty} \frac{2n}{10n + 7}$.

(b) The sequence $\left\{\dfrac{2n}{10n + 7}\right\}$.

4. Compute the value of the following improper integral. If it converges, enter its value. Enter *infinity* if it diverges to ∞, and *-infinity* if it diverges to $-\infty$. Otherwise, enter *diverges*.

$$\int_1^{\infty} \frac{3\,dx}{x^2 + 1} = $$

Does the series $\displaystyle\sum_{n=1}^{\infty} \frac{3}{n^2 + 1}$ converge or diverge? [Choose: converges | diverges to +infinity | diverges to -infinity | diverges]

5. In this exercise we investigate the sequence $\left\{\frac{b^n}{n!}\right\}$ for any constant b.

a. Use the Ratio Test to determine if the series $\sum \frac{10^k}{k!}$ converges or diverges.

b. Now apply the Ratio Test to determine if the series $\sum \frac{b^k}{k!}$ converges for any constant b.

c. Use your result from (b) to decide whether the sequence $\left\{\frac{b^n}{n!}\right\}$ converges or diverges. If the sequence $\left\{\frac{b^n}{n!}\right\}$ converges, to what does it converge? Explain your reasoning.

6. There is a test for convergence similar to the Ratio Test called the *Root Test*. Suppose we have a series $\sum a_k$ of positive terms so that $a_n \to 0$ as $n \to \infty$.

a. Assume

$$\sqrt[n]{a_n} \to r$$

as n goes to infinity. Explain why this tells us that $a_n \approx r^n$ for large values of n.

b. Using the result of part (a), explain why $\sum a_k$ looks like a geometric series when n is big. What is the ratio of the geometric series to which $\sum a_k$ is comparable?

c. Use what we know about geometric series to determine that values of r so that $\sum a_k$ converges if $\sqrt[n]{a_n} \to r$ as $n \to \infty$.

7. The associative and distributive laws of addition allow us to add finite sums in any order we want. That is, if $\sum_{k=0}^{n} a_k$ and $\sum_{k=0}^{n} b_k$ are finite sums of real numbers, then

$$\sum_{k=0}^{n} a_k + \sum_{k=0}^{n} b_k = \sum_{k=0}^{n} (a_k + b_k).$$

However, we do need to be careful extending rules like this to infinite series.

a. Let $a_n = 1 + \frac{1}{2^n}$ and $b_n = -1$ for each nonnegative integer n.

- Explain why the series $\sum_{k=0}^{\infty} a_k$ and $\sum_{k=0}^{\infty} b_k$ both diverge.
- Explain why the series $\sum_{k=0}^{\infty} (a_k + b_k)$ converges.
- Explain why

$$\sum_{k=0}^{\infty} a_k + \sum_{k=0}^{\infty} b_k \neq \sum_{k=0}^{\infty} (a_k + b_k).$$

This shows that it is possible to have to two divergent series $\sum_{k=0}^{\infty} a_k$ and $\sum_{k=0}^{\infty} b_k$ but yet have the series $\sum_{k=0}^{\infty} (a_k + b_k)$ converge.

b. While part (a) shows that we cannot add series term by term in general, we can under reasonable conditions. The problem in part (a) is that we tried to add divergent series. In this exercise we will show that if $\sum a_k$ and $\sum b_k$ are convergent series, then $\sum (a_k + b_k)$ is a convergent series and

$$\sum (a_k + b_k) = \sum a_k + \sum b_k.$$

- Let A_n and B_n be the nth partial sums of the series $\sum_{k=1}^{\infty} a_k$ and $\sum_{k=1}^{\infty} b_k$, respectively. Explain why

$$A_n + B_n = \sum_{k=1}^{n} (a_k + b_k).$$

- Use the previous result and properties of limits to show that

$$\sum_{k=1}^{\infty} (a_k + b_k) = \sum_{k=1}^{\infty} a_k + \sum_{k=1}^{\infty} b_k.$$

(Note that the starting point of the sum is irrelevant in this problem, so it doesn't matter where we begin the sum.)

c. Use the prior result to find the sum of the series $\sum_{k=0}^{\infty} \frac{2^k + 3^k}{5^k}$.

8. In the Limit Comparison Test we compared the behavior of a series to one whose behavior we know. In that test we use the limit of the ratio of corresponding terms of the series to determine if the comparison is valid. In this exercise we see how we can compare two series directly, term by term, without using a limit of sequence. First we consider an example.

a. Consider the series

$$\sum \frac{1}{k^2} \text{ and } \sum \frac{1}{k^2 + k}.$$

We know that the series $\sum \frac{1}{k^2}$ is a p-series with $p = 2 > 1$ and so $\sum \frac{1}{k^2}$ converges. In this part of the exercise we will see how to use information about $\sum \frac{1}{k^2}$ to determine information about $\sum \frac{1}{k^2+k}$. Let $a_k = \frac{1}{k^2}$ and $b_k = \frac{1}{k^2+k}$.

   i) Let S_n be the nth partial sum of $\sum \frac{1}{k^2}$ and T_n the nth partial sum of $\sum \frac{1}{k^2+k}$. Which is larger, S_1 or T_1? Why?

   ii) Recall that

$$S_2 = S_1 + a_2 \text{ and } T_2 = T_1 + b_2.$$

   Which is larger, a_2 or b_2? Based on that answer, which is larger, S_2 or T_2?

   iii) Recall that

$$S_3 = S_2 + a_3 \text{ and } T_3 = T_2 + b_3.$$

   Which is larger, a_3 or b_3? Based on that answer, which is larger, S_3 or T_3?

   iv) Which is larger, a_n or b_n? Explain. Based on that answer, which is larger, S_n or T_n?

   v) Based on your response to the previous part of this exercise, what relationship do you expect there to be between $\sum \frac{1}{k^2}$ and $\sum \frac{1}{k^2+k}$? Do you expect $\sum \frac{1}{k^2+k}$ to converge or diverge? Why?

b. The example in the previous part of this exercise illustrates a more general result. Explain why the Direct Comparison Test, stated here, works.

The Direct Comparison Test

If

$$0 \le b_k \le a_k$$

for every k, then we must have

$$0 \le \sum b_k \le \sum a_k$$

a. If $\sum a_k$ converges, then $\sum b_k$ converges.

b. If $\sum b_k$ diverges, then $\sum a_k$ diverges.

Note 8.3.9. This comparison test applies only to series with nonnegative terms.

   i) Use the Direct Comparison Test to determine the convergence or divergence of the series $\sum \frac{1}{k-1}$. Hint: Compare to the harmonic series.

   ii) Use the Direct Comparison Test to determine the convergence or divergence of the series $\sum \frac{k}{k^3+1}$.

8.4 Alternating Series

Motivating Questions

- What is an alternating series?

- What does it mean for an alternating series to converge?

- Under what conditions does an alternating series converge? Why?

- How well does the nth partial sum of a convergent alternating series approximate the actual sum of the series? Why?

- What is the difference between absolute convergence and conditional convergence?

In our study of series so far, almost every series that we've considered has exclusively non-negative terms. Of course, it is possible to consider series that have some negative terms. For instance, if we consider the geometric series

$$2 - \frac{4}{3} + \frac{8}{9} - \cdots + 2\left(-\frac{2}{3}\right)^n + \cdots,$$

which has $a = 2$ and $r = -\frac{2}{3}$, we see that not only does every other term alternate in sign, but also that this series converges to

$$S = \frac{a}{1-r} = \frac{2}{1-\left(-\frac{2}{3}\right)} = \frac{6}{5}.$$

In Preview Activity 8.4.1 and our following discussion, we investigate the behavior of similar series where consecutive terms have opposite signs.

> **Preview Activity 8.4.1.** Preview Activity 8.3.1 showed how we can approximate the number e with linear, quadratic, and other polynomial approximations. We use a similar approach in this activity to obtain linear and quadratic approximations to $\ln(2)$. Along the way, we encounter a type of series that is different than most of the ones we have seen so far. Throughout this activity, let $f(x) = \ln(1 + x)$.
>
> a. Find the tangent line to f at $x = 0$ and use this linearization to approximate $\ln(2)$. That is, find $L(x)$, the tangent line approximation to $f(x)$, and use the fact that $L(1) \approx f(1)$ to estimate $\ln(2)$.
>
> b. The linearization of $\ln(1 + x)$ does not provide a very good approximation to $\ln(2)$ since 1 is not that close to 0. To obtain a better approximation, we alter our approach; instead of using a straight line to approximate $\ln(2)$, we use a quadratic function to account for the concavity of $\ln(1 + x)$ for x close to 0. With the lin-

earization, both the function's value and slope agree with the linearization's value and slope at $x = 0$. We will now make a quadratic approximation $P_2(x)$ to $f(x) = \ln(1 + x)$ centered at $x = 0$ with the property that $P_2(0) = f(0)$, $P'_2(0) = f'(0)$, and $P''_2(0) = f''(0)$.

i. Let $P_2(x) = x - \frac{x^2}{2}$. Show that $P_2(0) = f(0)$, $P'_2(0) = f'(0)$, and $P''_2(0) = f''(0)$. Use $P_2(x)$ to approximate $\ln(2)$ by using the fact that $P_2(1) \approx f(1)$.

ii. We can continue approximating $\ln(2)$ with polynomials of larger degree whose derivatives agree with those of f at 0. This makes the polynomials fit the graph of f better for more values of x around 0. For example, let $P_3(x) = x - \frac{x^2}{2} + \frac{x^3}{3}$. Show that $P_3(0) = f(0)$, $P'_3(0) = f'(0)$, $P''_3(0) = f''(0)$, and $P'''_3(0) = f'''(0)$. Taking a similar approach to preceding questions, use $P_3(x)$ to approximate $\ln(2)$.

iii. If we used a degree 4 or degree 5 polynomial to approximate $\ln(1 + x)$, what approximations of $\ln(2)$ do you think would result? Use the preceding questions to conjecture a pattern that holds, and state the degree 4 and degree 5 approximation.

8.4.1 The Alternating Series Test

Preview Activity 8.4.1 gives us several approximations to $\ln(2)$, the linear approximation is 1 and the quadratic approximation is $1 - \frac{1}{2} = \frac{1}{2}$. If we continue this process we will obtain approximations from cubic, quartic (degree 4), quintic (degree 5), and higher degree polynomials giving us the approximations to $\ln(2)$ in Table 8.4.1.

The pattern here shows the fact that the number $\ln(2)$ can be approximated by the partial sums of the infinite series

$$\sum_{k=1}^{\infty} (-1)^{k+1} \frac{1}{k} \qquad (8.4.1)$$

where the alternating signs are determined by the factor $(-1)^{k+1}$.

linear	1	1
quadratic	$1 - \frac{1}{2}$	0.5
cubic	$1 - \frac{1}{2} + \frac{1}{3}$	$0.8\overline{3}$
quartic	$1 - \frac{1}{2} + \frac{1}{3} - \frac{1}{4}$	$0.58\overline{3}$
quintic	$1 - \frac{1}{2} + \frac{1}{3} - \frac{1}{4} + \frac{1}{5}$	$0.78\overline{3}$

Table 8.4.1

Using computational technology, we find that 0.6881721793 is the sum of the first 100 terms in this series. As a comparison, $\ln(2) \approx 0.6931471806$. This shows that even though the series (8.4.1) converges to $\ln(2)$, it must do so quite slowly, since the sum of the first 100 terms isn't particularly close to $\ln(2)$. We will investigate the issue of how quickly an alternating series converges later in this section. Again, note particularly that the series (8.4.1) is different from the series we have consider earlier in that some of the terms are negative. We call such a series an *alternating series*.

Definition 8.4.2. An alternating series is a series of the form

$$\sum_{k=0}^{\infty} (-1)^k a_k,$$

where $a_k > 0$ for each k.

We have some flexibility in how we write an alternating series; for example, the series

$$\sum_{k=1}^{\infty} (-1)^{k+1} a_k,$$

whose index starts at $k = 1$, is also alternating. As we will soon see, there are several very nice results that hold about alternating series, while alternating series can also demonstrate some unusual behaivior.

It is important to remember that most of the series tests we have seen in previous sections apply only to series with nonnegative terms. Thus, alternating series require a different test. To investigate this idea, we return to the example in Preview Activity 8.4.1.

Activity 8.4.2. Remember that, by definition, a series converges if and only if its corresponding sequence of partial sums converges.

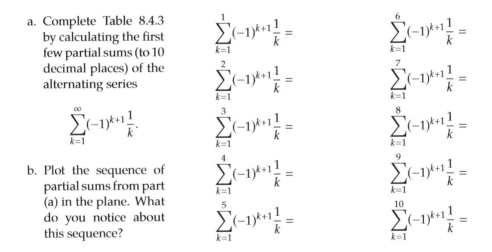

a. Complete Table 8.4.3 by calculating the first few partial sums (to 10 decimal places) of the alternating series

$$\sum_{k=1}^{\infty} (-1)^{k+1} \frac{1}{k}.$$

b. Plot the sequence of partial sums from part (a) in the plane. What do you notice about this sequence?

$$\sum_{k=1}^{1} (-1)^{k+1} \frac{1}{k} =$$

$$\sum_{k=1}^{2} (-1)^{k+1} \frac{1}{k} =$$

$$\sum_{k=1}^{3} (-1)^{k+1} \frac{1}{k} =$$

$$\sum_{k=1}^{4} (-1)^{k+1} \frac{1}{k} =$$

$$\sum_{k=1}^{5} (-1)^{k+1} \frac{1}{k} =$$

$$\sum_{k=1}^{6} (-1)^{k+1} \frac{1}{k} =$$

$$\sum_{k=1}^{7} (-1)^{k+1} \frac{1}{k} =$$

$$\sum_{k=1}^{8} (-1)^{k+1} \frac{1}{k} =$$

$$\sum_{k=1}^{9} (-1)^{k+1} \frac{1}{k} =$$

$$\sum_{k=1}^{10} (-1)^{k+1} \frac{1}{k} =$$

Table 8.4.3: Partial sums of the alternating series $\sum_{k=1}^{\infty} (-1)^{k+1} \frac{1}{k}$

Activity 8.4.2 exemplifies the general behavior that any convergent alternating series will demonstrate. In this example, we see that the partial sums of the alternating harmonic series oscillate around a fixed number that turns out to be the sum of the series.

Recall that if $\lim_{k \to \infty} a_k \neq 0$, then the series $\sum a_k$ diverges by the Divergence Test. From this point forward, we will thus only consider alternating series

$$\sum_{k=1}^{\infty} (-1)^{k+1} a_k$$

in which the sequence a_k consists of positive numbers that decrease to 0. For such a series, the nth partial sum S_n satisfies

$$S_n = \sum_{k=1}^{n} (-1)^{k+1} a_k.$$

Notice that

- $S_2 = a_1 - a_2$, and since $a_1 > a_2$ we have $0 < S_2 < S_1$.

- $S_3 = S_2 + a_3$ and so $S_2 < S_3$. But $a_3 < a_2$, so $S_3 < S_1$. Thus, $0 < S_2 < S_3 < S_1$.

- $S_4 = S_3 - a_4$ and so $S_4 < S_3$. But $a_4 < a_3$, so $S_2 < S_4$. Thus, $0 < S_2 < S_4 < S_3 < S_1$.

- $S_5 = S_4 + a_5$ and so $S_4 < S_5$. But $a_5 < a_4$, so $S_5 < S_3$. Thus, $0 < S_2 < S_4 < S_5 < S_3 < S_1$.

This pattern continues as illustrated in Figure 8.4.4 (with n odd) so that each partial sum lies between the previous two partial sums.

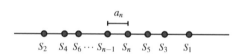

Figure 8.4.4: Partial sums of an alternating series

Note further that the absolute value of the difference between the $(n-1)$st partial sum S_{n-1} and the nth partial sum S_n is

$$|S_n - S_{n-1}| = a_n.$$

Since the sequence $\{a_n\}$ converges to 0, the distance between successive partial sums becomes as close to zero as we'd like, and thus the sequence of partial sums converges (even though we don't know the exact value to which the sequence of partial sums converges).

The preceding discussion has demonstrated the truth of the Alternating Series Test.

The Alternating Series Test

Given an alternating series

$$\sum (-1)^k a_k,$$

if the sequence $\{a_k\}$ of positive terms decreases to 0 as $k \to \infty$, then the alternating series converges.

Note particularly that if the limit of the sequence $\{a_k\}$ is not 0, then the alternating series diverges.

Activity 8.4.3. Which series converge and which diverge? Justify your answers.

a. $\sum_{k=1}^{\infty} \frac{(-1)^k}{k^2+2}$
b. $\sum_{k=1}^{\infty} \frac{(-1)^{k+1}2k}{k+5}$
c. $\sum_{k=2}^{\infty} \frac{(-1)^k}{\ln(k)}$

8.4.2 Estimating Alternating Sums

The argument for the Alternating Series Test also provides us with a method to determine how close the nth partial sum S_n is to the actual sum of a convergent alternating series. To see how this works, let S be the sum of a convergent alternating series, so

$$S = \sum_{k=1}^{\infty} (-1)^k a_k.$$

Recall that the sequence of partial sums oscillates around the sum S so that

$$|S - S_n| < |S_{n+1} - S_n| = a_{n+1}.$$

Therefore, the value of the term a_{n+1} provides an error estimate for how well the partial sum S_n approximates the actual sum S. We summarize this fact in the statement of the Alternating Series Estimation Theorem.

> **Alternating Series Estimation Theorem**
>
> If the alternating series
>
> $$\sum_{k=1}^{\infty} (-1)^{k+1} a_k$$
>
> converges and has sum S, and
>
> $$S_n = \sum_{k=1}^{n} (-1)^{k+1} a_k$$
>
> is the nth partial sum of the alternating series, then
>
> $$\left| \sum_{k=1}^{\infty} (-1)^{k+1} a_k - S_n \right| \le a_{n+1}.$$

Example 8.4.5. Let's determine how well the 100th partial sum S_{100} of

$$\sum_{k=1}^{\infty} \frac{(-1)^{k+1}}{k}$$

approximates the sum of the series.

Solution. If we let S be the sum of the series $\sum_{k=1}^{\infty} \frac{(-1)^{k+1}}{k}$, then we know that

$$|S_{100} - S| < a_{101}.$$

Now

$$a_{101} = \frac{1}{101} \approx 0.0099,$$

so the 100th partial sum is within 0.0099 of the sum of the series. We have discussed the fact (and will later verify) that

$$S = \sum_{k=1}^{\infty} \frac{(-1)^{k+1}}{k} = \ln(2),$$

and so $S \approx 0.693147$ while

$$S_{100} = \sum_{k=1}^{100} \frac{(-1)^{k+1}}{k} \approx 0.6881721793.$$

We see that the actual difference between S and S_{100} is approximately 0.0049750013, which is indeed less than 0.0099.

Activity 8.4.4. Determine the number of terms it takes to approximate the sum of the convergent alternating series

$$\sum_{k=1}^{\infty} \frac{(-1)^{k+1}}{k^4}$$

to within 0.0001.

8.4.3 Absolute and Conditional Convergence

A series such as

$$1 - \frac{1}{4} - \frac{1}{9} + \frac{1}{16} + \frac{1}{25} + \frac{1}{36} - \frac{1}{49} - \frac{1}{64} - \frac{1}{81} - \frac{1}{100} + \cdots \qquad (8.4.2)$$

whose terms are neither all nonnegative nor alternating is different from any series that we have considered to date. The behavior of these series can be rather complicated, but there is an important connection between these arbitrary series that have some negative terms and series with all nonnegative terms that we illustrate with the next activity.

Activity 8.4.5.

a. Explain why the series

$$1 - \frac{1}{4} - \frac{1}{9} + \frac{1}{16} + \frac{1}{25} + \frac{1}{36} - \frac{1}{49} - \frac{1}{64} - \frac{1}{81} - \frac{1}{100} + \cdots$$

must have a sum that is less than the series

$$\sum_{k=1}^{\infty} \frac{1}{k^2}.$$

b. Explain why the series

$$1 - \frac{1}{4} - \frac{1}{9} + \frac{1}{16} + \frac{1}{25} + \frac{1}{36} - \frac{1}{49} - \frac{1}{64} - \frac{1}{81} - \frac{1}{100} + \cdots$$

must have a sum that is greater than the series

$$\sum_{k=1}^{\infty} -\frac{1}{k^2}.$$

c. Given that the terms in the series

$$1 - \frac{1}{4} - \frac{1}{9} + \frac{1}{16} + \frac{1}{25} + \frac{1}{36} - \frac{1}{49} - \frac{1}{64} - \frac{1}{81} - \frac{1}{100} + \cdots$$

converge to 0, what do you think the previous two results tell us about the convergence status of this series?

As the example in Activity 8.4.5 suggests, if we have a series $\sum a_k$, (some of whose terms may be negative) such that $\sum |a_k|$ converges, it turns out to always be the case that the original series, $\sum a_k$, must also converge. That is, if $\sum |a_k|$ converges, then so must $\sum a_k$.

As we just observed, this is the case for the series (8.4.2), since the corresponding series of the absolute values of its terms is the convergent p-series $\sum \frac{1}{k^2}$. At the same time, there are series like the alternating harmonic series $\sum (-1)^{k+1} \frac{1}{k}$ that converge, while the corresponding series of absolute values, $\sum \frac{1}{k}$, diverges. We distinguish between these behaviors by introducing the following language.

Definition 8.4.6. Consider a series $\sum a_k$.

a. The series $\sum a_k$ *converges absolutely* (or is *absolutely convergent*) provided that $\sum |a_k|$ converges.

b. The series $\sum a_k$ *converges conditionally* (or is *conditionally convergent*) provided that $\sum |a_k|$ diverges and $\sum a_k$ converges.

In this terminology, the series (8.4.2) converges absolutely while the alternating harmonic series is conditionally convergent.

> **Activity 8.4.6.**
>
> a. Consider the series $\sum(-1)^k \frac{\ln(k)}{k}$.
>
>   i. Does this series converge? Explain.
>
>   ii. Does this series converge absolutely? Explain what test you use to determine your answer.
>
> b. Consider the series $\sum(-1)^k \frac{\ln(k)}{k^2}$.
>
>   i. Does this series converge? Explain.
>
>   ii. Does this series converge absolutely? Hint: Use the fact that $\ln(k) < \sqrt{k}$ for large values of k and then compare to an appropriate p-series.

Conditionally convergent series turn out to be very interesting. If the sequence $\{a_n\}$ decreases to 0, but the series $\sum a_k$ diverges, the conditionally convergent series $\sum(-1)^k a_k$ is right on the borderline of being a divergent series. As a result, any conditionally convergent series converges very slowly. Furthermore, some very strange things can happen with conditionally convergent series, as illustrated in some of the exercises.

8.4.4 Summary of Tests for Convergence of Series

We have discussed several tests for convergence/divergence of series in our sections and in exercises. We close this section of the text with a summary of all the tests we have encountered, followed by an activity that challenges you to decide which convergence test to apply to several different series.

Geometric Series The geometric series $\sum ar^k$ with ratio r converges for $-1 < r < 1$ and diverges for $|r| \geq 1$.

The sum of the convergent geometric series $\sum\limits_{k=0}^{\infty} ar^k$ is $\frac{a}{1-r}$.

Divergence Test If the sequence a_n does not converge to 0, then the series $\sum a_k$ diverges.

This is the first test to apply because the conclusion is simple. However, if $\lim_{n\to\infty} a_n = 0$, no conclusion can be drawn.

Integral Test Let f be a positive, decreasing function on an interval $[c, \infty)$ and let $a_k = f(k)$ for each positive integer $k \geq c$.

- If $\int_c^{\infty} f(t)\, dt$ converges, then $\sum a_k$ converges.
- If $\int_c^{\infty} f(t)\, dt$ diverges, then $\sum a_k$ diverges.

Use this test when $f(x)$ is easy to integrate.

Direct Comparision Test (see Ex 4 in Section 8.3)

Let $0 \le a_k \le b_k$ for each positive integer k.

- If $\sum b_k$ converges, then $\sum a_k$ converges.
- If $\sum a_k$ diverges, then $\sum b_k$ diverges.

Use this test when you have a series with known behavior that you can compare to — this test can be difficult to apply.

Limit Comparison Test Let a_n and b_n be sequences of positive terms. If

$$\lim_{k \to \infty} \frac{a_k}{b_k} = L$$

for some positive finite number L, then the two series $\sum a_k$ and $\sum b_k$ either both converge or both diverge.

Easier to apply in general than the comparison test, but you must have a series with known behavior to compare. Useful to apply to series of rational functions.

Ratio Test Let $a_k \ne 0$ for each k and suppose

$$\lim_{k \to \infty} \frac{|a_{k+1}|}{|a_k|} = r.$$

- If $r < 1$, then the series $\sum a_k$ converges absolutely.
- If $r > 1$, then the series $\sum a_k$ diverges.
- If $r = 1$, then test is inconclusive.

This test is useful when a series involves factorials and powers.

Root Test (see Exercise 2 in Section 8.3)

Let $a_k \ge 0$ for each k and suppose

$$\lim_{k \to \infty} \sqrt[k]{a_k} = r.$$

- If $r < 1$, then the series $\sum a_k$ converges.
- If $r > 1$, then the series $\sum a_k$ diverges.
- If $r = 1$, then test is inconclusive.

In general, the Ratio Test can usually be used in place of the Root Test. However, the Root Test can be quick to use when a_k involves kth powers.

Alternating Series Test If a_n is a positive, decreasing sequence so that $\lim_{n \to \infty} a_n = 0$, then the alternating series $\sum (-1)^{k+1} a_k$ converges.

This test applies only to alternating series — we assume that the terms a_n are all positive and that the sequence $\{a_n\}$ is decreasing.

Alternating Series Estimation Let $S_n = \sum_{k=1}^{n} (-1)^{k+1} a_k$ be the nth partial sum of the alternat-

ing series $\sum_{k=1}^{\infty} (-1)^{k+1} a_k$. Assume $a_n > 0$ for each positive integer n, the sequence a_n

decreases to 0 and $\lim_{n\to\infty} S_n = S$. Then it follows that $|S - S_n| < a_{n+1}$.

This bound can be used to determine the accuracy of the partial sum S_n as an approximation of the sum of a convergent alternating series.

Activity 8.4.7. For (a)-(j), use appropriate tests to determine the convergence or divergence of the following series. Throughout, if a series is a convergent geometric series, find its sum.

a. $\sum_{k=3}^{\infty} \frac{2}{\sqrt{k-2}}$

b. $\sum_{k=1}^{\infty} \frac{k}{1+2k}$

c. $\sum_{k=0}^{\infty} \frac{2k^2+1}{k^3+k+1}$

d. $\sum_{k=0}^{\infty} \frac{100^k}{k!}$

e. $\sum_{k=1}^{\infty} \frac{2^k}{5^k}$

f. $\sum_{k=1}^{\infty} \frac{k^3-1}{k^5+1}$

g. $\sum_{k=2}^{\infty} \frac{3^{k-1}}{7^k}$

h. $\sum_{k=2}^{\infty} \frac{1}{k^k}$

i. $\sum_{k=1}^{\infty} \frac{(-1)^{k+1}}{\sqrt{k+1}}$

j. $\sum_{k=2}^{\infty} \frac{1}{k\ln(k)}$

k. Determine a value of n so that the nth partial sum S_n of the alternating series $\sum_{n=2}^{\infty} \frac{(-1)^n}{\ln(n)}$ approximates the sum to within 0.001.

Summary

- An alternating series is a series whose terms alternate in sign. In other words, an alternating series is a series of the form

$$\sum (-1)^k a_k$$

where a_k is a positive real number for each k.

- An alternating series $\sum_{k=1}^{\infty} (-1)^k a_k$ converges if and only if its sequence $\{S_n\}$ of partial sums converges, where

$$S_n = \sum_{k=1}^{n} (-1)^k a_k.$$

- The sequence of partial sums of a convergent alternating series oscillates around and converge to the sum of the series if the sequence of nth terms converges to 0. That is why the Alternating Series Test shows that the alternating series $\sum_{k=1}^{\infty}(-1)^k a_k$ converges whenever the sequence $\{a_n\}$ of nth terms decreases to 0.

- The difference between the $n-1$st partial sum S_{n-1} and the nth partial sum S_n of a convergent alternating series $\sum_{k=1}^{\infty}(-1)^k a_k$ is $|S_n - S_{n-1}| = a_n$. Since the partial sums oscillate around the sum S of the series, it follows that

$$|S - S_n| < a_n.$$

So the nth partial sum of a convergent alternating series $\sum_{k=1}^{\infty}(-1)^k a_k$ approximates the actual sum of the series to within a_n.

Exercises

1. *(a)* Carefully determine the convergence of the series $\sum_{n=1}^{\infty} \frac{(-1)^n}{3^n}$. The series is

(b) Carefully determine the convergence of the series $\sum_{n=1}^{\infty} \frac{(-1)^n}{3n}$. The series is

2. For the following alternating series,

$$\sum_{n=1}^{\infty} a_n = 0.45 - \frac{(0.45)^3}{3!} + \frac{(0.45)^5}{5!} - \frac{(0.45)^7}{7!} + \ldots$$

how many terms do you have to compute in order for your approximation (your partial sum) to be within 0.0000001 from the convergent value of that series?

3. For the following alternating series,

$$\sum_{n=1}^{\infty} a_n = 1 - \frac{(0.4)^2}{2!} + \frac{(0.4)^4}{4!} - \frac{(0.4)^6}{6!} + \frac{(0.4)^8}{8!} - \ldots$$

how many terms do you have to go for your approximation (your partial sum) to be within 0.0000001 from the convergent value of that series?

4. For the following alternating series,

$$\sum_{n=1}^{\infty} a_n = 1 - \frac{1}{10} + \frac{1}{100} - \frac{1}{1000} + \ldots$$

how many terms do you have to go for your approximation (your partial sum) to be within 1e-07 from the convergent value of that series?

5. Conditionally convergent series converge very slowly. As an example, consider the famous formula[1]

$$\frac{\pi}{4} = 1 - \frac{1}{3} + \frac{1}{5} - \frac{1}{7} + \cdots = \sum_{k=0}^{\infty} (-1)^k \frac{1}{2k+1}. \tag{8.4.3}$$

In theory, the partial sums of this series could be used to approximate π.

---

[1]We will derive this formula in upcoming work.

a. Show that the series in (8.4.3) converges conditionally.

b. Let S_n be the nth partial sum of the series in (8.4.3). Calculate the error in approximating $\frac{\pi}{4}$ with S_{100} and explain why this is not a very good approximation.

c. Determine the number of terms it would take in the series (8.4.3) to approximate $\frac{\pi}{4}$ to 10 decimal places. (The fact that it takes such a large number of terms to obtain even a modest degree of accuracy is why we say that conditionally convergent series converge very slowly.)

6. We have shown that if $\sum(-1)^{k+1}a_k$ is a convergent alternating series, then the sum S of the series lies between any two consecutive partial sums S_n. This suggests that the average $\frac{S_n+S_{n+1}}{2}$ is a better approximation to S than is S_n.

a. Show that $\frac{S_n+S_{n+1}}{2} = S_n + \frac{1}{2}(-1)^{n+2}a_{n+1}$.

b. Use this revised approximation in (a) with $n = 20$ to approximate $\ln(2)$ given that

$$\ln(2) = \sum_{k=1}^{\infty}(-1)^{k+1}\frac{1}{k}.$$

Compare this to the approximation using just S_{20}. For your convenience, $S_{20} = \frac{155685007}{232792560}$.

7. In this exercise, we examine one of the conditions of the Alternating Series Test. Consider the alternating series

$$1 - 1 + \frac{1}{2} - \frac{1}{4} + \frac{1}{3} - \frac{1}{9} + \frac{1}{4} - \frac{1}{16} + \cdots,$$

where the terms are selected alternately from the sequences $\left\{\frac{1}{n}\right\}$ and $\left\{-\frac{1}{n^2}\right\}$.

a. Explain why the nth term of the given series converges to 0 as n goes to infinity.

b. Rewrite the given series by grouping terms in the following manner:

$$(1-1) + \left(\frac{1}{2}-\frac{1}{4}\right) + \left(\frac{1}{3}-\frac{1}{9}\right) + \left(\frac{1}{4}-\frac{1}{16}\right) + \cdots.$$

Use this regrouping to determine if the series converges or diverges.

c. Explain why the condition that the sequence $\{a_n\}$ *decreases* to a limit of 0 is included in the Alternating Series Test.

8. Conditionally convergent series exhibit interesting and unexpected behavior. In this exercise we examine the conditionally convergent alternating harmonic series $\sum_{k=1}^{\infty}\frac{(-1)^{k+1}}{k}$ and discover that addition is not commutative for conditionally convergent series. We will also encounter Riemann's Theorem concerning rearrangements of conditionally convergent series. Before we begin, we remind ourselves that

$$\sum_{k=1}^{\infty}\frac{(-1)^{k+1}}{k} = \ln(2),$$

a fact which will be verified in a later section.

a. First we make a quick analysis of the positive and negative terms of the alternating harmonic series.

    i. Show that the series $\sum_{k=1}^{\infty} \frac{1}{2k}$ diverges.

    ii. Show that the series $\sum_{k=1}^{\infty} \frac{1}{2k+1}$ diverges.

    iii. Based on the results of the previous parts of this exercise, what can we say about the sums $\sum_{k=C}^{\infty} \frac{1}{2k}$ and $\sum_{k=C}^{\infty} \frac{1}{2k+1}$ for any positive integer C? Be specific in your explanation.

b. Recall addition of real numbers is commutative; that is

$$a + b = b + a$$

for any real numbers a and b. This property is valid for any sum of finitely many terms, but does this property extend when we add infinitely many terms together?

The answer is no, and something even more odd happens. Riemann's Theorem (after the nineteenth-century mathematician Georg Friedrich Bernhard Riemann) states that a conditionally convergent series can be rearranged to converge to *any* prescribed sum. More specifically, this means that if we choose any real number S, we can rearrange the terms of the alternating harmonic series $\sum_{k=1}^{\infty} \frac{(-1)^{k+1}}{k}$ so that the sum is S. To understand how Riemann's Theorem works, let's assume for the moment that the number S we want our rearrangement to converge to is positive. Our job is to find a way to order the sum of terms of the alternating harmonic series to converge to S.

    i. Explain how we know that, regardless of the value of S, we can find a partial sum P_1

$$P_1 = \sum_{k=1}^{n_1} \frac{1}{2k+1} = 1 + \frac{1}{3} + \frac{1}{5} + \cdots + \frac{1}{2n_1+1}$$

of the positive terms of the alternating harmonic series that equals or exceeds S. Let

$$S_1 = P_1.$$

    ii. Explain how we know that, regardless of the value of S_1, we can find a partial sum N_1

$$N_1 = -\sum_{k=1}^{m_1} \frac{1}{2k} = -\frac{1}{2} - \frac{1}{4} - \frac{1}{6} - \cdots - \frac{1}{2m_1}$$

so that

$$S_2 = S_1 + N_1 \leq S.$$

    iii. Explain how we know that, regardless of the value of S_2, we can find a partial sum P_2

$$P_2 = \sum_{k=n_1+1}^{n_2} \frac{1}{2k+1} = \frac{1}{2(n_1+1)+1} + \frac{1}{2(n_1+2)+1} + \cdots + \frac{1}{2n_2+1}$$

of the remaining positive terms of the alternating harmonic series so that

$$S_3 = S_2 + P_2 \geq S.$$

iv. Explain how we know that, regardless of the value of S_3, we can find a partial sum

$$N_2 = - \sum_{k=m_1+1}^{m_2} \frac{1}{2k} = -\frac{1}{2(m_1+1)} - \frac{1}{2(m_1+2)} - \cdots - \frac{1}{2m_2}$$

of the remaining negative terms of the alternating harmonic series so that

$$S_4 = S_3 + N_2 \leq S.$$

v. Explain why we can continue this process indefinitely and find a sequence $\{S_n\}$ whose terms are partial sums of a rearrangement of the terms in the alternating harmonic series so that $\lim_{n \to \infty} S_n = S$.

8.5 Taylor Polynomials and Taylor Series

Motivating Questions

- What is a Taylor polynomial? For what purposes are Taylor polynomials used?

- What is a Taylor series?

- How are Taylor polynomials and Taylor series different? How are they related?

- How do we determine the accuracy when we use a Taylor polynomial to approximate a function?

In our work to date in Chapter 8, essentially every sum we have considered has been a sum of numbers. In particular, each infinite series that we have discussed has been a series of real numbers, such as

$$1 + \frac{1}{2} + \frac{1}{4} + \cdots + \frac{1}{2^k} + \cdots = \sum_{k=0}^{\infty} \frac{1}{2^k}. \tag{8.5.1}$$

In the remainder of this chapter, we will expand our notion of series to include series that involve a variable, say x. For instance, if in the geometric series in Equation (8.5.1) we replace the ratio $r = \frac{1}{2}$ with the variable x, then we have the infinite (still geometric) series

$$1 + x + x^2 + \cdots + x^k + \cdots = \sum_{k=0}^{\infty} x^k. \tag{8.5.2}$$

Here we see something very interesting: since a geometric series converges whenever its ratio r satisfies $|r| < 1$, and the sum of a convergent geometric series is $\frac{a}{1-r}$, we can say that for $|x| < 1$,

$$1 + x + x^2 + \cdots + x^k + \cdots = \frac{1}{1-x}. \tag{8.5.3}$$

Note well what Equation (8.5.3) states: the non-polynomial function $\frac{1}{1-x}$ on the right is equal to the infinite polynomial expresssion on the left. Moreover, it appears natural to truncate the infinite sum on the left (whose terms get very small as k gets large) and say, for example, that

$$1 + x + x^2 + x^3 \approx \frac{1}{1-x}$$

for small values of x. This shows one way that a polynomial function can be used to approximate a non-polynomial function; such approximations are one of the main themes in this section and the next.

In Preview Activity 8.5.1, we begin our explorations of approximating non-polynomial functions with polynomials, from which we will also develop ideas regarding infinite series that involve a variable, x.

Preview Activity 8.5.1. Preview Activity 8.3.1 showed how we can approximate the number e using linear, quadratic, and other polynomial functions; we then used similar ideas in Preview Activity 8.4.1 to approximate $\ln(2)$. In this activity, we review and extend the process to find the "best" quadratic approximation to the exponential function e^x around the origin. Let $f(x) = e^x$ throughout this activity.

a. Find a formula for $P_1(x)$, the linearization of $f(x)$ at $x = 0$. (We label this linearization P_1 because it is a first degree polynomial approximation.) Recall that $P_1(x)$ is a good approximation to $f(x)$ for values of x close to 0. Plot f and P_1 near $x = 0$ to illustrate this fact.

b. Since $f(x) = e^x$ is not linear, the linear approximation eventually is not a very good one. To obtain better approximations, we want to develop a different approximation that "bends" to make it more closely fit the graph of f near $x = 0$. To do so, we add a quadratic term to $P_1(x)$. In other words, we let

$$P_2(x) = P_1(x) + c_2 x^2$$

for some real number c_2. We need to determine the value of c_2 that makes the graph of $P_2(x)$ best fit the graph of $f(x)$ near $x = 0$.

Remember that $P_1(x)$ was a good linear approximation to $f(x)$ near 0; this is because $P_1(0) = f(0)$ and $P_1'(0) = f'(0)$. It is therefore reasonable to seek a value of c_2 so that

$$P_2(0) = f(0), \qquad P_2'(0) = f'(0), \qquad \text{and } P_2''(0) = f''(0).$$

Remember, we are letting $P_2(x) = P_1(x) + c_2 x^2$.

  i. Calculate $P_2(0)$ to show that $P_2(0) = f(0)$.

  ii. Calculate $P_2'(0)$ to show that $P_2'(0) = f'(0)$.

  iii. Calculate $P_2''(x)$. Then find a value for c_2 so that $P_2''(0) = f''(0)$.

  iv. Explain why the condition $P_2''(0) = f''(0)$ will put an appropriate "bend" in the graph of P_2 to make P_2 fit the graph of f around $x = 0$.

8.5.1 Taylor Polynomials

Preview Activity 8.5.1 illustrates the first steps in the process of approximating complicated functions with polynomials. Using this process we can approximate trigonometric, exponential, logarithmic, and other nonpolynomial functions as closely as we like (for certain values of x) with polynomials. This is extraordinarily useful in that it allows us to calculate values of these functions to whatever precision we like using only the operations of addition, subtraction, multiplication, and division, which are operations that can be easily programmed in a computer.

We next extend the approach in Preview Activity 8.5.1 to arbitrary functions at arbitrary points. Let f be a function that has as many derivatives at a point $x = a$ as we need. Since first learning it in Section 1.8, we have regularly used the linear approximation $P_1(x)$ to f at $x = a$, which in one sense is the best linear approximation to f near a. Recall that $P_1(x)$ is the tangent line to f at $(a, f(a))$ and is given by the formula

$$P_1(x) = f(a) + f'(a)(x - a).$$

If we proceed as in Preview Activity 8.5.1, we then want to find the best quadratic approximation

$$P_2(x) = P_1(x) + c_2(x - a)^2$$

so that $P_2(x)$ more closely models $f(x)$ near $x = a$. Consider the following calculations of the values and derivatives of $P_2(x)$:

$$P_2(x) = P_1(x) + c_2(x - a)^2 \qquad P_2(a) = P_1(a) = f(a)$$
$$P_2'(x) = P_1'(x) + 2c_2(x - a) \qquad P_2'(a) = P_1'(a) = f'(a)$$
$$P_2''(x) = 2c_2 \qquad\qquad\qquad P_2''(a) = 2c_2.$$

To make $P_2(x)$ fit $f(x)$ better than $P_1(x)$, we want $P_2(x)$ and $f(x)$ to have the same concavity at $x = a$. That is, we want to have

$$P_2''(a) = f''(a).$$

This implies that

$$2c_2 = f''(a)$$

and thus

$$c_2 = \frac{f''(a)}{2}.$$

Therefore, the quadratic approximation $P_2(x)$ to f centered at $x = 0$ is

$$P_2(x) = f(a) + f'(a)(x - a) + \frac{f''(a)}{2!}(x - a)^2.$$

This approach extends naturally to polynomials of higher degree. In this situation, we define polynomials

$$P_3(x) = P_2(x) + c_3(x - a)^3,$$
$$P_4(x) = P_3(x) + c_4(x - a)^4,$$
$$P_5(x) = P_4(x) + c_5(x - a)^5,$$

and so on, with the general one being

$$P_n(x) = P_{n-1}(x) + c_n(x - a)^n.$$

The defining property of these polynomials is that for each n, $P_n(x)$ must have its value and all its first n derivatives agree with those of f at $x = a$. In other words we require that

$$P_n^{(k)}(a) = f^{(k)}(a)$$

for all k from 0 to n.

To see the conditions under which this happens, suppose

$$P_n(x) = c_0 + c_1(x - a) + c_2(x - a)^2 + \cdots + c_n(x - a)^n.$$

Then

$$P_n^{(0)}(a) = c_0$$
$$P_n^{(1)}(a) = c_1$$
$$P_n^{(2)}(a) = 2c_2$$
$$P_n^{(3)}(a) = (2)(3)c_3$$
$$P_n^{(4)}(a) = (2)(3)(4)c_4$$
$$P_n^{(5)}(a) = (2)(3)(4)(5)c_5$$

and, in general,

$$P_n^{(k)}(a) = (2)(3)(4)\cdots(k-1)(k)c_k = k!c_k.$$

So having $P_n^{(k)}(a) = f^{(k)}(a)$ means that $k!c_k = f^{(k)}(a)$ and therefore

$$c_k = \frac{f^{(k)}(a)}{k!}$$

for each value of k. In this expression for c_k, we have found the formula for the degree n polynomial approximation of f that we seek.

Taylor Polynomials

The nth *order Taylor polynomial* of f centered at $x = a$ is given by

$$P_n(x) = f(a) + f'(a)(x - a) + \frac{f''(a)}{2!}(x - a)^2 + \cdots + \frac{f^{(n)}(a)}{n!}(x - a)^n$$
$$= \sum_{k=0}^{n} \frac{f^{(k)}(a)}{k!}(x - a)^k.$$

This degree n polynomial approximates $f(x)$ near $x = a$ and has the property that $P_n^{(k)}(a) = f^{(k)}(a)$ for $k = 0 \ldots n$.

Example 8.5.1. Determine the third order Taylor polynomial for $f(x) = e^x$, as well as the general nth order Taylor polynomial for f centered at $x = 0$.

Solution. We know that $f'(x) = e^x$ and so $f''(x) = e^x$ and $f'''(x) = e^x$. Thus,

$$f(0) = f'(0) = f''(0) = f'''(0) = 1.$$

So the third order Taylor polynomial of $f(x) = e^x$ centered at $x = 0$ is

$$P_3(x) = f(0) + f'(0)(x - 0) + \frac{f''(0)}{2!}(x - 0)^2 + \frac{f'''(0)}{3!}(x - 0)^3$$

$$= 1 + x + \frac{x^2}{2} + \frac{x^3}{6}.$$

In general, for the exponential function f we have $f^{(k)}(x) = e^x$ for every positive integer k. Thus, the kth term in the nth order Taylor polynomial for $f(x)$ centered at $x = 0$ is

$$\frac{f^{(k)}(0)}{k!}(x - 0)^k = \frac{1}{k!}x^k.$$

Therefore, the nth order Taylor polynomial for $f(x) = e^x$ centered at $x = 0$ is

$$P_n(x) = 1 + x + \frac{x^2}{2!} + \cdots + \frac{1}{n!}x^n = \sum_{k=0}^{n} \frac{x^k}{k!}.$$

Activity 8.5.2. We have just seen that the nth order Taylor polynomial centered at $a = 0$ for the exponential function e^x is

$$\sum_{k=0}^{n} \frac{x^k}{k!}.$$

In this activity, we determine small order Taylor polynomials for several other familiar functions, and look for general patterns that will help us find the Taylor series expansions a bit later.

a. Let $f(x) = \frac{1}{1-x}$.

   i. Calculate the first four derivatives of $f(x)$ at $x = 0$. Then find the fourth order Taylor polynomial $P_4(x)$ for $\frac{1}{1-x}$ centered at 0.

   ii. Based on your results from part (i), determine a general formula for $f^{(k)}(0)$.

b. Let $f(x) = \cos(x)$.

   i. Calculate the first four derivatives of $f(x)$ at $x = 0$. Then find the fourth order Taylor polynomial $P_4(x)$ for $\cos(x)$ centered at 0.

   ii. Based on your results from part (i), find a general formula for $f^{(k)}(0)$. (Think about how k being even or odd affects the value of the kth derivative.)

c. Let $f(x) = \sin(x)$.

    i. Calculate the first four derivatives of $f(x)$ at $x = 0$. Then find the fourth order Taylor polynomial $P_4(x)$ for $\sin(x)$ centered at 0.

    ii. Based on your results from part (i), find a general formula for $f^{(k)}(0)$. (Think about how k being even or odd affects the value of the kth derivative.)

It is possible that an nth order Taylor polynomial is not a polynomial of degree n; that is, the order of the approximation can be different from the degree of the polynomial. For example, in Activity 8.5.3 we found that the second order Taylor polynomial $P_2(x)$ centered at 0 for $\sin(x)$ is $P_2(x) = x$. In this case, the second order Taylor polynomial is a degree 1 polynomial.

8.5.2 Taylor Series

In Activity 8.5.2 we saw that the fourth order Taylor polynomial $P_4(x)$ for $\sin(x)$ centered at 0 is

$$P_4(x) = x - \frac{x^3}{3!}.$$

The pattern we found for the derivatives $f^{(k)}(0)$ describe the higher-order Taylor polynomials, e.g.,

$$P_5(x) = x - \frac{x^3}{3!} + \frac{x^{(5)}}{5!},$$

$$P_7(x) = x - \frac{x^3}{3!} + \frac{x^{(5)}}{5!} - \frac{x^{(7)}}{7!},$$

$$P_9(x) = x - \frac{x^3}{3!} + \frac{x^{(5)}}{5!} - \frac{x^{(7)}}{7!} + \frac{x^{(9)}}{9!},$$

and so on. It is instructive to consider the graphical behavior of these functions; the following figure shows the graphs of a few of the Taylor polynomials centered at 0 for the sine function.

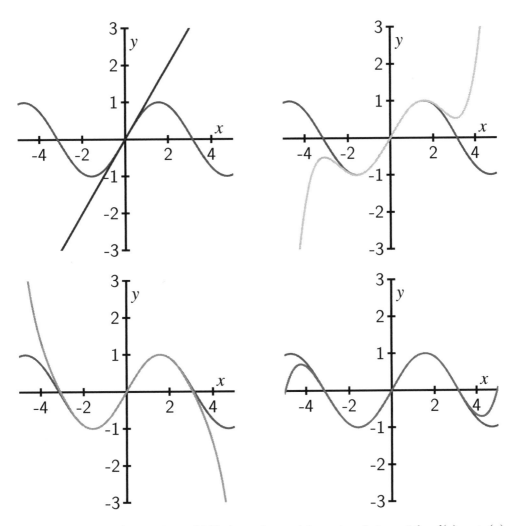

Figure 8.5.2: The order 1, 5, 7, and 9 Taylor polynomials centered at $x = 0$ for $f(x) = \sin(x)$.

Notice that $P_1(x)$ is close to the sine function only for values of x that are close to 0, but as we increase the degree of the Taylor polynomial the Taylor polynomials provide a better fit to the graph of the sine function over larger intervals. This illustrates the general behavior of Taylor polynomials: for any sufficiently well-behaved function, the sequence $\{P_n(x)\}$ of Taylor polynomials converges to the function f on larger and larger intervals (though those intervals may not necessarily increase without bound). If the Taylor polynomials ultimately converge to f on its entire domain, we write

$$f(x) = \sum_{k=0}^{\infty} \frac{f^{(k)}(a)}{k!}(x - a)^k.$$

Definition 8.5.3. Let f be a function all of whose derivatives exist at $x = a$. The **Taylor series** for f centered at $x = a$ is the series $T_f(x)$ defined by

$$T_f(x) = \sum_{k=0}^{\infty} \frac{f^{(k)}(a)}{k!}(x - a)^k.$$

In the special case where $a = 0$ in Definition 8.5.3, the Taylor series is also called the *Maclaurin* series for f. From Example 8.5.1 we know the nth order Taylor polynomial centered at 0 for the exponential function e^x; thus, the Maclaurin series for e^x is

$$\sum_{k=0}^{\infty} \frac{x^k}{k!}.$$

Activity 8.5.3. In Activity 8.5.2 we determined small order Taylor polynomials for a few familiar functions, and also found general patterns in the derivatives evaluated at 0. Use that information to write the Taylor series centered at 0 for the following functions.

  a. $f(x) = \frac{1}{1-x}$

  b. $f(x) = \cos(x)$ (You will need to carefully consider how to indicate that many of the coefficients are 0. Think about a general way to represent an even integer.)

  c. $f(x) = \sin(x)$ (You will need to carefully consider how to indicate that many of the coefficients are 0. Think about a general way to represent an odd integer.)

  d. Determine the n order Taylor polynomial for $f(x) = \frac{1}{1-x}$ centered at $x = 0$.

The next activity further considers the important issue of the x-values for which the Taylor series of a function converges to the function itself.

Activity 8.5.4.

  a. Plot the graphs of several of the Taylor polynomials centered at 0 (of order at least 5) for e^x and convince yourself that these Taylor polynomials converge to e^x for every value of x.

  b. Draw the graphs of several of the Taylor polynomials centered at 0 (of order at least 6) for $\cos(x)$ and convince yourself that these Taylor polynomials converge to $\cos(x)$ for every value of x. Write the Taylor series centered at 0 for $\cos(x)$.

  c. Draw the graphs of several of the Taylor polynomials centered at 0 for $\frac{1}{1-x}$. Based on your graphs, for what values of x do these Taylor polynomials appear to converge to $\frac{1}{1-x}$? How is this situation different from what we observe with e^x and cos(x)? In addition, write the Taylor series centered at 0 for $\frac{1}{1-x}$.

The Maclaurin series for e^x, $\sin(x)$, $\cos(x)$, and $\frac{1}{1-x}$ will be used frequently, so we should be certain to know and recognize them well.

8.5.3 The Interval of Convergence of a Taylor Series

In the previous section (in Figure 8.5.2 and Activity 8.5.4) we observed that the Taylor polynomials centered at 0 for e^x, $\cos(x)$, and $\sin(x)$ converged to these functions for all values of x in their domain, but that the Taylor polynomials centered at 0 for $\frac{1}{1-x}$ converged to $\frac{1}{1-x}$ for only some values of x. In fact, the Taylor polynomials centered at 0 for $\frac{1}{1-x}$ converge to $\frac{1}{1-x}$ on the interval $(-1, 1)$ and diverge for all other values of x. So the Taylor series for a function $f(x)$ does not need to converge for all values of x in the domain of f.

Our observations to date suggest two natural questions: can we determine the values of x for which a given Taylor series converges? Moreover, given the Taylor series for a function f, does it actually converge to $f(x)$ for those values of x for which the Taylor series converges?

Example 8.5.4. Graphical evidence suggests that the Taylor series centered at 0 for e^x converges for all values of x. To verify this, use the Ratio Test to determine all values of x for which the Taylor series

$$\sum_{k=0}^{\infty} \frac{x^k}{k!} \tag{8.5.4}$$

converges absolutely.

Solution. In previous work, we used the Ratio Test on series of numbers that did not involve a variable; recall, too, that the Ratio Test only applies to series of nonnegative terms. In this example, we have to address the presence of the variable x. Because we are interested in absolute convergence, we apply the Ratio Test to the series

$$\sum_{k=0}^{\infty} \left| \frac{x^k}{k!} \right| = \sum_{k=0}^{\infty} \frac{|x|^k}{k!}.$$

Now, observe that

$$\lim_{k \to \infty} \frac{a_{k+1}}{a_k} = \lim_{k \to \infty} \frac{\frac{|x|^{k+1}}{(k+1)!}}{\frac{|x|^k}{k}}$$

$$= \lim_{k \to \infty} \frac{|x|^{k+1} k!}{|x|^k (k+1)!}$$

$$= \lim_{k \to \infty} \frac{|x|}{k+1}$$

$$= 0$$

for any value of x. So the Taylor series (8.5.4) converges absolutely for every value of x, and thus converges for every value of x.

One key question remains: while the Taylor series for e^x converges for all x, what we have done does not tell us that this Taylor series actually converges to e^x for each x. We'll return to this question when we consider the error in a Taylor approximation near the end of this section.

We can apply the main idea from Example 8.5.4 in general. To determine the values of x for which a Taylor series

$$\sum_{k=0}^{\infty} c_k(x-a)^k,$$

centered at $x = a$ will converge, we apply the Ratio Test with $a_k = |c_k(x-a)^k|$ and recall that the series to which the Ratio Test is applied converges if $\lim_{k \to \infty} \frac{a_{k+1}}{a_k} < 1$.

Observe that

$$\frac{a_{k+1}}{a_k} = |x-a|\frac{|c_{k+1}|}{|c_k|},$$

so when we apply the Ratio Test, we get that

$$\lim_{k \to \infty} \frac{a_{k+1}}{a_k} = \lim_{k \to \infty} |x-a|\frac{c_{k+1}}{c_k}.$$

Note further that $c_k = \frac{f^{(k)}(a)}{k!}$, and say that

$$\lim_{k \to \infty} \frac{c_{k+1}}{c_k} = L.$$

Thus, we have found that

$$\lim_{k \to \infty} \frac{a_{k+1}}{a_k} = |x-a| \cdot L.$$

There are three important possibilities for L: L can be 0, a finite positive value, or infinite. Based on this value of L, we can therefore determine for which values of x the original Taylor series converges.

- If $L = 0$, then the Taylor series converges on $(-\infty, \infty)$.

- If L is infinite, then the Taylor series converges only at $x = a$.

- If L is finite and nonzero, then the Taylor series converges absolutely for all x that satisfy

$$|x-a| \cdot L < 1.$$

In other words, the series converges absolutely for all x such that

$$|x-a| < \frac{1}{L},$$

which is also the interval

$$\left(a - \frac{1}{L}, a + \frac{1}{L}\right).$$

Because the Ratio Test is inconclusive when the $|x - a| \cdot L = 1$, the endpoints $a \pm \frac{1}{L}$ have to be checked separately.

It is important to notice that the set of x values at which a Taylor series converges is always an interval centered at $x = a$. For this reason, the set on which a Taylor series converges is called the *interval of convergence*. Half the length of the interval of convergence is called the *radius of convergence*. If the interval of convergence of a Taylor series is infinite, then we say that the radius of convergence is infinite.

> **Activity 8.5.5.**
>
> a. Use the Ratio Test to explicitly determine the interval of convergence of the Taylor series for $f(x) = \frac{1}{1-x}$ centered at $x = 0$.
>
> b. Use the Ratio Test to explicitly determine the interval of convergence of the Taylor series for $f(x) = \cos(x)$ centered at $x = 0$.
>
> c. Use the Ratio Test to explicitly determine the interval of convergence of the Taylor series for $f(x) = \sin(x)$ centered at $x = 0$.

The Ratio Test tells us how we can determine the set of x values for which a Taylor series converges absolutely. However, just because a Taylor series for a function f converges, we cannot be certain that the Taylor series actually converges to $f(x)$ on its interval of convergence. To show why and where a Taylor series does in fact converge to the function f, we next consider the error that is present in Taylor polynomials.

8.5.4 Error Approximations for Taylor Polynomials

We now know how to find Taylor polynomials for functions such as $\sin(x)$, as well as how to determine the interval of convergence of the corresponding Taylor series. We next develop an error bound that will tell us how well an nth order Taylor polynomial $P_n(x)$ approximates its generating function $f(x)$. This error bound will also allow us to determine whether a Taylor series on its interval of convergence actually equals the function f from which the Taylor series is derived. Finally, we will be able to use the error bound to determine the order of the Taylor polynomial $P_n(x)$ for a function f that we need to ensure that $P_n(x)$ approximates $f(x)$ to any desired degree of accuracy.

In all of this, we need to compare $P_n(x)$ to $f(x)$. For this argument, we assume throughout that we center our approximations at 0 (a similar argument holds for approximations centered at a). We define the exact error, $E_n(x)$, that results from approximating $f(x)$ with $P_n(x)$ by

$$E_n(x) = f(x) - P_n(x).$$

We are particularly interested in $|E_n(x)|$, the distance between P_n and f. Note that since

$$P_n^{(k)}(0) = f^{(k)}(0)$$

for $0 \le k \le n$, we know that

$$E_n^{(k)}(0) = 0$$

for $0 \le k \le n$. Furthermore, since $P_n(x)$ is a polynomial of degree less than or equal to n, we know that

$$P_n^{(n+1)}(x) = 0.$$

Thus, since $E_n^{(n+1)}(x) = f^{(n+1)}(x) - P_n^{(n+1)}(x)$, it follows that

$$E_n^{(n+1)}(x) = f^{(n+1)}(x)$$

for all x.

Suppose that we want to approximate $f(x)$ at a number c close to 0 using $P_n(c)$. If we assume $|f^{(n+1)}(t)|$ is bounded by some number M on $[0, c]$, so that

$$\left| f^{(n+1)}(t) \right| \le M$$

for all $0 \le t \le c$, then we can say that

$$\left| E_n^{(n+1)}(t) \right| = \left| f^{(n+1)}(t) \right| \le M$$

for all t between 0 and c. Equivalently,

$$- M \le E_n^{(n+1)}(t) \le M \tag{8.5.5}$$

on $[0, c]$. Next, we integrate the three terms in Inequality (8.5.5) from $t = 0$ to $t = x$, and thus find that

$$\int_0^x -M \, dt \le \int_0^x E_n^{(n+1)}(t) \, dt \le \int_0^x M \, dt$$

for every value of x in $[0, c]$. Since $E_n^{(n)}(0) = 0$, the First FTC tells us that

$$-Mx \le E_n^{(n)}(x) \le Mx$$

for every x in $[0, c]$.

Integrating the most recent inequality, we obtain

$$\int_0^x -Mt \, dt \le \int_0^x E_n^{(n)}(t) \, dt \le \int_0^x Mt \, dt$$

and thus

$$-M\frac{x^2}{2} \le E_n^{(n-1)}(x) \le M\frac{x^2}{2}$$

for all x in $[0, c]$.

Integrating n times, we arrive at

$$-M\frac{x^{n+1}}{(n+1)!} \le E_n(x) \le M\frac{x^{n+1}}{(n+1)!}$$

for all x in $[0, c]$. This enables us to conclude that

$$|E_n(x)| \leq M \frac{|x|^{n+1}}{(n+1)!}$$

for all x in $[0, c]$, which shows an important bound on the approximation's error, E_n.

Our work above was based on the approximation centered at $a = 0$; the argument may be generalized to hold for any value of a, which results in the following theorem.

The Lagrange Error Bound for $P_n(x)$. Let f be a continuous function with $n + 1$ continuous derivatives. Suppose that M is a positive real number such that $|f^{(n+1)}(x)| \leq M$ on the interval $[a, c]$. If $P_n(x)$ is the nth order Taylor polynomial for $f(x)$ centered at $x = a$, then

$$|P_n(c) - f(c)| \leq M \frac{|c - a|^{n+1}}{(n+1)!}.$$

This error bound may now be used to tell us important information about Taylor polynomials and Taylor series, as we see in the following examples and activities.

Example 8.5.5. Determine how well the 10th order Taylor polynomial $P_{10}(x)$ for $\sin(x)$, centered at 0, approximates $\sin(2)$.

Solution. To answer this question we use $f(x) = \sin(x)$, $c = 2$, $a = 0$, and $n = 10$ in the Lagrange error bound formula. To use the bound, we also need to find an appropriate value for M. Note that the derivatives of $f(x) = \sin(x)$ are all equal to $\pm \sin(x)$ or $\pm \cos(x)$. Thus,

$$\left| f^{(n+1)}(x) \right| \leq 1$$

for any n and x. Therefore, we can choose M to be 1. Then

$$|P_{10}(2) - f(2)| \leq (1) \frac{|2 - 0|^{11}}{(11)!} = \frac{2^{11}}{(11)!} \approx 0.00005130671797.$$

So $P_{10}(2)$ approximates $\sin(2)$ to within at most 0.00005130671797. A computer algebra system tells us that

$$P_{10}(2) \approx 0.9093474427 \quad \text{and} \quad \sin(2) \approx 0.9092974268$$

with an actual difference of about 0.0000500159.

Activity 8.5.6. Let $P_n(x)$ be the nth order Taylor polynomial for $\sin(x)$ centered at $x = 0$. Determine how large we need to choose n so that $P_n(2)$ approximates $\sin(2)$ to 20 decimal places.

Example 8.5.6. Show that the Taylor series for $\sin(x)$ actually converges to $\sin(x)$ for all x.

Solution. Recall from the previous example that since $f(x) = \sin(x)$, we know

$$\left| f^{(n+1)}(x) \right| \leq 1$$

for any n and x. This allows us to choose $M = 1$ in the Lagrange error bound formula. Thus,

$$|P_n(x) - \sin(x)| \leq \frac{|x|^{n+1}}{(n+1)!} \tag{8.5.6}$$

for every x.

We showed in earlier work with the Taylor series $\sum_{k=0}^{\infty} \frac{x^k}{k!}$ converges for every value of x. Since the terms of any convergent series must approach zero, it follows that

$$\lim_{n \to \infty} \frac{x^{n+1}}{(n+1)!} = 0$$

for every value of x. Thus, taking the limit as $n \to \infty$ in the inequality (8.5.6), it follows that

$$\lim_{n \to \infty} |P_n(x) - \sin(x)| = 0.$$

As a result, we can now write

$$\sin(x) = \sum_{n=0}^{\infty} \frac{(-1)^n x^{2n+1}}{(2n+1)!}$$

for every real number x.

Activity 8.5.7.

a. Show that the Taylor series centered at 0 for $\cos(x)$ converges to $\cos(x)$ for every real number x.

b. Next we consider the Taylor series for e^x.

    i. Show that the Taylor series centered at 0 for e^x converges to e^x for every non-negative value of x.

    ii. Show that the Taylor series centered at 0 for e^x converges to e^x for every negative value of x.

    iii. Explain why the Taylor series centered at 0 for e^x converges to e^x for every real number x. Recall that we earlier showed that the Taylor series centered at 0 for e^x converges for all x, and we have now completed the argument that the Taylor series for e^x actually converges to e^x for all x.

c. Let $P_n(x)$ be the nth order Taylor polynomial for e^x centered at 0. Find a value of n so that $P_n(5)$ approximates e^5 correct to 8 decimal places.

Summary

- We can use Taylor polynomials to approximate complicated functions. This allows us to approximate values of complicated functions using only addition, subtraction, multiplication, and division of real numbers. The nth order Taylor polynomial centered at $x = a$ of a function f is

$$P_n(x) = f(a) + f'(a)(x - a) + \frac{f''(a)}{2!}(x - a)^2 + \cdots + \frac{f^{(n)}(a)}{n!}(x - a)^n$$
$$= \sum_{k=0}^{n} \frac{f^{(k)}(a)}{k!}(x - a)^k.$$

- The Taylor series centered at $x = a$ for a function f is

$$\sum_{k=0}^{\infty} \frac{f^{(k)}(a)}{k!}(x - a)^k.$$

- The nth order Taylor polynomial centered at a for f is the nth partial sum of its Taylor series centered at a. So the nth order Taylor polynomial for a function f is an approximation to f on the interval where the Taylor series converges; for the values of x for which the Taylor series converges to f we write

$$f(x) = \sum_{k=0}^{\infty} \frac{f^{(k)}(a)}{k!}(x - a)^k.$$

- The Lagrange Error Bound shows us how to determine the accuracy in using a Taylor polynomial to approximate a function. More specifically, if $P_n(x)$ is the nth order Taylor polynomial for f centered at $x = a$ and if M is an upper bound for $\left|f^{(n+1)}(x)\right|$ on the interval $[a, c]$, then

$$\left|P_n(c) - f(c)\right| \leq M\frac{|c - a|^{n+1}}{(n + 1)!}.$$

Exercises

1. Find the Taylor polynomials of degree n approximating $\cos(3x)$ for x near 0:

For $n = 2$, $P_2(x) =$

For $n = 4$, $P_4(x) =$

For $n = 6$, $P_6(x) =$

2. Suppose g is a function which has continuous derivatives, and that $g(7) = -1$, $g'(7) = -1$, $g''(7) = 1$, $g'''(7) = -4$.

(*a*) What is the Taylor polynomial of degree 2 for g near 7?

$P_2(x) = $ [_____]

(b) What is the Taylor polynomial of degree 3 for g near 7?

$P_3(x) = $ [_____]

(c) Use the two polynomials that you found in parts (a) and (b) to approximate $g(7.1)$.

With P_2, $g(7.1) \approx $ [_____]

With P_3, $g(7.1) \approx $ [_____]

3. Find the first four terms of the Taylor series for the function $\dfrac{5}{x}$ about the point $a = -2$. (Your answers should include the variable x when appropriate.)

$\dfrac{5}{x} = $ [_____] $+$ [_____] $+$ [_____] $+$

[_____] $+ \ldots$

4. Find the first four terms of the Taylor series for the function $\cos(x)$ about the point $a = -\pi/4$. (Your answers should include the variable x when appropriate.)

$\cos(x) = $ [_____] $+$ [_____] $+$

[_____] $+$ [_____] $+ \ldots$

5. Find the first five terms of the Taylor series for the function $f(x) = \ln(x)$ about the point $a = 3$. (Your answers should include the variable x when appropriate.)

$\ln(x) = $ [_____] $+$ [_____] $+$ [_____]

$+$ [_____] $+$ [_____] $+ \ldots$

6. In this exercise we investigation the Taylor series of polynomial functions.

a. Find the 3rd order Taylor polynomial centered at $a = 0$ for $f(x) = x^3 - 2x^2 + 3x - 1$. Does your answer surprise you? Explain.

b. Without doing any additional computation, find the 4th, 12th, and 100th order Taylor polynomials (centered at $a = 0$) for $f(x) = x^3 - 2x^2 + 3x - 1$. Why should you expect this?

c. Now suppose $f(x)$ is a degree m polynomial. Completely describe the nth order Taylor polynomial (centered at $a = 0$) for each n.

7. The examples we have considered in this section have all been for Taylor polynomials and series centered at 0, but Taylor polynomials and series can be centered at any value of a. We look at examples of such Taylor polynomials in this exercise.

a. Let $f(x) = \sin(x)$. Find the Taylor polynomials up through order four of f centered at $x = \frac{\pi}{2}$. Then find the Taylor series for $f(x)$ centered at $x = \frac{\pi}{2}$. Why should you have expected the result?

b. Let $f(x) = \ln(x)$. Find the Taylor polynomials up through order four of f centered at $x = 1$. Then find the Taylor series for $f(x)$ centered at $x = 1$.

c. Use your result from (b) to determine which Taylor polynomial will approximate ln(2) to two decimal places. Explain in detail how you know you have the desired accuracy.

8. We can use known Taylor series to obtain other Taylor series, and we explore that idea in this exercise, as a preview of work in the following section.

a. Calculate the first four derivatives of $\sin(x^2)$ and hence find the fourth order Taylor polynomial for $\sin(x^2)$ centered at $a = 0$.

b. Part (a) demonstrates the brute force approach to computing Taylor polynomials and series. Now we find an easier method that utilizes a known Taylor series. Recall that the Taylor series centered at 0 for $f(x) = \sin(x)$ is

$$\sum_{k=0}^{\infty} (-1)^k \frac{x^{2k+1}}{(2k+1)!}. \tag{8.5.7}$$

i. Substitute x^2 for x in the Taylor series (8.5.7). Write out the first several terms and compare to your work in part (a). Explain why the substitution in this problem should give the Taylor series for $\sin(x^2)$ centered at 0.

ii. What should we expect the interval of convergence of the series for $\sin(x^2)$ to be? Explain in detail.

9. Based on the examples we have seen, we might expect that the Taylor series for a function f always converges to the values $f(x)$ on its interval of convergence. We explore that idea in more detail in this exercise. Let $f(x) = \begin{cases} e^{-1/x^2} & \text{if } x \neq 0, \\ 0 & \text{if } x = 0. \end{cases}$

a. Show, using the definition of the derivative, that $f'(0) = 0$.

b. It can be shown that $f^{(n)}(0) = 0$ for all $n \geq 2$. Assuming that this is true, find the Taylor series for f centered at 0.

c. What is the interval of convergence of the Taylor series centered at 0 for f? Explain. For which values of x the interval of convergence of the Taylor series does the Taylor series converge to $f(x)$?

8.6 Power Series

Motivating Questions

- What is a power series?

- What are some important uses of power series?

- What is the connection between power series and Taylor series?

We have noted at several points in our work with Taylor polynomials and Taylor series that polynomial functions are the simplest possible functions in mathematics, in part because they essentially only require addition and multiplication to evaluate. Moreover, from the point of view of calculus, polynomials are especially nice: we can easily differentiate or integrate any polynomial. In light of our work in Section 8.5, we now know that several important non-polynomials have polynomial-like expansions. For example, for any real number x,

$$e^x = 1 + x + \frac{x^2}{2!} + \frac{x^3}{3!} + \cdots + \frac{x^n}{n!} + \cdots.$$

As we continue our study of infinite series, there are two settings where other series like the one for e^x arise: one is where we are simply given an expression like

$$1 + 2x + 3x^2 + 4x^3 + \cdots$$

and we seek the values of x for which the expression makes sense, while another is where we are trying to find an unknown function f, and we think about the possibility that the function has expression

$$f(x) = a_0 + a_1 x + a_2 x^2 + \cdots + a_k x^k + \cdots,$$

and we try to determine the values of the constants $a_0, a_1, \ldots$. The latter situation is explored in Preview Activity 8.6.1.

> **Preview Activity 8.6.1.** In Chapter 7, we learned some of the many important applications of differential equations, and learned some approaches to solve or analyze them. Here, we consider an important approach that will allow us to solve a wider variety of differential equations.
>
> Let's consider the familiar differential equation from exponential population growth given by
>
> $$y' = ky, \tag{8.6.1}$$
>
> where k is the constant of proportionality. While we can solve this differential equation using methods we have already learned, we take a different approach now that can be applied to a much larger set of differential equations. For the rest of this activity, let's assume that $k = 1$. We will use our knowledge of Taylor series to find a solution to the

differential equation (8.6.1).

To do so, we assume that we have a solution $y = f(x)$ and that $f(x)$ has a Taylor series that can be written in the form

$$y = f(x) = \sum_{k=0}^{\infty} a_k x^k,$$

where the coefficients a_k are undetermined. Our task is to find the coefficients.

a. Assume that we can differentiate a power series term by term. By taking the derivative of $f(x)$ with respect to x and substituting the result into the differential equation (8.6.1), show that the equation

$$\sum_{k=1}^{\infty} k a_k x^{k-1} = \sum_{k=0}^{\infty} a_k x^k$$

must be satisfied in order for $f(x) = \sum_{k=0}^{\infty} a_k x^k$ to be a solution of the DE.

b. Two series are equal if and only if they have the same coefficients on like power terms. Use this fact to find a relationship between a_1 and a_0.

c. Now write a_2 in terms of a_1. Then write a_2 in terms of a_0.

d. Write a_3 in terms of a_2. Then write a_3 in terms of a_0.

e. Write a_4 in terms of a_3. Then write a_4 in terms of a_0.

f. Observe that there is a pattern in (b)-(e). Find a general formula for a_k in terms of a_0.

g. Write the series expansion for y using only the unknown coefficient a_0. From this, determine what familiar functions satisfy the differential equation (8.6.1). (*Hint*: Compare to a familiar Taylor series.)

8.6.1 Power Series

As Preview Activity 8.6.1 shows, it can be useful to treat an unknown function as if it has a Taylor series, and then determine the coefficients from other information. In other words, we define a function as an infinite series of powers of x and then determine the coefficients based on something besides a formula for the function. This method of using series illustrated in Preview Activity 8.6.1 to solve differential equations is a powerful and important one that allows us to approximate solutions to many different types of differential equations even if we cannot explicitly solve them. This approach is different from defining a Taylor series based on a given function, and these functions we define with arbitrary coefficients are given a special name.

Definition 8.6.1. A **power series** centered at $x = a$ is a function of the form

$$\sum_{k=0}^{\infty} c_k(x - a)^k \qquad (8.6.2)$$

where $\{c_k\}$ is a sequence of real numbers and x is an independent variable.

We can substitute different values for x and test whether the resulting series converges or diverges. Thus, a power series defines a function f whose domain is the set of x values for which the power series converges. We therefore write

$$f(x) = \sum_{k=0}^{\infty} c_k(x - a)^k.$$

It turns out that[1], on its interval of convergence, a power series is the Taylor series of the function that is the sum of the power series, so all of the techniques we developed in the previous section can be applied to power series as well.

Example 8.6.2. Consider the power series defined by

$$f(x) = \sum_{k=0}^{\infty} \frac{x^k}{2^k}.$$

What are $f(1)$ and $f\left(\frac{3}{2}\right)$? Find a general formula for $f(x)$ and determine the values for which this power series converges.

Solution. If we evaluate f at $x = 1$ we obtain the series

$$\sum_{k=0}^{\infty} \frac{1}{2^k}$$

which is a geometric series with ratio $\frac{1}{2}$. So we can sum this series and find that

$$f(1) = \frac{1}{1 - \frac{1}{2}} = 2.$$

Similarly,

$$f(3/2) = \sum_{k=0}^{\infty} \left(\frac{3}{4}\right)^k = \frac{1}{1 - \frac{3}{4}} = 4.$$

In general, $f(x)$ is a geometric series with ratio $\frac{x}{2}$ and

$$f(x) = \sum_{k=0}^{\infty} \left(\frac{x}{2}\right)^k = \frac{1}{1 - \frac{x}{2}} = \frac{2}{2 - x}$$

---

[1]See Exercise 8.6.4.4 in this section.

provided that $-1 < \frac{x}{2} < 1$ (so that the ratio is less than 1 in absolute value). Thus, the power series that defines f converges for $-2 < x < 2$.

As with Taylor series, we define the interval of convergence of a power series (8.6.2) to be the set of values of x for which the series converges. In the same way as we did with Taylor series, we typically use the Ratio Test to find the values of x for which the power series converges absolutely, and then check the endpoints separately if the radius of convergence is finite.

Example 8.6.3. Let $f(x) = \sum_{k=1}^{\infty} \frac{x^k}{k^2}$. Determine the interval of convergence of this power series.

Solution. First we will draw graphs of some of the partial sums of this power series to get an idea of the interval of convergence. Let

$$S_n(x) = \sum_{k=1}^{n} \frac{x^k}{k^2}$$

for each $n \geq 1$. Figure 8.6.4 shows plots of $S_{10}(x)$ (in red), $S_{25}(x)$ (in blue), and $S_{50}(x)$ (in green).

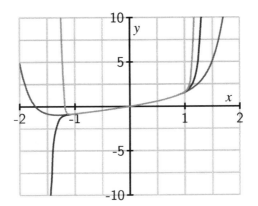

Figure 8.6.4: Graphs of partial sums of the power series $\sum_{k=1}^{\infty} \frac{x^k}{k^2}$

The behavior of S_{50} particularly highlights that it appears to be converging to a particular curve on the interval $(-1, 1)$, while growing without bound outside of that interval. This suggests that the interval of convergence might be $-1 < x < 1$. To more fully understand this power series, we apply the Ratio Test to determine the values of x for which the power series converges absolutely. For the given series, we have

$$a_k = \frac{x^k}{k^2},$$

so

$$\lim_{k \to \infty} \frac{|a_{k+1}|}{|a_k|} = \lim_{k \to \infty} \frac{\frac{|x|^{k+1}}{(k+1)^2}}{\frac{|x|^k}{k^2}}$$

$$= \lim_{k \to \infty} |x| \left(\frac{k}{k+1} \right)^2$$

$$= |x| \lim_{k \to \infty} \left(\frac{k}{k+1} \right)^2$$

$$= |x|.$$

Therefore, the Ratio Test tells us that the given power series $f(x)$ converges absolutely when $|x| < 1$ and diverges when $|x| > 1$. Since the Ratio Test is inconclusive when $|x| = 1$, we need to check $x = 1$ and $x = -1$ individually.

When $x = 1$, observe that

$$f(1) = \sum_{k=1}^{\infty} \frac{1}{k^2}.$$

This is a *p*-series with $p > 1$, which we know converges. When $x = -1$, we have

$$f(-1) = \sum_{k=1}^{\infty} \frac{(-1)^k}{k^2}.$$

This is an alternating series, and since the sequence $\left\{ \frac{1}{n^2} \right\}$ decreases to 0, the power series converges when $x = -1$ by the Alternating Series Test. Thus, the interval of convergence of this power series is $-1 \le x \le 1$.

Activity 8.6.2. Determine the interval of convergence of each power series.

a. $\sum_{k=1}^{\infty} \frac{(x-1)^k}{3k}$

b. $\sum_{k=1}^{\infty} kx^k$

c. $\sum_{k=1}^{\infty} \frac{k^2(x+1)^k}{4^k}$

d. $\sum_{k=1}^{\infty} \frac{x^k}{(2k)!}$

e. $\sum_{k=1}^{\infty} k!x^k$

8.6.2 Manipulating Power Series

Recall that we know several power series expressions for important functions such as $\sin(x)$ and e^x. Often, we can take a known power series expression for such a function and use that series expansion to find a power series for a different, but related, function. The next activity demonstrates one way to do this.

Activity 8.6.3. Our goal in this activity is to find a power series expansion for $f(x) = \frac{1}{1+x^2}$ centered at $x = 0$.

While we could use the methods of Section 8.5 and differentiate $f(x) = \frac{1}{1+x^2}$ several times to look for patterns and find the Taylor series for $f(x)$, we seek an alternate approach because of how complicated the derivatives of $f(x)$ quickly become.

    a. What is the Taylor series expansion for $g(x) = \frac{1}{1-x}$? What is the interval of convergence of this series?

    b. How is $g(-x^2)$ related to $f(x)$? Explain, and hence substitute $-x^2$ for x in the power series expansion for $g(x)$. Given the relationship between $g(-x^2)$ and $f(x)$, how is the resulting series related to $f(x)$?

    c. For which values of x will this power series expansion for $f(x)$ be valid? Why?

In a previous section we determined several important Maclaurin series and their intervals of convergence. Here, we list these key functions and remind ourselves of their corresponding expansions.

$$\sin(x) = \sum_{k=0}^{\infty} \frac{(-1)^k x^{2k+1}}{(2k+1)!} \qquad \text{for } -\infty < x < \infty$$

$$\cos(x) = \sum_{k=0}^{\infty} \frac{(-1)^k x^{2k}}{(2k)!} \qquad \text{for } -\infty < x < \infty$$

$$e^x = \sum_{k=0}^{\infty} \frac{x^k}{k!} \qquad \text{for } -\infty < x < \infty$$

$$\frac{1}{1-x} = \sum_{k=0}^{\infty} x^k \qquad \text{for } -1 < x < 1$$

As we saw in Activity 8.6.3, we can use these known series to find other power series expansions for related functions such as $\sin(x^2)$, e^{5x^3}, and $\cos(x^5)$. Another important way that we can manipulate power series is illustrated in the next activity.

Activity 8.6.4. Let f be the function given by the power series expansion

$$f(x) = \sum_{k=0}^{\infty} (-1)^k \frac{x^{2k}}{(2k)!}.$$

a. Assume that we can differentiate a power series term by term, just like we can differentiate a (finite) polynomial. Use the fact that

$$f(x) = 1 - \frac{x^2}{2!} + \frac{x^4}{4!} - \frac{x^6}{6!} + \cdots + (-1)^k \frac{x^{2k}}{(2k)!} + \cdots$$

to find a power series expansion for $f'(x)$.

b. Observe that $f(x)$ and $f'(x)$ have familiar Taylor series. What familiar functions are these? What known relationship does our work demonstrate?

c. What is the series expansion for $f''(x)$? What familiar function is $f''(x)$?

It turns out that our work in Activity 8.6.3 holds more generally. The corresponding theorem, which we will not prove, states that we can differentiate a power series for a function f term by term and obtain the series expansion for f', and similarly we can integrate a series expansion for a function f term by term and obtain the series expansion for $\int f(x)\, dx$. For both, the radius of convergence of the resulting series is the same as the original, though it is possible that the convergence status of the resulting series may differ at the endpoints. The formal statement of the Power Series Differentiation and Integration Theorem follows.

Power Series Differentiation and Integration Theorem

Suppose $f(x)$ has a power series expansion

$$f(x) = \sum_{k=0}^{\infty} c_k x^k$$

so that the series converges absolutely to $f(x)$ on the interval $-r < x < r$. Then, the power series $\sum_{k=1}^{\infty} k c_k x^{k-1}$ obtained by differentiating the power series for $f(x)$ term by term converges absolutely to $f'(x)$ on the interval $-r < x < r$. That is,

$$f'(x) = \sum_{k=1}^{\infty} k c_k x^{k-1}, \text{ for } |x| < r.$$

Similarly, the power series $\sum_{k=0}^{\infty} c_k \frac{x^{k+1}}{k+1}$ obtained by integrating the power series for $f(x)$ term by term converges absolutely on the interval $-r < x < r$, and

$$\int f(x)\, dx = \sum_{k=0}^{\infty} c_k \frac{x^{k+1}}{k+1} + C, \text{ for } |x| < r.$$

This theorem validates the steps we took in Activity 8.6.4. It is important to note that this result about differentiating and integrating power series tells us that we can differentiate and integrate term by term on the interior of the interval of convergence, but it does not tell us what happens at the endpoints of this interval. We always need to check what happens at the endpoints separately. More importantly, we can use use the approach of differentiating or integrating a series term by term to find new series.

Example 8.6.5. Find a series expansion centered at $x = 0$ for $\arctan(x)$, as well as its interval of convergence.

Solution. While we could differentiate $\arctan(x)$ repeatedly and look for patterns in the derivative values at $x = 0$ in an attempt to find the Maclaurin series for $\arctan(x)$ from the definition, it turns out to be far easier to use a known series in an insightful way. In Activity 8.6.3, we found that

$$\frac{1}{1 + x^2} = \sum_{k=0}^{\infty} (-1)^k x^{2k}$$

for $-1 < x < 1$. Recall that

$$\frac{d}{dx}[\arctan(x)] = \frac{1}{1 + x^2},$$

and therefore

$$\int \frac{1}{1 + x^2}\, dx = \arctan(x) + C.$$

It follows that we can integrate the series for $\frac{1}{1+x^2}$ term by term to obtain the power series expansion for $\arctan(x)$. Doing so, we find that

$$\arctan(x) = \int \left(\sum_{k=0}^{\infty} (-1)^k x^{2k} \right) dx$$

$$= \sum_{k=0}^{\infty} \left(\int (-1)^k x^{2k}\, dx \right)$$

$$= \left(\sum_{k=0}^{\infty} (-1)^k \frac{x^{2k+1}}{2k + 1} \right) + C.$$

The Power Series Differentiation and Integration Theorem tells us that this equality is valid for at least $-1 < x < 1$.

To find the value of the constant C, we can use the fact that $\arctan(0) = 0$. So

$$0 = \arctan(0) = \left(\sum_{k=0}^{\infty} (-1)^k \frac{0^{2k+1}}{2k + 1} \right) + C = C,$$

and we must have $C = 0$. Therefore,

$$\arctan(x) = \sum_{k=0}^{\infty} (-1)^k \frac{x^{2k+1}}{2k + 1} \tag{8.6.3}$$

for at least $-1 < x < 1$.

It is a straightforward exercise to check that the power series

$$\sum_{k=0}^{\infty} (-1)^k \frac{x^{2k+1}}{2k + 1}$$

converges both when $x = -1$ and when $x = 1$; in each case, we have an alternating series with terms $\frac{1}{2k+1}$ that decrease to 0, and thus the interval of convergence for the series expansion for $\arctan(x)$ in Equation (8.6.3) is $-1 \leq x \leq 1$.

Activity 8.6.5. Find a power series expansion for $\ln(1 + x)$ centered at $x = 0$ and determine its interval of convergence.

Summary

- A power series is a series of the form

$$\sum_{k=0}^{\infty} a_k x^k.$$

- We can often assume a solution to a given problem can be written as a power series, then use the information in the problem to determine the coefficients in the power series. This method allows us to approximate solutions to certain problems using partial sums of the power series; that is, we can find approximate solutions that are polynomials.

- The connection between power series and Taylor series is that they are essentially the same thing: on its interval of convergence a power series is the Taylor series of its sum.

Exercises

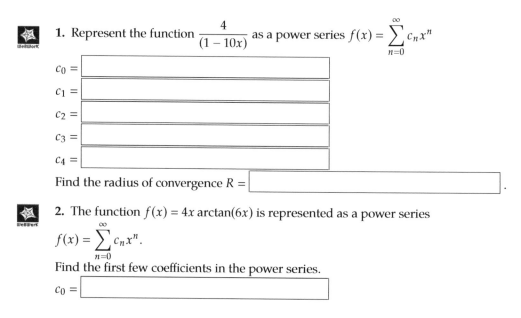

1. Represent the function $\dfrac{4}{(1 - 10x)}$ as a power series $f(x) = \displaystyle\sum_{n=0}^{\infty} c_n x^n$

$c_0 =$

$c_1 =$

$c_2 =$

$c_3 =$

$c_4 =$

Find the radius of convergence $R =$.

2. The function $f(x) = 4x \arctan(6x)$ is represented as a power series

$$f(x) = \sum_{n=0}^{\infty} c_n x^n.$$

Find the first few coefficients in the power series.

$c_0 =$

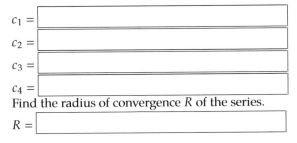

$c_1 = $

$c_2 = $

$c_3 = $

$c_4 = $

Find the radius of convergence R of the series.

$R = $

3. We can use power series to approximate definite integrals to which known techniques of integration do not apply. We will illustrate this in this exercise with the definite integral $\int_0^1 \sin(x^2)\,ds$.

    a. Use the Taylor series for $\sin(x)$ to find the Taylor series for $\sin(x^2)$. What is the interval of convergence for the Taylor series for $\sin(x^2)$? Explain.

    b. Integrate the Taylor series for $\sin(x^2)$ term by term to obtain a power series expansion for $\int \sin(x^2)\,dx$.

    c. Use the result from part (b) to explain how to evaluate $\int_0^1 \sin(x^2)\,dx$. Determine the number of terms you will need to approximate $\int_0^1 \sin(x^2)\,dx$ to 3 decimal places.

4. There is an important connection between power series and Taylor series. Suppose f is defined by a power series centered at 0 so that

$$f(x) = \sum_{k=0}^{\infty} a_k x^k.$$

    a. Determine the first 4 derivatives of f evaluated at 0 in terms of the coefficients a_k.

    b. Show that $f^{(n)}(0) = n!a_n$ for each positive integer n.

    c. Explain how the result of (b) tells us the following:

        On its interval of convergence, a power series is the Taylor series of its sum.

5. In this exercise we will begin with a strange power series and then find its sum. The Fibonacci sequence $\{f_n\}$ is a famous sequence whose first few terms are

$$f_0 = 0,\ f_1 = 1,\ f_2 = 1,\ f_3 = 2,\ f_4 = 3,\ f_5 = 5,\ f_6 = 8,\ f_7 = 13, \cdots,$$

where each term in the sequence after the first two is the sum of the preceding two terms. That is, $f_0 = 0$, $f_1 = 1$ and for $n \geq 2$ we have

$$f_n = f_{n-1} + f_{n-2}.$$

Now consider the power series

$$F(x) = \sum_{k=0}^{\infty} f_k x^k.$$

We will determine the sum of this power series in this exercise.

a. Explain why each of the following is true.

    i. $xF(x) = \sum_{k=1}^{\infty} f_{k-1}x^k$

    ii. $x^2F(x) = \sum_{k=2}^{\infty} f_{k-2}x^k$

b. Show that

$$F(x) - xF(x) - x^2F(x) = x.$$

c. Now use the equation

$$F(x) - xF(x) - x^2F(x) = x$$

to find a simple form for $F(x)$ that doesn't involve a sum.

d. Use a computer algebra system or some other method to calculate the first 8 derivatives of $\frac{x}{1-x-x^2}$ evaluated at 0. Why shouldn't the results surprise you?

6. Airy's equation[2]

$$y'' - xy = 0, \qquad (8.6.4)$$

can be used to model an undamped vibrating spring with spring constant x (note that y is an unknown function of x). So the solution to this differential equation will tell us the behavior of a spring-mass system as the spring ages (like an automobile shock absorber). Assume that a solution $y = f(x)$ has a Taylor series that can be written in the form

$$y = \sum_{k=0}^{\infty} a_k x^k,$$

where the coefficients are undetermined. Our job is to find the coefficients.

(a) Differentiate the series for y term by term to find the series for y'. Then repeat to find the series for y''.

(b) Substitute your results from part (a) into the Airy equation and show that we can write Equation (8.6.4) in the form

$$\sum_{k=2}^{\infty}(k-1)ka_k x^{k-2} - \sum_{k=0}^{\infty} a_k x^{k+1} = 0. \qquad (8.6.5)$$

(c) At this point, it would be convenient if we could combine the series on the left in (8.6.5), but one written with terms of the form x^{k-2} and the other with terms in the form x^{k+1}. Explain why

$$\sum_{k=2}^{\infty}(k-1)ka_k x^{k-2} = \sum_{k=0}^{\infty}(k+1)(k+2)a_{k+2} x^k. \qquad (8.6.6)$$

(d) Now show that

$$\sum_{k=0}^{\infty} a_k x^{k+1} = \sum_{k=1}^{\infty} a_{k-1} x^k. \qquad (8.6.7)$$

---

[2]The general differential equations of the form $y'' \pm k^2 xy = 0$ is called Airy's equation. These equations arise in many problems, such as the study of diffraction of light, diffraction of radio waves around an object, aerodynamics, and the buckling of a uniform column under its own weight.

(e) We can now substitute (8.6.6) and (8.6.7) into (8.6.5) to obtain

$$\sum_{n=0}^{\infty}(n+1)(n+2)a_{n+2}x^n - \sum_{n=1}^{\infty}a_{n-1}x^n = 0. \tag{8.6.8}$$

Combine the like powers of x in the two series to show that our solution must satisfy

$$2a_2 + \sum_{k=1}^{\infty}[(k+1)(k+2)a_{k+2} - a_{k-1}]x^k = 0. \tag{8.6.9}$$

(f) Use equation (8.6.9) to show the following:

  i. $a_{3k+2} = 0$ for every positive integer k,

  ii. $a_{3k} = \frac{1}{(2)(3)(5)(6)\cdots(3k-1)(3k)}a_0$ for $k \geq 1$,

  iii. $a_{3k+1} = \frac{1}{(3)(4)(6)(7)\cdots(3k)(3k+1)}a_1$ for $k \geq 1$.

(g) Use the previous part to conclude that the general solution to the Airy equation (8.6.4) is

$$y = a_0\left(1 + \sum_{k=1}^{\infty}\frac{x^{3k}}{(2)(3)(5)(6)\cdots(3k-1)(3k)}\right)$$

$$+ a_1\left(x + \sum_{k=1}^{\infty}\frac{x^{3k+1}}{(3)(4)(6)(7)\cdots(3k)(3k+1)}\right).$$

Any values for a_0 and a_1 then determine a specific solution that we can approximate as closely as we like using this series solution.

A Short Table of Integrals

a. $\int \frac{du}{a^2+u^2} = \frac{1}{a} \arctan \frac{u}{a} + C$

b. $\int \frac{du}{\sqrt{u^2 \pm a^2}} = \ln|u + \sqrt{u^2 \pm a^2}| + C$

c. $\int \sqrt{u^2 \pm a^2}\, du = \frac{u}{2}\sqrt{u^2 \pm a^2} \pm \frac{a^2}{2} \ln|u + \sqrt{u^2 \pm a^2}| + C$

d. $\int \frac{u^2 du}{\sqrt{u^2 \pm a^2}} = \frac{u}{2}\sqrt{u^2 \pm a^2} \mp \frac{a^2}{2} \ln|u + \sqrt{u^2 \pm a^2}| + C$

e. $\int \frac{du}{u\sqrt{u^2+a^2}} = -\frac{1}{a} \ln\left|\frac{a+\sqrt{u^2+a^2}}{u}\right| + C$

f. $\int \frac{du}{u\sqrt{u^2-a^2}} = \frac{1}{a} \sec^{-1} \frac{u}{a} + C$

g. $\int \frac{du}{\sqrt{a^2-u^2}} = \arcsin \frac{u}{a} + C$

h. $\int \sqrt{a^2 - u^2}\, du = \frac{u}{2}\sqrt{a^2 - u^2} + \frac{a^2}{2} \arcsin \frac{u}{a} + C$

i. $\int \frac{u^2}{\sqrt{a^2-u^2}}\, du = -\frac{u}{2}\sqrt{a^2 - u^2} + \frac{a^2}{2} \arcsin \frac{u}{a} + C$

j. $\int \frac{du}{u\sqrt{a^2-u^2}} = -\frac{1}{a} \ln\left|\frac{a+\sqrt{a^2-u^2}}{u}\right| + C$

k. $\int \frac{du}{u^2\sqrt{a^2-u^2}} = -\frac{\sqrt{a^2-u^2}}{a^2 u} + C$

Index